Partial Differential Equations of MATHEMATICAL PHYSICS

SECOND CORRECTED EDITION

Arthur Gordon Webster
A. B. (Harv.), Ph.D. (Berol.)

Edited by Samuel J. Plimpton
Ph.D. (Yale)

Dover Publications, Inc.
Mineola, New York

Bibliographical Note

This Dover edition, first published in 1966 and reissued in 2016, is an un-abridged republication of the second edition of the work, originally published by G. E. Stechert, New York, and B. G. Teubner, Leipzig, in 1933. The work was first published as a Dover edition in 1955, and reprinted in 1966 with a new Index prepared by Mr. Galoust M. Elgal.

Library of Congress Cataloging-in-Publication Data

Names: Webster, Arthur Gordon, 1863–1923. | Plimpton, Samuel James, 1883– editor.
Title: Partial differential equations of mathematical physics / Arthur Gordon Webster ; edited by Samuel J. Plimpton.
Description: Dover edition, second edition. | Mineola, New York : Dover Publications, Inc., 2016. | Originally published: New York : G.E. Stechert, and Leipzig : B.G. Teubner, 1927; Dover edition, published in 1955, reprinted in 1966, and reissued in 2016.
Identifiers: LCCN 2016009483 | ISBN 9780486805153 | ISBN 0486805158
Subjects: LCSH: Differential equations, Partial. | Mathematical physics.
Classification: LCC QA371 .W4 2016 | DDC 515/.353—dc23
LC record available at http://lccn.loc.gov/2016009483

Manufactured in the United States by RR Donnelley
80515801   2016
www.doverpublications.com

NOTE BY THE EDITOR

As the author did not write a preface to this volume, I venture to quote from memory his explanation to me of his purpose in writing it, and of the correlation covering a very broad field which he hoped to indicate.

"It seems to be the impression among students that mathematical physics consists in deriving a large number of partial differential equations and then solving them, individually, by an assortment of special mutually unrelated devices. It has not been made clear that there is any underlying unity of method and one has often been left entirely in the dark as to what first suggested a particular device to the mind of its inventor."

"I am attempting to show that there are only a few fundamental partial differential equations ordinarily required in physics and that these are all special cases of a general equation, which I will write down and that the method to be described for solving this includes all the other special methods."

At this time the first four chapters had been written in outline and I was asked to read these over and to suggest corrections or changes tending to clearness of presentation. During the remainder of the work it became the habit of the author to turn his manuscript over to me in this way and minor changes were frequently made. I mention this in justification for the slight liberties which I have taken with the text. At the time of the author's death more than one half of the work was in the hands of the publisher, in the form of uncorrected proof. The remainder was completely outlined in manuscript, lacking only in details to be carried out as indicated.

The numbered equations were all given, but it was found necessary to carry through the mathematical steps from the beginning of the book, adding here and there an equation or a sentence in order to make the connections sufficiently obvious. Such additions when in words were put in curly brackets, { }. It was not found feasible to add complication to the mathematics in this way. The only case in which there was a complete gap in the manuscript was in paragraph 101, where a considerable amount of mathematics was inserted, as indicated in a foot note.

I think Professor Webster must have been aware of his unusual success in correlating for the reader such an enormous amount of material as is here presented in a single volume. He several times, however, expressed a slight dissatisfaction with his result in one particular: "The work showed a tendency to fall into two parts." It seemed to be difficult to bind together into complete unity methods associated with Cauchy's use of the Fourier integral on the one hand and those employing normal functions on the other. It occurred to me that this was due to the fact that the two methods are really alternative methods, their relative convenience in any physical problem depending on the character of the known initial and boundary conditions, and I suggested to him that it would perhaps be best to point out this parallelism at the beginning of the book and carry the two developments along side by side. This plan he adopted and as it involved a slight revision of the text he asked me to suggest changes or additions tending to make this point clear. It is a little difficult to discover just which of these suggestions he intended to adopt. The correlation given in paragraphs 47 and 48 was taken from my notes. This has evidently been approved as it was included in the manuscript already sent to the printer. In all doubtful cases I have enclosed my comments in curly brackets.

In all respects the work has been unambiguously indicated by its author, and now stands practically as it would have come from his hands. Notwithstanding the intricate mathematical character of the text, it is not intended as merely a work of reference but rather as a complete course of reading in what has been called the "mathematical physics of the continuum."

I wish to thank Mr. A. Pen-Tung Sah for carrying through the arduous work of reading the proofs.

Worcester Mass U. S. A., June 1927.

S. J. PLIMPTON

No important changes have been made in the second edition, but numerous minor improvements and corrections have been introduced.

Worcester Mass U. S. A., May 1933.

S. J. PLIMPTON

CONTENTS

CHAPTER IX
THEORY OF INTEGRAL EQUATIONS

APPENDIX

CHAPTER I.

DEDUCTION OF THE DIFFERENTIAL EQUATIONS.

1. Scope of Mathematical Physics. Time and Space. It is the lofty aim of mathematical or theoretical physics to describe the universe in the most accurate manner. This manner must be by means of mathematics. Only those phenomena are amenable to it which are susceptible of measurement, but these include a great variety, viz.:

Mechanics (the laws of motion),

Sound (included under mechanics),

Heat,

Light,

Electricity and magnetism.

With some aspects of all these we shall here be concerned.

The phenomena of the physical universe take place in space and in time, the nature of which we must suppose known. Time is simpler than space in requiring but a single given number, t, to specify it. Space as we know it has three dimensions, that is, to specify the position of a point, we require three numerical data, most simply its Cartesian rectangular coordinates, x, y, z. Both space and time are continuously variable, that is, if ε be a number given at pleasure no matter how small, for every instant of time t there are as many others t' as we please such that $|t - t'| < \varepsilon$,[1]) and to every point x, y, z there are as many others as we please x', y', z' such that

$$|x - x'| < \varepsilon, \quad |y - y'| < \varepsilon, \quad |z - z'| < \varepsilon.$$

Thus the whole universe of space and time is characterized by four variables x, y, z, t.

A point may be supposed to be limited by a relation $\varphi(x, y, z) = 0$, so that it moves on a surface, which constitutes a portion of space of two dimensions, or by two relations $\varphi_1(x, y, z)$ and $\varphi_2(x, y, z) = 0$, so that it moves on a line, constituting a space of one dimension. For the characterization of a single point then, we require two, three or four variables.

---

1) By $|x|$ we mean the positive arithmetical or *absolute* value of x.

Mechanics has to do with the motions of points, finite or infinite in number, and considers the coordinates x, y, s of a point as functions of the time t. Its differential equations, involving the derivatives by the single variable t, are accordingly, for a finite number of points, ordinary differential equations. The other portions of mathematical physics consider the properties of substances occupying various portions of space at various times, and the dependence of these properties on the position and on the time. Thus we treat of the density, pressure, temperature, velocity, stress, electrical and magnetic polarization, and other properties of substances, which may all vary from one point to another, and from one instant to another. For all these quantities we have derivatives according to the time, x, y, s being constant, or according to either x, y, or s when t and the remaining coordinates are constant. Thus the derivatives are partial, and the differential equations are partial. Among these questions we must include the mechanics of a substance which continuously fills a portion of space. In mathematical physics, we assume that substances do continuously fill space, that is, to every point in the substance there are as many others as we please, as near as we please, at which the substance exists.

2. Continuity. Analytic Functions. We have, according to the point of view, to assume that matter is discontinuous, since we suppose it to be composed of atoms, which again may be composed of smaller separate parts, or that it is continuous, which is the point of view of mathematical physics. For instance, let us consider a solid body, which is bounded by a surface S. We may assume either that as we go along a continuous curve approaching the surface S on one side 1 and cross to the side 2, the density changes from a finite value to the value zero, thus being a discontinuous function of the distance s travelled (Fig. 1),

Fig. 1. d Fig. 2.

or we may suppose the discontinuity obviated by supposing the change to be made very suddenly, but not discontinuously, as in Fig. 2. The discontinuity may be treated as the limit of such a sudden variation. We may express this by the equation

$$F = \frac{F_1 + e^{ns} F_2}{1 + e^{ns}}$$

giving a function that is continuous when n is finite, but in the limit for $n = \infty$ passes discontinuously from the form F_1 to F_2 as s passes through zero.[1]

---

1) Maxwell, Electricity and Magnetism, Vol. I, § 8.

For example, if we look at a cumulus cloud from a distance, it appears to have a sharp boundary (the density of the water vapor discontinuous) but as we approach it we find it impossible to say where the boundary is, and it seems to shade off gradually, while again if we examine it with a microscope we find it is composed of minute drops, each of which we suppose to be composed of millions of atoms, each composed of many smaller parts. We have here an extreme example of the possible variety of points of view. We shall here adopt the hypothesis of general continuity of properties, with occasional discontinuity on surfaces, lines, or points, which may be removed as above.

A function f of a single variable x is termed continuous at a value x if, after a positive number ε is given as small as we please, we may find a number δ such that, for *all* values x such that $|x' - x| < \delta$, we have

$$|f(x') - f(x)| < \varepsilon.$$

A function which increases beyond any finite number at any point x is accordingly discontinuous.

If the function has at any point x_0 a single finite value, and has derivatives of *all* orders $f'(x_0), f''(x_0), \ldots$ each of a single finite value it is generally developable in a Taylor's series[1])

$$f(x) = f(x_0) + (x - x_0)f'(x_0) + \frac{(x - x_0)^2}{2!}f''(x_0) + \cdots$$

in some region including x_0, and is said to be an *analytic* function.

We are often obliged to deal with functions that are not analytic in some region. For instance, suppose that a harp string is drawn aside at a point so that it takes the shape Fig. 3. Then the line giving the shape of the string is represented by a continuous function of the distance from

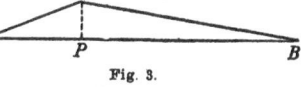

Fig. 3.

one end $y = f(x)$ and the derivative $\frac{dy}{dx} = f'(x)$ exists and is continuous except at P, where the derivative has one value on the left, and a different one on the right. The function y is not analytic in a region including P.

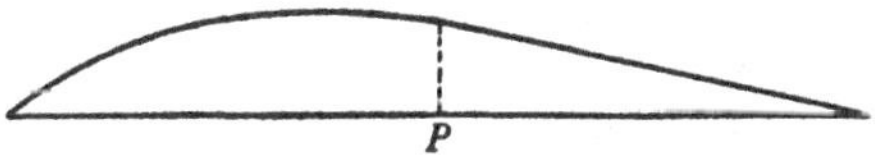

We may have the derivative continuous, as in Fig. 4,

Fig. 4.

where the line is an arc of a circle on the left of P, straight on the right. Then $f'(x)$ is continuous but $f''(x)$ changes abruptly as we pass P, — again

1) A curious case in which derivatives of all orders are finite is furnished by the function $e^{-\frac{1}{(x-x_0)^2}}$, and yet the development in Taylor's series would *not* represent the function. See Osgood, Lehrbuch der Funktionentheorie. Bd. I p. 90., also Pierpont, Theory of Functions of Real Variables, Vol. II , p. 176 *et seq*

y is *not* analytic. In fact if the function is analytic on one side, and developable in one Taylor's series, and analytic on the other side with a *different* development, it is *non*-analytic in the region including *P*. An analytic function is given at every point of its region if its values are given in a region however short, for this region suffices to find the derivatives of *all* orders, which determine the development. The function of Fig. 3 may have the discontinuity of the derivative removed by the physical consideration that it would be impossible to bend a real string

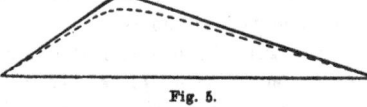

Fig. 5.

sharply, and substituting for it the function, full line Fig. 5, which is physically more nearly realizable, but still not analytic, or it may be replaced by an analytic function (dotted, Fig. 5) which differs from it at every point by less than ε, given however small. Thus by means of analytic functions and their limits we can suit all physical circumstances.

3. Vectors. Scalar and Vector Product. Some of the properties mentioned above are completely specified by a single number together with the unit of the property, for example, density, pressure, temperature. Others on the other hand, are specified only when their direction, as well as magnitude is given, such as velocity, acceleration, magnetic polarization. Quantities of the former category are called *scalars*, as susceptible of being measured off on a single arithmetical scale, quantities of the directed class *vectors*. This name arises from the consideration that the vector *A B carries* one from *A* to *B*. To specify it we must know the distance *A B*, denoted by $|AB|$, as well as the direction. Direction may be specified by two angular coordinates, such as the latitude and longitude of a point on a sphere such that a radius to it from the center has the given direction. Or the direction of a line may be given by the three direction cosines, between which there is the identical relation

$$(1) \qquad \cos^2(rx) + \cos^2(ry) + \cos^2(rz) = 1,$$

leaving two independent. To specify a vector quantity we occordingly need three independent data, one for the length, and two for the angle. A more convenient way of specifying a vector is to give its three *projections* on the rectangular coordinate axes. We shall here denote vectors by black letters A, the projections being given by suffixes. For instance we shall write

$$A \equiv A_x, \; A_y, \; A_z,$$

which may be read: the vector A has the projections A_x, A_y, A_z. The length, magnitude or absolute value of the vector will be denoted by $|A|$, so that evidently

$$(2) \qquad |A|^2 = A_x^2 + A_y^2 + A_z^2.$$

The direction cosines of the vector are evidently

(3) $\qquad \cos(Ax) = \dfrac{A_x}{|A|}, \qquad \cos(Ay) = \dfrac{A_y}{|A|}, \qquad \cos(Az) = \dfrac{A_z}{|A|},$

squaring and adding which, and using (2), they are found to satisfy (1). A vector of length unity in the direction of A will be denoted by A_1, so that we may write for the direction cosines of A

$$\cos (Ax) = A_{1x}, \quad \cos (Ay) = A_{1y}, \quad \cos (Az) = A_{1z}.$$

(The length, the projection, and the direction cosines of a vector are all merely numbers, that is scalar quantities.)

By a vector equation

(4) $\qquad\qquad\qquad\qquad A + B = C$

we mean, C, the *resultant* or vector sum of the components A and B, is the vector obtained by placing the initial point of the vector B in coincidence with the terminal point of the vector A, and taking C with its initial point coinciding with that of A, and its terminal point with that of B, Fig. 6. Evidently $A + B = B + A$. If $C = 0$, that is, its initial and terminal points coincide,

$$A + B = 0, \ A = -B,$$

showing that the negative of a vector has the same length but the *opposite* direction.

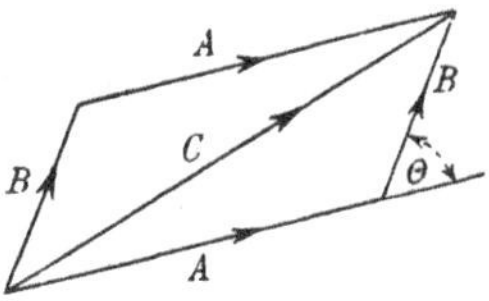

If we form the components of the vectors A, B, C, considering their algebraic signs, it is evident that the components of the resultants are equal respectively to the sums of the respective components, so that the single vector equation (4) may be replaced by the three scalar equations

(5) $\qquad A_x + B_x = C_x, \quad A_y + B_y = C_y, \quad A_z + B_z = C_z.$

Whether we write (4) or (5) is accordingly merely a matter of convenience and clearness. We shall make use of both. Evidently the addition of vectors may be extended to any number, and from it follows the idea of subtraction,

$$A - B = D. \ \text{(Fig. 6a.)}$$

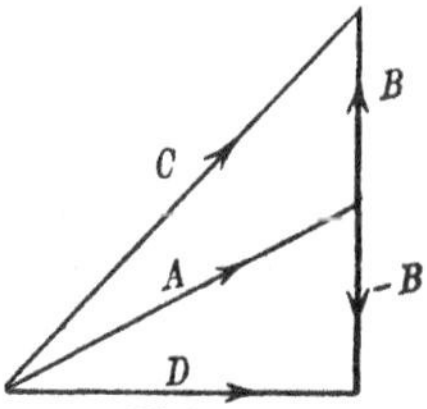

If we call θ the angle between the (positive) directions of A and B we have from the fundamental formula of trigonometry (Fig. 6).

$$|C|^2 = |A|^2 + |B|^2 + 2|A||B| \cos \theta,$$

or by equations (2) and (5)

$$A_x^2 + B_x^2 + 2A_xB_x + A_y^2 + B_y^2 + 2A_yB_y + A_z^2 + B_z^2 + 2A_zB_z$$
$$= A_x^2 + A_y^2 + A_z^2 + B_x^2 + B_y^2 + B_z^2 + 2|A||B| \cos \theta,$$

Fig. 6.

Fig. 6a.

from which

(6) $$|A||B|\cos\theta = A_x B_x + A_y B_y + A_z B_z$$

This expression, symmetrical in both vectors, and equal to the product of the lengths of both by the cosine of the included angle, is called the *scalar product* of the two vectors and is denoted by AB. Evidently

$$AB = BA.$$

If A and B are *unit* vectors $|A_1| = 1$ and $|B_1| = 1$, and (6) reduces to the familiar formula giving the cosine of the angle between the directions as the sum of the three products of their corresponding direction cosines. If two vectors are mutually perpendicular, their scalar product vanishes.

The *projection* of any vector A on any direction B, given by its direction cosines $B_{1x} = \cos(Bx)$, $B_{1y} = \cos(By)$, $B_{1z} = \cos(Bz)$, being defined as the product of its length by the cosine of the angle made by the vector with the direction of projection, (6) gives,

(7) $A_B = |A|\cos\theta = A_x \cos(Bx) + A_y \cos(By) + A_z \cos(Bz)$.

Thus the scalar product of two vectors may be defined as the length of either multiplied by the projection on its direction of the other. Beside the scalar product, defined as above, we define a vector, which we call the *vector* product of AB, and denote by [AB]. This we define as

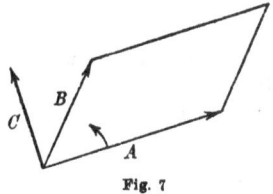

Fig. 7

a vector perpendicular to the plane of the vectors A and B, and drawn toward that side of their plane such that turning about the vector product as an axis in the direction required to turn a right handed screw advancing in the direction of the vector product will cause the first vector A to rotate towards the second B, Fig. 7. The magnitude of the vector product is defined as equal to the area of the parallelogram enclosed by A and B,

(8) $$|[AB]| = |A||B|\sin\theta_{AB}.$$

The vector product is thus in a sense the complement of the scalar product, but is not, like it, a symmetrical function, for

(9) $$[AB] = -[BA].$$

If C denote the vector product [AB], since by definition it is perpendicular to A and B,

(10) $$\begin{aligned} A_x C_x + A_y C_y + A_z C_z &= 0, \\ B_x C_x + B_y C_y + B_z C_z &= 0 \end{aligned}$$

Accordingly C_x, C_y, C_z are respectively proportional to the determinants of the other components, or

$$(11) \qquad \frac{C_x}{\begin{vmatrix} A_y, & A_z \\ B_y, & B_z \end{vmatrix}} = \frac{C_y}{\begin{vmatrix} A_z, & A_x \\ B_z, & B_x \end{vmatrix}} = \frac{C_z}{\begin{vmatrix} A_x, & A_y \\ B_x, & B_y \end{vmatrix}}$$

If we call the common ratio λ we have,

$$(12) \qquad \begin{aligned} C_x &= \lambda(A_y B_z - A_z B_y), \\ C_y &= \lambda(A_z B_x - A_x B_z). \\ C_z &= \lambda(A_x B_y - A_y B_x). \end{aligned}$$

Squaring and adding, we may determine the value of λ from the equation (8), having previously replaced $\sin^2 \theta$ by $1 - \cos^2 \theta$, where $\cos \theta$ is determined from (6).

Thus

$$\begin{aligned} |C|^2 &= C_x^2 + C_y^2 + C_z^2 \\ &= \lambda^2(A_y^2 B_z^2 + A_z^2 B_y^2 - 2A_y A_z B_y B_z \\ &\quad + A_z^2 B_x^2 + A_x^2 B_z^2 - 2A_z A_x B_z B_x \\ &\quad + A_x^2 B_y^2 + A_y^2 B_x^2 - 2A_x A_y B_x B_y) \end{aligned}$$

$$(13) \qquad = |A|^2 |B|^2 (1 - \cos^2 \theta)$$

$$= (A_x^2 + A_y^2 + A_z^2)(B_x^2 + B_y^2 + B_z^2) - (A_x B_x + A_y B_y + A_z B_z)^2.$$

Accordingly we must have $\lambda^2 = 1$, and the convention as to the direction of the vector product, shows that we must take $\lambda = +1$, as may be shown by taking A and B in the direction of two of the coordinate axes, say by putting all components except A_x, B_y, equal to zero. Accordingly the vector product is defined by the components,

$$(14) \qquad \begin{aligned} [AB]_x &= A_y B_z - A_z B_y, \\ [AB]_y &= A_z B_x - A_x B_z, \\ [AB]_z &= A_x B_y - A_y B_x. \end{aligned}$$

If two vectors are parallel, their vector product vanishes, thus all three expressions (14) vanish, and

$$\frac{A_x}{B_x} = \frac{A_y}{B_y} = \frac{A_z}{B_z},$$

which, by reference to (3) shows that the direction cosines are the same for both vectors. The volume of a parallelepiped bounded by three vectors A, B, C, being equal to the product of the area of the parallelogram included by two by the projection of the third on the direction perpendicular to their plane Fig. 8, is the *scalar* product of either by the *vector* product of the two others (in the proper order)

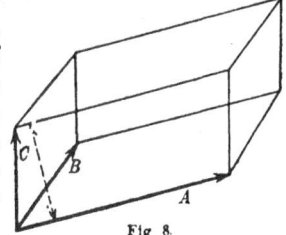

Fig 8.

$$A[BC] = B[CA] = C[AB]$$

$$(15) \quad = A_x(B_yC_z - B_zC_y) + A_y(B_zC_x - B_xC_z) + A_z(B_xC_y - B_yC_x) = \text{etc.}$$

$$= \begin{vmatrix} A_x, & A_y, & A_z \\ B_x, & B_y, & B_z \\ C_x, & C_y, & C_z \end{vmatrix}$$

If the three vectors are parallel to the same plane, this determinant vanishes.

Contrasted with this scalar product we may consider the vector product of A and [BC]. Since it is perpendicular to [BC], it is in the plane of B and C, and has a component parallel to each. Accordingly

$$[A[BC]] = xB + yC,$$

where x, y are scalars. Since the vector on the left is perpendicular to A,

$$A[A[BC]] = 0 = xBA + yCA,$$

$$\frac{x}{CA} = -\frac{y}{BA} = \lambda,$$

$$[A[BC]] = \lambda(B \cdot CA - C \cdot BA).$$

Writing out the x-component,

$$A_y(B_xC_y - B_yC_x) - A_z(B_zC_x - B_xC_z)$$

$$= B_x(A_xC_x + A_yC_y + A_zC_z) - C_x(A_xB_x + A_yB_y + A_zB_z),$$

showing that $\lambda = 1$. Accordingly

$$(16) \qquad [A[BC]] = B \cdot CA - C \cdot BA.$$

4. Scalar and Vector Functions. Gradient. We are accustomed in analysis to functions depending on the position of a point in space, or as we shall call them, functions of a point, or point functions. These are scalar quantities, for instance consider the temperature of a solid, varying from one point to another. We may graphically represent such a scalar function by a *level surface*, namely such a surface that on it the function is constant. Let $\varphi(x, y, z)$ be a point-function, then $\varphi = c$ is the equation of one of its level surfaces. On one side of the surface φ will be less than c, on the other greater. On the surface

$$(17) \qquad d\varphi \equiv \frac{\partial \varphi}{\partial x}dx + \frac{\partial \varphi}{\partial y}dy + \frac{\partial \varphi}{\partial z}dz = 0.$$

Now we may consider dx, dy, dz as the components of a vector of infinitesimal length ds, where

$$(18) \qquad ds^2 = dx^2 + dy^2 + dz^2$$

and

(19) $\quad \cos (ds, x) = \dfrac{dx}{ds}, \quad \cos (ds, y) = \dfrac{dy}{ds}, \quad \cos (ds, z) = \dfrac{dz}{ds},$

lying in the level surface, and if we form a vector **P** with the components

(20) $\qquad\qquad P_x = \dfrac{\partial \varphi}{\partial x}, \ P_y = \dfrac{\partial \varphi}{\partial y}, \ P_z = \dfrac{\partial \varphi}{\partial z},$

since by (17) the scalar product of **P** with ds vanishes, **P** is perpendicular to any vector lying in the tangent plane, and is therefore in the direction of the normal to $\varphi = c$. Accordingly by (3) the direction cosines of the normal to $\varphi = c$ are

(21)
$$\cos (nx) = \frac{\partial \varphi}{\partial x} \Big/ \sqrt{\left(\frac{\partial \varphi}{\partial x}\right)^2 + \left(\frac{\partial \varphi}{\partial y}\right)^2 + \left(\frac{\partial \varphi}{\partial z}\right)^2}$$
$$\cos (ny) = \frac{\partial \varphi}{\partial y} \Big/ \sqrt{\left(\frac{\partial \varphi}{\partial x}\right)^2 + \left(\frac{\partial \varphi}{\partial y}\right)^2 + \left(\frac{\partial \varphi}{\partial z}\right)^2}$$
$$\cos (nz) = \frac{\partial \varphi}{\partial z} \Big/ \sqrt{\left(\frac{\partial \varphi}{\partial x}\right)^2 + \left(\frac{\partial \varphi}{\partial y}\right)^2 + \left(\frac{\partial \varphi}{\partial z}\right)^2}$$

If ds is not in the tangent plane to the level surface, from (17) and (19) we obtain

$$d\varphi = ds \left(\frac{\partial \varphi}{\partial x} \cos (ds, x) + \frac{\partial \varphi}{\partial y} \cos (ds, y) + \frac{\partial \varphi}{\partial z} \cos (ds, z)\right).$$

We shall write the limit

$$\operatorname*{Lim}_{ds = 0} \frac{d\varphi}{ds} = \frac{\partial \varphi}{\partial s},$$

and call it the derivative of φ in the *direction* of ds. Accordingly

(22) $\quad \dfrac{\partial \varphi}{\partial s} = \dfrac{\partial \varphi}{\partial x} \cos (ds, x) + \dfrac{\partial \varphi}{\partial y} \cos (ds, y) + \dfrac{\partial \varphi}{\partial z} \cos (ds, z),$

which by (7) and (20) shows that the derivative in any direction is equal to the projection of the vector **P** on that direction. The vector **P** connected with the scalar function φ by the equations (20) is called the first or the *vector* differential parameter (Lamé) or *gradient* of φ, and we shall write

(23) $\qquad\qquad \nabla\varphi \equiv \operatorname{grad} \varphi \equiv \dfrac{\partial \varphi}{\partial x}, \ \dfrac{\partial \varphi}{\partial y}, \ \dfrac{\partial \varphi}{\partial z}.$

Its absolute value

(24) $\qquad\qquad |\operatorname{grad} \varphi| = \sqrt{\left(\dfrac{\partial \varphi}{\partial x}\right)^2 + \left(\dfrac{\partial \varphi}{\partial y}\right)^2 + \left(\dfrac{\partial \varphi}{\partial z}\right)^2}$

being greater than any of its projections, represents the *fastest* rate of increase of the scalar function φ for all directions at the point in question, and we have already seen that its direction is normal to the level surface $\varphi = c$.

It is convenient to observe that if we treat the operator $\nabla \equiv \dfrac{\partial}{\partial x}$, $\dfrac{\partial}{\partial y}$, $\dfrac{\partial}{\partial z}$ like a vector, and form the scalar product $A\nabla$ applied to a function φ, we have

(25)
$$A\nabla\varphi \equiv A_x\frac{\partial \varphi}{\partial x} + A_y\frac{\partial \varphi}{\partial y} + A_z\frac{\partial \varphi}{\partial z}$$
$$= |A| P_A = |A|\frac{\partial \varphi}{\partial s},$$

where ds is in the direction of A.[1]

Not only a scalar, but a vector, may be a function of a point. By a vector point-function, we mean one whose projections are themselves functions of a point. An example of such a vector point-function has been furnished in the above gradient of a scalar function. But there are an infinite number of others. Let A be such a vector,

$$A \equiv A_x(x, y, z), \quad A_y(x, y, z), \quad A_z(x, y, z).$$

We speak of the *field* of a vector function as a portion of space where the vector is defined. Such a field may be graphically represented by drawing at every point of it a curve whose tangent has at every point the direction of the vector. If ds represent the arc of such a curve, we have then

(26)
$$\frac{dx}{A_x} = \frac{dy}{A_y} = \frac{dz}{A_z}$$

as the differential equations of the curves, and the field is in a way represented by the curves, which we call curves of the vector A. As an example we may consider A to be the velocity of a fluid, then the vector lines are the stream-lines.

A vector may depend not only on space, but on the time. We may differentiate a vector by a scalar variable and we mean by the derivative

$$\frac{\partial A}{\partial t} \equiv \frac{\partial A_x}{\partial t}, \quad \frac{\partial A_y}{\partial t}, \quad \frac{\partial A_z}{\partial t}.$$

5. Invariants. Suppose we transform from axes x, y, z to a new set x', y', z', connected with the old by the following table representing the direction cosines of each with each,

(27)

	X	Y	Z
X'	α_1	β_1	γ_1
Y'	α_2	β_2	γ_2
Z'	α_3	β_3	γ_3

---

1) The operator ∇ is Hamilton's celebrated operator, called Nabla, ($\nu\acute{\alpha}\beta\lambda\alpha$, harp).

Then by successive application of the equation (7) we find for the components of a vector A in the new system.

$$
\begin{aligned}
(28) \quad & A_{x'} = A_x \alpha_1 + A_y \beta_1 + A_z \gamma_1, \\
& A_{y'} = A_x \alpha_2 + A_y \beta_2 + A_z \gamma_2, \\
& A_{z'} = A_x \alpha_3 + A_y \beta_3 + A_z \gamma_3.
\end{aligned}
$$

If for A we take the position vector whose components are x, y, z, x', y', z' we obtain the ordinary equations for transformation of coordinates,

$$
\begin{aligned}
(29) \quad & x' = \alpha_1 x + \beta_1 y + \gamma_1 z, \\
& y' = \alpha_2 x + \beta_2 y + \gamma_2 z, \\
& z' = \alpha_3 x + \beta_3 y + \gamma_3 z.
\end{aligned}
$$

Reversing the relations of the coordinates we have

$$
\begin{aligned}
(29') \quad & x = \alpha_1 x' + \alpha_2 y' + \alpha_3 z', \\
& y = \beta_1 x' + \beta_2 y' + \beta_3 z', \\
& z = \gamma_1 x' + \gamma_2 y' + \gamma_3 z'.
\end{aligned}
$$

Since $\alpha_1, \alpha_2, \alpha_3$ are the direction cosines of the X-axis with respect to x', y', z', we have by (1)

$$
\alpha_1^2 + \alpha_2^2 + \alpha_3^2 = 1,
$$

and in like manner,

$$
\begin{aligned}
(30) \quad & \beta_1^2 + \beta_2^2 + \beta_3^2 = 1, \\
& \gamma_1^2 + \gamma_2^2 + \gamma_3^2 = 1.
\end{aligned}
$$

Since the axes YZ, are perpendicular, we have

$$
\beta_1 \gamma_1 + \beta_2 \gamma_2 + \beta_3 \gamma_3 = 0
$$

and in like manner

$$
\begin{aligned}
(31) \quad & \gamma_1 \alpha_1 + \gamma_2 \alpha_2 + \gamma_3 \alpha_3 = 0 \\
& \alpha_1 \beta_1 + \alpha_2 \beta_2 + \alpha_3 \beta_3 = 0.
\end{aligned}
$$

Making use of these six equations we see that although the change of axes changes the *components*, the absolute value $|A'|$ is unchanged,

$$
(32) \quad |A'|^2 = A_{x'}^2 + A_{y'}^2 + A_{z'}^2 = A_x^2 + A_y^2 + A_z^2 = |A|^2.
$$

Such a function, which maintains its value when the coordinates are changed, is called an *invariant*. The scalar product AB is a second example. If Φ is a scalar function of x, y, z, it becomes a different function of x', y', z',

$$
\varphi(x, y, z) \equiv \varphi'(x', y', z'),
$$

and differentiating

$$
\frac{\partial \varphi'}{\partial x'} = \frac{\partial \varphi}{\partial x} \frac{\partial x}{\partial x'} + \frac{\partial \varphi}{\partial y} \frac{\partial y}{\partial x'} + \frac{\partial \varphi}{\partial z} \frac{\partial z}{\partial x'}.
$$

Now since by (29)

$$\frac{\partial x}{\partial x'} = \alpha_1, \quad \frac{\partial y}{\partial x'} = \beta_1, \quad \frac{\partial z}{\partial x'} = \gamma_1, \quad \text{etc.,}$$

we have

(33)
$$\frac{\partial \varphi'}{\partial x'} = \alpha_1 \frac{\partial \varphi}{\partial x} + \beta_1 \frac{\partial \varphi}{\partial y} + \gamma_1 \frac{\partial \varphi}{\partial z},$$

$$\frac{\partial \varphi'}{\partial y'} = \alpha_2 \frac{\partial \varphi}{\partial x} + \beta_2 \frac{\partial \varphi}{\partial y} + \gamma_2 \frac{\partial \varphi}{\partial z},$$

$$\frac{\partial \varphi'}{\partial z'} = \alpha_3 \frac{\partial \varphi}{\partial x} + \beta_3 \frac{\partial \varphi}{\partial y} + \gamma_3 \frac{\partial \varphi}{\partial z}.$$

Squaring and adding we find as before

(34)
$$\left(\frac{\partial \varphi'}{\partial x'}\right)^2 + \left(\frac{\partial \varphi'}{\partial y'}\right)^2 + \left(\frac{\partial \varphi'}{\partial z'}\right)^2 = \left(\frac{\partial \varphi}{\partial x}\right)^2 + \left(\frac{\partial \varphi}{\partial y}\right)^2 + \left(\frac{\partial \varphi}{\partial z}\right)^2.$$

The gradient is accordingly a *differential invariant*. In like manner, if φ and ψ are two scalar functions, the scalar product of their gradients

(35)
$$\nabla \varphi \nabla \psi \equiv \frac{\partial \varphi}{\partial x} \frac{\partial \psi}{\partial x} + \frac{\partial \varphi}{\partial y} \frac{\partial \psi}{\partial y} + \frac{\partial \varphi}{\partial z} \frac{\partial \psi}{\partial z}$$

may be shown to be a differential invariant.

6. Linear infinitesimal Deformation. Divergence. Curl.
One of the most important vector fields is that in which the vector A represents the velocity of the points of a continuous medium, fluid or solid, which is in motion It will be necessary to examine the changes of size, shape, and position of a small region which take place during a time dt. Let us consider a portion of space in which every point x, y, z, moves to a point $x + \xi$, $y + \eta$, $z + \zeta$, so that the displacement vector ξ, η, ζ is determined by the equations,

(36)
$$\xi = a_1 x + a_2 y + a_3 z,$$

$$\eta = b_1 x + b_2 y + b_3 z,$$

$$\zeta = c_1 x + c_2 y + c_3 z,$$

where the nine quantities a, b, c are constants. Such a motion is called a homogeneous strain, and has the property that any linear relation between the coordinates x, y, z, before the motion gives rise to a linear relation between the new coordinates $x + \xi$, $y + \eta$, $z + \zeta$ after the motion. Accordingly every plane remains a plane, parallel planes remain parallel, and every parallelopiped remains a parallelopiped, its angles being in general changed. A strain is considered small (infinitesimal) if all its nine constants are small, that is, if we may neglect their squares and products.

It is easily seen that all similar figures have their dimensions changed in the same ratio, likewise their volumes. To find the change of volume consider a rectangular parallelopiped whose sides are origi-

nally in the three coordinate axes, of lengths x, y, z, Fig. 9. These sides
are changed into vectors which we will call $\mathbf{A}$, $\mathbf{B}$ and $\mathbf{C}$, whose projections are found by adding $x, 0, 0$; $0, y, 0$; $0, 0, z$ to the three rows of
(36) after putting y, z, z, x and x, y, respectively equal to zero.

$$
\begin{aligned}
\mathbf{A}_x &= (1 + a_1)x, \\
\mathbf{A}_y &= b_1 x, \\
\mathbf{A}_z &= c_1 x, \\
\mathbf{B}_x &= a_2 y, \\
(37) \qquad \mathbf{B}_y &= (1 + b_2)y, \\
\mathbf{B}_z &= c_2 y, \\
\mathbf{C}_x &= a_3 z, \\
\mathbf{C}_y &= b_3 x, \\
\mathbf{C}_z &= (1 + c_3)z.
\end{aligned}
$$

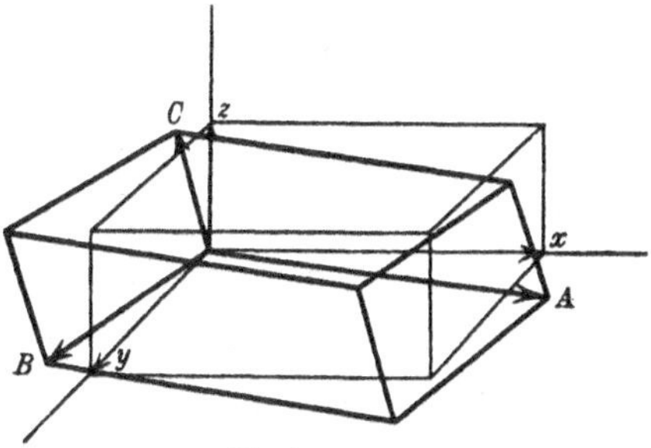

Fig. 9.

The volume of the new parallelopiped is by (15) equal to

$$
(38) \qquad V' = xyz \begin{vmatrix} (1 + a_1), & b_1 & , & c_1 \\ a_2 & (1 + b_2), & c_2 \\ a_3 & b_3 & , & 1 + c_3 \end{vmatrix},
$$

or expanding the determinant and neglecting all but first powers of the
small quantities,

$$
(39) \qquad V' = (1 + a_1 + b_2 + c_3)xyz = (1 + a_1 + b_2 + c_3)V,
$$

where V is the original volume. The relative increase of volume

$$
(40) \qquad \sigma = \frac{V' - V}{V} = a_1 + b_2 + c_3
$$

is called the *cubical dilatation* of the region, and is equal to the sum
of the diagonal coefficients in (36).

The simplest strain is one in which all constants but a_1 vanish.
Then every point moves in the direction of the x-axis, and the rectangular parallelopiped above is *stretched* in the x-direction. The strain
composed of three such simple stretches in three perpendicular directions
(in which all but a_1, b_2, c_3 are zero) is called a *pure* strain, and has the
property that the three mutually perpendicular edges of the parallelopiped retain their directions unchanged. These directions are called the
axes of the strain. Suppose that we consider a different rectangular
parallelopiped whose edges we take as new axes of x', y', z', connected
with x, y, z, by equations (29) (29'). Let us suppose that this parallelopiped is given stretches denoted by the equations

$$
(41) \qquad \xi' = s_1 x', \quad \eta' = s_2 y', \quad \zeta' = s_3 z'.
$$

Comparing the new coordinates, $x + \xi$, $y + \eta$, $z + \zeta$ with the old x, y, z in the first system with the same in the second system by means of (29) (29') (41), we have

$$\begin{aligned}
(42) \quad x + \xi &= \alpha_1(x' + \xi') + \alpha_2(y' + \eta') + \alpha_3(z' + \zeta') \\
&= \alpha_1(1 + s_1)x' + \alpha_2(1 + s_2)y' + \alpha_3(1 + s_3)z' \\
&= \alpha_1(1 + s_1)(\alpha_1 x + \beta_1 y + \gamma_1 z) + \alpha_2(1 + s_2)(\alpha_2 x + \beta_2 y + \gamma_2 z) \\
&\quad + \alpha_3(1 + s_3)(\alpha_3 x + \beta_3 y + \gamma_3 z),
\end{aligned}$$

which by means of (30), (31) becomes

$$\begin{aligned}
\xi &= (s_1\alpha_1^2 + s_2\alpha_2^2 + s_3\alpha_3^2)x + (s_1\alpha_1\beta_1 + s_2\alpha_2\beta_2 + s_3\alpha_3\beta_3)y \\
&\quad + (s_1\gamma_1\alpha_1 + s_2\gamma_2\alpha_2 + s_3\gamma_3\alpha_3)z,
\end{aligned}$$

and in like manner,

$$\begin{aligned}
(43) \quad \eta &= (s_1\alpha_1\beta_1 + s_2\alpha_2\beta_2 + s_3\alpha_3\beta_3)x + (s_1\beta_1^2 + s_2\beta_2^2 + s_3\beta_3^2)y \\
&\quad + (s_1\beta_1\gamma_1 + s_2\beta_2\gamma_2 + s_3\beta_3\gamma_3)z, \\
\zeta &= (s_1\gamma_1\alpha_1 + s_2\gamma_2\alpha_2 + s_3\gamma_3\alpha_3)x + (s_1\beta_1\gamma_1 + s_2\beta_2\gamma_2 + s_3\beta_3\gamma_3)y \\
&\quad + (s_1\gamma_1^2 + s_2\gamma_2^2 + s_3\gamma_3^2)z.
\end{aligned}$$

If we write

$$\begin{aligned}
(44) \quad s_x &= s_1\alpha_1^2 + s_2\alpha_2^2 + s_3\alpha_3^2, \\
s_y &= s_1\beta_1^2 + s_2\beta_2^2 + s_3\beta_3^2, \\
s_z &= s_1\gamma_1^2 + s_2\gamma_2^2 + s_3\gamma_3^2, \\
g_x &= s_1\beta_1\gamma_1 + s_2\beta_2\gamma_2 + s_3\beta_3\gamma_3, \\
g_y &= s_1\gamma_1\alpha_1 + s_2\gamma_2\alpha_2 + s_3\gamma_3\alpha_3, \\
g_z &= s_1\alpha_1\beta_1 + s_2\alpha_2\beta_2 + s_3\alpha_3\beta_3,
\end{aligned}$$

we may write (43)

$$\begin{aligned}
(45) \quad \xi &= s_x x + g_z y + g_y z, \\
\eta &= g_z x + s_y y + g_x z, \\
\zeta &= g_y x + g_x y + s_z z.
\end{aligned}$$

Thus the matrix of coefficients of a pure strain is symmetrical with respect to the principal diagonal. The dilatation is, by (40), $\sigma = s_x + s_y + s_z = s_1 + s_2 + s_3$, and is an invariant for the change of axes. In this case if

$$(46) \quad \varphi \equiv \tfrac{1}{2}s_x x^2 + \tfrac{1}{2}s_y y^2 + \tfrac{1}{2}s_z z^2 + g_x yz + g_y zx + g_z xy,$$

we have

$$(47) \quad \xi = \frac{\partial \varphi}{\partial x}, \qquad \eta = \frac{\partial \varphi}{\partial y}, \qquad \zeta = \frac{\partial \varphi}{\partial z},$$

or the displacement vector is the gradient of the scalar function φ. The quadric $\varphi = 1$, when referred to its principal axes, will have the form,

$$(48) \quad \varphi' \equiv \tfrac{1}{2}\lambda_1 x'^2 + \tfrac{1}{2}\lambda_2 y'^2 + \tfrac{1}{2}\lambda_3 z'^2,$$

from which, if we write

$$(49) \qquad \xi' = \frac{\partial \varphi'}{\partial x'} = \lambda_1 x', \qquad \eta' = \frac{\partial \varphi'}{\partial y'} = \lambda_2 y', \qquad \zeta' = \frac{\partial \varphi'}{\partial z'} = \lambda_3 z',$$

we get the strain as in (41), where $s_1 = \lambda_1$, $s_2 = \lambda_2$, $s_3 = \lambda_3$. Thus in a pure strain we can always find three mutually perpendicular lines which maintain their directions after the strain. These are the axes of the strain. In an impure strain the axes rotate.

The impure strain ξ, η, ζ, (36) may be represented as the resultant of two ξ_1, η_1, ζ_1, and ξ_2, η_2, ζ_2,

$$(50) \qquad \begin{aligned} \xi_1 &= a_1 x + \tfrac{1}{2}(a_2 + b_1)y + \tfrac{1}{2}(a_3 + c_1)z, \\ \eta_1 &= \tfrac{1}{2}(a_2 + b_1)x + b_2 y + \tfrac{1}{2}(b_3 + c_2)z, \\ \zeta_1 &= \tfrac{1}{2}(a_3 + c_1)x + \tfrac{1}{2}(b_3 + c_2)y + c_3 z, \end{aligned}$$

$$(51) \qquad \begin{aligned} \xi_2 &= 0 \cdot x - \tfrac{1}{2}(b_1 - a_2)y + \tfrac{1}{2}(a_3 - c_1)z, \\ \eta_2 &= \tfrac{1}{2}(b_1 - a_2)x + 0 \quad y - \tfrac{1}{2}(c_2 - b_3)z, \\ \zeta_2 &= -\tfrac{1}{2}(a_3 - c_1)x + \tfrac{1}{2}(c_2 - b_3)y + 0 \cdot z, \end{aligned}$$

the first of which is pure, the second has zero dilatation.

If we write

$$\begin{aligned} s_x &= a_1, & s_y &= b_2, & s_z &= c_3, \\ (52)\quad g_x &= \tfrac{1}{2}(c_2 + b_3), & g_y &= \tfrac{1}{2}(a_3 + c_1), & g_z &= \tfrac{1}{2}(b_1 + a_2), \\ \omega_x &= \tfrac{1}{2}(c_2 - b_3), & \omega_y &= \tfrac{1}{2}(a_3 - c_1), & \omega_z &= \tfrac{1}{2}(b_1 - a_2), \end{aligned}$$

(51) are

$$(53) \qquad \begin{aligned} \xi_2 &= \omega_y z - \omega_z y, \\ \eta_2 &= \omega_z x - \omega_x z, \\ \zeta_2 &= \omega_x y - \omega_y x. \end{aligned}$$

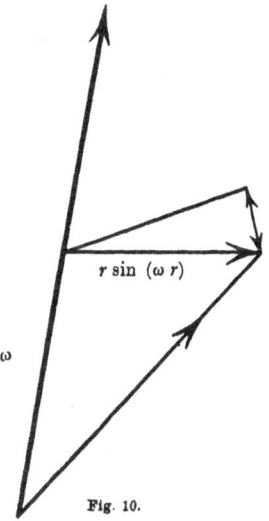

Thus the displacement (51) being the vector product of a vector $\omega \equiv \omega_x$, ω_y, ω_z and the position vector $r \equiv x, y, z$ is of magnitude $|\omega| \, |r| \sin(\omega r)$ and perpendicular to ω and r. It accordingly represents a displacement of the region in the form of a *rotation* of a rigid body through the small angle $|\omega|$ about an axis in the direction of the vector ω, Fig. 10. We accordingly see that every small strain may be considered as the resultant of a pure strain and a rotation without change of shape. The pure strain in which all constants are zero except one of the g's is called a simple *shear*.

$r \sin(\omega r)$

ω

Fig. 10.

Let us now examine the infinitesimal strain produced in the time dt in a vector field where the velocity of a point is the vector **A**, a func-

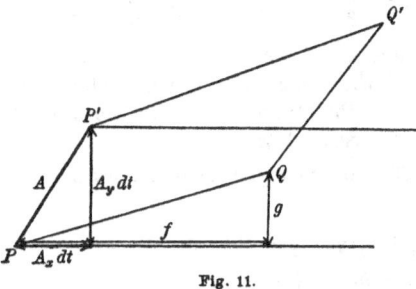

Fig. 11.

tion of x, y, z. A point P, Fig. 11, whose coordinates are x, y, z, will move to a point P' with coordinates $x + \mathbf{A}_x dt$, $y + \mathbf{A}_y dt$, $z + \mathbf{A}_z dt$. A neighboring point Q, $x + f$, $y + g$, $z + h$, will however, not have the same velocity but instead of $\mathbf{A}_x$ it will have, (if $\mathbf{A}_x$ is continuous)

$$\mathbf{A}_x + f\frac{\partial \mathbf{A}_x}{\partial x} + g\frac{\partial \mathbf{A}_x}{\partial y} + h\frac{\partial \mathbf{A}_x}{\partial z}.$$

Consequently the x-projection of PQ will have changed from f to

$$f + \left(\mathbf{A}_x + f\frac{\partial \mathbf{A}_x}{\partial x} + g\frac{\partial \mathbf{A}_x}{\partial y} + h\frac{\partial \mathbf{A}_x}{\partial z}\right)dt - \mathbf{A}_x\,dt.$$

If we call the new *relative* coordinates of Q with respect to P, $f + \xi$ $g + \eta$, $h + \zeta$, we accordingly have

$$\begin{aligned}
(54) \qquad
\xi &= \left(f\frac{\partial \mathbf{A}_x}{\partial x} + g\frac{\partial \mathbf{A}_x}{\partial y} + h\frac{\partial \mathbf{A}_x}{\partial z}\right)dt,\\
\eta &= \left(f\frac{\partial \mathbf{A}_y}{\partial x} + g\frac{\partial \mathbf{A}_y}{\partial y} + h\frac{\partial \mathbf{A}_y}{\partial z}\right)dt,\\
\zeta &= \left(f\frac{\partial \mathbf{A}_z}{\partial x} + g\frac{\partial \mathbf{A}_z}{\partial y} + h\frac{\partial \mathbf{A}_z}{\partial z}\right)dt.
\end{aligned}$$

We accordingly see that a small region in which f, g, h, are the coordinates of a variable point Q with respect to a given point P has received the homogeneous strain (54) in which the coefficients are the nine derivatives of the components of **A** multiplied by dt. Comparing with the general strain (36) we obtain the dilatation σ and the rotation ω from (40), (52),

$$(55) \qquad \sigma = dt\left(\frac{\partial \mathbf{A}_x}{\partial x} + \frac{\partial \mathbf{A}_y}{\partial y} + \frac{\partial \mathbf{A}_z}{\partial z}\right),$$

$$(56) \qquad
\begin{aligned}
\omega_x &= \tfrac{1}{2}dt\left(\frac{\partial \mathbf{A}_z}{\partial y} - \frac{\partial \mathbf{A}_y}{\partial z}\right), & g_x &= \tfrac{1}{2}dt\left(\frac{\partial \mathbf{A}_z}{\partial y} + \frac{\partial \mathbf{A}_y}{\partial z}\right),\\
\omega_y &= \tfrac{1}{2}dt\left(\frac{\partial \mathbf{A}_x}{\partial z} - \frac{\partial \mathbf{A}_z}{\partial x}\right), & g_y &= \tfrac{1}{2}dt\left(\frac{\partial \mathbf{A}_x}{\partial z} + \frac{\partial \mathbf{A}_z}{\partial x}\right),\\
\omega_z &= \tfrac{1}{2}dt\left(\frac{\partial \mathbf{A}_y}{\partial x} - \frac{\partial \mathbf{A}_x}{\partial y}\right), & g_z &= \tfrac{1}{2}dt\left(\frac{\partial \mathbf{A}_y}{\partial x} + \frac{\partial \mathbf{A}_x}{\partial y}\right).
\end{aligned}$$

Thus the time-rate of dilatation, of shear, and the angular velocity of rotation, of the infinitesimal region, are given by one-half the expressions in parentheses in (55) (56) respectively.

A scalar quantity derived from a vector A by the operation of (55) is called the *divergence* of the vector.

(57) $$\text{div } A \equiv \frac{\partial A_x}{\partial x} + \frac{\partial A_y}{\partial y} + \frac{\partial A_z}{\partial z}.$$

If $A = \lambda B$, where λ is any scalar function, we find

(58) $$\text{div } \lambda B = \lambda \text{ div } B + B \nabla \lambda,$$

using the notation (25).

If we consider the operator ∇ (note, p. 10) and form its symbolic scalar product with the vector A, we have

(59) $$\text{div } A = \nabla A.$$

On the contrary, if we form the symbolic *vector* product $[\nabla A]$ we obtain

(60) $$[\nabla A]_x = \frac{\partial A_z}{\partial y} - \frac{\partial A_y}{\partial z},$$
$$[\nabla A]_y = \frac{\partial A_x}{\partial z} - \frac{\partial A_z}{\partial x},$$
$$[\nabla A]_z = \frac{\partial A_y}{\partial x} - \frac{\partial A_x}{\partial y}.$$

The vector thus formed from A is called its *curl*, and we write

(61) $$[\nabla A] = \text{curl } A.$$

The appropriateness of the terms divergence and curl is seen from the example in which A represented the field of velocity of a continuous medium. If we change the axes of coordinates the projections A_x, A_y, A_z will change, but we find both div A and $|\text{curl } A|$ to be differential invariants.

7. Lamellar and Solenoidal Vector Fields. Two particular cases of vector fields are of importance. If the curl of the vector vanishes, the equations curl $A = 0$, are the conditions that the expression

(62) $$A_x dx + A_y dy + A_z dz = d\varphi$$

is a perfect differential. In that case, we have

(63) $$A_x = \frac{\partial \varphi}{\partial x}, \qquad A_y = \frac{\partial \varphi}{\partial y}, \qquad A_z = \frac{\partial \varphi}{\partial z},$$

and since the order of differentiation is immaterial, we have

$$\frac{\partial A_x}{\partial y} = \frac{\partial^2 \varphi}{\partial y \partial x} = \frac{\partial^2 \varphi}{\partial x \partial y} = \frac{\partial A_y}{\partial x}, \quad \text{etc.}$$

Accordingly, if the expression (62) is a perfect differential, the three expressions in (60) will vanish, or curl $A = 0$ is a necessary condition for the existence of the function. It can also be shown that the condition is sufficient. In this case the vector A which is derived from the scalar function φ by the equations (63) or $A = \operatorname{grad} \varphi$ is called a *lamellar* vector, for the reason that at every point the vector is normal to the level surface $\varphi = $ const, and by the property of the gradient, its magnitude is inversely proportional to the thickness of the level sheet, composed of two very near level surfaces. The scalar function φ is called the *potential* of the vector function A.

Let us suppose that the expression

$$A_x dx + A_y dy + A_z dz$$

is not a perfect differential, but that on multiplying by an integrating factor $\frac{1}{\lambda}$ it becomes so, that is

(64) $$A_x dx + A_y dy + A_z dz = \lambda d\varphi,$$

then we have

(65) $$A_x = \lambda \frac{\partial \varphi}{\partial x}, \quad A_y = \lambda \frac{\partial \varphi}{\partial y}, \quad A_z = \lambda \frac{\partial \varphi}{\partial z}, \quad \text{or} \quad A = \lambda \operatorname{grad} \varphi,$$

and although the vector is still normal to the level surface of φ it is not inversely proportional to the thickness of the sheet. The vector is then said to be *compound lamellar*.

In the case of the lamellar vector, the divergence has an important connection with the potential φ.

We have

(66) $$\operatorname{div} A = \operatorname{div} \operatorname{grad} \varphi = \frac{\partial^2 \varphi}{\partial x^2} + \frac{\partial^2 \varphi}{\partial y^2} + \frac{\partial^2 \varphi}{\partial z^2}.$$

The operator

$$\frac{\partial^2}{\partial x^2} + \frac{\partial^2}{\partial y^2} + \frac{\partial^2}{\partial z^2} \equiv \operatorname{div} \operatorname{grad}$$

is called Laplace's Operator and will be denoted by Δ, and the scalar expression $\Delta \varphi$, called by Lamé the second differential parameter of φ, is easily shown to be a differential invariant.

In the example in which A represents the velocity of the fluid, the condition curl $A = 0$ shows that a minute portion of the fluid does not rotate, but is simply carried along and deformed by a pure strain. The motion is said to be irrotational and φ is called the velocity-potential.

In the case of a compound lamellar vector, taking divergence, we have

(67) $$\operatorname{div}(\lambda \operatorname{grad} \varphi) \equiv \lambda \Delta \varphi + \operatorname{grad} \lambda \cdot \operatorname{grad} \varphi,$$

which according to (35) is again a differential invariant.

Taking the curl of (65)

$$\frac{\partial A_z}{\partial y} - \frac{\partial A_y}{\partial z} = \lambda \frac{\partial^2 \varphi}{\partial y \partial z} + \frac{\partial \lambda}{\partial y} \frac{\partial \varphi}{\partial z} - \lambda \frac{\partial^2 \varphi}{\partial z \partial y} - \frac{\partial \lambda}{\partial z} \frac{\partial \varphi}{\partial y}, \text{ etc.,}$$

and accordingly

(68) $$\operatorname{curl}(\lambda \operatorname{grad} \varphi) \equiv [\operatorname{grad} \lambda, \operatorname{grad} \varphi].$$

Since the vector product is perpendicular to both its factors,

$$A \operatorname{curl} A = \lambda \operatorname{grad} \varphi [\operatorname{grad} \lambda, \operatorname{grad} \varphi] = 0$$

so that the equation

(69) $$A \operatorname{curl} A = 0$$

is the necessary condition for a compound lamellar vector, including that for an ordinary lamellar vector.

Of equal importance with vectors which have zero curl are those with zero divergence. Such are typified by the velocity of an incompressible fluid. Suppose that anywhere in the field we draw a small closed curve, and through each of its points we draw a line of the vector A, thus forming a *tube* of the vector, Fig. 12. All the field may be thus divided up into vector tubes, whose walls have the same property as if they were material, that no fluid traverses them. Whatever flows in across one cross-section S_1, will then flow out at another S_2. If these are orthogonal to the tube, the volume of fluid flowing across

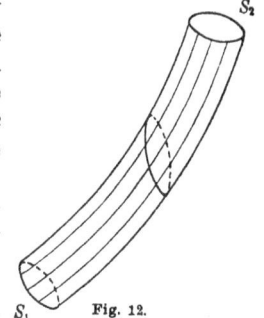

Fig. 12.

S_1, in a time dt will be contained in a cylinder of base S_1 and height $|A|_1 dt$, of volume $S_1 |A|_1 dt$. In like manner that flowing out at S_2 will be $S_2 |A|_2 dt$, and accordingly

$$S_1 |A|_1 = S_2 |A|_2.$$

Accordingly a vector for which $\operatorname{div} A = 0$, which will be called a *solenoidal* vector (σωλήν, a tube), has the property that its magnitude varies inversely as the cross-section of a tube of infinitesimal section. Tubes for which the product $S|A| = 1$ are called unit tubes.

If a vector is both lamellar and solenoidal,

(70) $$A = \operatorname{grad} \varphi, \quad \operatorname{div} A = \Delta \varphi = 0.$$

The equation $\Delta \varphi = 0$ is called Laplace's equation, and is one of the most important of the equations that we have to study.

8. Divergence Theorem. Suppose that we consider any closed surface S fixed in space, and find the amount of fluid that flows in through it in the time dt. From an element of surface dS at a point where the velocity is A there flows in the fluid filling a prism of slant height $|A| dt$ on the base dS (Fig. 13) of volume $dS |A| dt \cos(A, n) = A_n dS dt$, where n is the inward normal to the surface. Thus we have for the whole surface the integral,

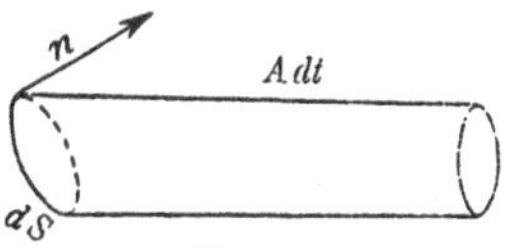

$$dt \int_S \int A_n dS.$$

Fig. 13.

Such a surface integral, in which the integrand is the normal component of a vector, is called the *flux* of that vector through the surface, whether closed or unclosed.

If we consider the meaning of the divergence, (55) we find that an element of volume $d\tau = dx\,dy\,dz$ gains in volume the amount $dt \cdot \operatorname{div} \mathbf{A} \cdot d\tau$ so that all the elements of volume inside the surface S gain the amount

$$dt \iiint \operatorname{div} \mathbf{A} \cdot d\tau$$

which must accordingly be equal to the amount that *passes out*. Equating this to minus the above we have

$$(71) \qquad \iiint \operatorname{div} \mathbf{A} \cdot d\tau = -\iint \mathbf{A}_n \, dS,$$

where n is, as will always be the case with us, the *interior* normal. This result, converting a volume integral into a surface integral, is called the Divergence Theorem, and will be proved in a more rigorous manner in chapter V. In the meantime we shall make use of it to deduce some of the differential equations of physics. In particular if a vector is solenoidal, $\operatorname{div} \mathbf{A} = 0$, and since the volume integral vanishes, the surface integral of flux through any closed surface is zero. Conversely if the flux through every closed surface in the field is zero, the vector is solenoidal throughout the field. For the theorem may be applied to any volume however small, and an integral cannot vanish in this case unless the integrand vanishes. Applying the divergence theorem to any tube of flow, and any cross-sections, whether or not orthogonal, we find the fundamental property of solenoidal vectors, for there is no flow across the walls of the tubes, hence the flow across one section is the same as for any other.

If the flow is not solenoidal, more fluid will flow out from S than in, or more in than out. In the first the fluid within S must have expanded, on the whole, in the second contracted. If we consider, not volumes, but masses, the mass in an element of volume is related to the volume by the relation $dm = \varrho \, d\tau$ where ϱ, the density, may vary from point to point and from instant to instant. Then the mass inside any surface is

$$(72) \qquad m = \iiint \varrho \, d\tau,$$

where the integral is extended over the volume inside S. The mass flowing in in unit time is found by multiplying the velocity $\mathbf{A}$ by the density ϱ. Thus since the gain of mass is equal to what comes in through the surface, we have for the rate of increase

$$(73) \qquad \frac{\partial m}{\partial t} = \frac{\partial}{\partial t} \iiint \varrho \, d\tau = -\iint \varrho \mathbf{A}_n \, dS.$$

We proceed as follows: since the surface S is fixed, and does not depend on the time, we may differentiate under the integral sign, then applying the divergence theorem to the surface integral we convert it into a volume integral over the same volume as the volume-integral, and transposing we have

$$(74) \qquad \int\int\int \left\{ \frac{\partial \varrho}{\partial t} + \text{div}\,(\varrho \mathbf{A}) \right\} d\tau = 0.$$

As before, since the reasoning applies to any volume whatever in the field, the integrand vanishes and we have

$$(75) \qquad \frac{\partial \varrho}{\partial t} + \text{div}\,(\varrho \mathbf{A}) = 0.$$

This is known as the *equation of continuity.* Using equation (58) this becomes

$$(76) \qquad \frac{\partial \varrho}{\partial t} + \mathbf{A}\nabla\varrho + \varrho\,\text{div}\,\mathbf{A} = 0.$$

The operation $\frac{\partial}{\partial t} + \mathbf{A}\nabla$, which occurs in (76), has a definite physical significance. If we fix our attention, not on a point fixed in space, but upon a point moving with velocity $\mathbf{A}$, any function of this moving point $f(x, y, z, t)$ will have the total differential

$$df = \frac{\partial f}{\partial t}dt + \frac{\partial f}{\partial x}dx + \frac{\partial f}{\partial y}dy + \frac{\partial f}{\partial z}dz,$$

and dividing by dt, since $\frac{dx}{dt} = \mathbf{A}_x, \quad \frac{dy}{dt} = \mathbf{A}_y, \quad \frac{dz}{dt} = \mathbf{A}_z,$

$$(77) \qquad \frac{df}{dt} = \frac{\partial f}{\partial t} + \mathbf{A}\nabla f$$

so that this operator denotes the time rate of change of the function of the point *as it moves.*

If we consider a small region of mass m and volume v, $m = \varrho v$, and since its mass does not change as it moves,

$$0 = \frac{dm}{dt} = \varrho\frac{dv}{dt} + v\frac{d\varrho}{dt}, \quad \frac{1}{\varrho}\frac{d\varrho}{dt} = -\frac{1}{v}\frac{dv}{dt}$$

and (76) becomes

$$\frac{1}{v}\frac{dv}{dt} = \text{div}\,\mathbf{A},$$

corresponding with the definition of div $\mathbf{A}$ in (55).

Suppose that we imagine fluid to be created or introduced into a portion of space without flowing in from outside, and that e represents the density of *strength of source,* or time-rate of creation per unit volume. Then we must add to the right hand member of (73) the integral representing the total strength of source, $Q = \int\int\int e\, d\tau$ so that instead of (75) we obtain

(75′) $\dfrac{\partial \varrho}{\partial t} + \operatorname{div}(\varrho A) = e.$

If ϱ is independent of the time we then interpret $\operatorname{div}(\varrho A)$ as the source-density.

Applying the results of this section to any vector, the flux $\int\int A_n\, dS$ represents the number of unit tubes crossing the surface, and in a region where $\operatorname{div} A > 0$ more tubes are issuing than entering, where $\operatorname{div} A < 0$ more are entering, hence the name divergence.

9. Notation of Hyperspace. As in space of three dimensions we call a set of quantities x, y, z the coordinates of a point so we may speak of the set of quantities

$$x_1, x_2, \ldots x_n$$

as the coordinates of a point in space of n-dimensions. This mode of speaking (for it is no more) enables us to transfer to algebra the language of geometry. For instance we speak of an equation

$$\varphi(x_1, x_2, \cdots x_n) = c$$

as the equation of a *hypersurface*, and a set of equations,

$$x_1 = f_1(t),\ x_2 = f_2(t),\ \cdots x_n = f_n(t),$$

as the equations of a *line* in n-dimensional space. We speak of the distance r of the point from the origin

$$r^2 = x_1^2 + x_2^2 + \cdots + x_n^2,$$

and its direction cosines

$$\frac{x_1}{r},\ \frac{x_2}{r},\ \cdots \frac{x_n}{r},$$

satisfying the relation

$$\cos^2(r x_1) + \cos^2(r x_2) + \cdots + \cos^2(r x_n) = 1.$$

We may speak of an n-dimensional vector

$$A \equiv A_{x_1}, A_{x_2}, \cdots A_{x_n},$$

of absolute value $|A|$, where,

$$|A|^2 = A_{x_1}^2 + A_{x_2}^2 + \cdots + A_{x_n}^2.$$

The gradient of a scalar function φ will be

$$\nabla\varphi \equiv \frac{\partial\varphi}{\partial x_1},\ \frac{\partial\varphi}{\partial x_2},\ \cdots \frac{\partial\varphi}{\partial x_n},$$

the divergence

$$\operatorname{div} A \equiv \frac{\partial A_{x_1}}{\partial x_1} + \frac{\partial A_{x_2}}{\partial x_2} + \cdots + \frac{\partial A_{x_n}}{\partial x_n},$$

and in a similar way all expressions may be generalized.

In the definition of a continuous function $f(x_1, x_2, \ldots x_n)$ we have merely to put for the change of the variable $|x - x'|$ the *remove* (*écart*) or distance r between the original point $x_1, x_2, \ldots x_n$ and the final point $x'_1, x'_2, \ldots x'_n$

$$r^2 = (x'_1 - x_1)^2 + (x'_2 - x_2)^2 + \cdots + (x'_n - x_n)^2,$$

and write

$$|f' - f| < \varepsilon \text{ for all points for which } r < \delta.$$

10. Flow of Heat and Electricity. In the example of the flow of fluid the vector $\varrho A = C$, defined as the amount of matter crossing unit area normal to the flow in unit time, is called the *current-density*. In the conduction of heat or electricity, formerly spoken of as imponderable fluids, although we may not speak of velocity, we may likewise define the current-density as a vector denoting the quantity crossing a surface normal to the direction of flow per unit of area and unit of time. Thus the amount flowing into any region in unit of time is (71)

$$\iint C_n \, dS = -\iiint \operatorname{div} C \cdot d\tau.$$

If we are dealing with flow of electricity, we obtain, as in (75)

$$(78) \qquad \frac{\partial \varrho}{\partial t} + \operatorname{div} C = 0,$$

where ϱ is the density of electrification.

In case we are dealing with flow of heat by conduction, the effect of the inflow is to heat the substance according to the law that the increase of temperature is proportional to the amount of heat flowing in. If φ denotes the temperature, Q, quantity of heat, m, the mass heated, then $dQ = cm\, d\varphi$ where c denotes the specific heat of the substance, then equation (73) becomes

$$(79) \qquad \frac{\partial Q}{\partial t} = \iiint \frac{\partial \varphi}{\partial t} c \varrho \, d\tau = \iint C_n \, dS = -\iiint \operatorname{div} C \cdot d\tau,$$

and (78) becomes

$$(80) \qquad c\varrho \frac{\partial \varphi}{\partial t} + \operatorname{div} C = 0.$$

We must now consider the law of flow, which is the same both for heat and electricity. The law was enunciated by Fourier in 1822 for heat and applied to electricity by Ohm in 1827. This law states that the current density is proportional to the gradient of the temperature in the case of heat or of the electric potential for electricity. Thus we have

$$(81) \qquad C = -\varkappa \operatorname{grad} \varphi$$

where $\varkappa$ denotes a physical property of the medium, called the conductivity for heat or electricity respectively. Introducing the equation (81) in (80) we obtain

$$(82) \qquad c\varrho\frac{\partial\varphi}{\partial t} = \operatorname{div}(\varkappa \operatorname{grad}\varphi)$$

which accordingly, (if $\varkappa$ is constant), necessitates the differential equation

$$(83) \qquad c\varrho\frac{\partial\varphi}{\partial t} = \varkappa\Delta\varphi.$$

This is the equation given by Fourier for the conduction of heat. In the case the flow is steady, that is to say, independent of the time, it reduces to Laplace's equation. If there are sources of heat, we add e on the right of (83). In order to get a similar equation for electricity, we require a relation between the electrical density ϱ and the potential φ, which will be given in Chapter V. We do *not* find ϱ proportional to φ.

11. Principles of Mechanics. We shall now give a very brief statement of the laws of mechanics in order to deduce some of its differential equations. According to the laws of Newton, the force applied to any particle is proportional to the product of its mass and acceleration. the mass being the scalar quantity characterizing the body and the acceleration and force being vectors. If **A** represents the acceleration, it is defined by the components

$$(84) \qquad \mathbf{A}_x = \frac{d^2x}{dt^2}, \quad \mathbf{A}_y = \frac{d^2y}{dt^2}, \quad \mathbf{A}_z = \frac{d^2z}{dt^2}, \quad \text{or} \quad \mathbf{A} = \frac{d^2\mathbf{r}}{dt^2},$$

where $\mathbf{r} \equiv x, y, z$, is the position vector of the point. In the case of a system of a number of points not free to move in all directions, but connected together by certain constraints, it is most convenient to use the method introduced by Lagrange known as that of generalized coordinates. By this we mean a number of independent parameters q_1, $q_2, \ldots q_n$ in terms of which we are able to specify the positions of all the points of the system. The time derivatives $q_r' = \dfrac{dq_r}{dt}$ of these parameters we term the corresponding velocities. We have then to consider the function T called the kinetic energy which is defined as the sum for all points of the product of one half the mass by the square of the velocity,

$$(85) \qquad T = \frac{1}{2}\sum mv^2.$$

If we have $x_r = f_r(q_1, q_2, \ldots q_n)$, $y_r = g_r(q_1, q_2, \ldots q_n)$, $z_r = h_r(q_1, q_2, \ldots q_n)$

$$(86) \qquad \begin{aligned}
\frac{dx_r}{dt} &= \frac{\partial f_r}{\partial q_1}q_1' + \frac{\partial f_r}{\partial q_2}q_2' + \cdots + \frac{\partial f_r}{\partial q_n}q_n', \\[4pt]
\frac{dy_r}{dt} &= \frac{\partial g_r}{\partial q_1}q_1' + \frac{\partial g_r}{\partial q_2}q_2' + \cdots + \frac{\partial g_r}{\partial q_n}q_n', \\[4pt]
\frac{dz_r}{dt} &= \frac{\partial h_r}{\partial q_1}q_1' + \frac{\partial h_r}{\partial q_2}q_2' + \cdots + \frac{\partial h_r}{\partial q_n}q_n'.
\end{aligned}$$

We find the velocities are expressed linearly in terms of the generalized velocities $q_1', q_2', \ldots q_n'$. Squaring, multiplying by the masses and summing, we find for T a *homogeneous quadratic function* of the *velocities* $q_1', q_2', \ldots q_n'$

$$(87) \qquad T = \frac{1}{2} \sum_{r=1}^{r=n} \sum_{s=1}^{s=n} Q_{rs} q_r' q_s',$$

the coefficients Q_{rs} of which are functions of the *coordinates* $q_1, q_2, \ldots q_n$.

Besides T we deal with a function of the coordinates q alone, termed the potential energy W, which is defined by the property that the force acting to change any coordinate q_s is

$$(88) \qquad P_r = - \frac{\partial W}{\partial q_r}.$$

We may in both T and W distinguish between the self- and mutual energies. If the system moves so that only one coordinate q_r changes, the kinetic energy reduces to $\frac{1}{2} Q_{rr} q_r'^2$. This we will call the self-energy belonging to this coordinate. If q_r and q_s change simultaneously, besides the two self-energies $\frac{1}{2} Q_{rr} q_r'^2 + \frac{1}{2} Q_{ss} q_s'^2$, there is the energy $Q_{rs} q_r' q_s'$, which vanishes with either velocity, and which we may call the *mutual* energy of the two motions. Such a decomposition of the potential energy will be possible in particular cases to be mentioned later. Now we are able to sum up all the laws of mechanics in a single formula known as Hamilton's Principle which reads as follows:

$$(89) \qquad \delta \int_{t_0}^{t_1} (T - W) dt = 0.$$

The symbol δ is to be taken in the sense of the Calculus of Variations, which is as follows: suppose that we know the path actually taken by every point of the system in a given motion, that is to say we have

$$(90) \qquad \begin{aligned} q_1 &= \varphi_1(t), \quad q_2 = \varphi_2(t), \ldots q_n = \varphi_n(t) \\ q_1' &= \frac{d\varphi_1}{dt}, \quad q_2' = \frac{d\varphi_2}{dt}, \ldots q_n' = \frac{d\varphi_n}{dt}. \end{aligned}$$

We can then calculate for every instant of time the kinetic and potential energies of the system, form their differences, and integrate throughout the motion. If now we suppose that the paths described by the particles are changed for each value of t by an infinitesimal amount, by changing the form of the functions $\varphi_r(t)$ to $\varphi_r + \delta q_r$ where the quantities δq_r are *arbitrary* functions of t, multiplied by an infinitesimal constant ε, then the change in the integral thus brought about and denoted by δ must vanish as far as the terms in the first power of ε. It is to be understood that the limits of integration denoting the time of start and finish of the motion as well as the positions at those instants are definitely given. Instead of proving this formula, we will show that it leads to the laws of Newton.

In the case of a single free point whose motion is described in terms of rectangular coordinates x, y, z, the velocity-square is $\left(\frac{dx}{dt}\right)^2 + \left(\frac{dy}{dt}\right)^2 + \left(\frac{dz}{dt}\right)^2$, and we have

$$(91) \qquad T = \frac{1}{2}\,m\left[\left(\frac{dx}{dt}\right)^2 + \left(\frac{dy}{dt}\right)^2 + \left(\frac{dz}{dt}\right)^2\right] = \frac{1}{2}\,m(x'^2 + y'^2 + z'^2).$$

Then Hamilton's Principle is

$$(92) \qquad \delta\int_{t_0}^{t_1}\left\{\frac{m}{2}(x'^2 + y'^2 + z'^2) - W\right\}dt = 0,$$

and using the first principle of the Calculus of Variations, namely that the operations of integration and variation may be commuted,

$$(93) \qquad \int_{t_0}^{t_1}\left\{m(x'\,\delta x' + y'\,\delta y' + z'\,\delta z') - \delta W\right\}dt = 0.$$

We now use the second principle, that the operations of differentiation by the independent variable t and variation may be commuted, that is

$$(94) \qquad \delta x' = \delta\frac{dx}{dt} = \frac{d}{dt}\,\delta x.$$

The term

$$\int_{t_0}^{t_1} x'\,\frac{d}{dt}\,\delta x\,dt$$

can then be integrated by parts,

$$(95) \qquad \int_{t_0}^{t_1} x'\,\frac{d}{dt}\,\delta x\,dt = x'\,\delta x\Big|_{t_0}^{t_1} - \int_{t_0}^{t_1}\delta x\,\frac{d^2x}{dt^2}\,dt.$$

Now at the limits t_0, t_1 the positions are given, so that δx is zero. Thus the integrated part vanishes. We proceed thus in the first three terms. In the next place, the potential energy W depending only on the coordinates x, y, z, we have

$$(96) \qquad \delta W = \frac{\partial W}{\partial x}\,\delta x + \frac{\partial W}{\partial y}\,\delta y + \frac{\partial W}{\partial z}\,\delta z.$$

$$(97) \quad \int_{t_0}^{t_1}\left[\left(m\frac{d^2x}{dt^2} + \frac{\partial W}{\partial x}\right)\delta x + \left(m\frac{d^2y}{dt^2} + \frac{\partial W}{\partial y}\right)\delta y + \left(m\frac{d^2z}{dt^2} + \frac{\partial W}{\partial z}\right)\delta z\right]dt = 0.$$

Now since the variations δx, δy, δz are perfectly *arbitrary* functions of t, it is impossible that the integral should vanish for all possible choices unless the parentheses multiplying them vanish, giving

$$(98) \qquad m \frac{d^2 x}{dt^2} = - \frac{\partial W}{\partial x}, \; m \frac{d^2 y}{dt^2} = - \frac{\partial W}{\partial y}, \; m \frac{d^2 z}{dt^2} = - \frac{\partial W}{\partial z}.$$

But by definition, the derivatives of W are the components of the forces acting, so that these are Newton's equations of motion.

Suppose now we have the system specified in terms of generalized coordinates, all independent. Then, since W depends only on the q's

$$(99) \qquad \delta W = \frac{\partial W}{\partial q_1} \delta q_1 + \frac{\partial W}{\partial q_2} \delta q_2 + \cdots + \frac{\partial W}{\partial q_n} \delta q_n.$$

On the contrary T depends both on the q's and q''s.

$$\delta T = \frac{\partial T}{\partial q_1} \delta q_1 + \cdots + \frac{\partial T}{\partial q_n} \delta q_n + \frac{\partial T}{\partial q_1'} \delta q_1' + \cdots + \frac{\partial T}{\partial q_n'} \delta q_n'.$$

Hamilton's equation then reads,

$$(100) \qquad \int_{t_0}^{t_1} \sum_{r=1}^{r=n} \left\{ \frac{\partial T}{\partial q_r'} \delta q_r' + \left(\frac{\partial T}{\partial q_r} - \frac{\partial W}{\partial q_r} \right) \delta q_r \right\} dt = 0.$$

As before we have

$$(101) \qquad \delta q_r' = \delta \frac{dq_r}{dt} = \frac{d}{dt} \delta q_r,$$

and integrating by parts, the integrated parts vanishing for the same reason as before,

$$(102) \qquad \int_{t_0}^{t_1} \sum_{r=1}^{r=n} \delta q_r \left\{ \frac{\partial T}{\partial q_r} - \frac{\partial W}{\partial q_r} - \frac{d}{dt} \left(\frac{\partial T}{\partial q_r'} \right) \right\} dt = 0.$$

Since the δq's are all arbitrary, as before, the coefficients must vanish, and we have

$$(103) \qquad \frac{d}{dt} \left(\frac{\partial T}{\partial q_r'} \right) - \frac{\partial T}{\partial q_r} = - \frac{\partial W}{\partial q_r} = P_r.$$

These equations, one for each coordinate q, are known as Lagrange's equations for generalized coordinates. In the case where the q's are Cartesian coordinates, the terms $\frac{\partial T}{\partial q_r}$ do not appear. From analogy with Cartesian coordinates, we call the quantities $p_r = \frac{\partial T}{\partial q_r'}$, the *momenta* corresponding to coordinate q_r, velocity q_r', and force P_r.

12. Equation of Hydrodynamics and Sound. If we consider the motion of the fluid contained in an infinitesimal rectangular parallelopiped of dimensions dx, dy, dz, and of mass $\rho\,dx\,dy\,dz$, the forces acting on it are not only those depending on the presence of a field of force, such as gravity, and proportional to the mass acted on, equal to F per unit of mass of the fluid, but also to the pressure acting normally on the six faces. On the face perpendicular to the x-axis, and nearest the origin, of area $dy\,dz$, the pressure p (force per unit area) produces the force $p\,dy\,dz$ toward the right. On the opposite face instead of p we have $p + \frac{\partial p}{\partial x}dx$ multiplied by $dy\,dz$, acting toward the left. The resultant toward the right is thus $-\frac{\partial p}{\partial x}dx\,dy\,dz$. Adding this to the component $F_x\varrho\,dx\,dy\,dz$ we have the total force acting on the mass $\varrho\,dx\,dy\,dz$, which is to be equated to the mass multiplied by the acceleration. Since acceleration is the rate of change of velocity of a particular particle as it moves, we have by (77) if $\mathbf{v}$ is the vector velocity of the fluid,

$$(104) \qquad \varrho\,dx\,dy\,dz\left(\frac{\partial \mathbf{v}}{\partial t} + \mathbf{v}\nabla\mathbf{v}\right)_x = F_x\varrho\,dx\,dy\,dz - \frac{\partial p}{\partial x}dx\,dy\,dz.$$

In like manner we have the other components, so that dividing by $\varrho\,dx\,dy\,dz$ we have the vector equation

$$(105) \qquad \frac{\partial \mathbf{v}}{\partial t} + \mathbf{v}\nabla\mathbf{v} = \mathbf{F} - \frac{1}{\varrho}\operatorname{grad} p$$

or written out,

$$(106) \qquad \begin{aligned}
\frac{\partial v_x}{\partial t} + v_x\frac{\partial v_x}{\partial x} + v_y\frac{\partial v_x}{\partial y} + v_z\frac{\partial v_x}{\partial z} &= F_x - \frac{1}{\varrho}\frac{\partial p}{\partial x}, \\
\frac{\partial v_y}{\partial t} + v_x\frac{\partial v_y}{\partial x} + v_y\frac{\partial v_y}{\partial y} + v_z\frac{\partial v_y}{\partial z} &= F_y - \frac{1}{\varrho}\frac{\partial p}{\partial y}, \\
\frac{\partial v_z}{\partial t} + v_x\frac{\partial v_z}{\partial x} + v_y\frac{\partial v_y}{\partial y} + v_z\frac{\partial v_z}{\partial z} &= F_z - \frac{1}{\varrho}\frac{\partial p}{\partial z}.
\end{aligned}$$

With this we combine the equation of continuity (76)

$$(107) \qquad \frac{\partial \varrho}{\partial t} + \mathbf{v}\nabla\varrho + \varrho\operatorname{div}\mathbf{v} = 0.$$

The term $\mathbf{v}\nabla\mathbf{v}$, as we see, involves components of $\mathbf{v}$ multiplied by their derivatives. The equation is much simplified if the velocities are so small that their products with each other and with other derivatives, being of the second order, may be neglected. We shall consider two such cases, the first being the motion of sound in air. We then put,

$$(108) \qquad \frac{\partial \mathbf{v}}{\partial t} = \mathbf{F} - \frac{1}{\varrho}\operatorname{grad} p.$$

Under the circumstances assumed, the changes in density are small, so that we may put

$$(109) \qquad \varrho = \varrho_0(1 + s),$$

where ϱ_0 is a constant, and s is a small quantity called the *compression*. Differentiating logarithmically,

$$(110) \qquad \frac{d\varrho}{\varrho} = \frac{ds}{1+s} = ds,$$

if we neglect s in comparison with unity. From this and (107) we find

$$(111) \qquad -\frac{\partial s}{\partial t} = \operatorname{div} \mathbf{v}$$

as the equation of continuity.

Let us call q the displacement of the particle of air, so that $\mathbf{v} = \frac{d\mathbf{q}}{dt}$. Then (108) is

$$(112) \qquad \frac{\partial^2 \mathbf{q}}{\partial t^2} = \mathbf{F} - \frac{1}{\varrho}\operatorname{grad} p.$$

The equation of continuity becomes $\left(\text{since } \frac{\partial}{\partial t}\operatorname{div} = \operatorname{div}\frac{\partial}{\partial t}\right).$ [1])

$$(113) \qquad -\frac{\partial s}{\partial t} = \frac{\partial}{\partial t}\operatorname{div} \mathbf{q}, \quad \text{or} \quad s = -\operatorname{div} \mathbf{q}.$$

We now need a physical relation between the density and pressure, and we assume $dp = eds$, where e is the coefficient of elasticity of the fluid. Accordingly

$$\operatorname{grad} p = e \operatorname{grad} s.$$

Taking the divergence of (112),

$$\frac{\partial^2}{\partial t^2}\operatorname{div} q = \operatorname{div} \mathbf{F} - \frac{e}{\varrho}\operatorname{div}\operatorname{grad} s,$$

and by (113),

$$(114) \qquad \frac{\partial^2 s}{\partial t^2} = \frac{e}{\varrho}\Delta s - \operatorname{div} \mathbf{F}.$$

---

[1] Since we have $\frac{\partial^2 u}{\partial x \partial y} = \frac{\partial^2 u}{\partial y \partial x}$, all linear operations involving differentiation and multiplication by constants are commutative, e. g., if D_1 stands for the operator $D_1 = a\frac{\partial}{\partial x} + b\frac{\partial}{\partial y} + c\frac{\partial}{\partial z} + \cdots$, and D_2 for the operator $D_2 = \alpha\frac{\partial}{\partial x} + \beta\frac{\partial}{\partial y} + \gamma\frac{\partial}{\partial z} + \cdots$, we have $D_1 D_2 u = D_2 D_1 u$, which is an abbreviation for

$$\left(a\frac{\partial}{\partial x} + b\frac{\partial}{\partial y} + c\frac{\partial}{\partial z} + \cdots\right)\left(\alpha\frac{\partial u}{\partial x} + \beta\frac{\partial u}{\partial y} + \gamma\frac{\partial u}{\partial z} + \cdots\right)$$

$$= a\alpha\frac{\partial^2 u}{\partial x^2} + (a\beta + b\alpha)\frac{\partial^2 u}{\partial x \partial y} + \cdots$$

$$= \left(\alpha\frac{\partial}{\partial x} + \beta\frac{\partial}{\partial y} + \gamma\frac{\partial}{\partial z} + \cdots\right)\left(a\frac{\partial u}{\partial x} + b\frac{\partial u}{\partial y} + c\frac{\partial u}{\partial z} + \cdots\right).$$

If there are no applied forces, the equation

(115) $$\frac{\partial^2 s}{\partial t^2} = a^2 \Delta s, \quad a^2 = \frac{e}{\varrho},$$

is the ordinary equation for the propagation of sound, and we shall refer to it as the *wave-equation*. A region where div F is not zero is called a source of sound.

If the applied force F vanishes, or is lamellar, it may be shown that v is also lamellar, or derived from a velocity potential φ. Let us write

$$\mathbf{F} = -\operatorname{grad} V, \quad \mathbf{v} = \frac{\partial \mathbf{q}}{\partial t} = \operatorname{grad}\varphi, \quad \frac{\partial s}{\partial t} = -\operatorname{div}\frac{\partial \mathbf{q}}{\partial t} = -\Delta\varphi,$$

then (112) becomes

(112') $$\frac{\partial}{\partial t}\operatorname{grad}\varphi = -\operatorname{grad} V - \frac{e}{\varrho}\operatorname{grad} s.$$

Differentiating by t, this may be written

$$\operatorname{grad}\left\{\frac{\partial^2\varphi}{\partial t^2} + \frac{\partial V}{\partial t} - \frac{e}{\varrho}\Delta\varphi\right\} = 0.$$

A quantity whose gradient vanishes everywhere is constant in space, we therefore have, putting the parenthesis equal to zero,

(115') $$\frac{\partial^2\varphi}{\partial t^2} = a^2\Delta\varphi - \frac{\partial V}{\partial t},$$

or the velocity potential satisfies the same equation as the compression, $-\partial V/\partial t$ being proportional to the source-density.

In the simplest case, that of propagation in one dimension, put

$$\mathbf{q} = \mathbf{q}_x = u, \quad s = -\frac{\partial u}{\partial x}, \quad \frac{\partial p}{\partial x} = e\frac{\partial s}{\partial x} = -e\frac{\partial^2 u}{\partial x^2}$$

and equation (112) becomes

$$\frac{\partial^2 u}{\partial t^2} = a^2\frac{\partial^2 u}{\partial x^2} + \mathbf{F},$$

so that in this case the displacement itself satisfies the same wave-equation. Since $s = -\partial u/\partial x$, and we may commute the order of differentiation, a derivative of a solution is a solution of the linear equation, so that (if F = 0) s satisfies the same equation (115), as already proved.

The equation of sound thus differs from that of heat in having the second time derivative instead of the first. A source of sound corresponds exactly to a source of heat. If the phenomena of heat or sound depend only on x, or on x and y, the Laplacian Δ reduces to $\frac{\partial^2}{\partial x^2}$ or $\frac{\partial^2}{\partial x^2} + \frac{\partial^2}{\partial y^2}$ respectively.

In connection with the sound problem let us consider the motion of a stretched flexible string or membrane. In either case let us consider the weight to be negligible in comparison with the tension τ.

Consider a portion of string of length ds, and mass $\rho\, ds$, and let it be displaced at each point by an amount y so small that we may neglect the square of the angle made by the tangent with the x-axis taken as the equilibrium position of the string. Then we may put

$$ds = dx\left\{1 + \left(\frac{\partial y}{\partial x}\right)^2\right\}^{\frac{1}{2}} = dx,\quad \sin\theta = \frac{dy}{ds} = \frac{\partial y}{\partial x},$$ where θ is the angle made by the tangent with the x-axis, Fig. 14. Since we neglect stretching of the string and its weight, the tension τ is constant. Suppose that there is a transverse force Y per unit of length acting on the string. There is at the left hand end of the element ds the y-component of the tension

$$\tau \sin\theta = \tau\frac{\partial y}{\partial x}$$

pulling downwards, and at the right

$$\tau\sin(\theta + d\theta) = \tau\left[\frac{\partial y}{\partial x} + dx\frac{\partial^2 y}{\partial x^2}\right]$$

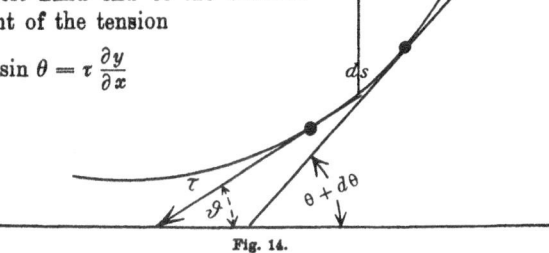

Fig. 14.

pulling upwards. The difference $\tau\, dx\dfrac{\partial^2 y}{\partial x^2}$, added to the force $Y\,dx$ is thus to be equated to the mass $\rho\, dx$ times the acceleration $\dfrac{\partial^2 y}{\partial t^2}$, giving, on dividing by dx,

$$(116)\qquad \rho\frac{\partial^2 y}{\partial t^2} = \tau\frac{\partial^2 y}{\partial x^2} + Y.$$

Thus the displacement of the string, if $Y = 0$, satisfies the same equation as the motion of sound depending on one coordinate. A region where Y does not vanish may be called a *source* of motion.

In the case of a membrane we have to find the total force normal to the membrane due to the tension. Again assuming the displacement z to be small, let the equation of the surface into which the plane membrane is distorted be $z = ax^2 + by^2$ as far as terms of the second order, the axes of x, y, being in the tangent plane in the direction of the lines of curvature. In polar coordinates

$$x = r\cos\varphi,\quad y = r\sin\varphi,$$

$$z = r^2(a\cos^2\varphi + b\sin^2\varphi) = \frac{r^2}{2}\{(a+b) + (a-b)\cos 2\varphi\},$$

$$\frac{\partial z}{\partial r} = \tan\theta = r\{(a+b) + (a-b)\cos 2\varphi\},$$

where θ represents the slope of a normal section of the surface with the horizontal, Fig. 15. The tension, in the case of a surface, is the force exerted across unit length of a curve. Calling this τ, let us consider the component in the z-direction of the tension on the boundary of a small circular area of radius r with center at the origin. This will be $\tau\,ds\,\sin\theta = \tau r\,d\varphi\dfrac{\partial z}{\partial r}$ on each element of the circumference, and

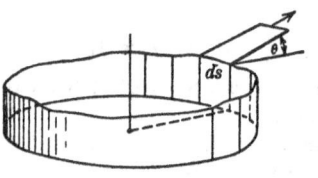

on the whole

$$Z' = \int_0^{2\pi} \tau r^2 [(a+b) + (a-b)\cos 2\varphi]\,d\varphi$$

$$= 2\pi\tau r^2(a+b) = 2\tau A(a+b),$$

where A is the area of the circle. But from the equation of the surface this gives

Fig. 15.

$$Z' = \tau A\left[\frac{\partial^2 z}{\partial x^2} + \frac{\partial^2 z}{\partial y^2}\right]_{\substack{x=0 \\ y=0}}$$

so that the effect of the tension is proportional to the Laplacian for two dimensions, as in the string for one. If in addition there is an applied force Z per unit of area, we must add ZA to Z' and equate to the mass σA times the acceleration. Accordingly we obtain the equation

(117)
$$\sigma\frac{\partial^2 z}{\partial t^2} = \tau\left[\frac{\partial^2 z}{\partial x^2} + \frac{\partial^2 z}{\partial y^2}\right] + Z,$$

where σ is the *surface* density, τ the *surface* tension. A region of Z not equal to zero is a source. Thus the motion of strings and membranes is governed by the same equation as that of sound in one or two dimensions.

The motion of waves on the surface of water is more complicated than any of the motions already treated, but is simplified if we assume that the displacements are small in comparison with the depth of the water, and that the slope is small. This is the case in tidal waves in the ocean. We shall again neglect the terms of the second order, and write, if q is the displacement, $F_z = -g$,

(118)
$$\varrho\frac{\partial^2 q_x}{\partial t^2} = -\frac{\partial p}{\partial x} + \varrho F_x,$$

$$\varrho\frac{\partial^2 q_y}{\partial t^2} = -\frac{\partial p}{\partial y} + \varrho F_y,$$

$$\varrho\frac{\partial^2 q_z}{\partial t^2} = -\frac{\partial p}{\partial z} - \varrho g.$$

We shall neglect the vertical acceleration, so that

(119)
$$\frac{\partial p}{\partial z} = -\varrho g,$$

that is we assume the pressure to be due merely to the depth of the water. If P is the pressure at the surface, h the undisturbed depth of the water, ζ the elevation of the surface above the undisturbed level, integrating (119) from the bottom up to the point in question we have

(120) $$p = P + g\varrho(h + \zeta - z)$$

(121) $$\frac{\partial p}{\partial x} = \frac{\partial P}{\partial x} + g\varrho\frac{\partial\zeta}{\partial x}, \quad \frac{\partial p}{\partial y} = \frac{\partial P}{\partial y} + g\varrho\frac{\partial\zeta}{\partial y}.$$

Since P and ζ depend only on x and y, if we make the assumption that F_x, F_y do likewise, we find that the horizontal acceleration depends only on x and y, so that a vertical line of particles remains vertical, and we have to consider only the motion at the surface.

Inserting these values in (118) we have at the surface,

(122)
$$\varrho\frac{\partial^2 q_x}{\partial t^2} = -g\varrho\frac{\partial\zeta}{\partial x} - \frac{\partial P}{\partial x} + \varrho F_x,$$
$$\varrho\frac{\partial^2 q_y}{\partial t^2} = -g\varrho\frac{\partial\zeta}{\partial y} - \frac{\partial P}{\partial y} + \varrho F_y.$$

The liquid being incompressible, the equation of continuity is

$$\operatorname{div} q = \frac{\partial q_x}{\partial x} + \frac{\partial q_y}{\partial y} + \frac{\partial q_z}{\partial z} = 0,$$

from which we get the value of ζ,

(123) $$\zeta = -\int_0^h \left[\frac{\partial q_x}{\partial x} + \frac{\partial q_y}{\partial y}\right] dz = -h\left[\frac{\partial q_x}{\partial x} + \frac{\partial q_y}{\partial y}\right].$$

Differentiating (122) respectively by x and y, multiplying by $-h$, adding and using (123) we find

(124) $$\varrho\frac{\partial^2\zeta}{\partial t^2} = g\varrho h\left[\frac{\partial^2\zeta}{\partial x^2} + \frac{\partial^2\zeta}{\partial y^2}\right] + h\left[\frac{\partial^2 P}{\partial x^2} + \frac{\partial^2 P}{\partial y^2}\right] + \varrho h\left[\frac{\partial F_x}{\partial x} + \frac{\partial F_y}{\partial y}\right].$$

Thus the equation is the same as for sound in two dimensions or for a membrane. A region where $\Delta P + \varrho \operatorname{div} F$ is not zero is a source.

13. Equations of Elasticity. In the case of a solid body the forces developed by the relative motion of its parts are of a more complicated nature than in the case of a fluid, since the forces across a plane element separating two portions of the solid are not normal to the element, as in the case of fluid pressure.

The stress, or system of forces developed, is known if we know at each point the resultant action across any small plane element dS between the matter on its opposite sides. If we call this force $P dS$, P is called the stress-vector, and depends on the orientation of the element dS. We may characterize the stress by giving the stress vector on each side of an infinitesimal rectangular parallelopiped, Fig. 16. Let us suppose that at the point x, y, z, the stress-vector on the side normal to the x-axis is A, on that normal to y is B, and on that normal to z is C, and it is to be understood that these are the forces acting in the positive direction on the matter on the negative sides of the three

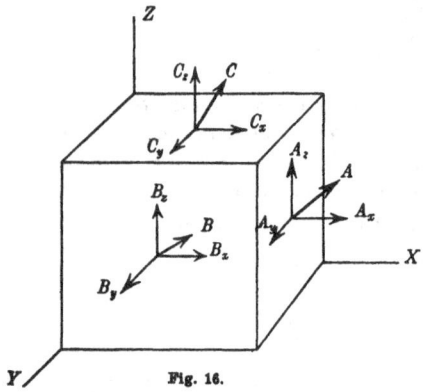

Fig. 16.

planes. The strain is characterized by the nine components

$$A_x, A_y, A_z, B_x, B_y, B_z, C_x, C_y, C_z$$

and it can be easily shown that the lack of tendency to turn the cube entails the equalities,

$$(125) \quad B_z = C_y, \quad C_x = A_z,$$
$$A_y = B_x.$$

Of these A_x, B_y, C_z are the normal tractions in the three coordinate directions. The others are tangential or *shearing* forces. They are zero for a perfect fluid.

Consider the resultant of all the X forces acting on the cube. On the side normal to x nearest the origin, Fig. 17, we have $A_x dy dz$, on the opposite side $(A_x + \frac{\partial A_x}{\partial x} dx) dy dz$, pulling in the opposite direction, the resultant of both being $\frac{\partial A_x}{\partial x} dx dy dz$.

On the side normal to y nearest the origin we have the tangential force $(B_x = A_y) dz dx$ and on the opposite side $(A_y + \frac{\partial A_y}{\partial y} dy) dz dx$ in

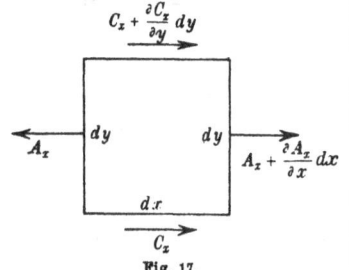

Fig. 17.

the opposite direction, the resultant being $\frac{\partial A_y}{\partial y} dx dy dz$. From the other two sides we get the resultant $\frac{\partial A_z}{\partial z} dx dy dz$. In all then we have in the X direction,

$$\operatorname{div} A \, dx dy dz,$$

and this is the force tending to move the cube in the X direction, beside the force due to an external field $\varrho F_x dx dy dz$ as in the case of a fluid.

The mass of the cube is $\varrho dx dy dz$, if ϱ is the density. The displacement is q_x, hence the equation of motion

$$\varrho \frac{\partial^2 q_x}{\partial t^2} = \operatorname{div} A + \varrho F_x.$$

Similarly in the other directions

$$(126) \quad \varrho \frac{\partial^2 q_y}{\partial t^2} = \operatorname{div} B + \varrho F_y$$
$$\varrho \frac{\partial^2 q_z}{\partial t^2} = \operatorname{div} C + \varrho F_z.$$

We must now find expressions for the stress in terms of the three linear expansions s_x, s_y, s_s and the three shears g_x, g_y, g_s specifying the strain. It is found by experiment that for small relative displacements the six stress-components may be put equal to linear functions of these six quantities with constant coefficients, known as the elastic constants of the body.

If the body is elastically isotropic, that is alike in all directions, their number is reduced to two. First, if the body receives upon the two faces normal to the x-axis equal and opposite normal stresses A_x, it will experience a *stretch* $s_x = aA_x$ in the x-direction, but will be contracted in the y and s directions by a different amount $s_y = s_s = -bA_x$, where a and b are the two elastic constants. Supposing normal stresses B_y and C_s on the other faces, we get for the resultant stretches,

$$(127) \qquad \begin{aligned} s_x &= aA_x - b(B_y + C_s)\,, \\ s_y &= aB_y - b(C_s + A_x)\,, \\ s_s &= aC_s - b(A_x + B_y)\,, \end{aligned}$$

giving the cubical dilatation

$$(128) \qquad \sigma = s_x + s_y + s_s = (a - 2b)(A_x + B_y + C_s).$$

Secondly suppose the cube is deformed without change of volume so that every plane parallel to the xz-plane is shoved in the x-direction by an amount proportional to its distance from that plane, so that

$$(129) \qquad q_x = 2gy\,, \quad q_y = q_s = 0.$$

The square side in the xy-plane is distorted into a rhombus, Fig. 18, the original right angle being changed by an angle $\dfrac{\partial q_x}{\partial y} = 2g$. Such a strain is called a shear of angle $2g$, and is caused by a tangential stress $B_x = T$ on the top-side, with an equal and opposite on the bottom, proportional to the angle of shear,

$$(130) \qquad 2g\mu = T$$

where μ is called the modulus of rigidity. (This is zero for perfect fluids, which can be sheared without stress). We will relate μ to the constants a and b.

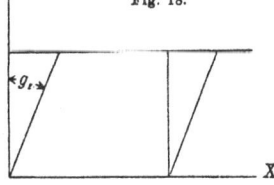

Fig. 18.

The above shear has rotated the diagonal of the square into the diagonal of the rhombus, but we easily see that if we turn it back, through the angle g the same strain is given by the equations

$$(131) \qquad q_x = g_s y\,, \quad q_y = g_s x\,, \quad q_s = 0.$$

The square $OACB$ (Fig. 19) becomes the rhombus $OA'C'B'$. The diagonals AB and OC maintain their directions unchanged, and are accordingly two of the axes of the pure strain, the axis of z being the third. The stretch-ratio along OC is

$$\frac{OC' - OC}{OC} = \frac{CC'}{OC} = \frac{C''C'}{BC} = \frac{BB'}{OB} = g_s,$$

as may be seen from the figure. The stretch along the perpendicular axis OE is negative.

$$\frac{OE' - OE}{OE} = -\frac{EE'}{OE} = -\frac{E'E''}{AE} = -g_s.$$

Consequently, a shear may be defined as a stretch along one axis combined with an equal negative stretch, or *squeeze*, along a perpen-

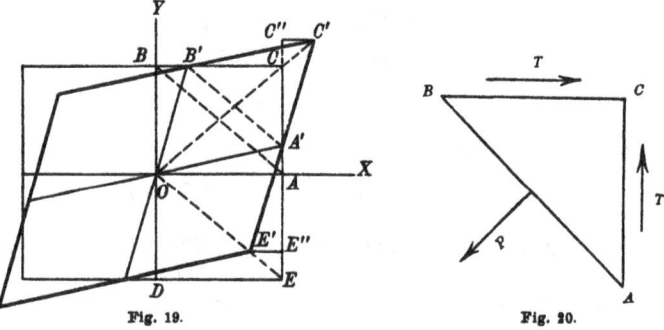

Fig. 19.   Fig. 20.

dicular axis, and no stretch along an axis perpendicular to both. It is evident that it may be produced by a normal stress, or traction P in the first direction, combined with an equal negative stress or pressure $-P$ in the second. Taking these for new axes x', y' we have by (127)

$$(132) \qquad \begin{aligned} s'_x &= aP - b(-P) = (a + b)P = g_s, \\ s'_y &= a(-P) - bP = -(a+b)P = -g_s. \end{aligned}$$

It remains only to connect the normal stress P with the tangential stress T above. This we may do by considering the equilibrium of the triangular prism ABC, Fig. 20, whose sides are of lengths 1, 1, $\sqrt{2}$. Evidently the shear is produced by two equal tangential forces on BC, AC, $T = 2\mu g_s$, but these are opposed by the normal traction P on the face AB. Resolving the two former by multiplying by

$$\cos 45^\circ = 1/\sqrt{2}$$

we have for the total component,

$$\frac{2T}{\sqrt{2}} = \sqrt{2} \cdot 2\mu g_s.$$

But P acts in the face of area $\sqrt{2}$, accordingly

$$P\sqrt{2} = \sqrt{2} \cdot 2\mu g,$$

and comparing with (132) we have $1/2\mu = a + b$.

Substituting $1/2\mu$ for $a + b$ in (127) we have, by (128)

(133)
$$s_x = \frac{A_x}{2\mu} - \frac{b}{a-2b}\,\sigma,$$
$$s_y = \frac{B_y}{2\mu} - \frac{b}{a-2b}\,\sigma,$$
$$s_z = \frac{C_z}{2\mu} - \frac{b}{a-2b}\,\sigma.$$

Solving for A_x, B_y, C_z, and writing $\lambda = 2\mu b/(a-2b)$,

(134)
$$A_x = \lambda\sigma + 2\mu s_x = \lambda\,\mathrm{div}\,q + 2\mu\frac{\partial q_x}{\partial x},$$
$$B_y = \lambda\sigma + 2\mu s_y = \lambda\,\mathrm{div}\,q + 2\mu\frac{\partial q_y}{\partial y},$$
$$C_z = \lambda\sigma + 2\mu s_z = \lambda\,\mathrm{div}\,q + 2\mu\frac{\partial q_z}{\partial z},$$

for the normal stresses, while for the tangential components, we have 2μ multiplied by the corresponding g's, and from (56)

(135)
$$B_z = C_y = 2\mu g_x = \mu\left\{\frac{\partial q_z}{\partial y} + \frac{\partial q_y}{\partial z}\right\},$$
$$C_x = A_z = 2\mu g_y = \mu\left\{\frac{\partial q_x}{\partial z} + \frac{\partial q_z}{\partial x}\right\},$$
$$A_y = B_x = 2\mu g_z = \mu\left\{\frac{\partial q_y}{\partial x} + \frac{\partial q_x}{\partial y}\right\}.$$

Substituting these values of the stress in the first equation of motion, (126)

(136) $$\varrho\frac{\partial^2 q_x}{\partial t^2} = \lambda\frac{\partial}{\partial x}\,\mathrm{div}\,q + 2\mu\frac{\partial^2 q_x}{\partial x^2} + \mu\left\{\frac{\partial^2 q_y}{\partial x\,\partial y} + \frac{\partial^2 q_x}{\partial y^2} + \frac{\partial^2 q_x}{\partial z^2} + \frac{\partial^2 q_z}{\partial x\,\partial z}\right\} + \varrho F_x$$
$$= (\lambda + \mu)\frac{\partial}{\partial x}\,\mathrm{div}\,q + \mu\Delta q_x + \varrho F_x.$$

This is one component of the vector equation

(136) $$\varrho\frac{\partial^2 q}{\partial t^2} = (\lambda + \mu)\,\mathrm{grad}\,\mathrm{div}\,q + \mu\Delta q + \varrho F,$$

understanding by Δq a vector whose components are the Laplacians of the respective components of q.

Taking the divergence of (136) since $\mathrm{div}\,\Delta = \Delta\,\mathrm{div}$,

(137) $$\varrho\frac{\partial^2}{\partial t^2}\,\mathrm{div}\,q = (\lambda + 2\mu)\Delta\,\mathrm{div}\,q + \mathrm{div}\,(\varrho F).$$

Consequently the dilatation obeys the equation of sound (114), and a region where $\mathrm{div}(\varrho F)$ is not zero is a source.

Secondly, taking the curl of (136), since curl grad $= 0$, curl $\Delta \equiv \Delta$ curl,

$$(138) \qquad \varrho \frac{\partial^2}{\partial t^2} \operatorname{curl} q = \mu \Delta \operatorname{curl} q + \operatorname{curl}(\varrho F)$$

so that the molecular rotation is propagated according to the same equation, but with the constant μ instead of $\lambda + 2\mu$. We shall see that (137) and (138) correspond to longitudinal and transverse waves with different velocities of propagation.

13a. Viscous Fluids. An ideal or perfect fluid differs from a solid in not being able to support tangential stress, that is, $\mu = 0$. In such a fluid curl q can be only a linear function of the time, that is rotation is not propagated. But in actual fluids which are all somewhat *viscous*, while no tangential stresses can exist in a state of rest, during motion such stresses can exist, and their relation to the velocities is of precisely the same character as that of the stresses to the *displacements* in a solid. We may therefore put, instead of (134)

$$(134')\qquad \begin{aligned} A_x &= -p + \lambda\sigma + 2\mu\frac{\partial v_x}{\partial x}, \\ B_y &= -p + \lambda\sigma + 2\mu\frac{\partial v_y}{\partial y}, \\ C_z &= -p + \lambda\sigma + 2\mu\frac{\partial v_z}{\partial z}, \end{aligned}$$

$$(135')\qquad \begin{aligned} B_z &= C_y = \mu\left\{\frac{\partial v_z}{\partial y} + \frac{\partial v_y}{\partial z}\right\}, \\ C_x &= A_z = \mu\left\{\frac{\partial v_x}{\partial z} + \frac{\partial v_z}{\partial x}\right\}, \\ A_y &= B_x = \mu\left\{\frac{\partial v_y}{\partial x} + \frac{\partial v_x}{\partial y}\right\}, \end{aligned}$$

where
$$\sigma = \operatorname{div} v,$$

and λ and μ are two *new* constants. The terms in p are inserted because, even when $v = 0$ there may be a uniform pressure. Adding (134′)

$$A_x + B_y + C_z = -3p + (3\lambda + 2\mu)\sigma.$$

Now there seems to be no reason why a uniform dilatation should give rise to any viscous forces, or should change the pressure. Accordingly we put $3\lambda + 2\mu = 0$. Inserting the values from (134′) in (126) but writing the accelerations in terms of the velocities as in (106) we have

$$(106')\qquad \frac{\partial v}{\partial t} + v\nabla v = F - \frac{1}{\varrho}\left(\operatorname{grad} p + 2\mu\Delta v - \frac{2\mu}{3}\operatorname{grad} \operatorname{div} v\right).$$

The non-linear term $v\nabla v$ may be neglected when the velocities are small, as in the case of sound. Proceeding as in § 12, taking divergence, and using (111), we get, instead of (113)

$$(114') \qquad \frac{\partial^2 s}{\partial t^2} = - \operatorname{div} \mathbf{F} + \frac{e}{\varrho} \Delta s + \frac{4}{3} \frac{\mu}{\varrho} \frac{\partial}{\partial t} \Delta s$$

as the equation of sound as affected by viscosity.

14. Equation for Transverse Vibrations of Rod. If we consider a solid of cylindrical form whose thickness is negligible in comparison with its length, which we will call a rod or wire, we may find its equation of motion under the same restrictions as in the case of a string, namely that the displacement is small, and that the slope is small at each point. If x be measured in the direction of the rod, u be the displacement in the y direction, u and $\frac{\partial u}{\partial x}$ are to be small.

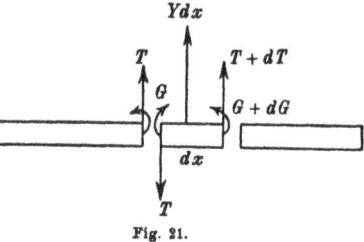

Fig. 21.

Let us consider the equilibrium of a portion of the rod of length dx, the forces on which are shown in Fig. 21. If the resultant of the shearing forces A_y on one end be T and the other $T + \frac{\partial T}{\partial x} dx$, and if there be applied an external force Y per unit of length, we have, putting the resultant equal to zero,

$$- T + \left(T + \frac{\partial T}{\partial x} dx \right) + Y dx = 0,$$

$$(139) \qquad \frac{\partial T}{\partial x} = - Y.$$

The forces in the x-direction balance, and we do not need to consider them, as we have already considered the matter of *longitudinal* motion in connection with sound. But we have another condition in expressing the fact that there is no tendency to *turn* the element about an axis parallel to the z-axis. Suppose that the moment of all the normal forces A_x Fig. 21 a on one end is $- G$ and at the other $G + \frac{\partial G}{\partial x} dx$. The tangential force at one end has a moment about a point in the other. Let

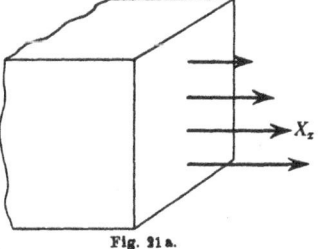

Fig. 21 a.

us take moments about an axis parallel to the z-axis through a point in the x-axis at the point x. Then the force T has no moment, while the force $T + dT$ has the moment $(T + dT)dx$. Adding this to the resultant moment we have, in the limit,

(140) $$\frac{\partial G}{\partial x} = -T.$$

Differentiating, and using (139) we eliminate T,

(141) $$\frac{\partial^2 G}{\partial x^2} = Y.$$

The moment G may be considered as due to the stretching of the longitudinal fibres of the rod in bending. If R is the radius of curvature of the axis of the rod, Fig. 22, which before bending coincided with the x-axis, we find the ratio of the stretched to the unstretched length $\dfrac{ds'}{ds} = \dfrac{R-y}{R}$

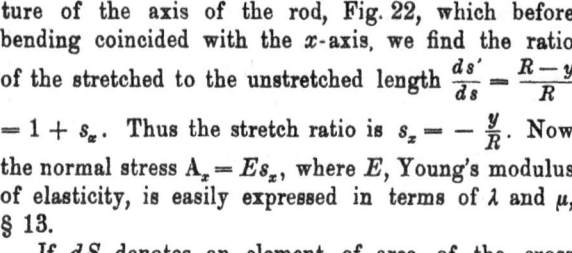

Fig. 22.

$= 1 + s_x$. Thus the stretch ratio is $s_x = -\dfrac{y}{R}$. Now the normal stress $A_x = Es_x$, where E, Young's modulus of elasticity, is easily expressed in terms of λ and μ, § 13.

If dS denotes an element of area of the cross section, the force on it $A_x dS$ has about a point on the x-axis a moment which is found by multiplying it by the distance y of its point of application from the x-axis, so that adding for all points in the cross-section we have

(142) $$G = -\iint y A_x dS = -E \iint s_x y\, dS = \frac{E}{R} \iint y^2 dS.$$

The integral $\iint y^2 dS$ is called the moment of inertia of the section, and will be denoted by I. Since the slope of the bent axis is small, the curvature is

(143) $$\frac{1}{R} = \frac{\partial^2 u}{\partial x^2}.$$

Consequently

(144) $$G = EI \frac{\partial^2 u}{\partial x^2},$$

and inserting this value in (141)

(145) $$EI \frac{\partial^4 u}{\partial x^4} = Y.$$

If there is no applied force Y, so that the bar is not in equilibrium, we have the unbalanced force of magnitude equal to $-Y dx$ acting to produce acceleration, and if ϱ is the density, S the cross-section,

(146) $$\varrho S dx \frac{\partial^2 u}{\partial t^2} = -Y dx.$$

Introducing the value of Y from (145) we have the equation

(147) $$\varrho S \frac{\partial^2 u}{\partial t^2} + E I \frac{\partial^4 u}{\partial x^4} = 0.$$

A source of motion may be introduced on the right, as in (116).

In a similar manner, but with more complication, we may deduce the equation for the transverse displacement or motion of a thin *plate*, or body one of whose dimensions is small in comparison with the other two, slightly bent from a plane. Let us take the xy-plane as parallel to the surfaces of the undeformed plate, and half-way between them, so that the equations of the top and bottom surfaces are $s = \pm h$. Consider the equilibrium of a small rectangular portion of the plate of dimensions dx and dy. Let the resultant of the shearing forces on the face normal to the X-axis and on the side nearest the origin be $- T_1 \, dy$, and on the side farthest from the origin $\left(T_1 + \frac{\partial T_1}{\partial x} \, dx\right) dy$, Fig. 23. The resultant of the two is then $\frac{\partial T_1}{\partial x} dx dy$. Similarly for the two faces normal to the Y-axis the resultant is $\frac{\partial T_2}{\partial y} dx dy$, and if the whole external force applied to the plate per unit area and in the Z-direction is Z we have the equation of equilibrium corresponding to (139),

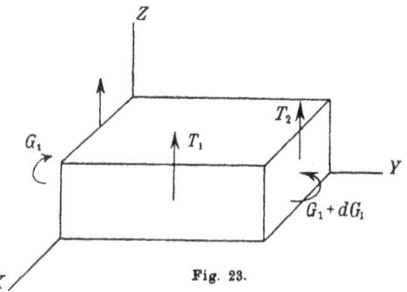

Fig. 23.

(139′) $$\frac{\partial T_1}{\partial x} + \frac{\partial T_2}{\partial y} + Z = 0.$$

Consider now the moments of the stresses applied to these four faces. Beside the flexural moment corresponding to what we had for the bar, $- G_1 dy$ on the first face and $\left(G_1 + \frac{\partial G_1}{\partial x} \, dx\right) dy$ on the second, with the resultant moment as before $\frac{\partial G_1}{\partial x} \, dx dy$, we now have to consider the *torsional* moment of the forces about a normal to the respective faces, which we will call $- H_1 dy$ and $\left(H_1 + \frac{\partial H_1}{\partial x} \, dx\right) dy$ for the faces normal to the X-axis, $- H_2 dx$ and $\left(H_2 + \frac{\partial H_2}{\partial y} \, dy\right) dx$ for those normal to the Y-axis. Corresponding to (140) we have then

(140′)
$$\frac{\partial G_1}{\partial x} + \frac{\partial H_2}{\partial y} - T_1 = 0,$$
$$\frac{\partial H_1}{\partial x} + \frac{\partial G_2}{\partial y} + T_2 = 0.$$

(It is to be noticed that the moments are positive if they tend to turn about the X-axis from Y to Z, and similarly in cyclic order.)

Differentiating by x and y respectively, and eliminating T_1 and T_2 by (139'),

(141')
$$\frac{\partial^2 G_1}{\partial x^2} + \frac{\partial^2 H_2}{\partial x \partial y} - \frac{\partial^2 H_1}{\partial x \partial y} - \frac{\partial^2 G_2}{\partial y^2} + Z = 0.$$

We have now to find the moments G_1, G_2, H_1, H_2 in terms of the bending. We shall assume, as we did for the bar, that all points lying on any normal to the suface $s = 0$ remain on a normal to the central surface into which that plane is deformed. If u denote, as for the bar, the z-displacement of the central surface, we have, as in Fig. 22, for the horizontal components of the displacement

$$\mathfrak{q}_x = - z \frac{\partial u}{\partial x}, \qquad \mathfrak{q}_y = - z \frac{\partial u}{\partial y},$$

so that as before,

$$s_x = \frac{\partial \mathfrak{q}_x}{\partial x} = - z \frac{\partial^2 u}{\partial x^2}, \qquad s_y = \frac{\partial \mathfrak{q}_y}{\partial y} = - z \frac{\partial^2 u}{\partial y^2}.$$

We also have for the shear,

$$2 g_s = \frac{\partial \mathfrak{q}_y}{\partial x} + \frac{\partial \mathfrak{q}_x}{\partial y} = - 2 z \frac{\partial^2 u}{\partial x \partial y}.$$

We shall assume that the stress $C_z = 0$. We accordingly find from (127),

$$s_x = a A_x - b B_y, \qquad s_y = a B_y - b A_x,$$

which being solved give

$$A_x = \frac{a s_x + b s_y}{a^2 - b^2}, \qquad B_y = \frac{b s_x + a s_y}{a^2 - b^2}.$$

Using the values of s_x, s_y, g_s we accordingly have

$$A_x = - \frac{z}{a^2 - b^2} \left(a \frac{\partial^2 u}{\partial x^2} + b \frac{\partial^2 u}{\partial y^2} \right), \qquad B_y = - \frac{z}{a^2 - b^2} \left(b \frac{\partial^2 u}{\partial x^2} + a \frac{\partial^2 u}{\partial y^2} \right)$$

$$A_y = B_x = - 2 \mu z \frac{\partial^2 u}{\partial x \partial y} = - \frac{z}{a + b} \frac{\partial^2 u}{\partial x \partial y},$$

all of which depend on the bending of the central surface, and in addition are proportional to the distance of the point in question from it s.

We now have according to the definition of the flexural and torsional couples

$$G_1 \, dy = \int_{-h}^{h} z \mathrm{A}_z \, dz \, dy = -\frac{2}{3} \frac{h^3}{a^2 - b^2} \left(a \frac{\partial^2 u}{\partial x^2} + b \frac{\partial^2 u}{\partial y^2} \right) dy,$$

$$G_2 \, dx = -\int_{-h}^{h} z \mathrm{B}_y \, dz \, dx = \frac{2}{3} \frac{h^3}{a^2 - b^2} \left(b \frac{\partial^2 u}{\partial x^2} + a \frac{\partial^2 u}{\partial y^2} \right) dx,$$

(142′)

$$H_1 \, dy = -\int_{-h}^{h} z \mathrm{A}_y \, dz \, dy = \frac{2}{3} \frac{h^3}{a+b} \frac{\partial^2 u}{\partial x \partial y} \, dy,$$

$$H_2 \, dx = \int_{-h}^{h} \mathrm{B}_x \, dz \, dy = -\frac{2}{3} \frac{h^3}{a+b} \frac{\partial^2 u}{\partial x \partial y} \, dx.$$

Introducing these values into (141′) we easily obtain

(145′)	$$D \left(\frac{\partial^4 u}{\partial x^4} + 2 \frac{\partial^4 u}{\partial x^2 \partial y^2} + \frac{\partial^4 u}{\partial y^4} \right) = Z,$$

where the constant D, called the flexural rigidity of the plate, has the value

$$D = \frac{2}{3} \frac{h^3 a}{a^2 - b^2} = \frac{2}{3} \frac{E h^3}{1 - \eta^2},$$

$E_1 = \frac{1}{a}$ being Young's modulus, and $\eta = \frac{b}{a}$ Poisson's coefficient representing the ratio of lateral contraction to longitudinal extension under the sole force A_x, as we see from equations (127).

If we denote the Laplacian in two dimensions by

$$\Delta \equiv \frac{\partial^2}{\partial x^2} + \frac{\partial^2}{\partial y^2},$$

we may write the equation (145′)

$$D \Delta \Delta u = Z.$$

The equation

$$\Delta \Delta u = 0$$

satisfied for a plate under the influence of no forces on its surface, but bent by couples at its edge, is called the double Laplace equation.

If the plate is not in equilibrium, but in motion, we have to replace the force Z by the reaction due to the acceleration produced (per unit area of the plate), and if ϱ is the density of the plate, $Z = -2 h \varrho \frac{\partial^2 u}{\partial t^2}$, so that

(148)	$$2 h \varrho \frac{\partial^2 u}{\partial t^2} + D \Delta \Delta u = 0$$

is the equation for the motion of the plate.

These equations for the bar and plate are the only equations that we have found (except 114′) introducing derivatives of order higher than the second.

15. Flow of Electricity in Wires. Suppose we consider a long straight wire, in which current flows depending on t, the time, and x, the distance along the wire. The electrical variables to be considered are the current and the potential. If C denote the current density, V the potential, we have by Ohm's Law (81)

$$(149) \qquad C = -\varkappa \operatorname{grad} V = -\varkappa \frac{\partial V}{\partial x}.$$

We define the total current I at any point as the total quantity of electricity passing per unit of time, and as C is the quantity per unit of area, if A is the cross-section of the wire

$$(150) \qquad I = CA = -\varkappa A \frac{\partial V}{\partial x} = -\frac{1}{R} \frac{\partial V}{\partial x},$$

where R is the *resistance* of the line per unit of length. Thus at the two ends of a portion of wire of length dx, we have the difference of potential

$$(151) \qquad dV = \frac{\partial V}{\partial x} dx = -R dx I.$$

Secondly, a portion of conductor receiving a charge dq has its potential increased so that $dq = K dV$, where K is called the electrostatic *capacity* of the conductor. This is Faraday's Law. If K refers to unit of length, the element dx of capacity $K dx$ receives in unit time the charge

$$(152) \qquad \frac{\partial q}{\partial t} = K dx \frac{\partial V}{\partial t}.$$

But there enters per unit time from the left the quantity I, and there goes out on the right $I + \frac{\partial I}{\partial x} dx$, the total gain is thus $-\frac{\partial I}{\partial x} dx$, equating which to (152) we have

$$(153) \qquad K \frac{\partial V}{\partial t} = -\frac{\partial I}{\partial x}.$$

If some of the current leaks off from the line to the earth, whose potential is zero, the amount lost will be proportional to V, so that from the length dx there is lost the amount per unit of time,

$$S dx V$$

where S is the *leakage conductance* per unit of length. Subtracting SV on the right of (153) we get

$$(154) \qquad K \frac{\partial V}{\partial t} + SV = -\frac{\partial I}{\partial x}.$$

Thirdly, it was discovered by Faraday that if the current vary with the time it reacts, or requires an additional potential difference to drive it, as the inertia of a body makes it react, and require a force to drive it proportional to its acceleration. The electrical analogue of the mass is called self-inductance, and if L represent the self-inductance per unit of length, the reaction is $Ldx\dfrac{\partial I}{\partial t}$. This must be subtracted from the dV in (151) so that we get the equation

$$(155) \qquad L\frac{\partial I}{\partial t} + RI = -\frac{\partial V}{\partial x},$$

we now have two equations connecting I and V, of similar form. If we call the operators

$$(156) \qquad K\frac{\partial}{\partial t} + S \equiv D_1, \quad L\frac{\partial}{\partial t} + R \equiv D_2, \quad -\frac{\partial}{\partial x} \equiv D_3.$$

We may write the equations (154), (155)

$$(157) \qquad D_1 V = D_3 I, \qquad D_2 I = D_3 V \qquad (158)$$

We may eliminate either I or V by differentiation. Apply the operator D_2 to (157),

$$(159) \qquad D_2 D_1 V = D_2 D_3 I = D_3 D_2 I = D_3^2 V.$$

Applying D_1 to (158),

$$(160) \qquad D_1 D_2 I = D_1 D_3 V = D_3 D_1 V = D_3^2 I.$$

Consequently both I and V satisfy the same equation

$$(161) \qquad D_1 D_2 \varphi = D_3^2 \varphi,$$

or written out,

$$(162) \qquad KL\frac{\partial^2 \varphi}{\partial t^2} + (RK + SL)\frac{\partial \varphi}{\partial t} + RS\varphi = \frac{\partial^2 \varphi}{\partial x^2}.$$

This is called the Telegraphist's Equation. It first appeared in a paper by Kirchhoff[1]), in 1857, next in one by Heaviside[2]) in 1876, but first attracted attention when treated by Poincaré[3]) in 1893.

In the case of a line of no resistance and no leakage, it reduces to

$$(163) \qquad \frac{\partial^2 \varphi}{\partial t^2} = \frac{1}{KL}\frac{\partial^2 \varphi}{\partial x^2},$$

which is the same as the equation of sound for one dimension. If on

---

1) Über die Bewegung der Elektrizität in Drähten. *Prgg. Ann.* Bd. 100, *Ges. Abh.* p. 131.

2) On the Extra Current. *Phil. Mag.*, Aug. 1876, S. 5, vol. 2. *Electrical Papers*, vol. I, p. 53.

3) Sur la propagation de l'électricité. *Comptes rendus*, 117, p. 1027, 1893.

the other hand the inductance L is negligible in comparison with the resistance, we have

$$(164) \qquad \frac{\partial \varphi}{\partial t} + \frac{S}{K}\, \varphi = \frac{1}{KR}\, \frac{\partial^2 \varphi}{\partial x^2},$$

which is the equation for the flow of heat in one dimension (83), the leakage term, for heat flowing along a bar, and losing heat by radiation, being of the same form as for electricity. It was the equation in this form that was first treated by Lord Kelvin[1]), in 1855, in connection with the Atlantic Cable. We shall see how the appearance of the term in the self-inductance completely changes the character of the propagation.

In order to make plain the analogy between the propagation of sound in one dimension and electrical actions in a line without resistance, let us consider the analogy in detail. The pressure p corresponds to the potential V, the flow being proportional to the negative gradient of either. The equations

$$(112) \qquad \varrho\, \frac{\partial^2 q}{\partial t^2} = -\frac{\partial p}{\partial x}, \ \ L\, \frac{\partial I}{\partial t} = -\frac{\partial V}{\partial x}, \qquad\qquad (155)$$

show that the current I corresponds to the velocity $\frac{\partial q}{\partial t}$, the inductance L to the density ϱ.

The equation

$$dp = eds$$

on being divided by dt, with (111), giving

$$\frac{\partial p}{\partial t} = -e\, \frac{\partial}{\partial x}\, \frac{\partial q}{\partial t},$$

corresponds to

$$(153) \qquad \frac{\partial V}{\partial t} = -\frac{1}{K}\, \frac{\partial I}{\partial x},$$

so that $\frac{1}{K}$ corresponds to the elasticity e.

In order to get a term in sound corresponding to the term RI in (155) we need to introduce a term showing a resistance to motion proportional to the velocity $\varkappa\, \frac{\partial q}{\partial t}$, such as might come from friction with the sides of the tube. The leakage term SV could also be introduced in the sound equation, in case of a porous tube. We then should have exact similarity.

16. Stokes's Curl Theorem. We find by differentiation of (60) that we have identically

$$\operatorname{div} \operatorname{curl} \mathbf{A} = 0,$$

---

1) On the Theory of the Electric Telegraph, *Proc. Roy. Soc.*, May 1855; *Math. and Phys. Papers*, vol. II, p. 61.

or any vector which is the curl of another is solenoidal. Consequently, the flux

$$\iint (\text{curl } \mathbf{A})_n \, dS$$

of curl through any *closed* surface is zero. If then we draw a closed curve on such a surface, the flux through the portion of surface on one side of the curve is the same as that on the other, but opposite in sign. We see then that the flux through any *unclosed* piece of surface depends not on the surface, but only on the contour, for considering a closed surface of any two sheets having a common contour, the flux out through both is zero.

Consider now the line integral around the contour, of the tangential component of a vector $\mathbf{A}$.

(165) $\int \mathbf{A}_s \, ds = \int (\mathbf{A}_x dx + \mathbf{A}_y dy + \mathbf{A}_z dz).$

Suppose first that the contour is a *plane* curve in the xy-plane. By drawing lines parallel to the axes, divide up the enclosed area into rectangles of sides dx, dy.

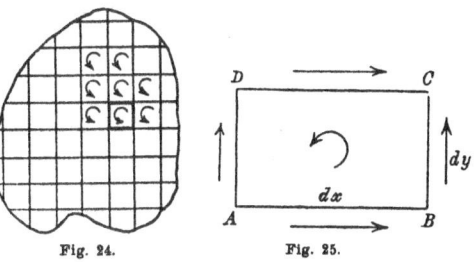

Then if we take the line-integral in turn around each of the infinitesimal meshes $ABCD$ in the same direction of circulation, Fig. 24, the parts of the integral belonging to adjacent meshes will cancel, the integrand being the

Fig. 24.   Fig. 25.

same, but the signs of ds opposite. Thus the whole integral around the contour is the sum of all those around the meshes. Consider one mesh Fig. 25. Going around in the direction of the arrow, we have on the side AB, $\mathbf{A}_x dx$ and on the side CD, $-\left(A_x + \frac{\partial A_x}{\partial y} dy\right) dx.$ On the side AD we have $-A_y dy$ and on BC, $\left(A_y + \frac{\partial A_y}{\partial x} dx\right) dy.$ Adding these four we have

(166) $\left(\frac{\partial A_y}{\partial x} - \frac{\partial A_x}{\partial y}\right) dx \, dy = (\text{curl } \mathbf{A})_z \, dS,$

where dS is the area of the mesh. Consequently, we have, adding together the result for all the meshes, the equality of the contour integral to a surface-integral,

(167) $\int \mathbf{A}_s \, ds = \iint (\text{curl } \mathbf{A})_z \, dS.$

If the surface is not plane, let us divide it into infinitesimal strips by planes parallel to the xy-plane, and then divide each of these strips into triangles by drawing planes parallel to the yz- and zx-planes Fig. 26. Let us then consider the integral around one triangular mesh. Consider the

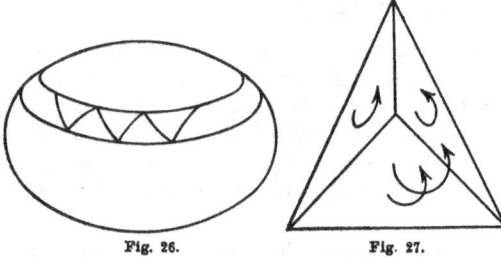

tetrahedron made by the mesh and three right triangles parallel to the coordinate planes, Fig. 27. It is obvious that the line-integral around the oblique triangle is the sum of those about the three right triangles.

Fig. 26.   Fig. 27.

But if the areas of these are dS, dS_x, dS_y, dS_z respectively, since the others are the projections of dS, we have

(168) $\quad dS_x = dS \cos (nx), \; dS_y = dS \cos (ny), \; dS_z = dS \cos (nz),$

and

(169) $\quad \iint (\text{curl } A)_n \, dS = \iint [(\text{curl } A)_x \cos (nx) + (\text{curl } A)_y \cos (ny)$

$+ (\text{curl } A)_z \cos (nz)] dS = \iint (\text{curl} A)_x dS_x + \iint (\text{curl } A)_y dS_y + \iint (\text{curl } A)_z dS_z.$

Now integrating over a plane area, we have by (167) and the two similar formulae, the three surface-integrals equal to the line integral (165) around the three plane triangles. Consequently we have for the whole surface integral

(170) $\qquad\qquad \iint (\text{curl } A)_n \, dS = \int A_s \, ds.$

This is known as Stokes's curl theorem.[1]

The normal is to be drawn toward the side toward which a right-handed screw would move when turning in the direction of integration around the contour.

17. Maxwell's Equations of the Electromagnetic Field. In the theory of electricity we have to consider two pairs of vectors, the electric and magnetic *field-strengths* E and H, and the corresponding *inductions* D, B, connected with them by the equations,

(171) $\qquad\qquad D = \varepsilon E, \quad B = \mu H,$

where ε and μ denote physical properties of the substance in question and are called the electrical and magnetic inductivities respectively. The vector $\frac{D}{4\pi}$ is called by Maxwell the electric displacement. The divergence of $\frac{D}{4\pi}$ is the density of electricity ϱ,

(172) $\qquad\qquad \varrho = \frac{1}{4\pi} \text{ div } D.$

---

1) For an entirely different proof see the author's *Dynamics*, p. 84.

This may be taken at present merely as a definition of **D**, though it will be shown in Chapter V why the density is so connected with the field. The corresponding magnetic density is zero, so that we are to have

(173) $$\text{div } \mathbf{B} = 0.$$

If the electrical and magnetic quantities are independent of the time, it is found that curl $\mathbf{E} = 0$, $\mathbf{E}$ is a lamellar vector, and we have $\mathbf{E} = - \text{grad } V$, where V is called the electric potential. If however, the magnetic field is changing, it was found by Faraday from experiments with closed wire circuits, that there is produced around any closed circuit an *electromotive force* defined by the line-integral of the electric field around the circuit,

$$\int \mathbf{E}_s \, ds,$$

and of magnitude equal to the time-rate of diminution of the flux of magnetic induction through the circuit, the direction of normal and direction of integration being as defined above. Thus Faraday's law is

(174) $$\int \mathbf{E}_s \, ds = - \frac{\partial}{\partial t} \int\int \mathbf{B}_n \, dS.$$

The contour being fixed, we may differentiate under the integral, and applying Stokes's theorem,

(175) $$\int\int (\text{curl } \mathbf{E})_n \, dS = - \int\int \frac{\partial \mathbf{B}_n}{\partial t} \, dS.$$

The equation being true for *any* closed contour, the integrands must be equal, and since n may have any direction, we must have

(176) $$- \frac{\partial \mathbf{B}}{\partial t} = \text{curl } \mathbf{E}.$$

This is the first of Maxwell's equations, and agrees with the statement made above that if **B** is independent of the time **E** is lamellar, since its curl vanishes.

The second equation follows from Oersted's discovery of electromagnetism, namely, that a current of electricity gives rise to a magnetic field. This is proportional to the current, and from considerations of the work required to take a magnetic pole around a closed path, it is shown that the magneto-motive force, defined as the line-integral of magnetic field, around any closed curve is proportional to the total current flowing through the curve. If **C** is the current-density

(177) $$\int \mathbf{H}_s \, ds = 4\pi \int\int \mathbf{C}_n \, dS$$

the factor 4π depending on the units used. Applying Stokes's theorem,

(178) $$\int\int (\text{curl } \mathbf{H})_n \, dS = 4\pi \int\int \mathbf{C}_n \, dS,$$

and by the same reasoning as above,

(179) $$\text{curl } \mathbf{H} = 4\pi\mathbf{C}.$$

In case there is no current at a point of the field, curl $\mathbf{H} = 0$, and $\mathbf{H}$ is lamellar, $\mathbf{H} = -\text{ grad } \Omega$, where Ω is the magnetic potential. Before Oersted's discovery, it was supposed that $\mathbf{H}$ was always lamellar.

The equation (179) was the one used before Maxwell's ideas became known. It was Maxwell's brilliant suggestion, that even in insulators, where there is no conduction current $\mathbf{C}$, a changing electric displacement acts like a current, and produces a magnetic field. The manner of introduction of this displacement current is as follows. The density of electricity is connected, on the one hand, with $\mathbf{D}$ (172), on the other with the conduction current $\mathbf{C}$ by (78). Comparing these two, we get

(180) $$\text{div } \frac{1}{4\pi} \frac{\partial \mathbf{D}}{\partial t} + \text{div } \mathbf{C} = 0.$$

Hence the vector

$$\frac{1}{4\pi} \frac{\partial \mathbf{D}}{\partial t} + \mathbf{C}$$

is solenoidal, and Maxwell calls it the *total* current, $\frac{1}{4\pi} \frac{\partial \mathbf{D}}{\partial t} = \mathbf{C}_2$ being called the *displacement* current. Introducing this in (179) in place of $\mathbf{C}$

(181) $$\frac{\partial \mathbf{D}}{\partial t} + 4\pi\,\mathbf{C} = \text{curl } \mathbf{H}.$$

This is Maxwell's second relation. We make use of (171) and the equivalent of (81) to express all in terms of $\mathbf{E}$ and $\mathbf{H}$.

(182) $$\mathbf{D} = \varepsilon\mathbf{E}, \quad \mathbf{B} = \mu\mathbf{H}, \quad \mathbf{C} = \varkappa\mathbf{E},$$

when our two equations become, ε, μ, $\varkappa$ being constants,

(183) $$-\mu\,\frac{\partial \mathbf{H}}{\partial t} = \text{curl } \mathbf{E},$$

(184) $$\varepsilon\,\frac{\partial \mathbf{E}}{\partial t} + 4\pi\varkappa\mathbf{E} = \text{curl } \mathbf{H}.$$

Calling the operators

$$-\mu\,\frac{\partial}{\partial t} \equiv D_1, \quad \varepsilon\,\frac{\partial}{\partial t} + 4\pi\varkappa \equiv D_2, \quad \text{curl} \equiv D_3,$$

these are

$$D_1 H = D_3 E, \quad D_2 E = D_3 H,$$

from which we obtain, as in (159) and (160),

(185)
$$D_2 D_1 H = D_2 D_3 E = D_3 D_2 E = D_3^2 H,$$
$$D_1 D_2 E = D_1 D_3 H = D_3 D_1 H = D_3^2 E,$$

so that both vectors satisfy the same differential equation. The significance of the operator $D_3^2 = \operatorname{curl}(\operatorname{curl}) \equiv \operatorname{curl}^2$, is easily found from the x-component,

$$(\operatorname{curl}^2 A)_x = \frac{\partial}{\partial y}\left(\frac{\partial A_y}{\partial x} - \frac{\partial A_x}{\partial y}\right) - \frac{\partial}{\partial z}\left(\frac{\partial A_x}{\partial z} - \frac{\partial A_z}{\partial x}\right)$$

$$(186) \qquad = \frac{\partial}{\partial x}\left(\frac{\partial A_x}{\partial x} + \frac{\partial A_y}{\partial y} + \frac{\partial A_z}{\partial z}\right) - \left(\frac{\partial^2 A_x}{\partial x^2} + \frac{\partial^2 A_x}{\partial y^2} + \frac{\partial^2 A_x}{\partial z^2}\right)$$

$$= (\operatorname{grad}\,\operatorname{div}\,A)_x - \Delta A_x,$$

so that we may write

$$(187) \qquad\qquad \operatorname{curl}^2 A \equiv \operatorname{grad}\,\operatorname{div}\,A - \Delta\,A$$

meaning by ΔA a vector whose components are the Laplacians of the corresponding components of A.

Now we have (173)

$$\operatorname{div}\,B = \mu\,\operatorname{div}\,H = 0,$$

and by (184) taking divergence,

$$(188) \quad \operatorname{div}\left(\varepsilon\frac{\partial E}{\partial t} + 4\pi\varkappa E\right) = \varepsilon\frac{\partial}{\partial t}\operatorname{div}\,E + 4\pi\varkappa\,\operatorname{div}\,E = 0,$$

an equation for div E, integrating which

$$(189) \qquad\qquad \operatorname{div}\,E = \varepsilon^{-\frac{4\pi\varkappa}{\varepsilon}t}(\operatorname{div}\,E)_{t=0}.$$

Even if, then, E has any divergence, it dies away as the time goes on, or if the substance is an insulator, $\varkappa = 0$, and div E is independent of the time, that is, the electricity remains in place unchanged, and does not concern us. Let us then put div $E = 0$. The first term in *curl²* then disappears, and changing all signs, we get for both vectors E and H the equation

$$(190) \qquad\qquad \mu\varepsilon\frac{\partial^2 A}{\partial t^2} + 4\pi\varkappa\mu\frac{\partial A}{\partial t} = \Delta A.$$

If $\varkappa = 0$, each of the three components of E or of H respectively satisfies the wave-equation, and we are thus led to the idea of electromagnetic waves in an insulator, which Maxwell identifies with lightwaves. If $\varkappa$ is not zero, (190) is the general telegraphist's equation (162), while if ε is negligible in comparison with $\varkappa$, we have the same equation as for the flow of heat (83).

By a celebrated experiment made by Rowland, of Baltimore, it was shown that a charge of electricity in motion produces a magnetic field, so that to the conduction and displacement currents we now add a term called the *convection* current, of density $C_3 = \varrho v$, where v is the velocity of motion of the charge. We accordingly write, instead of (181)

$$(191) \qquad\qquad \frac{\partial D}{\partial t} + 4\pi(C + \varrho v) = \operatorname{curl}\,H.$$

According to the theory of H. A. Lorentz, of Leyden, all conduction currents are due to the flow of electrons, or minute grains of electricity so that if ϱ and v be their electrical density and velocity respectively, we may include C in ϱv. Let us, for simplicity, consider a medium for which $\varepsilon = \mu = 1$, $D = E$, $B = H$. Proceeding as before, taking the time derivative of (191),

$$(192) \quad \frac{\partial^2 D}{\partial t^2} + 4\pi \frac{\partial}{\partial t}(\varrho v) = \text{curl} \frac{\partial H}{\partial t} = - \text{curl}^2 D = \Delta D - \text{grad div } D.$$

Thus we obtain, since div $D = 4\pi\varrho$,

$$(193) \quad \Delta D - \frac{\partial^2 D}{\partial t^2} = 4\pi \left\{ \text{grad } \varrho + \frac{\partial}{\partial t}(\varrho v) \right\}.$$

Taking the curl of (191)

$$\text{curl} \frac{\partial D}{\partial t} = \frac{\partial}{\partial t} \text{curl } D = - \frac{\partial^2 B}{\partial t^2} = \text{curl}^2 B - 4\pi \text{ curl } (\varrho v),$$

and since div $B = 0$,

$$(194) \quad \Delta B - \frac{\partial^2 B}{\partial t^2} = - 4\pi \text{ curl } (\varrho v).$$

The operator $\Delta - \frac{1}{a^2} \frac{\partial^2}{\partial t^2}$ is called by Lorentz the d'Alembertian (because in one dimension its vanishing gives the equation of the string as found by d'Alembert) and denoted by $\square$.

Accordingly when electricity (electrons) is moved about, both fields obey equations of the form

$$(195) \quad \square \; \varphi = f,$$

where f is a given function of space and time. This is the same equation as that for sound where there are sources (114), (115'), (117), and was originally given by L. V. Lorenz of Copenhagen.[1]) It was treated by Lord Rayleigh, and later by Beltrami and Poincaré.

The equations become somewhat simpler if we assume, since div $B = 0$,

$$(196) \quad B = \text{curl } A,$$

where A is a vector known as the vector-potential. Introducing this into (191), (183)

$$(197) \quad \frac{\partial D}{\partial t} + 4\pi \varrho v = \text{curl}^2 A = \text{grad div } A - \Delta A,$$

$$(198) \quad - \frac{\partial B}{\partial t} = - \text{curl} \frac{\partial A}{\partial t} = \text{curl } D,$$

which latter being written

$$(199) \quad \text{curl} \left(D + \frac{\partial A}{\partial t} \right) = 0,$$

---

1) Mémoire sur la théorie de l'élasticité des corps homogènes à élasticité constante. Crelle's Journal 58, pp. 328—351. Oeuvres scientifiques, t. 2, p. 3.

shows that the vector in parenthesis is lamellar, and is accordingly derived from a potential,

$$(200) \qquad \mathbf{D} + \frac{\partial \mathbf{A}}{\partial t} = - \operatorname{grad} V$$

where V is the ordinary or scalar potential, for in the unvarying state, $\mathbf{D} = - \operatorname{grad} V$.

Taking the divergence of (200)

$$(201) \qquad 4\pi\varrho + \frac{\partial}{\partial t} \operatorname{div} \mathbf{A} = - \Delta V.$$

The vector $\mathbf{A}$ is not yet fully determined, only its curl being given. If we assume

$$(202) \qquad \operatorname{div} \mathbf{A} = - \frac{\partial V}{\partial t},$$

equation (201) becomes

$$(203) \qquad \Delta V - \frac{\partial^2 V}{\partial t^2} = - 4\pi\varrho.$$

Differentiating (200) by t, using (202),

$$(204) \qquad \frac{\partial \mathbf{D}}{\partial t} + \frac{\partial^2 \mathbf{A}}{\partial t^2} = - \operatorname{grad} \frac{\partial V}{\partial t} = \operatorname{grad} \operatorname{div} \mathbf{A},$$

and subtracting this from (197),

$$(205) \qquad \Delta \mathbf{A} - \frac{\partial^2 \mathbf{A}}{\partial t^2} = - 4\pi\varrho\mathbf{v}.$$

Thus both the scalar and vector-potentials satisfy the Lorenz-Beltrami equation.

The equations (196) and (200) were given by Maxwell, but the utility of the two potentials in considering propagation was shown by Liénard and Wiechert.

18. Recapitulation of Equations. We will now collect the various equations that we have deduced. Laplace's equation (70)

$$\Delta \varphi = 0,$$

satisfied by the velocity-potential of the irrotational flow of an incompressible fluid, and by the steady flow of heat and electricity in conductors. If there are sources, we have Poisson's equation,

$$\Delta \varphi = - e,$$

where e, the source-density is a given point-function.

The equation of heat-conduction

$$(83) \qquad \frac{\partial \varphi}{\partial t} - a^2 \Delta \varphi = 0,$$

or

$$(83) \qquad \frac{\partial \varphi}{\partial t} - a^2 \Delta \varphi = a^2 e$$

if sources are present.

The wave-equation,

$$(115, 116, 117, 124, 190) \quad \frac{\partial^2 \varphi}{\partial t^2} - a^2 \Delta \varphi = 0,$$

or

$$(114, 116, 117, 124) \quad \frac{\partial^2 \varphi}{\partial t^2} - a^2 \Delta \varphi = a^2 e,$$

satisfied by the motion of sound, or the displacement of a string or membrane, long low water-waves, and electro-magnetic waves.

The equation

$$(147, 148) \quad \frac{\partial^2 \varphi}{\partial t^2} + a^2 \Delta \Delta \varphi = 0, \quad \text{or} \quad = a^2 e$$

for the transverse motion of a bar or plate.

The telegraphist's equation

$$(162) \quad a \frac{\partial^2 \varphi}{\partial t^2} + b \frac{\partial \varphi}{\partial t} + c \Delta \varphi = -ce,$$

includes all except the last preceding, by giving appropriate values to a, b, c.

These are the chief equations of mathematical physics, and are alike in all being linear, of the second order (with one exception), the coefficients of the derivatives being constants. A more general equation, containing all these, with the same exception, is

$$\sum_{r=1}^{r=n} \sum_{s=1}^{s=n} A_{rs} \frac{\partial^2 \varphi}{\partial x_r \partial x_s} + \sum_{r=1}^{r=n} B_r \frac{\partial \varphi}{\partial x_r} + C\varphi = F$$

where all the A's, B's, C, and F are functions of the independent variables x_1, x_2, ... x_n. In all the cases we have dealt with, n is equal to 2, 3 or 4.

CHAPTER II.

EQUATIONS OF THE FIRST ORDER.

19. Definition of Partial Differential Equation. A partial differential equation is a relation between any number of independent variables x_1, x_2, ... x_n, a dependent variable z depending upon them, and its partial derivatives with respect to them. We shall use the notation

$$p_1 = \frac{\partial z}{\partial x_1}, \quad p_2 = \frac{\partial z}{\partial x_2}, \quad \cdots p_n = \frac{\partial z}{\partial x_n},$$

$$p_{rs} = \frac{\partial^2 z}{\partial x_r \partial x_s}, \quad p_{rr} = \frac{\partial^2 z}{\partial x_r^2}, \quad p_{rst} = \frac{\partial^3 z}{\partial x_r \partial x_s \partial x_t}, \text{ etc.}$$

The order of the equation is that of the derivative of highest order contained in it. Thus the partial differential equation of the first order is

$$F(x_1, x_2, \ldots x_n, z, p_1, p_2, \ldots p_n) = 0.$$

We shall frequently deal with the simplest case, that of only two independent variables, in which case we may make use of the geometrical interpretation in space of three dimensions, which may be represented by perspective figures. All that we say may however be extended to space of n dimensions, and the same figures may be used by an effort of the imagination. Instead of x_1, x_2, z, p_1, p_2, we use the common notation x, y, z,

$$p = \frac{\partial z}{\partial x}, \quad q = \frac{\partial z}{\partial y},$$

so that

(1) $$F(x, y, z, p, q) = 0$$

is the differential equation. We shall consider F to be an algebraic function. If it is satisfied by $z = f(x, y)$, or $\varphi(x, y, z) = 0$, that is, if in (1) we put $z = f(x, y)$,

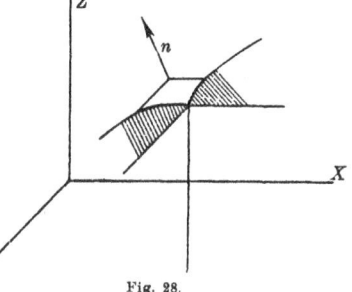

$p = \dfrac{\partial f}{\partial x}$, $q = \dfrac{\partial f}{\partial y}$, and thus reduce it to an identity for *all* values of x and y, we call z an integral or solution of the differential equation. Geometrically the integral is represented by a surface, and p and q are the slopes of the intersection of the tangent plane to the surface with the vertical planes though the point x, y, z, parallel

Fig. 28.

to the xz- and yz-planes respectively, Fig. 28. By § 4, (21), if we write

$$\varphi(x, y, z) \equiv f(x, y) - z = 0$$

the direction cosines of the normal are proportional to p, q, -1.

We shall see that the differential equation represents more than a single surface.

20. Geometrical Interpretation. We will now contrast the partial differential equation (1) with the ordinary differential equation

(2) $$F\left(x, y, \frac{dy}{dx}\right) = 0,$$

or

(3) $$\frac{dy}{dx} = f(x, y),$$

defining y as a function of x. At every point of the xy-plane the slope $\dfrac{dy}{dx}$ is given as $f(x, y)$ so that if we begin with an arbitrary point and go a small distance in the direction $f(x_0, y_0)$ to a neighboring point x_1, y_1, and at that point we move in the new direction $f(x_1, y_1)$ and so continue, we obtain in the limit as the points approach each other, a definite curve, called the integral curve of the differential equation (3), Fig. 29. There is such an integral curve through *every* point in the plane, and if we draw a straight line there is one curve through each of its points, there are thus as many integral curves as there are points in a line, that is a single infinity, ∞^1. A *particular* solution is completely specified by giving a particular point through which the integral curve shall pass.

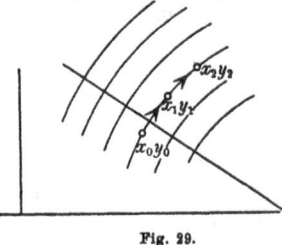

In the case of the partical differential equation (1), for every point x, y, z in space we have given not p and q but a *relation between p* and q, {i. e. between} the direction cosines of the normal to an integral sur face through the point. Accordingly the

Fig. 29.

normal, and with it the tangent plane, is not uniquely given, but may take any one of an infinite number of directions. If we draw all lines in these directions, any one of which has the equations, in running coordinates ξ, η, ζ,

$$(4) \qquad \frac{\xi - x}{p} = \frac{\eta - y}{q} = \frac{\zeta - z}{-1},$$

the y will generate a cone, which we shall call the normal cone N at x, y, z, whose equation is found by inserting the values of p, q from (4) in (1) to be

$$(5) \qquad F\left(x, y, z, -\frac{\xi - x}{\zeta - z}, -\frac{\eta - y}{\zeta - z}\right) = 0.$$

The planes normal to these rays, any one of which is a possible tangent plane to an integral surface, the equation of which plane is

$$(6) \qquad \zeta - z = p(\xi - x) + q(\eta - y),$$

envelope a cone which we shall call the tangent cone T. The line in which the plane (6) touches the cone T is given by the intersection of the plane (6) with the neighboring plane

$$(7) \qquad \zeta - z = (p + dp)(\xi - x) + (q + dq)(\eta - y),$$

in which dp and dq are related by

$$(8) \qquad \frac{\partial F}{\partial p} dp + \frac{\partial F}{\partial q} dq = 0.$$

It will be convenient to use the abbreviations

$$(9) \qquad X = \frac{\partial F}{\partial x}, \quad Y = \frac{\partial F}{\partial y}, \quad Z = \frac{\partial F}{\partial z}, \quad P = \frac{\partial F}{\partial p}, \quad Q = \frac{\partial F}{\partial q},$$

so that we have

$$(10) \qquad P dp + Q dq = 0.$$

Combining (6) and (7)

(11) $dp(\xi - x) + dq(\eta - y) = 0,$

from which and (10)

$$\frac{\xi - x}{P} = \frac{\eta - y}{Q}.$$

Calling this ratio λ, and putting

$$\xi - x = \lambda P, \quad \eta - y = \lambda Q$$

in (6) we obtain

$$\zeta - z = \lambda (Pp + Qq),$$

so that

(12) $\dfrac{\xi - x}{P} = \dfrac{\eta - y}{Q} = \dfrac{\zeta - z}{Pp + Qq}$

are the equations of the line of contact of the plane (6) with the cone T.

We accordingly see that the integral surface is far from being determined when we make it pass through a given point, but is susceptible of a far greater variety than in the case of the corresponding ordinary differential equation. We shall see that, as was shown by Cauchy, the integral surface may be determined by making it pass through a single infinity of points forming a continuum by lying on an arbitrarily given twisted curve in space, for instance, such that when $\varphi(x, y) = 0$, φ being a *given* function, then $z = \psi(x, y)$ where ψ is an *arbitrary function*[1]), or in other terms, when $y = \lambda(x)$, given, $z = \nu(x)$, arbitrary, or simultaneously, x, y, z are given in terms of a parameter u,

$$x = \lambda(u), \quad y = \mu(u), \quad z = \nu(u).$$

Thus instead of a single or double infinity of surfaces, the number is an infinity of the order of the number of points in a linear continuum. This is more simply expressed by saying that the solution contains an arbitrary *function*.

21. Cauchy's Problem.

That an integral surface may be determined by passing it through a given twisted curve (Cauchy's problem) may be made plausible by the following geometrical consideration.

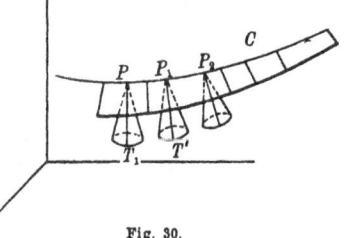

Fig. 30.

On any point on the given curve C, Fig. 30, draw the tangent cone T. There will be at P at least one plane tangent both to the curve

---

{ [1] We erect on the curve $\varphi(x, y) = 0$ a sort of cylindrical surface or fence whose height $z = \psi(x, y)$. }

C and to the cone T.[1]) An infinitesimal element of this plane then coincides with the integral surface at P. At a neighboring point P' there will be another cone T', differing slightly from that at P and furnishing another element of the surface. Thus we get a continuous set of plane elements, forming a strip of infinitesimal width, and passing through the curve C. Passing now to one edge of this strip treat it as a new curve C', when we get a new strip, and thus continuing we get a whole surface generated. That this process is possible may be rigidly proved analytically. We shall however see that the process fails for certain important curves.

22. Linear Equations, Characteristics. The simplest differential equations of any order are *linear*, that is those in which the derivatives appear only in the first power. Thus the linear equation of the first order is

$$P_1 p_1 + P_2 p_2 + \cdots + P_n p_n = Z,$$

where $P_1, P_2, \ldots P_n, Z$ are functions of $x_1, x_2, \ldots x_n, z$. If $Z = 0$ the equation is said to be homogeneous. In the case of two independent variables we shall write the non-homogeneous equation as

(13) $$Pp + Qq = R,$$

where P, Q, R, are functions of x, y, z. (Since we may write

$$F(x, y, z, p, q) \equiv Pp + Qq - R$$

the notation is the same as before in (9). In this case the normal cone, being of the first order, is a plane, and the cone T reduces to a line, which by (12) and (13) is

(14) $$\frac{\xi - x}{P} = \frac{\eta - y}{Q} = \frac{\zeta - z}{R}.$$

The simplest way in which z may be defined as a function of x and y involving one arbitrary function is by an equation between two *given* functions $u(x, y, z)$ and $v(x, y, z)$ such as,

$$u = \varphi(v) \quad \text{or} \quad \psi(u, v) = 0,$$

where φ or ψ is an *arbitrary* function. We shall see that such a relation makes z the solution of a linear partial differential equation of the first order. For differentiating the equation $\psi(u, v) = 0$ totally

(15)
$$\frac{\partial \psi}{\partial u} \left(\frac{\partial u}{\partial x} dx + \frac{\partial u}{\partial y} dy + \frac{\partial u}{\partial z} dz \right)$$
$$+ \frac{\partial \psi}{\partial v} \left(\frac{\partial v}{\partial x} dx + \frac{\partial v}{\partial y} dy + \frac{\partial v}{\partial z} dz \right) = 0$$

---

1) It may be that this tangent plane is *imaginary*, e. g., if the line in Fig. 30 should enter the cone. {In this case no *real* integral surface can be defined by the curve.}

and since z is a function of x and y,

$$(16) \qquad dz = \frac{\partial z}{\partial x} dx + \frac{\partial z}{\partial y} dy = p\,dx + q\,dy,$$

inserting in (15) and remembering that since x and y are independent dx and dy are totally unrelated, and therefore their coefficients in (15) must vanish independently, we find

$$(17) \qquad \begin{aligned} \frac{\partial \psi}{\partial u}\left(\frac{\partial u}{\partial x} + p\frac{\partial u}{\partial z}\right) + \frac{\partial \psi}{\partial v}\left(\frac{\partial v}{\partial x} + p\frac{\partial v}{\partial z}\right) &= 0, \\ \frac{\partial \psi}{\partial u}\left(\frac{\partial u}{\partial y} + q\frac{\partial u}{\partial z}\right) + \frac{\partial \psi}{\partial v}\left(\frac{\partial v}{\partial y} + q\frac{\partial v}{\partial z}\right) &= 0, \end{aligned}$$

expressing the vanishing of the partial derivatives by x and y.

Eliminating $\dfrac{\partial \psi}{\partial u}$, $\dfrac{\partial \psi}{\partial v}$, we get

$$\left(\frac{\partial u}{\partial x} + p\frac{\partial u}{\partial z}\right)\left(\frac{\partial v}{\partial y} + q\frac{\partial v}{\partial z}\right) = \left(\frac{\partial u}{\partial y} + q\frac{\partial u}{\partial z}\right)\left(\frac{\partial v}{\partial x} + p\frac{\partial v}{\partial z}\right),$$

or collecting terms in p and q,

$$p\,\frac{\partial(u, v)}{\partial(y, z)} + q\,\frac{\partial(u, v)}{\partial(z, x)} = \frac{\partial(u, v)}{\partial(x, y)}\,^{1)}$$

which is of the form

$$(13) \qquad Pp + Qq = R$$

where P, Q, R, are the known functions given by the three Jacobian determinants. Conversely we shall show that the most general solution of (13) is given by $\psi(u, v) = 0$ where ψ is arbitrary.

Let us first consider the homogeneous linear equation in *three* independent variables x, y, z,

$$(18) \qquad X\frac{\partial f}{\partial x} + Y\frac{\partial f}{\partial y} + Z\frac{\partial f}{\partial z} = 0$$

where X, Y, Z, are functions of x, y, z, only. The geometrical meaning of this equation is that, if $f(x, y, z) = 0$ is a solution, representing an integral surface, since the direction cosines of its normal are proportional to $\dfrac{\partial f}{\partial x}$, $\dfrac{\partial f}{\partial y}$, $\dfrac{\partial f}{\partial z}$, the normal to the integral surface is everywhere perpendicular to the vector X, Y, Z. In fact, if $f(x, y, z) = 0$ we have

$$(19) \qquad \frac{\partial f}{\partial x} dx + \frac{\partial f}{\partial y} dy + \frac{\partial f}{\partial z} dz = 0$$

---

1) We write

$$\frac{\partial(u, v)}{\partial(x, y)} = \begin{vmatrix} \dfrac{\partial u}{\partial x}, & \dfrac{\partial u}{\partial y} \\[2mm] \dfrac{\partial v}{\partial x}, & \dfrac{\partial v}{\partial y} \end{vmatrix}$$

and if this is to be compatible with (18) for all points x, y, z we must have

(20)
$$\frac{dx}{X} = \frac{dy}{Y} = \frac{dz}{Z}.$$

But these two simultaneous ordinary differential equations are the equations of the lines of the vector X, Y, Z (Chap. I, 26) one of which can be drawn through any point in space. In particular a different one may be drawn through each of the points of a plane, so that there are in all ∞^2 of these lines, and we have seen, the integral surface $f = 0$ is tangent to one of these at every one of its points.

Let two integrals of the equations (20) be

(21)
$$u(x, y, z) = a$$
$$v(x, y, z) = b,$$

where a and b are arbitrary constants. Then we may show that every integral of (18) is of the form $f \equiv \Phi(u, v)$. For since f, u, v are solutions of (18),

(22)
$$X \frac{\partial f}{\partial x} + Y \frac{\partial f}{\partial y} + Z \frac{\partial f}{\partial z} = 0,$$
$$X \frac{\partial u}{\partial x} + Y \frac{\partial u}{\partial y} + Z \frac{\partial u}{\partial z} = 0,$$
$$X \frac{\partial v}{\partial x} + Y \frac{\partial v}{\partial y} + Z \frac{\partial v}{\partial z} = 0,$$

rom which, eliminating X, Y, Z,

$$\frac{\partial(f, u, v)}{\partial(x, y, z)} = 0$$

which indicates a functional relation $f = \Phi(u, v)$. (See appendix).

From this relation Lagrange obtained the general integral of the linear non-homogeneous equation in two variables

(13)
$$Pp + Qq = R.$$

Suppose this is satisfied by $f(x, y, z) = 0$.

Differentiating partially, as above (z being a function of x and y),

(23)
$$\frac{\partial f}{\partial x} + p \frac{\partial f}{\partial z} = 0, \quad \frac{\partial f}{\partial y} + q \frac{\partial f}{\partial z} = 0.$$

Multiplying by P, Q, respectively, adding, and substituting in (13)

(24)
$$P \frac{\partial f}{\partial x} + Q \frac{\partial f}{\partial y} + R \frac{\partial f}{\partial z} = 0.$$

Now the curves given by the equations

(25)
$$\frac{dx}{P} = \frac{dy}{Q} = \frac{dz}{R}$$

have at every point the direction of the line (14) to which the cone T

reduces in this case, and on account of their important relations, are called the *characteristics* of the differential equation (13).

An integral $u(x, y, z) = a$ is a surface generated by a single infinity of characteristics. Changing the constant a, we get another surface, and giving a all values from $-\infty$ to $+\infty$ we get ∞^1 surfaces. Similarly $v(x, y, z) = b$ represents another surface, and giving b all possible values gives a second ∞^1 of surfaces. The intersections of the ∞^2 integral surfaces

$$u(x, y, z) = a, \quad v(x, y, z) = b$$

gives all the ∞^2 characteristic curves forming a so-called congruence, Fig. 31. The functional relation $\Phi(u, v) = 0$ denotes that to every value

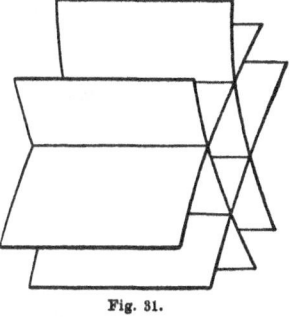

of u is associated one (or more) definite values of v, that is we pick out in an arbitrary manner ∞^1 intersections of the u and v surfaces. This may be done by choosing an arbitrary twisted curve C at each of whose points the u-surface and the v-surface will define the characteristic and all these characteristics sweep out an integral surface passing through the curve. It is obvious that if the given curve C is itself a characteristic, it does not define an integral surface, accordingly the char-

Fig. 31.

acteristics are curves of exception to the possible solution of Cauchy's problem, at the same time lying on integral surfaces.

Examples.

Equation of Cylinders.

If the generators are parallel to a line whose direction cosines are $a, b, 1$ the normals are perpendicular to it, accordingly

$$ap + bq = 1.$$

The equations (25) are

$$\frac{dx}{a} = \frac{dy}{b} = \frac{dz}{1},$$

of which two integrals are

$$x - az \equiv u = \text{const.,}$$
$$y - bz \equiv v = \text{const.,}$$

and an integral is

$$\Phi(x - az, y - bz) = 0.$$

The characteristics are the congruence of straight lines parallel to the given line, and any twisted curve determines a cylinder generated by them.

Equation of Cones.

The equation of surfaces whose tangent planes pass through a fixed point $a, b, c,$ is

$$(x - a)p + (y - b)q = z - c.$$

The equations (25) are

$$\frac{dx}{x - a} = \frac{dy}{y - b} = \frac{dz}{z - c},$$

of which two integrals are

$$\frac{x - a}{z - c} \equiv u = \text{const.}, \quad \frac{y - b}{z - c} \equiv v = \text{const},$$

and the desired integral,

$$\Phi\left(\frac{x - a}{z - c}, \frac{y - b}{z - c}\right) = 0.$$

The characteristics are all straight rays emanating from the point $a, b, c.$
Surfaces of Revolution.

These have the property that the normal always meets a fixed line. Let its equations be

$$\frac{X}{\alpha} = \frac{Y}{\beta} = \frac{Z}{\gamma} = \varrho.$$

Let the equations of the normal be

$$\frac{X' - x}{p} = \frac{Y' - y}{q} = \frac{Z' - z}{-1} = \varrho'.$$

If these meet there is a point for which

$$X = X', \quad Y = Y', \quad Z = Z',$$

then

$$x + p\varrho' = \alpha\varrho,$$
$$y + q\varrho' = \beta\varrho,$$
$$z - \varrho' = \gamma\varrho.$$

Eliminating $\varrho, \varrho',$

$$\begin{vmatrix} x, & \alpha, & p \\ y, & \beta, & q \\ z, & \gamma, & -1 \end{vmatrix} = 0,$$

which is the required differential equation, or expanded,

$$(\gamma y - \beta z)p + (\alpha z - \gamma x)q = \beta x - \alpha y.$$

Put

$$\frac{dx}{\gamma y - \beta z} = \frac{dy}{\alpha z - \gamma x} = \frac{dz}{\beta x - \alpha y} = dt,$$
$$dx = (\gamma y - \beta z)\, dt,$$
$$dy = (\alpha z - \gamma x)\, dt,$$
$$dz = (\beta x - \alpha y)\, dt,$$

from which we deduce the integrable combinations

$$x\,dx + y\,dy + z\,dz = 0,$$
$$\alpha\,dx + \beta\,dy + \gamma\,dz = 0,$$

giving

$$x^2 + y^2 + z^2 \equiv u = \text{const.},$$
$$\alpha x + \beta y + \gamma z \equiv v = \text{const.},$$
$$\Phi(x^2 + y^2 + z^2, \ \alpha x + \beta y + \gamma z) = 0.$$

The characteristics are all the circles in planes perpendicular to the given axis, and with their centers on it, and cut out by the spheres u and the planes v.

23. Classification of Integrals. It was proved by Cauchy, and by Mme. Kovalevski, that a partial differential equation of the first order can always be solved by a function which, for a particular value of one of the independent variables x, becomes an arbitrary function of all the others, that is when

$$x_1 = x_1^0, \quad z = \varphi(x_2, x_3, \ldots x_n).$$

In the case of two independent variables for

$$x = x_0, \quad z = \varphi(y).$$

This represents a plane curve, and the case of a twisted curve, mentioned in § 20 can be reduced to this by a suitable transformation. Such an integral is called the *general* solution, and every solution obtained therefrom by giving the arbitrary function a particular form is called a particular solution or integral. A solution which contains, not an arbitrary function, but as many arbitrary *constants* as there are independent variables, was called by Lagrange a *complete* integral.

Consider the equation

$$(26) \qquad\qquad V(x, y, z, a, b) = 0,$$

where a and b are arbitrary constants. This determines z as a function of x, y, and we may show that this is a complete integral of an equation of the first order. Let us indicate partial derivatives under the circumstances that z is a function of x and y by brackets. Then from the equation

$$dV = \left(\frac{\partial V}{\partial x}\right) dx + \left(\frac{\partial V}{\partial y}\right) dy = \frac{\partial V}{\partial x}\,dx + \frac{\partial V}{\partial y}\,dy + \frac{\partial V}{\partial z}\,dz$$

we get

$$(27) \qquad \begin{aligned} \left(\frac{\partial V}{\partial x}\right) &= \frac{\partial V}{\partial x} + p\,\frac{\partial V}{\partial z} = 0, \\ \left(\frac{\partial V}{\partial y}\right) &= \frac{\partial V}{\partial y} + q\,\frac{\partial V}{\partial z} = 0. \end{aligned}$$

From the three equations (26), (27) containing x, y, z, p, q, a, b, we may eliminate a, b obtaining an equation

$$(28) \qquad F(x, y, z, p, q) = 0.$$

Lagrange sought from a complete integral to deduce the general integral. Since the equation (28) is obtained by eliminating a, b between (26), (27) it expresses the condition that (26) and (27) define three functions

$$z = f_1(x, y), \quad a = f_2(x, y), \quad b = f_3(x, y)$$

satisfying (26) and (27). Let us differentiate (26) on the assumption that a, b are so defined.

$$(29) \qquad \frac{\partial V}{\partial x} dx + \frac{\partial V}{\partial y} dy + \frac{\partial V}{\partial z}(p\,dx + q\,dy) + \frac{\partial V}{\partial a} da + \frac{\partial V}{\partial b} db = 0,$$

and by reason of equations (27) which we have already used, this reduces to

$$(30) \qquad \frac{\partial V}{\partial a} da + \frac{\partial V}{\partial b} db = 0.$$

This is satisfied as follows:

1^0. If a and b are constants; this gives the complete integral, representing any one of a double infinity of surfaces.

2^0. If

$$\frac{\partial V}{\partial a} = 0, \quad \frac{\partial V}{\partial b} = 0.$$

If we eliminate a and b from these two equations and (26) we obtain a relation between x, y, z, without arbitrary constants or functions, and therefore not derivable from the general integral. It is called the *singular* integral, and represents a surface which is the *envelope* of the ∞^2 surfaces given by the complete integral.

3^0. More generally, since the equation (30) can be written,

$$\frac{\partial V}{\partial a}\left(\frac{\partial a}{\partial x} dx + \frac{\partial a}{\partial y} dy\right) + \frac{\partial V}{\partial b}\left(\frac{\partial b}{\partial x} dx + \frac{\partial b}{\partial y} dy\right) = 0,$$

we have, dx and dy being independent,

$$\frac{\partial V}{\partial a} \frac{\partial a}{\partial x} + \frac{\partial V}{\partial b} \frac{\partial b}{\partial x} = 0,$$

$$\frac{\partial V}{\partial a} \frac{\partial a}{\partial y} + \frac{\partial V}{\partial b} \frac{\partial b}{\partial y} = 0,$$

which by elimination of $\dfrac{\partial V}{\partial a}, \dfrac{\partial V}{\partial b}$ give

$$(31) \qquad \frac{\partial(a, b)}{\partial(x, y)} = 0$$

showing a functional relation between a and b, say $b = \varphi(a)$, $db = \varphi'(a)da$, and substituting in (30)

$$(32) \qquad \frac{\partial V}{\partial a} + \frac{\partial V}{\partial b} \varphi'(a) = 0.$$

Eliminating a between (32) and (26) gives a solution, containing no arbitrary constant, but an arbitrary function — it is accordingly the general integral. Of course, the elimination cannot be practically effected until the arbitrary function has been determined.

Geometrically, when we determine b as a function of a, we select a single infinity out of the double infinity of surfaces $V(x, y, z, a, b) = 0$ and eliminating between

$$(33) \qquad V(x, y, z, a, \varphi(a)) = 0$$

and its derivative according to a

$$(34) \qquad \frac{\partial V(x, y, z, a, \varphi(a))}{\partial a} = 0$$

we obtain the envelope of these ∞^1 surfaces, representing the general solution. It is to be noticed how this differs from the singular solution, which is the envelope of ∞^2 surfaces, while the general solution is the envelope of ∞^1 surfaces, arbitrarily selected.

Consider an example.

$$V \equiv ax + by + z - a^2 - b^2 = 0.$$

The complete integral gives us a set of planes. Equations (27) are

$$a + p = 0, \; b + q = 0,$$

and the equation (28) is

$$px + qy - z + p^2 + q^2 = 0,$$

which is not linear. The singular solution is obtained from

$$\frac{\partial V}{\partial a} = x - 2a = 0, \; \frac{\partial V}{\partial b} = y - 2b = 0$$

and eliminating a, b

$$x^2 + y^2 + 4z = 0,$$

a paraboloid of revolution, enveloping all the planes. The general solution is given by

$$ax + \varphi(a)y + z - a^2 - (\varphi(a))^2 = 0,$$
$$x + \varphi'(a)y \quad - 2a - 2\varphi(a)\varphi'(a) = 0.$$

In certain cases of the equation (28), the complete integral can be found by inspection. For instance,

$$pq = 1.$$

Evidently

$$z = ax + \frac{y}{a} + b$$

is a complete integral. More generally, if x, y, z, do not appear in F,

$$q = f(p),$$

$p = a$, gives a complete integral

$$z = ax + f(a)y + b.$$

24. Method of Characteristics. We shall now consider the method developed by Cauchy and Lie, which introduces us to the importance of the nature of the characteristics of a non-linear equation of the first order. We define as an *element* of space a point x, y, z and a plane passing through it with·slopes p, q. Thus the set of values x, y, z, p, q defines an element, and there are ∞^5 elements in space, for at each of the ∞^3 points there are ∞^2 directions possible for normals to the planes. Suppose we consider a continuous multiplicity of elements given by

$$(35) \quad x = f_1(u), \quad y = f_2(u), \quad z = f_3(u), \quad p = f_4(u), \quad q = f_5(u),$$

where u is a variable parameter.

This represents a curve, and at each point we may construct a definite plane element (Fig. 32). In general the consecutive planes will

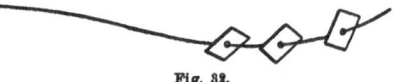

Fig. 32.

intersect in lines that do not pass through the curve. If they do, we shall call the elements *united* and they have an enveloping surface, which does pass through the curve. The condition for this is

$$(36) \quad z = \Phi(x, y), \quad p = \frac{\partial \Phi}{\partial x}, \quad q = \frac{\partial \Phi}{\partial y},$$

or the single equation

$$(37) \quad dz = p\,dx + q\,dy,$$

necessitating the relation

$$f_3'(u) = f_4(u)f_1'(u) + f_5(u)f_2'(u).$$

Suppose (36) denotes a particular integral of

$$(38) \quad F(x, y, z, p, q) = 0.$$

Let us begin with any element x_0, y_0, z_0, p_0, q_0, satisfying (38), and let us leave the point x_0, y_0, z_0, in such a direction that projected on the xy-plane,

$$(39) \quad \frac{dx}{P} = \frac{dy}{Q} = du,$$

where u is a variable which, {we will suppose,} vanishes for $x = x_0$, $y = y_0$.

Since P and Q are known functions (defined in 9) of x, y, z, p, q, and, making use of (36), of x, y alone, x and y will take on a definite series of values, and for each of these z, p, q will be known, so that we obtain a united continuum of elements, represented by a curve in space, and a strip of surface containing it, Fig. 33. This succession of elements is called a characteristic, in-cluding both the curve and the strip of surface. Through every point of the integral surface there goes a charac-teristic, there are therefore in the surface ∞^1 characteristics, and the surface is known if we know all its characteristics.

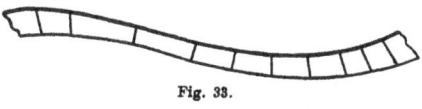

Fig. 33.

Let x, y, z, p, q be an element,

$$x + \delta x, \ y + \delta y, \ z + \delta z, \ p + \delta p, \ q + \delta q$$

any infinitely near element not on the same characteristic. If we write

$$r = \frac{\partial^2 \Phi}{\partial x^2}, \quad s = \frac{\partial^2 \Phi}{\partial x \partial y}, \quad t = \frac{\partial^2 \Phi}{\partial y^2},$$

we have by definition,

$$\begin{aligned} \delta z &= p\,\delta x + q\,\delta y, \\ \delta p &= r\,\delta x + s\,\delta y, \\ \delta q &= s\,\delta x + t\,\delta y. \end{aligned}$$

(40)

Let

$$x + dx, \ y + dy, \ z + dz, \ p + dp, \ q + dq$$

be another element on the same characteristic as x, y, z, p, q. Then

$$\begin{aligned} dz &= p\,dx + q\,dy, \\ dp &= r\,dx + s\,dy, \\ dq &= s\,dx + t\,dy. \end{aligned}$$

(41)

Differentiating (38) totally, by varying the characteristic,

$$X\delta x + Y\delta y + Z\delta z + P\delta p + Q\delta q = 0.$$

Replacing P and Q by their values from (39),

(39)
$$P = \frac{dx}{du}, \quad Q = \frac{dy}{du},$$

$\delta z, \delta p$ and δq by their values from (40),

(42) $\delta x\left(X + pZ + r\frac{dx}{du} + s\frac{dy}{du}\right) + \delta y\left(Y + qZ + s\frac{dx}{du} + t\frac{dy}{du}\right) = 0,$

or by the last two of (41)

$$\delta x\left(X + pZ + \frac{dp}{du}\right) + \delta y\left(Y + qZ + \frac{dq}{du}\right) = 0,$$

and since δx, δy are independent,

$$X + pZ + \frac{dp}{du} = 0,$$

(43)

$$Y + qZ + \frac{dq}{du} = 0.$$

Now differentiating (38) along the characteristic,

$$X dx + Y dy + Z dz + P dp + Q dq = 0,$$

and replacing dx by Pdu, dy by Qdu, dp by $-(X+pZ)du$, dq by $-(Y+qZ)du$ from (43), we obtain

$$\{PX + QY - P(X+pZ) - Q(Y+qZ)\}\,du + Z\,dz = 0,$$

or on dividing by Z,

(44) $$dz = (Pp + Qq)\,du.$$

We have accordingly the differential equations (39), (43), (44),

(45) $$\frac{dx}{P} = \frac{dy}{Q} = \frac{dz}{Pp+Qq} = -\frac{dp}{X+pZ} = -\frac{dq}{Y+qZ} = du,$$

of which the first two have already been given in (25) for the linear equation, (13). These simultaneous equations, whose denominators are all given functions of x, y, z, p, q, and nothing else, on being integrated, will give, together with the initial element $x_0. y_0, z_0, p_0, q_0$, perfectly definite values for the variables, x, y, z, p, q.

$$x = f_1(u, x_0, y_0, z_0, p_0, q_0),$$
$$y = f_2(u, x_0, y_0, z_0, p_0, q_0),$$
(46) $$z = f_3(u, x_0, y_0, z_0, p_0, q_0),$$
$$p = f_4(u, x_0, y_0, z_0, p_0, q_0),$$
$$q = f_5(u, x_0, y_0, z_0, p_0, q_0).$$

These are independent of Φ, hence we have the remarkable result: every integral which contains the element x_0, y_0, z_0, p_0, q_0, contains all the elements of the characteristic through it. Or, two surfaces representing integrals {and tangent at one point,} are tangent to each other along a characteristic.

Equations (46) represent a solution if

(47) $$F(x_0, y_0, z_0, p_0, q_0) = 0.$$

The values of x, y, z, p, q corresponding to the various elements of the different integrals, are accordingly expressed in terms of the six parameters $u, x_0, y_0, z_0, p_0, q_0$, of which the last five satisfy (47). Leaving these constant, and varying u, we obtain all the elements of one characteristic, which thus form what is called a continuous group of one parameter. By varying x_0, y_0, z_0, p_0, q_0, we pass from one characteristic to another, making an infinitesimal or finite *transformation* of the group.

Let

$$(48) \qquad z = \Phi(x, y), \quad p = \frac{\partial \Phi}{\partial x}, \quad q = \frac{\partial \Phi}{\partial y}$$

be any integral. In order that it may contain a given element x, y, z, p, q, it need only contain the initial element x_0, y_0, z_0, p_0, q_0 lying on the same characteristic, that is,

$$(49) \qquad z_0 = \Phi(x_0, y_0), \quad p_0 = \frac{\partial \Phi}{\partial x_0}, \quad q_0 = \frac{\partial \Phi}{\partial y_0}.$$

These three equations, with (47) and (46) characterize all the elements of the integral. Hence by the elimination from these nine of u, x_0, y_0, z_0, p_0, q_0 we get three equations, giving z, p, q in terms of x, y, that is the integral.

The number of possible characteristics may be considered as follows. From (47) we may express q_0, and from $x = f_1(u, x_0, y_0, z_0, p_0, q_0)$ we may express u, thus we get

$$(50) \qquad \begin{aligned} y &= \psi_1(x, x_0, y_0, z_0, p_0), \\ z &= \psi_2(x, x_0, y_0, z_0, p_0), \\ p &= \psi_3(x, x_0, y_0, z_0, p_0), \\ q &= \psi_4(x, x_0, y_0, z_0, p_0), \end{aligned}$$

and since for a given x_0 we may take arbitrarily y_0, z_0, p_0, we have ∞^3 characteristics.

Geometrically speaking, as we have seen that at every point we have a single infinity of directions, and as each determines a characteristic, there are at each point ∞^1 characteristics which are generators of a surface representing a particular solution, having a singular { or conical } point. As there are ∞^3 points in space, we might then expect ∞^4 characteristics, but as in each there are ∞^1 points there are only ∞^3 characteristics, composing what is called a *complex* of curves. In the case of a linear equation, we have seen that there are only ∞^2 characteristics, for at each point there is but *one* direction of a characteristic.

Suppose we represent the initial element in terms of a single parameter v,

$$(51) \quad x_0 = \varphi_1(v), \ y_0 = \varphi_2(v), \ z_0 = \varphi_3(v), \ p_0 = \varphi_4(v), \ q_0 = \varphi_5(v).$$

How shall we choose the five functions φ so that the surfaces generated by the characteristics shall be an integral?

The equations

$$(52) \qquad \begin{aligned} x &= f_1(u, \varphi_1(v), \varphi_2(v), \varphi_3(v), \varphi_4(v), \varphi_5(v)), \\ y &= f_2(u, \varphi_1(v), \varphi_2(v), \varphi_3(v), \varphi_4(v), \varphi_5(v)), \\ z &= f_3(u, \varphi_1(v), \varphi_2(v), \varphi_3(v), \varphi_4(v), \varphi_5(v)), \\ p &= f_4(u, \varphi_1(v), \varphi_2(v), \varphi_3(v), \varphi_4(v), \varphi_5(v)), \\ q &= f_5(u, \varphi_1(v), \varphi_2(v), \varphi_3(v), \varphi_4(v), \varphi_5(v)), \end{aligned}$$

in which u and v are independent parameters, represent a surface (by x, y, z) and a set of elements (by p, q) and for them to be united we must have

$$(53) \qquad \frac{\partial z}{\partial u} = p \frac{\partial x}{\partial u} + q \frac{\partial y}{\partial u},$$

$$(54) \qquad \frac{\partial z}{\partial v} = p \frac{\partial x}{\partial v} + q \frac{\partial y}{\partial v},$$

$$(55) \qquad F(x, y, z, p, q) = 0.$$

As the five functions f_1, f_2, f_3, f_4, f_5 are integrals of the equations (45) we have

$$F(x, y, z, p, q) = F(x_0, y_0, z_0, p_0, q_0),$$

hence we need have only

$$F(x_0, y_0, z_0, p_0, q_0) = 0.$$

Equation (53) is satisfied, by the differential equations (45) (in which v is constant). In order to satisfy (54) put

$$(56) \qquad H = \frac{\partial z}{\partial v} - p \frac{\partial x}{\partial v} - q \frac{\partial y}{\partial v}.$$

Differentiating by u,

$$(57) \qquad \frac{\partial H}{\partial u} = \frac{\partial^2 z}{\partial u \partial v} - p \frac{\partial^2 x}{\partial u \partial v} - q \frac{\partial^2 y}{\partial u \partial v} - \frac{\partial p}{\partial u} \frac{\partial x}{\partial v} - \frac{\partial q}{\partial u} \frac{\partial y}{\partial v}.$$

Differentiating (53) by v,

$$(58) \qquad 0 = \frac{\partial^2 z}{\partial u \partial v} - p \frac{\partial^2 x}{\partial u \partial v} - q \frac{\partial^2 y}{\partial u \partial v} - \frac{\partial p}{\partial v} \frac{\partial x}{\partial u} - \frac{\partial q}{\partial v} \frac{\partial y}{\partial u}.$$

Subtracting we obtain

$$(59) \qquad \frac{\partial H}{\partial u} = \frac{\partial p}{\partial v} \frac{\partial x}{\partial u} - \frac{\partial p}{\partial u} \frac{\partial x}{\partial v} + \frac{\partial q}{\partial v} \frac{\partial y}{\partial u} - \frac{\partial q}{\partial u} \frac{\partial y}{\partial v},$$

and replacing derivatives by u by their values from (45)

$$(60) \qquad \frac{\partial H}{\partial u} = P \frac{\partial p}{\partial v} + Q \frac{\partial q}{\partial v} + (X + pZ) \frac{\partial x}{\partial v} + (Y + qZ) \frac{\partial y}{\partial v}$$
$$= X \frac{\partial x}{\partial v} + Y \frac{\partial y}{\partial v} + P \frac{\partial p}{\partial v} + Q \frac{\partial q}{\partial v} + Z \left(p \frac{\partial x}{\partial v} + q \frac{\partial y}{\partial v} \right).$$

Now since

$$F(x, y, z, p, q) = 0$$

we have

$$(61) \qquad X \frac{\partial x}{\partial v} + Y \frac{\partial y}{\partial v} + Z \frac{\partial z}{\partial v} + P \frac{\partial p}{\partial v} + Q \frac{\partial q}{\partial v} = 0,$$

subtracting which from (60) gives

$$(62) \qquad \frac{\partial H}{\partial u} = Z \left(p \frac{\partial x}{\partial v} + q \frac{\partial y}{\partial v} - \frac{\partial z}{\partial v} \right) = -ZH.$$

This is a differential equation in u, integrating which,

$$(63) \qquad\qquad H = H_0 e^{-\int_{u_0}^{u} Z\, du}$$

If Z is a holomorphic function of u, the exponential function cannot vanish, so that for H to be zero, we must have $H_0 = 0$, that is

$$(64) \qquad\qquad \frac{\partial z_0}{\partial v} - p_0 \frac{\partial x_0}{\partial v} - q_0 \frac{\partial y_0}{\partial v} = 0 \,.$$

Consequently for an integral surface we need only put x_0, y_0, z_0, p_0, q_0 functions of a variable v satisfying (47) and (64). We may thus solve Cauchy's problem, putting x_0, y_0, z_0, in terms of v so as to lie on the given curve, when (64) and (47) determine p_0 and q_0. Cauchy obtained the general integral by putting

$$x_0 = \text{const.,}$$
$$z_0 = \varphi(y_0),$$
$$q_0 = \varphi'(y_0) \,.$$

In this case $v = y_0$, and (64) is satisfied. The determining curve is plane.[1])

We have then the practical rule: to integrate an equation of the first order $F(x, y, z, p, q) = 0$, integrate the simultaneous equations (45), obtaining (46) giving initial values that satisfy

$$F(x_0, y_0, z_0, p_0, q_0) = 0,$$

and putting

$$z_0 = \varphi(y_0), \quad q_0 = \varphi'(y_0) \,.$$

Eliminating from these eight equations in eleven variables u, y_0, z_0, p_0, q_0 we obtain the three

$$z = \varphi(x, y), \quad p = \psi(x, y), \quad q = \chi(x, y) \,.$$

Examples.

As a very simple example consider the equation

$$\left(\frac{\partial z}{\partial x}\right)^2 + \left(\frac{\partial z}{\partial y}\right)^2 = 1, \text{ or } p^2 + q^2 = 1, \quad F \equiv \tfrac{1}{2}(p^2 + q^2 - 1) = 0$$
$$X = Y = Z = 0, \quad P = p, \quad Q = q \,.$$

Our equations (45) then are

$$\frac{dx}{p} = \frac{dy}{q} = \frac{dz}{1} = \frac{dp}{0} = \frac{dq}{0}$$

From the last two terms we must have $dp = dq = 0$, $p = p_0$, $q = q_0$, accordingly

$$\frac{dx}{p_0} = \frac{dy}{q_0} = dz \,,$$

{1} By a suitable transformation a twisted curve may always be reduced to a plane curve. }

from which we obtain the integrals

$$q_0(x - x_0) = p_0(y - y_0), \quad x - x_0 = p_0(z - z_0)$$

which are the equations of a straight line inclined at an angle of 45^0 to the xy-plane, (as may be seen since $dx^2 + dy^2 = (p_0^2 + q_0^2)dz^2 = dz^2$). The complex of characteristics is composed of all such lines. Those passing through a given point x_0, y_0, z_0 form a circular cone

$$(\zeta - z_0)^2 = (\xi - x_0)^2 + (y - y_0)^2$$

as can at once be shown by finding the envelope of the tangent plane

$$\zeta - z_0 = p_0(\xi - x_0) + q_0(y - y_0)$$

by differentiating by p_0 after putting $q_0 = \sqrt{1 - p_0^2}$, and eliminating p_0.

The general integral is found from the equations

$$q_0(x - x_0) = p_0(y - y_0), \quad x - x_0 = p_0(z - z_0), \quad p = p_0, \quad q = q_0$$
$$p_0^2 + q_0^2 = 1, \quad z_0 = \varphi(y_0), \quad q_0 = \varphi'(y_0).$$

Consider the equation

$$z = \frac{\partial z}{\partial x} \frac{\partial z}{\partial y} \quad \text{or} \quad F \equiv z - pq = 0.$$

We have

$$X = Y = 0, \quad Z = 1, \quad P = -q, \quad Q = -p,$$

and equations (45) are

$$\frac{dx}{q} = \frac{dy}{p} = \frac{dz}{2pq} = \frac{dp}{p} = \frac{dq}{q},$$

from which we deduce the integrals

$$x - x_0 = q - q_0, \quad y - y_0 = p - p_0, \quad \frac{p}{p_0} = \frac{q}{q_0},$$

using which in

$$dz = 2p\,dx = 2\frac{p_0}{q_0}q\,dx = 2\frac{p_0}{q_0}(x - x_0 + q_0)dx$$

leads to the further integral

$$z = \frac{p_0}{q_0}(x - x_0 + q_0)^2$$

which with the three above and

$$p_0 q_0 = z_0, \quad z_0 = \varphi(y_0), \quad q_0 = \varphi'(y_0),$$

give the general integral. The characteristics are parabolas in vertical planes, those through a point x_0, y_0, z_0, having the equations

$$p_0(x - x_0) = q_0(y - y_0), \quad z = \frac{p_0}{q_0}(x - x_0 + q_0)^2, \quad p_0 q_0 = z_0.$$

Consider the equation

$$xy = \frac{\partial z}{\partial x}\frac{\partial z}{\partial y}, \quad \text{or} \quad F \equiv xy - pq = 0,$$

$$X = y, \quad Y = x, \quad Z = 0, \quad P = -q, \quad Q = -p,$$

$$\frac{dx}{q} = \frac{dy}{p} = \frac{dz}{2pq} = \frac{dp}{y} = \frac{dq}{x}.$$

Multiplying by $xy = pq$,

$$pdx = qdy = \frac{dz}{2} = xdp = ydq,$$

and integrating,

$$\frac{x}{x_0} = \frac{p}{p_0}, \quad \frac{y}{y_0} = \frac{q}{q_0}, \quad dz = 2pdx = 2\frac{p_0}{x_0}xdx,$$

$$z - z_0 = \frac{p_0}{x_0}(x^2 - x_0^2) = \frac{q_0}{y_0}(y^2 - y_0^2).$$

The characteristics are twisted curves.

Multiplying together the two values of $z - z_0$, since $x_0 y_0 = p_0 q_0$ we obtain the complete integral

$$(z - z_0)^2 = (x^2 - x_0^2)(y^2 - y_0^2)$$

and the general integral

$$[z - \varphi(y_0)]^2 = (x^2 - x_0^2)(y^2 - y_0^2)$$

$$z - \varphi(y_0) = \frac{\varphi'(y_0)}{y_0}(y^2 - y_0^2),$$

from which y_0 is to be eliminated.

Cauchy's method may be extended to equations in any number of independent variables. Suppose equation is

(38') $$F(x_1, x_2 \cdots x_n, z, p_1, p_2 \cdots p_n) = 0$$

where we make use of the extended notation,

$$p_r = \frac{\partial z}{\partial x_r}, \quad X_r = \frac{\partial F}{\partial x_r}, \quad P_r = \frac{\partial F}{\partial p_r}, \quad Z = \frac{\partial F}{\partial z}.$$

Then proceeding exactly as before we shall find corresponding to (45) the equations

(45') $$\frac{dx_1}{P_1} = \frac{dx_2}{P_2} = \cdots = \frac{dx_n}{P_n} = \frac{dz}{P_1 p_1 + P_2 p_2 \cdots + P_n p_n}$$

$$= -\frac{dp_1}{X_1 + p_1 Z} = \cdots - \frac{dp_n}{X_n + p_n Z} = du,$$

and the characteristics are defined as before by means of a continuous group of elements in the hyperspace. We shall need these characteristics in Chapter VI.

CHAPTER III.

WAVE EQUATION. VIBRATIONS AND NORMAL FUNCTIONS.

25. Euler's Equation. Three Types.

The solution of equations of orders higher than the first is reduced to the study of particular equations — of these the equations of mathematical physics are the most important, and will now be studied. They are linear and the simplest ones have constant coefficients.

The simplest equation in two independent variables of order higher than the first is

$$(1) \qquad \frac{\partial^{m+n} z}{\partial x^m \partial y^n} = 0.$$

Its general integral is easily found. Introduce as an auxiliary variable,

$$(2) \qquad u = \frac{\partial^n z}{\partial y^n},$$

then our equation is

$$(3) \qquad \frac{\partial^m u}{\partial x^m} = 0,$$

whose solution is

$$(4) \qquad u = A_0 + A_1 x + A_2 x^2 \cdots + A_{m-1} x^{m-1},$$

where the A's are independent of x, i. e., are arbitrary functions of y. Let

$$A_0 = \frac{d^n Y_0}{dy^n}, \quad A_1 = \frac{d^n Y_1}{dy^n}, \quad \ldots A_{m-1} = \frac{d^n Y_{m-1}}{dy^n}.$$

Integrating n times we obtain

$$(5) \qquad z_1 = Y_0 + Y_1 x + Y_2 x^2 \cdots + Y_{m-1} x^{m-1}$$

as a particular solution. Let

$$z = z_1 + \zeta,$$

then we have

$$(6) \qquad \frac{\partial^n z}{\partial y^n} = \frac{\partial^n z_1}{\partial y^n} + \frac{\partial^n \zeta}{\partial y^n} = u,$$

and therefore, since

$$\frac{\partial^n z_1}{\partial y^n} = u, \quad \frac{\partial^n \zeta}{\partial y^n} = 0,$$

whose integral is

$$(7) \qquad \zeta = X_0 + X_1 y + X_2 y^2 \cdots + X_{n-1} y^{n-1},$$

where the X's are arbitrary functions of x. Consequently we have for our general solution

$$(8) \qquad z = Y_0 + Y_1 x + Y_2 x^2 + \cdots + Y_{m-1} x^{m-1}$$
$$+ X_0 + X_1 y + X_2 y^2 + \cdots + X_{n-1} y^{n-1}.$$

Introducing $m + n$ arbitrary functions.

In particular the equation of the second order

$$(9) \qquad \frac{\partial^2 z}{\partial x \partial y} = 0$$

has the general integral

$$(10) \qquad z = X_0 + Y_0 = \varphi(x) + \psi(y).$$

A more general equation is that treated by Euler:

$$(11) \qquad a \frac{\partial^2 z}{\partial x^2} + 2b \frac{\partial^2 z}{\partial x \partial y} + c \frac{\partial^2 z}{\partial y^2} = 0,$$

where a, b, c, are constants.

Let us change the independent variables, by the linear transformation

$$(12) \qquad \alpha x + \beta y = \xi, \ \gamma x + \delta y = \eta,$$

$$\frac{\partial z}{\partial x} = \frac{\partial z}{\partial \xi} \frac{\partial \xi}{\partial x} + \frac{\partial z}{\partial \eta} \frac{\partial \eta}{\partial x} = \alpha \frac{\partial z}{\partial \xi} + \gamma \frac{\partial z}{\partial \eta},$$

$$\frac{\partial z}{\partial y} = \frac{\partial z}{\partial \xi} \frac{\partial \xi}{\partial y} + \frac{\partial z}{\partial \eta} \frac{\partial \eta}{\partial y} = \beta \frac{\partial z}{\partial \xi} + \delta \frac{\partial z}{\partial \eta},$$

$$(13) \qquad \frac{\partial^2 z}{\partial x^2} = \left(\alpha \frac{\partial}{\partial \xi} + \gamma \frac{\partial}{\partial \eta}\right)^2 z, \ \frac{\partial^2 z}{\partial y^2} = \left(\beta \frac{\partial}{\partial \xi} + \delta \frac{\partial}{\partial \eta}\right)^2 z,$$

$$\frac{\partial^2 z}{\partial x \partial y} = \left(\alpha \frac{\partial}{\partial \xi} + \gamma \frac{\partial}{\partial \eta}\right) \left(\beta \frac{\partial z}{\partial \xi} + \delta \frac{\partial z}{\partial \eta}\right)$$

so that the transformed equation is

$$(14) \qquad \begin{aligned} & (a\alpha^2 + 2b\alpha\beta + c\beta^2) \frac{\partial^2 z}{\partial \xi^2} + (a\gamma^2 + 2b\gamma\delta + c\delta^2) \frac{\partial^2 z}{\partial \eta^2} \\ & + 2[a\alpha\gamma + b(\alpha\delta + \beta\gamma) + c\beta\delta] \frac{\partial^2 z}{\partial \xi \partial \eta} = 0. \end{aligned}$$

We may make the first two terms disappear by a suitable choice of α, β, γ, δ.

Put

$$\alpha = \gamma = 1, \ \beta = \lambda_1^1, \ \delta = \lambda_2,$$

where λ_1, λ_2 are the roots of the quadratic

$$(15) \qquad a + 2b\lambda + c\lambda^2 = 0.$$

According as we have

$$(16) \qquad \begin{aligned} b^2 - ac &> 0, \text{ roots real,} \\ b^2 - ac &= 0, \text{ roots equal,} \\ b^2 - ac &< 0, \text{ roots imaginary,} \end{aligned}$$

we characterize the differential equations as hyperbolic, parabolic, or elliptic. The importance of this classification will appear in Chapter VI.

We have then

$$(17) \qquad \{a + b(\lambda_1 + \lambda_2) + c\lambda_1\lambda_2\} \frac{\partial^2 z}{\partial \xi \partial \eta} = 0$$

or since

$$\lambda_1 + \lambda_2 = -\frac{2b}{c}, \ \lambda_1 \lambda_2 = \frac{a}{c}$$

(18)
$$\frac{2}{c}(ac - b^2)\frac{\partial^2 z}{\partial \xi \partial \eta} = 0.$$

If then
$$ac \neq b^2,$$

(19)
$$\frac{\partial^2 z}{\partial \xi \partial \eta} = 0, \ z = \varphi(\xi) + \psi(\eta).$$

Accordingly we have

(20)
$$z = \varphi(x + \lambda_1 y) + \psi(x + \lambda_2 y).$$

This is the case for the hyperbolic and elliptic equations.

If $ac = b^2$, the roots λ_1, λ_2 are equal, and $\xi = \eta$. Therefore we must proceed otherwise.

In that case put
$$\alpha = 1, \ \beta = \lambda, \text{ with } \gamma, \ \delta$$

arbitrary. The third coefficient now becomes

$$a\gamma + b(\delta + \gamma\lambda) + c\lambda\delta = (a + b\lambda)\gamma + (b + c\lambda)\delta = 0,$$

since
$$-\lambda = \frac{a}{b} = \frac{b}{c}.$$

The second coefficient does not vanish, and we therefore have

(21)
$$\frac{\partial^2 z}{\partial \eta^2} = 0,$$

(22) $$z = \varphi(\xi) + \eta\psi(\xi) = \varphi(x + \lambda y) + (\gamma x + \delta y)\psi(x + \lambda y),$$

the general integral of the parabolic differential equation.

26. Equation of String. Waves. This method applies to the differential equation of a stretched string, Chap. I, (116) or of plane waves of sound (without impressed forces)

(23)
$$\frac{\partial^2 u}{\partial t^2} = a^2 \frac{\partial^2 u}{\partial x^2}, \ a^2 = \frac{\tau}{\varrho}.$$

Putting
$$\xi = x + \lambda_1 t, \ \eta = x + \lambda_2 t$$

where
$$\lambda^2 - a^2 = 0, \ \lambda_1 = a, \ \lambda_2 = -a,$$

we obtain

(24)
$$u = \varphi(x + at) + \psi(x - at).$$

This solution was given by d'Alembert in 1747.

Let us consider the meaning of the two terms separately.
$$u_1 = \psi(\eta) = \psi(x - at).$$

For each value of η, u has a definite value. If we represent the values of u_1 in terms of η by a surface, we have, along the line $\eta = x - at =$ const., u_1 constant, so that the surface is a cylinder, with generators parallel to the line $x = at$, $u = 0$, Fig. 34.

To find the values of u_1 at any time t, pass a plane at a distance t from and parallel to the plane $t = 0$. The section is the same as that by the plane $t = 0$, moved a distance at to the right.

Thus the initial *form* of the string is preserved, but travels to the right with velocity a in the form of a wave. The other term represents a wave travelling to the left with the same velocity.

We have now to determine the arbitrary functions arising in the solution. Suppose that for a given instant we give the displacement of

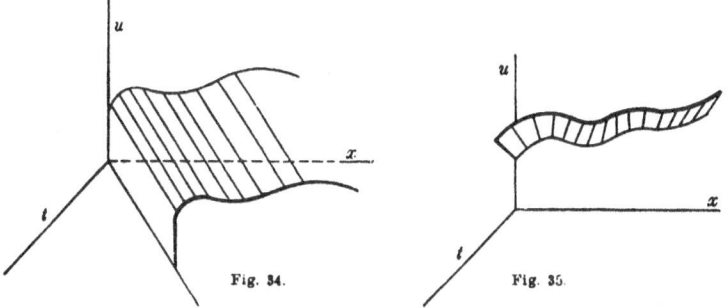

Fig. 34.   Fig. 35.

every particle of the string or of the air, and also its velocity. For instance, suppose for

$$t = 0; \quad u = F(x),$$
$$\frac{\partial u}{\partial t} = G(x),$$

and let us suppose that F and G are zero except between $x_0 < x < x_1$

This is a particular case of Cauchy's problem for the equation of the second order, in which we give along the *support*[1] $t = 0$ (Fig. 35), the value not only of u but also of the derivative $\frac{\partial u}{\partial t}$, the *slope* of the integral surface, so that we have not only a curve but a strip given, for the surface to coincide with.

Then since for all t and x,

(25)
$$u = \varphi(x + at) + \psi(x - at),$$
$$\frac{\partial u}{\partial t} = a[\varphi'(x + at) - \psi'(x - at)],$$

---

1) The curve which "supports" the data in Cauchy's problem will be so called.

putting $t = 0$ we have

$$F(x) = \varphi(x) + \psi(x)$$
$$G(x) = a[\varphi'(x) - \psi'(x)].$$

Integrating with respect to x from the value x_0,

$$\varphi(x) - \psi(x) = \frac{1}{a}\int_{x_0}^{x} G'(x)\,dx$$

and from this and the value of $\varphi + \psi$,

$$\varphi(x) = \tfrac{1}{2}\left[F(x) + \frac{1}{a}\int_{x_0}^{x} G(x)\,dx \right]$$

(26)

$$\psi(x) = \tfrac{1}{2}\left[F(x) - \frac{1}{a}\int_{x_0}^{x} G(x)\,dx \right]$$

for all values of x. Inserting the values $x + at$ and $x - at$ respectively we have

$$u = \tfrac{1}{2}\left[F(x+at) + F(x-at) + \frac{1}{a}\int_{x_0}^{x+at} G(x)\,dx - \frac{1}{a}\int_{x_0}^{x-at} G(x)\,dx \right]$$

(27)

$$= \tfrac{1}{2}\left[F(x+at) + F(x-at) + \frac{1}{a}\int_{x-at}^{x+at} G(x)\,dx \right].$$

We thus have an instance of Stokes's rule: The terms depending on the initial configuration are obtained from those depending on the initial velocity by differentiating according to the time and replacing the function giving the velocity by that giving the displacement. (The reason for this will appear later.)

Suppose the string distorted and let go without velocity, then $G(x) = 0$ and

$$u = \tfrac{1}{2}[F(x+at) + F(x-at)]$$

or the two waves are alike, and each one has half the original displacement. On the other hand suppose the string is started from zero displacement with arbitrary velocities.

Then

$$F(x) = 0,$$

and supposing

$$G(x) = \frac{dg}{dx}$$

$$u = \frac{1}{2a}\{g(x+at) - g(x-at)\}$$

so that in this case the two waves are opposite to each other. In the general case the displacement is dependent on both the original velocity and displacement.

From the symmetry of the differential equation with respect to x and at, it is evident that for any value of x we might give, say for

$$x = 0; \quad u = f(t),$$
$$\frac{\partial u}{\partial x} = g(t).$$

Thus we have two conditions for a point of time, or for a point of space, in either case involving two arbitrary functions.

27. String with Fixed Ends. The string or air has so far been supposed unlimited. Suppose we consider it to have an end at a point $x = l$, we may there impose conditions of a different kind. Suppose that the end is fixed, $u = 0$. Such a point is called a node. This is easily realized in the case of sound by bounding the air by a wall perpendicular to the x-axis, as in a closed pipe. Then for

$$x = l, \quad u = 0,$$
$$\frac{\partial u}{\partial t} = 0,$$
$$0 = F(l + at) + F(l - at) + \frac{1}{a}\int_{l-at}^{l+at} G(x)dx$$
$$0 = aF'(l + at) - aF'(l - at) + G(l + at) + G(l - at).$$

This is an ordinary differential equation in t connecting the functions F and G, and may be written

$$aF'(l + at) + G(l + at) = aF'(l - at) - G(l - at).$$

One way in which this may be solved for all values of t is by $F'(l + at)$ being an even, $G(l + at)$ an odd function of t. (By an even function

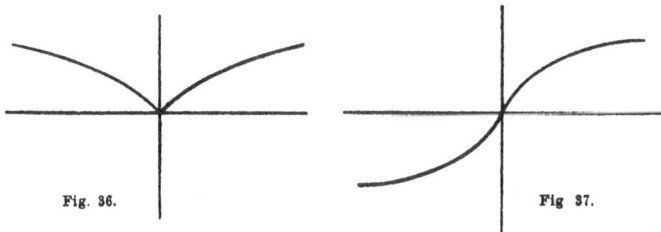

Fig. 36.   Fig 37.

of x we mean one in which $f(-x) = f(x)$, by an odd function one in which $f(-x) = -f(x)$. The graph of an even function is mirrored in the y-axis, Fig. 36, that of an odd function in the y and the x-axes, Fig. 37. The cosine and sine are familiar examples. Derivatives of even and odd functions are odd and even respectively.)

Suppose $G(x) = 0$. Then if we carry the values of $F(x)$ between 0 and l to the plane $x = l$ by means of the cylindrical generators parallel to $x = at$, $u = 0$, Fig. 38, we get the values of $F(l - at)$ from the

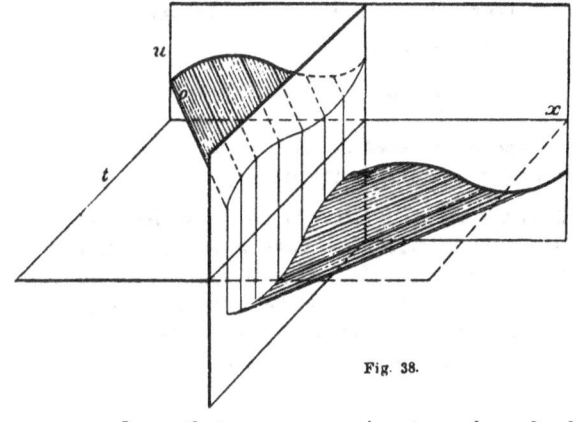

Fig. 38.

wave going to the right. These must be counteracted by the values of the wave going to the left, which are found by carrying the negative values to the plane $t = 0$ by the generators of the other cylinder parallel to $x = -at$, so that $F(l + x) = -F(l - x)$. Thus the form of the curve is *mirrored* about both axes at the point $x = l$, so that a wave running towards a fixed point is *reflected* with its sign changed. Thus an echo is explained. At the end $x = 0$ we have a similar refléction. The wave due to initial velocity may be similarly dealt with.

Accordingly the motion of a string with two fixed ends is composed of two waves continually travelling along it in opposite directions, the whole displacement being the resultant of the two waves and their reflections in the ends.

The motion is accordingly periodic, for after either wave has been reflected at both ends, that is after travelling a distance $2l$, the string has resumed its original form, and the motion is repeated over and over again.

Let us return to the air vibration. Since the dilatation $s = \dfrac{\partial u}{\partial x}$ satisfies the same equation as u,

$$(28) \qquad \frac{\partial^2 s}{\partial t^2} = a^2 \frac{\partial^2 s}{\partial x^2}$$

$$(29) \qquad s = \chi(x + at) + \omega(x - at)$$

and is likewise propagated in waves. At a wall normal to the direction of propagation, we have $u = 0$, but not $s = 0$.

Since

$$u = \varphi(x + at) + \psi(x - at)$$

$$\frac{1}{a}\frac{\partial u}{\partial t} = \varphi'(x + at) - \psi(x - at)$$

$$\frac{\partial u}{\partial x} = \varphi'(x + at) + \psi'(x - at) = s.$$

Comparing with (29)
$$\chi \equiv \varphi', \quad \omega \equiv \psi'.$$

If $x = l$ is a node, $u = 0$.

$$\varphi(l + at) = -\psi(l - at)$$

(30) $$\varphi'(l + at) = \psi'(l - at)$$

$$\chi(l + at) = \omega(l - at) \text{ for all values of } t, \text{ so that}$$

$$\chi(l + x) = \omega(l - x).$$

Consequently the dilatation is reflected, *without* change of sign. This is what happens at a closed end of an organ-pipe. Consequently the change in pressure at a node is *doubled*. At an *open* end, on the contrary, the condition is, since $p = p_0$, $s = 0$, so that there the dilatation s is reflected with sign changed, that is, as a compression, while the displacement is reflected without change of sign.

In the case of the telegraphist's equation, if we put the resistance $R = 0$, we have the same equation as for the string. If we have a pair of parallel wires, or a single wire such as a 'wireless' antenna, with a break in the circuit at the end, then the current I, is zero and is reflected with change of sign. On the contrary, if the circuit of parallel wires is closed, the potential difference V, is zero, and V is reflected with change in sign. In the first case, V is reflected without, and in the second I without change of sign.

Suppose we consider such a motion that at any point x we have u a *periodic* function of the time

$$u(t + T) = u(t).$$

Such a motion will be given by

$$\begin{array}{c}\cos\\\sin\end{array} k(x \pm at), \text{ (either function)}$$

the period being given by

(31) $$kaT = 2\pi, \quad T = \frac{2\pi}{ka},$$

but any multiple of a period is also a period, therefore whatever integer n may be, terms like

$$\begin{array}{c}\cos\\\sin\end{array} nk(x \pm at)$$

whose period is $\frac{T}{n}$, have also the period T.

Thus a periodic solution in t will be given by

(32) $$u = \sum_{n=1}^{n=\infty} \{A_n \cos nk(x + at) + B_n \sin nk(x + at) \\ + C_n \cos nk(x - at) + D_n \sin nk(x - at)\},$$

if the series is convergent.

Suppose we have nodes at $x = 0$, $x = l$. Then putting $x = 0$,

$$\sum_{n=1}^{n=\infty} \{A_n + C_n\} \cos nkat + (B_n - D_n) \sin nkat\} = 0.$$

Now it will be shown later that if such a series vanishes identically the coefficient of the sine or cosine of every multiple of t must vanish. Therefore

$$C_n = -A_n, \quad B_n = D_n, \quad n = 1, 2, 3 \ldots$$

and

$$u = \sum_{n=1}^{n=\infty} \{A_n(\cos nk(x + at) - \cos nk(x - at))$$

(33)
$$+ B_n(\sin nk(x + at) + \sin nk(x - at))\}$$

$$= 2 \sum_{n=1}^{n=\infty} \{-A_n \sin nkat + B_n \cos nkat\} \sin nkx.$$

Since $x = l$ is a node, we must have

(34)
$$\sum_{n=1}^{n=\infty} \{-A_n \sin nkat + B_n \cos nkat\} \sin nkl = 0.$$

As before we must have the coefficients of each term vanish, therefore either the $A_n's$ and $B_n's$ vanish, or

(35)
$$\sin nkl = 0.$$

This is a transcendental equation to determine k, and its roots are given by

$$nkl = m\pi$$

where m is any integer.

(36)
$$nk = \frac{m\pi}{l}, \quad T_m = \frac{2\pi}{nka} = \frac{2l}{ma}.$$

Putting $m = 1$ we obtain the longest period of the vibration, which is thus twice the length of the string or pipe divided by the velocity of sound, or of the transverse motion of the string.

The term

$$\frac{\sin}{\cos} \frac{m\pi}{l}(x \pm at)$$

is periodic not only in t but in x, so that when we increase x by the amount λ given by

$$\frac{m\pi\lambda}{l} = 2\pi,$$

the ordinates are repeated. The length

(37)
$$\lambda_m = \frac{2l}{m}$$

is called the *wave-length* of this vibration, and it is equal to the distance travelled by the wave during the time T_m. Thus the longest wave-length of a possible vibration is given by $\lambda = 2l$.

The term

(38)
$$\sin \frac{m\pi x}{l} \left(A_m \cos \frac{m\pi a t}{l} + B_m \sin \frac{m\pi a t}{l} \right)$$

has nodes at

$$x = \frac{l}{m}, \ \frac{2l}{m}, \ \frac{3l}{m}, \ \ldots l$$

or divides the string into m *ventral segments*. The extreme positions of the string are shown in Fig. 39. Its period and wavelength are $\frac{1}{m}$ of those of the slowest vibration, or *fundamental*, and its frequency m times as great. These various vibrations are called the *harmonics*, or *overtones* of the fundamental. The existence of music thus depends on the equation (35). Thus the most general motion of a string consists of the resultant of an infinite number of harmonics.

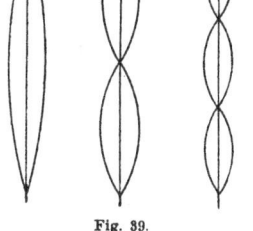

(39) $y = \displaystyle\sum_{m=1}^{m=\infty} \sin\frac{m\pi x}{l}\left(A_m\cos\frac{m\pi a t}{l} + B_m\sin\frac{m\pi a t}{l}\right).$

The method of building up from particular solutions is due to Daniel Bernoulli, 1775.

Fig. 39.

In order that the string may take a given form $u = f(x)$ when $t = 0$, we must take an infinite series of terms like the above, and putting $t = 0$

(40)
$$f(x) = \sum_{m=1}^{m=\infty} A_m \sin \frac{m\pi x}{l}.$$

Thus we are brought into contact with the question of development of functions in trigonometric series.

28. Theory of Small Vibrations. Before proceeding with the question of trigonometric series let us consider another way of looking at the matter of vibrations of a string. To that end we shall consider the small vibrations of any system, after the method of Lagrange (see chapter I, 11).

Suppose a system is defined by n parameters $q_1, q_2, \ldots q_n$. Its kinetic energy will be a quadratic function of their time-derivatives $q_1', q_2', \ldots q_n'$,

(41)
$$T = \frac{1}{2} \sum_{r=1}^{r=n} \sum_{s=1}^{s=n} a_{rs} q_r' q_s',$$

where the a's are functions of the coordinates q alone. The potential energy will depend only on the q's, and developing by Taylor's theorem

$$(42) \qquad W = W_0 + \sum_{r=1}^{r=n} q_r \left(\frac{\partial W}{\partial q_r}\right)_0 + \frac{1}{2}\sum_{r=1}^{r=n} \sum_{s=1}^{s=n} q_r q_s \left(\frac{\partial^2 W}{\partial q_r \partial q_s}\right)_0 + \cdots$$

where W_0 is the value when all the q's are zero. The forces acting are obtained by differentiation of W, so that

$$(43) \qquad P_r = -\frac{\partial W}{\partial q_r}.$$

Suppose that there is a configuration of equilibrium, for which every $P_r = 0$, then W is a minimum, and every $\left(\frac{\partial W}{\partial q_r}\right)_0 = 0$, if we take this as the zero configuration Thus, neglecting the constant W_0, which does not affect the motion, W begins with a *quadratic* function of the q's. If the motion is small enough, we may neglect the terms of higher orders — we shall accordingly put W a homogeneous quadratic function of the q's, with *constant* coefficients.

$$(44) \qquad W = \frac{1}{2}\sum_{r=1}^{r=n} \sum_{s=1}^{s=n} c_{rs} q_r q_s.$$

Developing the functions a_{rs} in series, the term

$$a_{rs} q_r' q_s' = q_r' q_s' \left[a_{rs}^{(0)} + \sum_{t=1}^{t=n} q_t \left(\frac{\partial a_{rs}}{\partial q_t}\right)_0 + \cdots \right],$$

and since the q''s are small together with the q's, we may neglect all but the term of lowest order, so that we may consider the a's as constants.

We thus have the two quadratic functions or *forms*.

$$(45) \qquad T = \frac{1}{2}\sum_{r=1}^{r=n} \sum_{s=1}^{s=n} a_{rs} q_r' q_s'$$

$$W = \frac{1}{2}\sum_{r=1}^{r=n} \sum_{s=1}^{s=n} c_{rs} q_r q_s$$

both of which are positive for all values of their variables, and from which we are to form the differential equations of motion,

$$(46) \qquad \frac{dp_r}{dt} - \frac{\partial T}{\partial q_r} = P_r = -\frac{\partial W}{\partial q_r},$$

with $\frac{\partial T}{\partial q_r} = 0$ and

$$(47) \qquad p_r = \frac{\partial T}{\partial q_r'} = a_{r1} q_1 + a_{r2} q_2' + \cdots + a_{rn} q_n'.$$

In the case of a single coordinate we have

$$T = \tfrac{1}{2}aq'^2, \quad W = \tfrac{1}{2}cq^2,$$

(48)
$$p = aq', \qquad P = -cq,$$

$$a\frac{d^2q}{dt^2} = -cq$$

where a is the inertia, and c the coefficient of stiffness that is the force of restitution per unit of displacement. Thus in general we may call the a's coefficients of inertia, the c's coefficients of stiffness.

The integral of the equation is found by putting

(49)
$$q = Ae^{mt}, \quad q'' = m^2Ae^{mt},$$

and inserting in the differential equation

(50)
$$am^2 + c = 0,$$

$$m = \pm i\sqrt{\frac{c}{a}}.$$

$$\text{Now} \quad e^{i\sqrt{\frac{c}{a}}t} \quad \text{and} \quad e^{-i\sqrt{\frac{c}{a}}t}$$

being particular solutions, so are their sum and difference, multiplied by any constant, or

$$\cos\sqrt{\frac{c}{a}}\cdot t, \quad \sin\sqrt{\frac{c}{a}}\cdot t.$$

We therefore have the general solution

(51)
$$q = A\cos\sqrt{\frac{c}{a}}\cdot t + B\sin\sqrt{\frac{c}{a}}\cdot t$$

$$= C\cos\left(\sqrt{\frac{c}{a}}\cdot t - \alpha\right), \quad C = \sqrt{A^2+B^2}, \quad \alpha = \tan^{-1}\frac{B}{A},$$

which represents a harmonic vibration of period

(52)
$$T = 2\pi\sqrt{\frac{a}{c}}$$

and frequency

(53)
$$\frac{1}{T} = \frac{1}{2\pi}\sqrt{\frac{c}{a}},$$

but of arbitrary amplitude C, and phase α.

If the motion in the last case is resisted by a force proportional to the velocity,

(54)
$$a\frac{d^2q}{dt^2} = -cq - \varkappa\frac{dq}{dt},$$

the equation for m is

(55)
$$a m^2 + \varkappa m + c = 0,$$

$$m = -\frac{\varkappa}{2a} \pm \sqrt{\frac{\varkappa^2}{4a^2} - \frac{c}{a}}.$$

If $\varkappa^2 < 4ac$, the radicand is negative, and both roots are complex, say

$$m = \mu \pm i\nu.$$

Then the solution is

(56)
$$q = C e^{-\frac{\varkappa t}{2a}} \cos\left(\sqrt{\frac{c}{a} - \frac{\varkappa^2}{4a^2}} \cdot t - \alpha\right)$$

denoting a harmonic oscillation with logarithmic decrement.

Return now to the general problem. We have the linear differential equations with constant coefficients

(57) $a_{r1}\dfrac{d^2 q_1}{dt^2} + a_{r2}\dfrac{d^2 q_2}{dt^2} + \cdots + a_{rn}\dfrac{d^2 q_n}{dt^2} + c_{r1} q_1 + c_{r2} q_2 + \cdots + c_{rn} q_n = 0,$

for $r = 1, 2, \ldots n$.

Let us put

(58)
$$q_r = A_r e^{mt},$$

m being the same for all the q's, and substituting, dividing by e^{mt},

$$A_1(m^2 a_{11} + c_{11}) + A_2(m^2 a_{12} + c_{12}) + \cdots + A_n(m^2 a_{1n} + c_{1n}) = 0,$$
$$A_1(m^2 a_{21} + c_{21}) + A_2(m^2 a_{22} + c_{22}) + \cdots + A_n(m^2 a_{2n} + c_{2n}) = 0,$$
(59) $\cdot \quad \cdot \quad \cdot \quad \cdot \quad \cdot \quad \cdot \quad \cdot \quad \cdot \quad \cdot \quad \cdot \quad \cdot \quad \cdot$
$\cdot \quad \cdot \quad \cdot \quad \cdot \quad \cdot \quad \cdot \quad \cdot \quad \cdot \quad \cdot \quad \cdot \quad \cdot \quad \cdot$
$$A_1(m^2 a_{n1} + c_{n1}) + A_2(m^2 a_{n2} + c_{n2}) + \cdots + A_n(m^2 a_{nn} + c_{nn}) = 0.$$

These are linear equations in the A's, and in order that they shall hold their determinant must vanish, namely

(60)
$$D(m^2) = \begin{vmatrix} m^2 a_{11} + c_{11}, & \ldots & m^2 a_{1n} + c_{1n} \\ m^2 a_{21} + c_{21}, & \ldots & m^2 a_{2n} + c_{2n} \\ \cdot & \cdot & \cdot \\ \cdot & \cdot & \cdot \\ m^2 a_{n1} + c_{n1}, & \ldots & m^2 a_{nn} + c_{nn} \end{vmatrix} = 0.$$

This is Lagrange's determinantal equation for m. It is of degree n in m^2. Thus we have the roots $m = \pm m_1, m_2, \ldots m_n$.

We have to prove that every root is a pure imaginary. After the roots are found, substituting a root, say m_r, in the linear equations (59) we get a set of values for the ratios of the A's. Call them

$$A_1^{(r)} : A_2^{(r)} : \ldots : A_n^{(r)}.$$

For $-m_r$ we get the same ratios in the A's as for m_r.

Substituting different roots we get n different sets of ratios. The general solution is the sum of the particular solutions

$$q_1 = A_1^{(1)} e^{m_1 t} + A_1^{(2)} e^{m_2 t} + \cdots + A_1^{(n)} e^{m_n t},$$

(61)

$$q_2 = A_2^{(1)} e^{m_1 t} + A_2^{(2)} e^{m_2 t} + \cdots + A_2^{(n)} e^{m_n t},$$

$$\cdots \cdots \cdots \cdots \cdots \cdots \cdots$$

It is to be observed that it is the ratios of the A's in the same column that are determined, those in the same row being arbitrary.

If we multiply our equations (59), having substituted m_r, respectively by

$$A_1^{(r)}, A_2^{(r)}, \ldots A_n^{(r)},$$

and add, we obtain

(62)

$$m_r^2 \sum_{s=1}^{s=n} \sum_{t=1}^{t=n} a_{st} A_s^{(r)} A_t^{(r)} + \sum_{s=1}^{s=n} \sum_{t=1}^{t=n} c_{st} A_s^{(r)} A_t^{(r)} = 0.$$

But the term multiplying m_r^2 and the one independent of it are respectively what we get if we substitute for the variables q_s' in T, q_s in W, the values $A_s^{(r)}$, say $TA^{(r)}$ and $WA^{(r)}$.

Thus

(63)

$$m_r^2 = -\frac{WA^{(r)}}{TA^{(r)}}.$$

But the functions T and W are from their physical properties essentially positive, whatever the values of the variables.

Therefore every m_r^2 is negative, and all the roots m_r are purely imaginary. If we put $m^2 = -\lambda$,

(64)

$$m_r^2 = -\lambda_r \quad m_r = i\nu_r = i\sqrt{\lambda_r},$$

we may then substitute trigonometric for exponential solutions, putting, with other constants, a, α,

(65) $q_r = a_r^{(1)} \cos(\nu_1 t - \alpha_1) + a_r^{(2)} \cos(\nu_2 t - \alpha_2) + \cdots + a_r^{(n)} \cos(\nu_n t - \alpha_n).$

We see that this is possible, since for the solution

$$A_r^{(s)}(e^{m_s t} \pm e^{-m_s t}) = A_r^{(s)}(e^{i\nu_s t} \pm e^{-i\nu_s t})$$

we may substitute $\cos \nu_s t$ or $\sin \nu_s t$, multiplied by constants,

$$A_r^{(s)} \cos \nu_s t + B_r^{(s)} \sin \nu_s t,$$

where

$$\frac{B_r^{(s)}}{A_r^{(s)}} = \frac{B_1^{(s)}}{A_1^{(s)}} = \tan \alpha_s, \quad r = 2, 3, \ldots n,$$

that is the B's satisfy the same equations (59) as the A's.

Then

$$q_1 = A_1^{(1)} \cos \nu_1 t + B_1^{(1)} \sin \nu_1 t + A_1^{(2)} \cos \nu_2 t + B_1^{(2)} \sin \nu_2 t + \cdots$$

(66) $q_2 = A_2^{(1)} \cos \nu_1 t + B_2^{(1)} \sin \nu_1 t + A_2^{(2)} \cos \nu_2 t + B_2^{(2)} \sin \nu_2 t + \cdots$

$$\cdots \cdots \cdots \cdots \cdots \cdots \cdots \cdots \cdots$$

It is to be noticed that the ratios of A's and B's in any column, or the relative amplitudes and phases of the harmonic terms of different q's of the *same period* are determined by the nature of the system, whereas the relative amplitudes of the terms of different period in a single coordinate are arbitrary. To determine the A's and B's suppose we have the initial values of the q's given. $q_r = g_r,\ q'_r = h_r$

Then

$$g_1 = A_1^{(1)} + A_1^{(2)} + \cdots + A_1^{(n)},$$
$$g_2 = A_2^{(1)} + A_2^{(2)} + \cdots + A_2^{(n)},$$
$$\cdot\ \ \cdot\ \ \cdot\ \ \cdot\ \ \cdot\ \ \cdot\ \ \cdot\ \ \cdot\ \ \cdot\ \ \cdot$$
$$\cdot\ \ \cdot\ \ \cdot\ \ \cdot\ \ \cdot\ \ \cdot\ \ \cdot\ \ \cdot\ \ \cdot\ \ \cdot$$

(67)

$$h_1 = \nu_1 B_1^{(1)} + \nu_2 B_1^{(2)} + \cdots + \nu_n B_1^{(n)},$$
$$h_2 = \nu_1 B_2^{(1)} + \nu_2 B_2^{(2)} + \cdots + \nu_n B_2^{(n)},$$
$$\cdot\ \ \cdot\ \ \cdot\ \ \cdot\ \ \cdot\ \ \cdot\ \ \cdot\ \ \cdot\ \ \cdot\ \ \cdot$$
$$\cdot\ \ \cdot\ \ \cdot\ \ \cdot\ \ \cdot\ \ \cdot\ \ \cdot\ \ \cdot\ \ \cdot\ \ \cdot$$

Thus the sums of A's in a row, and the ratios of those in a column being given, all are determined.

29. Normal Coordinates. There is one set of coordinates of peculiar importance. Suppose we make a linear transformation with constant coefficients.

$$q_1 = \alpha_{11}\psi_1 + \alpha_{12}\psi_2 + \cdots + \alpha_{1n}\psi_n,$$
$$q_2 = \alpha_{21}\psi_1 + \alpha_{22}\psi_2 + \cdots + \alpha_{2n}\psi_n,$$
$$\cdot\ \ \cdot\ \ \cdot\ \ \cdot\ \ \cdot\ \ \cdot\ \ \cdot\ \ \cdot\ \ \cdot\ \ \cdot$$

(68)

$$\cdot\ \ \cdot\ \ \cdot\ \ \cdot\ \ \cdot\ \ \cdot\ \ \cdot\ \ \cdot\ \ \cdot\ \ \cdot$$
$$q_n = \alpha_{n1}\psi_1 + \alpha_{n2}\psi_2 + \cdots + \alpha_{nn}\psi_n,$$

of such a nature that the two functions T and W are simultaneously transformed to sums of squares. It may be shown by algebra that this may be done in *one* way (see Chapter IX).

(69)
$$T = \tfrac{1}{2}(a_1\psi_1'^2 + a_2\psi_2'^2 + \cdots + a_n\psi_n'^2),$$
$$W = \tfrac{1}{2}(c_1\psi_1^2 + c_2\psi_2^2 + \cdots + c_n\psi_n^2).$$

Then our equations (57) are

(70)
$$a_r \frac{d^2\psi_r}{dt^2} + c_r\psi_r = 0,$$

(71)
$$\psi_r = A_r \cos\left(\sqrt{\frac{c_r}{a_r}}\,t - \alpha_r\right).$$

In other words each coordinate performs a harmonic vibration independent of the others, with its own period. The ψ's are called *normal coordinates*, and the q's being linear functions of the ψ's perform compound harmonic vibrations (65). A *normal vibration* is one where all

ψ's but one are zero. The effect of this on the q's is, if every ψ vanishes except ψ_s, to make

$$(72) \qquad q_1 = \alpha_{1s}\psi_s, \quad q_2 = \alpha_{2s}\psi_s, \; \cdots \; q_n = \alpha_{ns}\psi_s,$$

that is, in a normal vibration all the coordinates keep time, being in constant ratios with one another throughout the motion, as the solutions (65) reduce to a single column. Thus the general solution is the resultant of n normal vibrations. The normal coordinates have the property that the energy of any vibration is the sum of the energies of the separate normal vibrations, that is the mutual energies vanish.

$$(73) \qquad T + W = \tfrac{1}{2}(c_1 a_1^2 + c_2 a_2^2 + \cdots + c_n a_n^2).$$

30. Inverse Method. It will be advantageous for certain purposes to modify our treatment of equations (57). We have the equations for the force acting upon any particle in terms of the displacements of all the particles,

$$(74) \qquad - P_r = \frac{\partial W}{\partial q_r} = \sum_{s=1}^{s=n} c_{rs} \, q_s.$$

Let us solve these equations of the q's in terms of the P's and let the solutions be

$$(75) \qquad - q_s = \sum_{t=1}^{t=n} C_{st} P_t.$$

As we have called the coefficients c_{rs} coefficients of stiffness so we may call the coefficients C_{rs} coefficient of *yielding* or of *influence*. Any coefficient C_{rs} is the displacement a coordinate q_r will take when acted upon by an external force of amount unity, applied to another coordinate q_s. Since we have $c_{rs} = c_{sr}$, it is evident that we have $C_{rs} = C_{sr}$, so that we get the theorem due to Maxwell that a force applied to one coordinate produces a displacement in a second equal to the displacement produced in the first by the application of the same force to the second.

Since by Euler's theorem on homogeneous functions, or from (74)

$$W = \frac{1}{2} \sum_{r=1}^{r=n} q_r \frac{\partial W}{\partial q_r} = - \frac{1}{2} \sum_{r=1}^{r=n} P_r q_r,$$

using (75) we have

$$W = \frac{1}{2} \sum_{r=1}^{r=n} \sum_{s=1}^{s=n} C_{rs} P_r P_s,$$

$$- \frac{\partial W}{\partial P_s} = - \sum_{r=1}^{r=n} C_{rs} P_r = q_s.$$

This last equation expresses Castigliano's theorem.

If we substitute in (74) the values of q_s from (75) we obtain an identity, and by comparing the coefficients on both sides, we obtain the relations

$$(76) \qquad \sum_{s=1}^{s=n} c_{rs} C_{st} = \begin{matrix} 0, & r \neq t, \\ 1, & r = t, \end{matrix} \quad \text{or} \quad \sum_{r=1}^{r=n} c_{rs} C_{rt} = \begin{matrix} 0, & s \neq t, \\ 1, & s = t, \end{matrix}$$

which are well known relations concerning the minors of a determinant. If we now substitute for P_r in (75) its value $\frac{dp_r}{dt}$ from the dynamical equation, that is, if we multiply the r'th equation (59) by the constant C_{rt} and sum for all values of r we obtain in virtue of the identities (76), the equations

$$(77) \qquad \sum_{r=1}^{r=n}\sum_{s=1}^{s=n} C_{rt}(m^2 a_{rs} + c_{rs})A_s = m^2 \sum_{r=1}^{r=n}\sum_{s=1}^{s=n} C_{rt} a_{rs} A_s + A_t = 0.$$

If we put

$$(78) \qquad \sum_{r=1}^{r=n} C_{rt} a_{rs} = K_{st} = K_{ts}, \quad m^2 = -\lambda,$$

we finally have the equations

$$(79) \qquad \begin{aligned} (1 - \lambda K_{11})A_1 - \lambda K_{12} A_2 \ldots - \lambda K_{1n} A_n &= 0, \\ -\lambda K_{21} A_1 + (1 - \lambda K_{22})A_2 \ldots - \lambda K_{2n} A_n &= 0, \end{aligned}$$
$$\cdot \quad \cdot \quad \cdot \quad \cdot \quad \cdot \quad \cdot \quad \cdot$$

with the determinant

$$(80) \qquad d(\lambda) = \begin{vmatrix} 1 - \lambda K_{11}, & -\lambda K_{12}, & \ldots & -\lambda K_{1n}, \\ -\lambda K_{21}, & 1 - \lambda K_{22}, & \ldots & -\lambda K_{2n} \\ \cdot & \cdot & \cdot & \cdot \\ \cdot & \cdot & \cdot & \cdot \\ -\lambda K_{n1}, & -\lambda K_{n2}, & \ldots & 1 - \lambda K_{nn} \end{vmatrix} = 0.$$

As the equations (79) are equivalent to the equations (59), it is evident that the determinants $D(\lambda)$ and $d(\lambda)$ differ only by a constant factor, and consequently have the same roots in λ. The determinant $d(\lambda)$ is simpler for computation.

If the coordinates are normal, we have

$$(81) \qquad a_{rs} = c_{rs} = C_{rs} = K_{rs} = 0, \ r \neq s, \quad C_{rr} = \frac{1}{c_{rr}}, \quad K_{rr} = \frac{a_{rr}}{c_{rr}}$$

and both determinants reduce to their diagonal columns, so that

$$(82) \qquad \begin{aligned} D(\lambda) &= (-\lambda a_{11} + c_{11})(-\lambda a_{22} + c_{22}) \cdots (-\lambda a_{nn} + c_{nn}), \\ d(\lambda) &= \left(1 - \lambda \frac{a_{11}}{c_{11}}\right)\left(1 - \lambda \frac{a_{22}}{c_{22}}\right) \cdots \left(1 - \lambda \frac{a_{nn}}{c_{nn}}\right), \end{aligned}$$

whose roots are

$$\lambda_r = \frac{c_{rr}}{a_{rr}},$$

and we see that both have the same roots, which are directly visible, those for the different normal coordinates separating from each other.

31. Vibration of String of Beads. Let us now consider, after Lagrange, the motion of a massless string on which are fastened a number of massive equal and equidistant beads. Let their number be n and the mass of each m, and let the length of the string be l. Let the ends be fastened, and the distance of each bead from the next or from an end

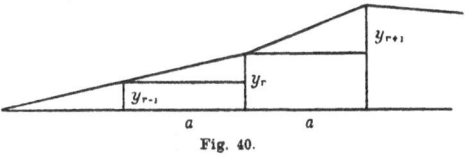

Fig. 40.

be $a = l/(n+1)$. Let the lateral displacements, all in the same plane, be $y_1, y_2 \cdots y_n$.

Then

$$(83) \qquad T = \frac{m}{2}\left(y_1'^2 + y_2'^2 + \cdots + y_n'^2\right).$$

The displacements being so small that the tension may be considered constant, say τ, the force on any particle r is, resolving parallel to y, Fig. 40,

$$(84) \qquad \tau\left\{-\frac{y_r - y_{r-1}}{a} + \frac{y_{r+1} - y_r}{a}\right\} = -\frac{\partial W}{\partial y_r},$$

from which

$$(85) \qquad W = \frac{\tau}{2a}\left\{y_1^2 + (y_2 - y_1)^2 + (y_3 - y_2)^2 + \cdots + y_n^2\right\}.$$

We neglect the action of gravity.

We accordingly have the equations of motion

$$(86) \qquad \begin{aligned} m\frac{d^2 y_1}{dt^2} + \frac{\tau}{a}(y_1 - 0 + y_1 - y_2) &= 0, \\ m\frac{d^2 y_2}{dt^2} + \frac{\tau}{a}(y_2 - y_1 + y_2 - y_3) &= 0 \\ \cdots \cdots \cdots \cdots \cdots \cdots \cdots \\ m\frac{d^2 y_n}{dt^2} + \frac{\tau}{a}(y_n - y_{n-1} + y_n - 0) &= 0. \end{aligned}$$

Now putting $y_r = A_r e^{\sqrt{-\lambda}t}$ we obtain

$$(87) \qquad \begin{aligned} \left[-m\lambda + \frac{2\tau}{a}\right]A_1 \qquad -\frac{\tau}{a}A_2 \qquad &= 0, \\ -\frac{\tau}{a}A_1 + \left[-m\lambda + \frac{2\tau}{a}\right]A_2 - \frac{\tau}{a}A_3 &= 0 \end{aligned}$$

$$\cdots \cdots \cdots \cdots \cdots \cdots \cdots \cdots$$

or dividing through by $\frac{\tau}{a}$ and putting $2 - ma\frac{\lambda}{\tau} = C$,

(88)
$$
\begin{aligned}
CA_1 - A_2 + 0 \cdots &= 0, \\
-A_1 + CA_2 - A_3 + 0 + \cdots &= 0, \\
0 - A_2 + CA_3 - A_4 + 0 + \cdots &= 0,
\end{aligned}
$$
$$\cdot \quad \cdot \quad \cdot \quad \cdot \quad \cdot \quad \cdot \quad \cdot \quad \cdot \quad \cdot \quad \cdot \quad \cdot$$

The determinantal equation is

(89)
$$
D_n(\lambda) = \begin{vmatrix}
C, & -1, & 0, & 0, & 0 \cdots \\
-1, & C, & -1, & 0, & 0 \cdots \\
0, & -1, & C, & -1, & 0 \cdots \\
0, & 0, & -1, & C, & -1 \cdots
\end{vmatrix} \quad n \text{ rows}
$$

But expanding in terms of first minors

$$D_n = CD_{n-1} - D_{n-2}.$$

If we put

$$C = 2\cos\theta,$$

the equation will be satisfied by

$$D_n = c\sin(n+1)\theta.$$

For, the c dividing out,

$$\sin(n+1)\theta = 2\cos\theta\sin n\theta - \sin(n-1)\theta.$$

To find c put $n = 1$,

$$D_1 = C = 2\cos\theta = \frac{\sin 2\theta}{\sin\theta}, \quad c = \frac{1}{\sin\theta}.$$

Thus

(90)
$$D_n = \frac{\sin(n+1)\theta}{\sin\theta},$$

and if this is to vanish we must have

$$(n+1)\theta = s\pi, \quad s = 1, 2, 3 \ldots n,$$

giving

$$C = 2 - ma\frac{\lambda}{\tau} = 2\cos\theta = 2\cos\frac{s\pi}{n+1},$$

(91)
$$\lambda_s = \nu_s^2 = \frac{2\tau}{ma}\left(1 - \cos\frac{s\pi}{n+1}\right)$$

$$\nu_s = 2\sqrt{\frac{\tau}{ma}}\,\sin\frac{s}{n+1}\frac{\pi}{2}.$$

Thus we obtain n different frequencies, varying as the ordinates of points dividing a quadrant into $n + 1$ parts, Fig. 41.

We may reach the same result by observing that the linear equations for the A's,

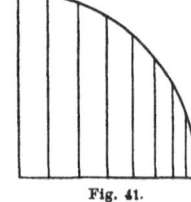

Fig. 41.

(92)
$$-A_{r-1} + CA_r - A_{r+1} = 0,$$

are satisfied by

(93)
$$A_s = P \sin s\theta \text{ with } P \text{ constant, since}$$
$$- \sin (r - 1)\theta + 2 \cos \theta \sin r\theta - \sin (r + 1)\theta = 0.$$

Accordingly let us substitute in the differential equation

(94)
$$- y_{r-1} + 2 y_r + \frac{m a}{\tau} \frac{d^2 y_r}{dt^2} - y_{r+1} = 0,$$

the solution

(95)
$$y_r = P \sin r\theta \cos(\nu t - \alpha).$$

Every term will contain the same cosine, so that dividing out,

(96)
$$- \sin (r - 1)\theta + 2\left(1 - \frac{m a \nu^2}{2\tau}\right) \sin r\theta - \sin(r + 1)\theta = 0,$$

which is an identity if

(97)
$$1 - \frac{m a \nu^2}{2\tau} = \cos\theta,$$

giving

$$\nu^2 = \frac{2\tau}{ma}(1 - \cos\theta),$$

as before.

If we put for $\theta_s = \dfrac{s\pi}{n+1}$,

$$\nu_s = 2\sqrt{\frac{\tau}{ma}} \sin \frac{s\pi}{2(n+1)}$$

(98)
$$y_r = \sum_{s=1}^{s=n} P_s \sin\frac{r s\pi}{n+1} \cos(\nu_s t - \alpha_s).$$

If x is the distance of the r'th particle from the end of the string,

$$x = ra = \frac{rl}{(n+1)},$$

from which

$$\frac{r}{n+1} = \frac{x}{l},$$

so that

(99)
$$y(x) = \sum_{s=1}^{s=n} P_s \sin\frac{s\pi x}{l} \cos(\nu_s t - \alpha_s).$$

32. Continuous String. Rayleigh's Principle. We shall now show that, according to the principle known by the name of Lord Rayleigh, all the properties here enumerated for discrete systems, particularly the properties of normal coordinates, persist when we allow n to increase without limit, so that we have a continuous distribution of mass, in this case the continuous string with which we began the chapter. The mass is distributed so that, ϱ being the line-density,

(100)
$$m = \varrho a = \frac{\varrho l}{(n+1)}, \quad ma = \varrho a^2 = \frac{\varrho l^2}{(n+1)^2},$$

and introducing this into the value of v_s

$$(101) \quad v_s = 2\sqrt{\frac{\tau}{\varrho}} \lim_{n=\infty}\left(\frac{n+1}{l}\sin\frac{s}{n+1}\frac{\pi}{2}\right) = \frac{s\pi}{l}\sqrt{\frac{\tau}{\varrho}}\lim_{n=\infty}\left(\frac{\sin\frac{s}{n+1}\frac{\pi}{2}}{\frac{s}{n+1}\frac{\pi}{2}}\right).$$

As n increases without limit y preserves its form, while v_s, approaches the limit

$$(102) \qquad\qquad v_s = \frac{s\pi}{l}\sqrt{\frac{\tau}{\varrho}}. \; ^{1)}$$

We have therefore for the continuous string,

$$(103) \qquad y = \sum_{s=1}^{s=\infty} P_s\sin\frac{s\pi x}{l}\cos\left(\frac{s\pi}{l}\sqrt{\frac{\tau}{\varrho}}t\cdot - \alpha_s\right),$$

which is the form already found, (39).

The frequencies of the various components are now all integral multiples of the lowest. In order to show how nearly this is the case for *finite* values of n, we may compare, after Lord Rayleigh, the values of $N = \frac{2(n+1)}{\pi}\sin\frac{\pi}{2(n+1)}$ with unity.

n	1	2	3	4	9	19	39
N	·9003	·9549	·9745	·9836	·9959	·9990	·9997

This gives the ratios of the fundamental, $s = 1$, and we see that the approach to unity is very rapid.

The differential equations

$$(86) \qquad \frac{m}{a}\frac{d^2y_r}{dt^2} + \frac{\tau}{a}\left(\frac{y_r - y_{r-1}}{a} - \frac{y_{r+1} - y_r}{a}\right) = 0,$$

become in the limit, replacing a by dx, since the difference of differences becomes a second derivative,

$$(104) \qquad \varrho\frac{\partial^2 y}{\partial t^2} - \tau\frac{\partial^2 y}{\partial x^2} = 0,$$

which is the same as (23).

The kinetic energy, instead of a sum, becomes the definite integral,

$$(105) \qquad T = \frac{1}{2}\int_0^l \varrho\left(\frac{\partial y}{\partial t}\right)^2 dx.$$

The potential energy also becomes the integral,

$$(106) \qquad W = \frac{1}{2}\int_0^l \tau\left(\frac{\partial y}{\partial x}\right)^2 dx,$$

---

1) Since $\lim_{x=0}\frac{\sin x}{x} = 1$.

and instead of Lagrange's equations we use Hamilton's priciple,

$$(107) \quad 0 = \delta \int_{t_0}^{t_1} (T-W)dt = \frac{1}{2}\int_{t_0}^{t_1} dt \int_0^l \delta \left\{ \varrho \left(\frac{\partial y}{\partial t}\right)^2 - \tau \left(\frac{\partial y}{\partial x}\right)^2 \right\} dx$$

$$= \int_{t_0}^{t_1} dt \int_0^l \left\{ \varrho \frac{\partial y}{\partial t}\frac{\partial \delta y}{\partial t} - \tau \frac{\partial y}{\partial x}\frac{\partial \delta y}{\partial x} \right\} dx.$$

Integrating the first term partially by the time, and the second by x,

$$(108) \int_0^l \left[\varrho \frac{\partial y}{\partial t}\delta y \right]_{t_0}^{t_1} dx - \int_{t_0}^{t_1} \left[\tau \frac{\partial y}{\partial x}\delta y \right]_0^l dt - \int_{t_0}^{t_1} dt \int_0^l dx \left\{ \varrho \frac{\partial^2 y}{\partial t^2} - \tau \frac{\partial^2 y}{\partial x^2} \right\} \delta y = 0.$$

Now since δy vanishes at t_0 and t_1, and since the ends of the string are fixed, $\delta y = 0$ at the limits, so that both single integrals vanish.

Consequently, we must have

$$(104) \quad \varrho \frac{\partial^2 y}{\partial t^2} - \tau \frac{\partial^2 y}{\partial x^2} = 0.$$

{We first obtained the force on each particle in terms of the displacement of all the particles.}

The inverse method, § 30, which consists in making the applied forces rather than the displacements the main subject of attention, is convenient in the case of the string of beads, and enables us to pass to the limit in a different manner If we apply a force P_r to a point whose coordinate is ra, the string is displaced into a broken line, Fig. 42, and resolving the two tensions we find

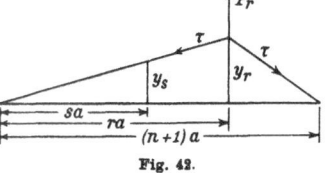

Fig. 42.

$$(109) \quad P_r = \tau \left[\frac{y_r}{ra} + \frac{y_r}{(n+1-r)a} \right] = \frac{\tau}{a}\frac{n+1}{r(n+1-r)}y_r,$$

from which we find for the point $x = sa$,

$$(110) \quad y_s = \frac{s}{r}y_r \text{ if } s < r; \ y_s = \frac{n+1-s}{n+1-r}y_r, \text{ if } s > r,$$

so that the coefficient of influence is

$$(111) \quad C_{rs} = \frac{y_s}{P_r} = \frac{as(n+1-r)}{\tau(n+1)}, \ s < r; \ C_{rs} = \frac{ar(n+1-s)}{\tau(n+1)}, \ s > r.$$

Using these values we obtain for the equations (75) (P_r is now changed in sign),

$$y_1 = \frac{a}{\tau(n+1)}\{1 \cdot n P_1 + 1(n-1)P_2 + 1(n-2)P_3 + \cdots + 1 \cdot 1 P_n\}$$

$$y_2 = \frac{a}{\tau(n+1)}\{2 \cdot n P_1 + 2(n-1)P_2 + 2(n-2)P_3 + \cdots + 2 \cdot 1 P_n\}$$

(112)

$$y_3 = \frac{a}{\tau(n+1)}\{3 \cdot n P_1 + 3(n-1)P_2 + 3(n-2)P_3 + \cdots + 3 \cdot 1 P_n\}$$

. . . .

$$y_n = \frac{a}{\tau(n+1)}\{1 \cdot 1 P_1 \quad + 2 \cdot 1 P_2 \quad + 3 \cdot 1 P_3 + \cdots + n \cdot 1 P_n\}$$

Since we have

(113) $T = \frac{1}{2}m y_r'^2,\ a_{rr} = m,\ a_{rs} = 0,\ r + s,\ K_{rs} = m C_{rs},$

we see that putting

(114) $K_{rs} = \frac{mas(n+1-r)}{\tau(n+1)},\ s < r;\ K_{rs} = \frac{mar(n+1-s)}{\tau(n+1)},\ s > r,$

$$l = \frac{ma\lambda}{\tau(n+1)},$$

we have the determinant, {corresponding to (80)},

(115)

$$d(l) = \begin{vmatrix} 1 - 1 \cdot nl, & -1(n-1)l, & -1(n-2)l, \ldots -1 \cdot 1 \cdot l \\ -2nl, & 1 - 2(n-1)l, & -2(n-2)l, \ldots -2 \cdot 1 \cdot l \\ -3nl, & -3(n-1)l, & 1 - 3(n-2)l, \ldots -3 \cdot 1 \cdot l \\ \cdot \ \cdot \ \cdot \ \cdot \ \cdot \ \cdot \ \cdot \ \cdot \ \cdot \ \cdot \\ \cdot \ \cdot \ \cdot \ \cdot \ \cdot \ \cdot \ \cdot \ \cdot \ \cdot \ \cdot \\ -1 \cdot 1 \cdot l, & -2 \cdot 1 \cdot l, & -3 \cdot 1 \cdot l, \ldots 1 - n \cdot 1 \cdot l \end{vmatrix}$$

We may use this determinant for computation instead of D and shall obtain for $n = 1, 2, 3, \ldots$ the values already found.

The equations (79) are now

(77) $$A_r = \lambda \sum_{s=1}^{s=n} K_{rs} A_s,\quad r = 1, 2 \ldots n,$$

with the values of K_{rs} in (114). These equations take the place of (87). Now letting n increase without limit, putting

$$x = ra,\ \xi = sa,\ A_r = \varphi(x),\ A_s = \varphi(\xi),$$

$$K_{rs} = \frac{m}{\tau}\frac{x(l-\xi)}{l} = \frac{\varrho\,d\xi}{\tau}\cdot\frac{x(l-\xi)}{l} = \frac{\varrho}{\tau}d\xi\,K(x,\xi),$$

(116)

$$K(x, \xi) = \frac{x(l-\xi)}{l},\ x < \xi,$$

$$K(x, \xi) = \frac{\xi(l-x)}{l},\ x > \xi,\ \frac{\lambda\varrho}{\tau} = k^2$$

Then passing to the limit, our linear equation (79) becomes

$$(117) \qquad \varphi(x) - k^2 \int_0^l K(x, \xi) \varphi(\xi) d\xi = 0,$$

which is called a homogeneous *integral* equation, to determine the unknown *function* φ. This has only the solution $\varphi = 0$ unless k is one of the characteristic numbers obtained from the roots of the determinantal equation

$$\lim_{n = \infty} d(\lambda) = 0 \text{ or } D(\lambda) = 0,$$

that is $\sin kl = 0$ in which case the integral equation has as a solution the corresponding normal function, $\varphi(x)$, representing a normal vibration or standing wave. {It will be seen that total motion of the string may be represented in the form $y = \sum_{n m} \varphi_n(x) \psi_m(t)$, each product term being an independent coordinate of the function y. (See § 29).}

33. Determination of Coefficients. Fourier's Series. Let us now determine the $2n$ arbitrary constants P_s, α_s, (103), in the case of n finite. Let the initial values of y_r be denoted by y_r^0.

Put

$$B_s = P_s \cos \alpha_s.$$

Then

$$(118) \quad \begin{aligned} y_1^0 &= B_1 \sin \frac{\pi a}{l} + B_2 \sin \frac{2\pi a}{l} + \cdots + B_n \sin \frac{n\pi a}{l}, \\ y_2^0 &= B_1 \sin \frac{2\pi a}{l} + B_2 \sin \frac{4\pi a}{l} + \cdots + B_n \sin \frac{2n\pi a}{l}, \\ & \;\cdot \quad \cdot \quad \cdot \quad \cdot \quad \cdot \quad \cdot \quad \cdot \quad \cdot \quad \cdot \quad \cdot \quad \cdot \\ & \;\cdot \quad \cdot \quad \cdot \quad \cdot \quad \cdot \quad \cdot \quad \cdot \quad \cdot \quad \cdot \quad \cdot \quad \cdot \\ y_n^0 &= B_1 \sin \frac{n\pi a}{l} + B_2 \sin \frac{2n\pi a}{l} + \cdots + B_n \sin \frac{n^2 \pi a}{l}. \end{aligned}$$

To find B_s, multiply the first by $\sin \frac{s\pi a}{l}$, the second by $\sin \frac{2s\pi a}{l}$ and so on, and add. Put for brevity $\theta = \frac{\pi a}{l}$.

The equations (118) are then

$$(119) \qquad y_r^0 = \sum_{t=1}^{t=n} B_t \sin tr\theta,$$

from which

$$(120) \qquad \sum_{r=1}^{r=n} y_r^0 \sin rs\theta = \sum_{r=1}^{r=n} \sum_{t=1}^{t=n} B_t \sin rs\theta \sin rt\theta.$$

Now

$$\sin a \sin b = \tfrac{1}{2}[\cos(a - b) - \cos(a + b)].$$

Hence the coefficient of B_t is

$$(121) \quad \tfrac{1}{2}[\cos(s-t)\theta + \cos 2(s-t)\theta + \cdots + \cos n(s-t)\theta]$$
$$- \tfrac{1}{2}[\cos(s+t)\theta + \cos 2(s+t)\theta + \cdots + \cos n(s+t)\theta].$$

We have therefore to sum two series of the form

$$(122) \qquad S = \cos\varphi + \cos 2\varphi + \cdots + \cos n\varphi.$$

Multiplying by $2\cos\varphi$,

$$2S\cos\varphi = 2(\cos\varphi\cos\varphi + \cos 2\varphi\cos\varphi + \cdots + \cos n\varphi\cos\varphi)$$

transforming which by means of the formula,

$$2\cos a\cos b = \cos(a-b) + \cos(a+b),$$

gives

$$2S\cos\varphi = 1 + \cos\varphi + \cos 2\varphi + \cdots + \cos(n-1)\varphi$$
$$+ \cos 2\varphi + \cos 3\varphi + \cdots + \cos(n-1)\varphi + \cos n\varphi + \cos(n+1)\varphi,$$
$$2S\cos\varphi = 1 + S - \cos n\varphi + S + \cos(n+1)\varphi - \cos\varphi.$$

Solving for S,

$$2S(1-\cos\varphi) = -(1-\cos\varphi) + \cos n\varphi - \cos(n+1)\varphi,$$

$$(123) \qquad S = -\tfrac{1}{2} + \frac{\sin(2n+1)\dfrac{\varphi}{2}}{2\sin\dfrac{\varphi}{2}}.$$

Accordingly the coefficient of B_t is the difference $S_1 - S_2$, where

$$S_1 = -\tfrac{1}{2} + \frac{\sin(2n+1)(s-t)\dfrac{\theta}{2}}{2\sin(s-t)\dfrac{\theta}{2}}, \qquad \theta = \frac{\pi a}{l} = \frac{\pi}{n+1},$$

$$(124)$$

$$S_2 = -\tfrac{1}{2} + \frac{\sin(2n+1)(s+t)\dfrac{\theta}{2}}{2\sin(s+t)\dfrac{\theta}{2}}, \qquad (s+t)\frac{\theta}{2} \leqq \frac{2n\pi}{2(n+1)} < \pi.$$

The value $0:0$ cannot occur, on account of the limits for s and t. Put

$$(s \pm t)\theta = \frac{h\pi}{n+1},$$

where h is an integer,

$$S_1 = -\tfrac{1}{2} + \frac{\sin\left(h\pi - \dfrac{h\pi}{2(n+1)}\right)}{2\sin\dfrac{h\pi}{2(n+1)}} = -\tfrac{1}{2}(1 + \cos h\pi).$$

This is 0 or -1, as h is *odd*, or *even*. But $s+t$ and $s-t$ are of the same parity, therefore

$$S_1 - S_2 = 0, \text{ if } s+t.$$

If $s = t$

$$S_1 = \cos 0 + \cos 2 \cdot 0 + \cdots + \cos n \cdot 0 = n,$$

$$S_2 = -1$$

$$S_1 - S_2 = n + 1.$$

Thus we have finally, since terms in (120) vanish except where $s = t$.

(125) $$B_s = \frac{2}{n+1} \sum_{r=1}^{r=n} y_r^0 \sin \frac{rs\pi a}{l}.$$

If now n increases without limit, putting $ra = x$, $y_r^0 = f(ra)$,

(126) $$y = f(x) = \sum_{s=1}^{s=\infty} B_s \sin \frac{s\pi x}{l},$$

and putting for a,

$$\frac{l}{(n+1)} = dx,$$

(127) $$B_s = \frac{2}{l} \int_0^l f(x) \sin \frac{s\pi x}{l} dx.$$

This is the form for the coefficients of Fourier's trigonometric series (as far as *sine* terms go).

Let us now approach the problem of the motion of a continuous string in a slightly different manner from that thus far employed. Let us seek the condition for a so-called *standing* wave, or one in which the displacement, y, is a function of x multiplied by a function of the time. A so called standing wave may be described as one whose form remains unchanged but whose magnitude changes with the time. That is, let us seek a particular solution of the equation

(23) $$\frac{\partial^2 y}{\partial t^2} = a^2 \frac{\partial^2 y}{\partial x^2},$$

of the form

(128) $$y = \psi(t)\varphi(x).$$

Substituting in the differential equation, we obtain

(129) $$\psi''(t)\varphi(x) \equiv a^2 \psi(t)\varphi''(x) \text{ or } \frac{\psi''(t)}{\psi(t)} = a^2 \frac{\varphi''(x)}{\varphi(x)}.$$

On one side we have a function of t alone, on the other one of x alone. These cannot be equal for all independent values of t and x, unless each side is constant. Put therefore the common value equal to $-m^2$, so that

(130) $$\frac{d^2\psi}{dt^2} + m^2\psi = 0,$$

(130') $$\frac{d^2\varphi}{dx^2} + k^2\varphi = 0, \text{ where } k = \frac{m}{a}.$$

Thus our partial differential equation breaks up into two ordinary differential equations (130) whose solutions are

(131)
$$\psi = A \cos mt + B \sin mt,$$
$$\varphi = C \cos kx + D \sin kx.$$

Accordingly

(132) $y = (A \cos mt + B \sin mt)(C \cos kx + D \sin kx)$

is a particular solution, where A, B, C, D, are arbitrary constants. The admissible values of m are determined by the terminal conditions. If $x = 0$, $x = l$ are to be nodes, we must have

(133) $C = 0$, $\sin kl = 0$, $kl = \dfrac{ml}{a} = s\pi$, $s = 1, 2, 3 \cdots$

Thus

(134) $y = \left(A \cos \dfrac{s\pi a t}{l} + B \sin \dfrac{s\pi a t}{l}\right) \sin \dfrac{s\pi x}{l}$

is a solution representing a standing wave. A sum of such solutions is a solution, (*not* representing a standing wave since the space and time functions do not separate), therefore

(135) $y = \sum\limits_{s=1}^{s=\infty}\left(A_s \cos \dfrac{s\pi a t}{l} + B_s \sin \dfrac{s\pi a t}{l}\right) \sin \dfrac{s\pi x}{l},$

if a convergent series, is a solution. For $t = 0$, let

$$u = F_{(x)}, \quad \frac{\partial u}{\partial t} = G(x),$$

that is

(136)
$$F(x) = \sum\limits_{s=1}^{s=\infty} A_s \sin \frac{s\pi x}{l},$$

$$G(x) = \sum\limits_{s=1}^{s=\infty} B_s \frac{sa\pi}{l} \sin \frac{s\pi x}{l}.$$

We thus again reach the problem of developing an arbitrary function in a series advancing by sines of integral multiples of a variable $\dfrac{\pi x}{l}$.

Let us consider the question of the development in a series of both sines and cosines of such multiples. Suppose (calling the variable x for simplicity),

(137) $f(x) = \frac{1}{2}A_0 + A_1 \cos x + A_2 \cos 2x + \cdots + A_s \cos sx + \cdots$
$$+ B_1 \sin x + B_2 \sin 2x + \cdots + B_s \sin sx + \cdots$$

Proceeding as in the case of the finite series, let us multiply by $\cos sx$ and (instead of summing), integrate between $x = -\pi$ and $x = \pi$.

The series can be integrated only if uniformly convergent, as shown in Appendix. If it is such,

$$(138)\quad \int_{-\pi}^{\pi} f(x)\cos sx\,dx = \tfrac{1}{2}A_0\int_{-\pi}^{\pi}\cos sx\,dx + \sum_{r=1}^{r=\infty}A_r\int_{-\pi}^{\pi}\cos rx\cos sx\,dx$$

$$+ \sum_{r=1}^{r=\infty}B_r\int_{-\pi}^{\pi}\sin rx\cos sx\,dx.$$

Now we have

$$\int_{-\pi}^{\pi}\cos rx\cos sx\,dx = 0,\quad r \neq s,$$

$$\int_{-\pi}^{\pi}\cos rx\sin sx\,dx = 0,\quad \begin{array}{l} r \neq s, \\ r = s, \end{array}$$

$$(139)$$

$$\int_{-\pi}^{\pi}\sin rx\sin sx\,dx = 0,\quad r \neq s,$$

$$\int_{-\pi}^{\pi}\cos^2 rx\,dx = \int_{-\pi}^{\pi}\sin^2 rx\,dx = \pi,\quad r \neq 0,\quad \int_{-\pi}^{\pi}dx = 2\pi.$$

In virtue of these equalities, all the terms except the one in A_s disappear, and we have accordingly

$$(140)\qquad\qquad A_s = \frac{1}{\pi}\int_{-\pi}^{\pi} f(x)\cos sx\,dx.$$

In like manner multiplying by $\sin sx$, we obtain the coefficient

$$(141)\qquad\qquad B_s = \frac{1}{\pi}\int_{-\pi}^{\pi} f(x)\sin sx\,dx,$$

as before (127). The integrals A_s and B_s are called the Fourier constants for the function $f(x)$. (Instead of the limits $-\pi, \pi$ we may take $0, 2\pi$.)

Every term in the series is periodic with the period 2π, consequently the whole series is periodic with the same period, and if a function is represented by the dotted line in Fig. 43, the

Fig. 43.

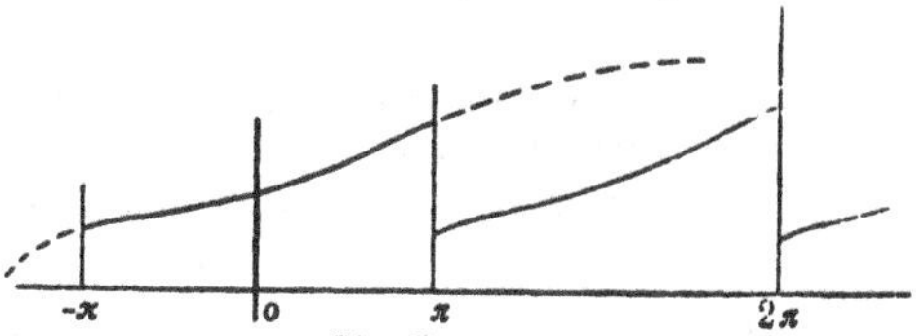

series instead of agreeing with the function, outside of the interval stated, repeats itself according to the full line in the figure rather than the dotted. We consequently find the very interesting phenomenon of two functions which agree in a certain portion of a continuous range, but totally disagree in the remaining portion.

The integral relations of (139) are not accidental. We may find them from the properties of normal coordinates. A normal coordinate was one which appeared in the differential equation containing the time unaccompanied by other coordinates. If we put

(128) $$y = \varphi(x)\psi(t),$$

where

(130) $$\frac{d^2\psi}{dt^2} + m^2\psi = 0,$$

ψ is a normal coordinate, and we may take either $\sin mt$ or $\cos mt$.

Substituting in the differential equation we obtained

(130') $$\frac{d^2\varphi}{dx^2} + k^2\varphi = 0.$$

The functions $\sin kx$, $\cos kx$ which now take the place of the arbitrary *constants* in the problem of n degrees of freedom, are called *normal functions*. We have then the question of developing a function of x in a series of normal functions. Thus

(142) $$y = \varphi_1\psi_1 + \varphi_2\psi_2 + \cdots$$

where $$\varphi_s = D_s \sin\frac{s\pi}{l}x \text{ or } C_s \cos\frac{s\pi}{l}x.$$

Now the other property of normal coordinates is that no product terms occur in the energy functions. But

(143) $$T = \frac{1}{2}\int_0^l \varrho\left(\frac{\partial y}{\partial t}\right)^2 dx = \frac{1}{2}\int_0^l \varrho\left(\sum_{r=0}^{r=\infty}\varphi_r\psi_r'\right)^2 dx$$

$$= \frac{1}{2}\sum_{r=1}^{r=\infty}\varrho\psi_r'^2\int_0^l\varphi_r^2 dx + \varrho\sum_{r=1}^{r=\infty}\sum_{s=1}^{s=\infty}\psi_r'\psi_s'\int_0^l\varphi_r\varphi_s dx.$$

Consequently we must have

(144) $$\int_0^l\varphi_r\varphi_s dx = 0 \text{ when } r \neq s.$$

This is the fundamental property of normal functions. Functions having it are said to be *orthogonal* to each other, for the reason that, since two vectors A, B are orthogonal if their scalar product vanishes,

$$A_x B_x + A_y B_y + A_z B_z = 0.$$

If we consider two n dimensional vectors

$$A_1, A_2, \ldots A_n,$$
$$B_1, B_2, \ldots B_n,$$

they are orthogonal if

$$A_1 B_1 + A_2 B_2 + \cdots + A_n B_n = 0.$$

Making n increase without limit, the scalar product becomes the definite integral

$$\int A(x) B(x) \, dx.$$

If we compare with the formulae (31) of Chapter I we see that the set of mutually orthogonal functions are analogous to the direction cosines of a vector. They will be completely so if in addition they satisfy equations corresponding to (30), that is if

$$(145) \qquad \int_0^l \varphi_r^2 \, dx = 1.$$

Since the properties of the normal functions are unaltered if they are multiplied by arbitrary constants, this may always be brought about. The functions φ_r are then said to be *normalized*.

{The convenience of this will appear later.}

We may now sum up the result of the limiting process of Rayleigh's Principle. Instead of the n ordinary differential equations of motion, we have a single partial differential equation. Instead of the algebraic equation for the frequencies $D(\lambda) = 0$, with n roots, we have the transcendental equation (35), with an infinite number of roots. The normal coordinates ψ_r are multiplied not by constants A_r, but by normal functions φ_r of the space variable x, and in a normal vibration the space factor is a normal function. There is one normal function corresponding to each characteristic number k_r, or root of the equation (35). The energy functions become, in the normal coordinates, sums of squares of an infinite number of variables. We shall find that the transformation of a quadratic form into normal coordinates is of fundamental importance for the theory of developments in infinite series, used in connection with partial differential equations in general. We shall see in the next section that the normal functions may be defined as solutions of the integral equation (117). The question of the validity of the passage to the limit involved in Rayleighs principle, which has been settled by Hilbert, may be left until Chapter IX. The question of the convergence of Fourier's series will be treated in Chapter IV.

Suppose that we wish a development of $f(x)$ valid in a different interval. We may obtain this by a change of variable. Put

$$x' = \frac{\pi x}{l}$$

and develop

$$f(x) = f\left(\frac{l x'}{\pi}\right)$$

by the preceding method, in sines and cosines of multiples of x':

$$(146) \qquad f(x) = \tfrac{1}{2} A_0 + \sum_{s=1}^{s=\infty} \left(A_s \cos \frac{s \pi x}{l} + B_s \sin \frac{s \pi x}{l} \right),$$

where

$$(147) \qquad A_s = \frac{1}{\pi} \int_{-\pi}^{\pi} f \left(\frac{l x'}{\pi} \right) \cos s x' \, d x' = \frac{1}{l} \int_{-l}^{l} f(x) \cos \frac{s \pi x}{l} \, dx,$$

$$B_s = \frac{1}{l} \int_{-l}^{l} f(x) \sin \frac{s \pi x}{l} \, dx.$$

As an illustration let us consider the problem of a string plucked at a point $x = \xi$, so that there $y = b$, and then let go. Then we have

$$f(x) = \frac{b x}{\xi} \quad \text{from} \quad 0 < x < \xi,$$

$$f(x) = b \frac{x - l}{\xi - l} \quad \text{from} \quad \xi < x < l.$$

Since the ends are nodes, we have only sines, and since on the left of the origin we may extend $f(x)$ by an odd function, we may integrate from 0 to l and multiply by two. Thus

$$(148) \quad B_s = \frac{2b}{l} \left[\frac{1}{\xi} \int_0^{\xi} x \sin \frac{s \pi x}{l} dx + \frac{1}{\xi - l} \int_{\xi}^{l} (x - l) \sin \frac{s \pi x}{l} dx \right].$$

Integrating by parts,

$$B_s = \frac{2b}{l} \left[-\frac{l}{s \pi \xi} x \cos \frac{s \pi x}{l} \Big|_0^{\xi} + \frac{l}{s \pi \xi} \int_0^{\xi} \cos \frac{s \pi x}{l} dx \right.$$

$$(149)$$

$$\left. -\frac{l(x - l)}{s \pi (\xi - l)} \cos \frac{s \pi x}{l} \Big|_{\xi}^{l} + \frac{l}{s \pi (\xi - l)} \int_{\xi}^{l} \cos \frac{s \pi x}{l} dx \right].$$

The integrated terms vanish at $x = 0$, and $x = l$ respectively, and cancel each other at $x = \xi$, while the integrals give

$$B_s = \frac{2b}{l} \left[\frac{l^2}{s^2 \pi^2 \xi} \sin \frac{s \pi \xi}{l} - \frac{l^2}{s^2 \pi^2 (\xi - l)} \sin \frac{s \pi \xi}{l} \right]$$

$$(150)$$

$$= \frac{2 b l^2}{\pi^2 \xi (l - \xi)} \frac{\sin \frac{s \pi \xi}{l}}{s^2}.$$

Accordingly we have for the initial form of the string

$$(151) \qquad f(x) = \frac{2 b l^2}{\pi^2 \xi (l - \xi)} \sum_{s=}^{s=\infty} \frac{1}{s^2} \sin \frac{s \pi \xi}{l} \sin \frac{s \pi x}{l},$$

which is symmetrical in x and ξ.

For the subsequent motion we have

$$(152) \qquad y = \frac{2bl^2}{\pi^2 \xi(l-\xi)} \sum_{s=1}^{s=\infty} \frac{1}{s^3} \sin\frac{s\pi\xi}{l} \sin\frac{s\pi x}{l} \cos\frac{s\pi a t}{l}.$$

We see that the amplitude of the harmonics falls off very rapidly as the order increases, being of the order $\frac{1}{s^3}$. All the amplitudes increase as the plucked point is nearer an end of the string. Any particular harmonic vanishes if $\sin\frac{s\pi\xi}{l} = 0$, that is, if the plucked point is a node for that harmonic. This was first stated by Thomas Young.

As a contrast to this problem, let us consider the case of a string initially straight, and struck at a certain point, or in a short region, in such a way as to receive there a constant velocity, while elsewhere the velocity is zero. Such would be the case in a string struck by a perfectly hard hammer. Let it extend from $\xi - \varepsilon$ to $\xi + \varepsilon$, and impart a velocity v. Then if $y = 0$ for $t = 0$,

$$(153) \qquad \begin{aligned} y &= \sum_{s=1}^{s=\infty} B_s \sin\frac{s\pi x}{l} \sin\frac{s\pi a t}{l} \\ \left(\frac{\partial y}{\partial t}\right)_{t=0} &= \sum_{s=1}^{s=\infty} B_s \frac{s\pi a}{l} \sin\frac{s\pi x}{l}, \end{aligned}$$

and we are to determine B_s so that

$$(154) \qquad B_s \frac{s\pi a}{l} = \frac{2}{l}\int_{\xi-\varepsilon}^{\xi+\varepsilon} v \sin\frac{s\pi x}{l}\, dx = \frac{4v}{s\pi} \sin\frac{s\pi\xi}{l} \sin\frac{s\pi\varepsilon}{l}.$$

Consequently we have

$$(155) \qquad y = \frac{4vl}{a\pi^2} \sum_{s=1}^{s=\infty} \frac{1}{s^2} \sin\frac{s\pi\xi}{l} \sin\frac{s\pi\varepsilon}{l} \sin\frac{s\pi x}{l} \sin\frac{s\pi a t}{l}.$$

We see that the higher harmonics do not fall off as rapidly as in the case of the plucked string, a phenomenon which is accentuated as ε is smaller, since in the limit for $\varepsilon = 0$ we have s only in the *first* power in the denominator. Young's theorem still holds. The theory of the string struck by a hammer has been extended by Helmholtz to cover the case of a soft hammer, which will be in contact with the string for a finite length of time depending on its mass and elasticity.

34. Forced Vibrations. Resonance. We shall now consider the case of a system executing *forced* vibrations, i. e., those caused by the action of periodically varying forces. It will be simplest to consider the system referred to normal coordinates,

$$(156) \qquad T = \frac{1}{2}\sum a_r q_r'^2, \quad W = \frac{1}{2}\sum c_r q_r^2,$$

and we have for each q,

(157)
$$a_r \frac{d^2 q_r}{dt^2} + c_r q_r = P_r.$$

Let us suppose that $P_r = F_r \cos pt$. Then we find the particular solution

$$q_r = A_r \cos pt, \text{ if}$$

(158)
$$A_r(c_r - p^2 a_r) = F_r,$$

$$q_r = \frac{F_r \cos pt}{c_r - p^2 a_r},$$

giving a definite amplitude for each normal coordinate. Such a vibration is called a *forced* vibration, since it has its period $\frac{2\pi}{p}$ imposed upon it by the force from without, instead of having it determined by the constitution of the system, as in the case of the free vibration. The free vibration has $\cos \nu_r t$, where $\nu_r = \sqrt{\frac{c_r}{a_r}}$ (71), so that if $p^2 = \nu_r^2$, that is, if the period of the force agrees with that of the free vibration the corresponding amplitude is infinite.[1] This is called *resonance*. We see that there are as many possible resonances as there are normal coordi-

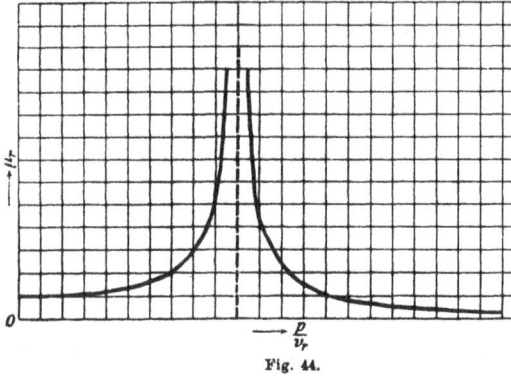

nates, that is, the number of degrees of freedom of the system. If we should neglect the inertia of the system, putting the a_r's equal to zero, we should have,

(159) $q_r = \frac{P_r}{c_r}.$

This is called the *equilibrium theory*, because it assumes that each displace-

Fig. 44.

ment is related to the force as in the case of equilibrium. If we call the ratio of the amplitude in the true dynamical theory to that in the equilibrium theory the dynamical magnification μ, we have

(160)
$$\mu_r = \frac{1}{1 - \dfrac{p^2}{\nu_r^2}}$$

In the case of resonance the dynamical magnification is infinite. The manner of its variation as a function of p is shown in Fig. 44). For small enough values of p the equilibrium theory is approximately true.

---

1) In reality the occurrence of an infinite amplitude is prevented by damping forces, as in (54), but for simplicity we shall neglect them.

Let us now consider the forced vibrations of a string. The force due to the tension has been found to be, on an element dx,

$$\tau\, dx\, \frac{\partial^2 y}{\partial x^2}.$$

Suppose that in addition we have the force $\tau Y(x, t)dx$, and suppose at first for simplicity that this is $\tau f(x)\cos pt\, dx$, the equation of motion will be

(161) $$\varrho\, \frac{\partial^2 y}{\partial t^2} = \tau\left(\frac{\partial^2 y}{\partial x^2} + f(x)\cos pt\right).$$

As before (128) we may put, for a standing wave,

$$y = \varphi(x)\cos pt,$$

obtaining

(162) $$-p^2\varphi = a^2\left\{\frac{d^2\varphi}{dx^2} + f(x)\right\}, \quad \text{where} \quad a^2 = \frac{\tau}{\varrho}$$

or with the usual notation, $\frac{p}{a} = k$,

(163) $$\frac{d^2\varphi}{dx^2} + k^2\varphi = -f(x).$$

This equation may be solved for any value of k and for any form of the function $f(x)$. It will be convenient to consider two extreme cases of distribution of *load*, as we may call the force per unit of length $\tau f(x)$, borrowing a word from the case of equilibrium, $k = 0$. First let the load be constant, $f(x) = R$. Then we may write

(164) $$\varphi_1 = \varphi + \frac{R}{k^2}, \quad \frac{d^2\varphi_1}{dx^2} + k^2\varphi_1 = 0,$$

so that we get a solution vanishing at 0 and l by putting

(165) $$\varphi = \frac{R}{k^2\sin kl}\{\sin kx + \sin k(l-x) - \sin kl\},$$

which is composed of the superposition of two waves each of the wavelength corresponding to the forced frequency, each with a node at one end, and the proper compensating constant. We find the maximum va¹ e of φ for varying x, to be

(166) $$\varphi_m = \frac{R}{k^2}\,\frac{1 - \cos\dfrac{kl}{2}}{\cos\dfrac{kl}{2}}.$$

We have resonance when $\sin kl = 0$.

On the equilibrium theory, $k = 0$, we should have

(167) $$\frac{d^2\varphi}{dx^2} + R = 0, \quad \varphi = \frac{R}{2}x(l-x),$$

(168) $$\varphi_m = \frac{Rl^2}{8}, \quad \varphi = 4\varphi_m\frac{x(l-x)}{l^2}.$$

Although the parabolic form of the equilibrium theory imitates the actual form well only for small values of p, it was shown by Lord Rayleigh that if we assume a free vibration having a distribution of amplitude like that of the equilibrium theory instead of a normal function, we shall get a period differing but slightly from that of the lowest normal vibration. This principle is often of value, enabling us to make an approximation without solving a differential equation. Let us then put

$$y = \frac{4x(l-x)}{l^2}\,\psi\,, \qquad \frac{\partial y}{\partial t} = \frac{4x(l-x)}{l^2}\,\psi'\,,$$

(169)
$$T = \frac{1}{2}\varrho\,\psi'^2\,\frac{16}{l^4}\int_0^l x^2(l-x)^2\,dx = \frac{1}{2}\,\varrho l\psi'^2\,\frac{16}{30}\,,$$

$$W = \frac{1}{2}\int_0^l \tau\left(\frac{\partial y}{\partial x}\right)^2 dx = \frac{\tau}{2}\,\frac{16\,\psi^2}{l^4}\int_0^l (l-2x)^2\,dx = \frac{1}{2}\,\tau\,\frac{16}{3}\,\frac{\psi^2}{l}\,.$$

If then $\psi'' + p_0^2\psi = 0$ we find

(170)
$$p_0 = \sqrt{10}\,\frac{a}{l} = 3.1623\,\frac{a}{l}\,.$$

while the true value is $\pi\,\frac{a}{l}$. The ratio of the period of this *quasi-equilibrium* theory to the true one is 1 to $\cdot 9936$. The work done by the load is

$$P\delta\psi = \int_0^l \tau R\delta y\,dx = \frac{2}{3}\,\tau Rl\delta\psi\,.$$

For the forced vibration on this quasi-equilibrium theory we have

(171)
$$\frac{16}{30}\,\varrho l\,\frac{d^2\psi}{dt^2} + \frac{16}{3}\,\frac{\tau}{l}\,\psi = \frac{2}{3}\,\tau Rl\cos pt\,,$$

from which we obtain

(172)
$$\psi = \frac{5}{4\varrho}\,\frac{\tau R\cos pt}{p_0^2 - p^2}\,.$$

From this we get the value of the maximum amplitude

(173)
$$y_m = \frac{5}{4}\,\frac{R}{k_0^2 - k^2}\,,$$

as compared with (166) in the dynamical theory. If we put

(174) $k = k_0 + \varepsilon = \dfrac{\pi}{l} + \varepsilon\,, \quad \dfrac{kl}{2} = \dfrac{\pi}{2} + \dfrac{\varepsilon l}{2}\,, \quad \cos\dfrac{kl}{2} = -\sin\dfrac{\varepsilon l}{2}\,.$

When ε is small,

(175)
$$\varphi_m = \frac{Rl^2}{\pi^2}\,\frac{2}{\varepsilon l}\,, \qquad y_m = \frac{5}{4}\,\frac{Rl}{2\pi\varepsilon}\,.$$

These two values are related to each other as $\cdot 2026$ to $\cdot 1989$. The values of φ_m from (166) and y_m from (173) are plotted in Fig. 45.

35. Point-source. Green's Function. We shall now consider the other extreme of distribution of load, namely concentration in a single point. We have to put $f(x) = 0$ except in a very short region from $x = \xi - \varepsilon$ to $x = \xi + \varepsilon$ in which $f(x)$ increases in such a way that the integral $\lim\limits_{\varepsilon = 0} \int_{\xi - \varepsilon}^{\xi + \varepsilon} f(x)\,dx$ is finite, say unity. Such a region we may call a point load, or point source of sound. The function, represented by the full line, Fig. 43, may be called the jag-function, and may be considered as the limit

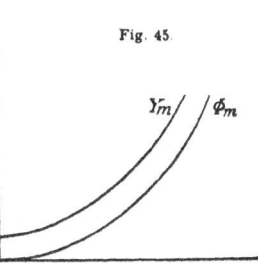

Fig. 45

of a continuous function, such as $e^{-\mu^2 x^2} \dfrac{\mu}{\sqrt{\pi}}$, represented by the dotted line, Fig. 43. (See Appendix).

We shall show that in the case of a string a point source will cause a discontinuity in the first derivative $\dfrac{\partial y}{\partial x}$. For integrating (163),

$$(176) \qquad \int_{\xi - \varepsilon}^{\xi + \varepsilon} \frac{d^2 \varphi}{dx^2}\,dx + k^2 \int_{\xi - \varepsilon}^{\xi + \varepsilon} \varphi\,dx = - \int_{\xi - \varepsilon}^{\xi + \varepsilon} f(x\,dx\,.$$

$$(177) \qquad \lim_{\varepsilon = 0} \frac{d\varphi}{dx} \Big|_{\xi - \varepsilon}^{\xi + \varepsilon} = - \lim_{\varepsilon = 0} \int_{\xi - \varepsilon}^{\xi + \varepsilon} f(x)\,dx$$

since φ is finite, and the range vanishes, the second integral vanishes. We shall call the strength of the source the value of the integral on the right. Accordingly in passing a source of value unity, the derivative falls off by the amount unity,

$$(178) \qquad \frac{d\varphi}{dx} \Big|_{\xi - 0}^{\xi + 0} = - 1\,.$$

This we also see by considering the resultant of the vertical components of the tension on the two sides, a unit point-source being produced by an applied force equal to the tension τ. See page 114.

Let us first consider the case where $k = 0$. This is the case $p = 0$, of zero frequency, or equilibrium.

Let us call the function satisfying the equation $\dfrac{d^2 \varphi}{dx^2} = 0$, vanishing at $x = 0$, $x = l$, and having a unit source at $x = \xi$, the Green's function for our equation and denote it by $K(x, \xi)$. {Mechanically this will denote the displacement of the string, under the influence of a unit force applied at ξ.} We have

$$K \equiv A_1 x + B_1, \quad x < \xi,$$
$$K \equiv A_2 x + B_2, \quad x > \xi$$

and evidently by the end conditions $B_1 = 0$, $A_2 l + B_2 = 0$, by the condition at the source $A_2 - A_1 = -1$, $A_2 \xi + B_2 = A_1 \xi$, so that

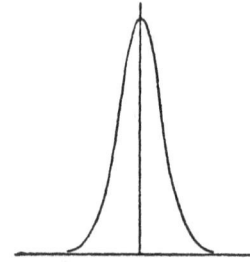

Fig. 46.

$$(179) \qquad \begin{aligned} K(x, \xi) &= \frac{x(l - \xi)}{l}, \quad x < \xi, \\ K(x, \xi) &= \frac{\xi(l - x)}{l}, \quad x > \xi, \end{aligned}$$

as we found in (116). We may call this the *influence*-function, corresponding by Rayleigh's Principle to the influence coefficients (111). We have $K(\xi, \xi) = \xi(l - \xi)/l$ which is a maximum for $\xi = \frac{1}{2}$. The function is evidently symmetrical in x and ξ, Fig. 46, and is represented by two hyperbolic paraboloids, intersecting in a parabola in the plane $x = \xi$. If we have a different source at $x = \eta$,

$$(180) \qquad \begin{aligned} K(x, \eta) &= \frac{x(l - \eta)}{l}, \quad x < \eta, \\ K(x, \eta) &= \frac{\eta(l - x)}{l}, \quad x > \eta. \end{aligned}$$

Now

$$(181) \qquad \begin{aligned} K(\xi, \eta) &= \frac{\xi(l - \eta)}{l}, \quad \xi < \eta, \\ K(\eta, \xi) &= \frac{\xi(l - \eta)}{l}, \quad \eta > \xi, \\ K(\xi, \eta) &= \frac{\eta(l - \xi)}{l}, \quad \xi > \eta, \\ K(\eta, \xi) &= \frac{\eta(l - \xi)}{l}, \quad \eta < \xi, \end{aligned}$$

so that in either case

$$(182) \qquad K(\xi, \eta) = K(\eta, \xi).$$

This symmetry of the Green's function is a theorem of great importance in mechanics and mathematical physics. Here it means that a source at a point ξ produces the same displacement at a point η that the same source at η produces at ξ, and corresponds to the fact that for a discrete system $C_{rs} = C_{sr}$.

Let us deduce the same result from the differential equation. Put

$$K(x, \xi) = u, \quad K(x, \eta) = v.$$

Then

$$\frac{d^2 u}{d x^2} = 0, \quad \frac{d^2 v}{d x^2} = 0,$$

from which

$$u \frac{d^2 v}{d x^2} - v \frac{d^2 u}{d x^2} = 0,$$

and integrating over the fundamental interval

$$(183) \qquad \int_0^l \left(u \frac{d^2 v}{dx^2} - v \frac{d^2 u}{dx^2} \right) dx = 0 .$$

The integrand is the derivative of $u \dfrac{dv}{dx} - v \dfrac{du}{dx}$, which has two discontinuities, at $x = \xi$, $x = \eta$. In order to consider the effect of discontinuities, let us suppose that $f(x)$ has a discontinuity at $x = \xi$. Then $\int_a^b f'(x) dx = f(b) - f(a)$, if ξ is not in the range. If it is we may remove from the field of integration a small region of width 2ε containing the point of discontinuity.

$$(184) \qquad \int_a^b f'(x) dx = \lim_{\varepsilon = 0} \left[\int_a^{\xi-\varepsilon} f'(x) dx + \int_{\xi+\varepsilon}^b f'(x) dx \right] = f(\xi - 0) - f(a)$$
$$+ f(b) - f(\xi + 0).$$

$$\int_a^b f'(x) dx = f(b) - f(a) + f(x) \Big|_{\xi+0}^{\xi-0}$$

Suppose $\xi < \eta$. Then we must take the integral (183) over the successive ranges of continuity, so that

$$(185) \qquad (uv' - vu') \Big|_0^{\xi-0} + (uv' - vu') \Big|_{\xi+0}^{\eta-0} + (uv' - vu') \Big|_{\eta+0}^l = 0 .$$

Now at $x = 0$, $x = l$ both u and v are zero. They are continuous throughout, but their derivatives have discontinuities at ξ and η respectively. We thus obtain

$$(186) \qquad u(\xi) v' \Big|_{\xi+0}^{\xi-0} - v(\xi) u' \Big|_{\xi+0}^{\xi-0} + u(\eta) v' \Big|_{\eta+0}^{\eta-0} - v(\eta) u' \Big|_{\eta+0}^{\eta-0} = 0 .$$

Only the second and third terms are different from zero, and considering the definitions of the discontinuities, we have

$$(187) \qquad u(\eta) - v(\xi) = 0 , \quad \text{or} \quad K(\eta, \xi) - K(\xi, \eta) = 0 ,$$

the result directly obtained.

Suppose we have the equation

$$(188) \qquad \frac{d^2 y}{dx^2} = - f(x)$$

representing the equilibrium of the string with a distribution of sources of density $f(x)$. We will suppose $f(x)$ is piecewise continuous, that is that in the fundamental interval it has only a finite number of points of discontinuity, and approaches a definite finite limit on both sides of each of these. Evidently since a unit source at $x = \xi$ produces the displacement $y = K(x, \xi)$ and since the equation is linear, the effect of each of the sources is added to that of the others, {i. e. the sum of solutions is a solution}, so that a source of strength $f(\xi)d\xi$ produces the displacement at x, equal to $K(x, \xi) f(\xi)$, and the whole effect is

$$(189) \qquad y(x) = \int_0^l K(x, \xi) f(\xi)\, d\xi.$$

A function $y(x)$ that may be thus represented is said to be representable *source-wise*.

Let us verify that the above definite integral satisfies the equation (188). Since K is a continuous function of x, we have

$$(190) \qquad \frac{dy}{dx} = \int_0^l \frac{\partial K}{\partial x} f(\xi)\, d\xi.$$

Now $\dfrac{\partial K}{\partial x}$ is discontinuous at $x = \xi$.

If we write

$$(191) \qquad \frac{dy}{dx} = \int_0^x \frac{\partial K}{\partial x} f(\xi)\, d\xi + \int_x^l \frac{\partial K}{\partial x} f(\xi)\, d\xi,$$

we have, at points where $f(\xi)$ is continuous (differentiating according to the limits),

$$(192) \qquad \frac{d^2 y}{dx^2} = \int_0^x \frac{\partial^2 K}{\partial x^2} f(\xi)\, d\xi + \frac{\partial K}{\partial x}(x, x - 0) f(x)$$

$$+ \int_x^l \frac{\partial^2 K}{\partial x^2} f(\xi)\, d\xi - \frac{\partial K}{\partial x}(x, x + 0) f(x),$$

and since

$$\frac{\partial^2 K}{\partial x^2} = 0, \qquad \frac{\partial K}{\partial x}\Big|_{\xi+0}^{\xi-0} = 1$$

and $\dfrac{\partial K}{\partial x}(x, x - 0)$ represents the slope just on the right of the source, $\dfrac{\partial K}{\partial x}(x, x + 0)$ the slope just on the left, we have

$$(188) \qquad \frac{d^2 y}{dx^2} = - f(x).$$

This process is true if $f(x)$ is only piece-wise continuous, but will be undetermined at the points of discontinuity.

36. Forced Vibration. General Source. Integral Equation.

Let us now return to the equation (163) where $k \neq 0$, and consider a forced vibration from a unit point-source at $x = \xi$. On each side we have

$$(130) \qquad \frac{d^2 \varphi}{dx^2} + k^2 \varphi = 0.$$

Since $\varphi(0) = \varphi(l) = 0$, we have

(194)
$$\varphi(x) = \varphi(\xi) \frac{\sin kx}{\sin k\xi}, \qquad x < \xi,$$

$$\varphi(x) = \varphi(\xi) \frac{\sin k(l - x)}{\sin k(l - \xi)}, \quad x > \xi.$$

If the motion is prescribed at $x = \xi$, that is, if $\varphi(\xi)$ is given, we see that if either $\sin k\xi$ or $\sin k(l - \xi)$ is zero, the motion is infinite (on one side) that is, if motion is applied at what would be a node for a motion with the prescribed period, there will be resonance.

On the contrary, if not the motion, but the strength of the source is given, since

(195)
$$\varphi'(x) = \varphi(\xi) \frac{k \cos kx}{\sin k\xi}, \quad x < \xi,$$

$$\varphi'(x) = -\varphi(\xi) \frac{k \cos k(l - x)}{\sin k(l - \xi)}, \quad x > \xi,$$

and by the definition of a unit source,

$$k\varphi(\xi) \left[\frac{\cos k\xi}{\sin k\xi} + \frac{\cos k(l - \xi)}{\sin k(l - \xi)} \right] = 1,$$

(196)
$$\varphi(\xi) = \frac{\sin k\xi \sin k(l - \xi)}{k \sin kl},$$

(197)
$$\varphi(x) = \frac{\sin kx \sin k(l - \xi)}{k \sin kl} = \Gamma(x, \xi), \quad x < \xi,$$

$$\varphi(x) = \frac{\sin k\xi \sin k(l - x)}{k \sin kl} = \Gamma(x, \xi), \quad x > \xi.$$

{ We shall call $\Gamma(x, \xi)$ the Green's function for the equation (130) while $K(x, \xi)$ is for the case $k = 0$.}

We now have resonance only if $\sin kl = 0$ (that is, if p and k correspond to one of the natural frequencies and wave-lengths respectively), no matter where the force be applied, while if the source be applied at a node for this period, i.e. if $\sin k\xi = 0$ when $x = \xi$, no motion is produced, *beyond the node*.

Suppose we have a continuous distribution of sources of density $f(x)$

(163)
$$\frac{d^2\varphi}{dx^2} + k^2\varphi = -f(x).$$

Then as before, considering all sources, taking the values both for $x < \xi$ and $x > \xi$,

(198) $\varphi(x) = \dfrac{\sin k(l - x)}{k \sin kl} \displaystyle\int_0^x \sin k\xi f(\xi) d\xi + \dfrac{\sin kx}{k \sin kl} \displaystyle\int_x^l \sin k(l - \xi) f(\xi) d\xi.$

This solution could also be obtained by the method of "variation of the constants".

Let us make use of the Green's function $K(x, \xi)$. We have $\dfrac{\partial^2 K}{\partial x^2} = 0$. Multiply our equation for φ by K, that for K by φ and take the difference. Then

$$(199) \qquad K\frac{d^2\varphi}{dx^2} - \varphi\frac{d^2 K}{dx^2} + k^2 K\varphi = -Kf.$$

Integrating, and considering the discontinuity, as before

$$(200) \quad \left(K\frac{d\varphi}{dx} - \varphi\frac{dK}{dx}\right)\Big|_{\xi+0}^{\xi-0} + k^2\int_0^l K(x, \xi)\varphi(x)\,dx = -\int_0^l K(x, \xi)f(x)\,dx,$$

$$(201) \qquad -\varphi(\xi) + k^2\int_0^l K(x, \xi)\,\varphi(x)\,dx = -\int_0^l K(x, \xi)f(x)\,dx.$$

The last integral contains only given functions, and is therefore a given function of ξ, say $-F(\xi)$, we may then write

$$(202) \qquad \varphi(\xi) - k^2\int_0^l K(x, \xi)\,\varphi(x)\,dx = F(\xi).$$

This equation, in which the unknown function φ appears both outside and under the integral sign is known as an *integral equation* of the second kind. We have found its solution, (198) by means of the differential equation, in case $F(\xi)$ can be represented source-wise with source-density f,

$$(203) \qquad F(\xi) = \int_0^l K(x, \xi)f(x)\,dx.$$

We find the density

$$f(x) = -\frac{d^2 F}{dx^2}$$

{since (189) is a solution of (188)}.

In case k is one of the characteristic numbers, that is, roots of the equation $\sin kl = 0$, the integral equation has no solution, for we then have resonance.

We get a physical notion of the genesis of the integral equation as follows. Since a force P applied to the string in equilibrium at the point $x = \xi$ produces such a discontinuity in $\dfrac{\partial y}{\partial x}$ that

$$\tau\frac{\partial y}{\partial x}\Big|_{\xi+0}^{\xi-0} = P,$$

a unit source is produced by a force τ. Suppose now the string is in motion. Then each portion of string of length $d\xi$ produces a reaction $-\varrho\,d\xi\dfrac{\partial^2 y}{\partial t^2}$, and this force at ξ produces at a point x the displacement

$$-\frac{K(x, \xi)}{\tau}\varrho\,d\xi\left(\frac{\partial^2 y}{\partial t^2}\right)_{x=\xi},$$

beside that impressed from outside. Accordingly the effect of the whole string is to produce at x the displacement

$$(204) \quad y(x) = -\frac{\varrho}{\tau}\int_0^l K(x,\xi)\left(\frac{\partial^2 y}{\partial t^2}\right)_{x=\xi}d\xi + \int_0^l K(x,\xi)f(\xi)\cos(pt)d\xi.$$

If we put $y = \varphi(x)\cos pt$, this reduces to

$$(205) \quad \varphi(x) = k^2\int_0^l K(x,\xi)\varphi(\xi)d\xi + F(x)$$

which is the integral equation (202) (with x and ξ interchanged, which does not affect K). This corresponds to the inverse method of treating vibrations.

37. Normal Functions. Bilinear Formula. Let us now return to the equation for the general source

$$(161) \quad \varrho\frac{\partial^2 y}{\partial t^2} = \tau\left[\frac{\partial^2 y}{\partial x^2} + Y(x,t)\right],$$

and consider the effect of putting in the normal functions,

$$(206) \quad y = \sum_{r=1}^{r=\infty}\varphi_r(x)\psi_r(t).$$

We must find the expressions for the kinetic and potential energies, as in (156), and of the forces P_r belonging to each normal coordinate, in order to find a differential equation like (157) for each ψ_r. Now using the definitions (105) and (106), and the properties expressed by (144)[1] and (145), we find

$$T = \frac{1}{2}\int_0^l \varrho\left(\frac{\partial y}{\partial t}\right)^2 dx = \frac{1}{2}\varrho\int_0^l\left(\sum_{r=1}^{r=\infty}\varphi_r\psi_r'\right)^2 dx = \frac{1}{2}\varrho\sum_{r=1}^{r=\infty}\psi_r'^2,$$

$$W = \frac{1}{2}\int_0^l \tau\left(\frac{\partial y}{\partial x}\right)^2 dx = \frac{1}{2}\tau\int_0^l\left(\sum_{r=1}^{r=\infty}\psi_r\frac{d\varphi_r}{dx}\right)^2 dx = \frac{1}{2}\tau\sum_{r=1}^{r=\infty}\psi_r^2\int_0^l\left(\frac{d\varphi_r}{dx}\right)^2 dx.$$

The expression for W may be transformed by an integration by parts,

$$\int_0^l\left(\frac{d\varphi_r}{dx}\right)^2 dx = \varphi_r\frac{d\varphi_r}{dx}\bigg|_0^l - \int_0^l\varphi_r\frac{d^2\varphi_r}{dx^2}dx,$$

---

1) The function W, like T, can contain no product terms if the ψ's are normal coordinates.

and since both ends of the string are fixed the integrated terms disappear, and we have, replacing $\frac{d^2\varphi_r}{dx^2}$ by $-k_r^2\,\varphi_r$, by (130),

$$W = \frac{1}{2}\tau \sum k_r^2\psi_r^2.$$

For the work of the sources in an arbitrary displacement we find

$$\int_0^l \tau\,Y\delta y\,dx = \tau \sum_{r=1}^{r=\infty} \delta\psi_r \int_0^l Y\varphi_r\,dx = \sum_{r=1}^{r=\infty} P_r\delta\psi_r,$$

which gives the value of the force,

$$(207)\qquad\qquad P_r = \tau \int_0^l Y\varphi_r\,dx.$$

{This corresponds to developing the function Y in the manner of Fourier, as in (140).}

Accordingly we find by Lagrange's method the equation for ψ_r,

$$(208)\qquad\qquad \varrho\,\frac{d^2\psi_r}{dt^2} + \tau k_r^2\psi_r = P_r = \tau \int_0^l Y\varphi_r\,dx,$$

which is of the same form as (157), and reduces to (130) when $P_r = 0$.

Let us, as before, begin with the equilibrium theory $\frac{d^2\psi_r}{dt^2} = 0$.
Then

$$(209)\qquad\qquad \psi_r = \frac{P_r}{\tau k_r^2} = \frac{1}{k_r^2}\int_0^l Y\varphi_r\,dx,$$

$$(210)\qquad\qquad y = \sum_{r=1}^{r=\infty} \frac{\varphi_r}{k_r^2}\int_0^l Y\varphi_r\,dx.$$

Let us now reduce the sources to a point-source at $x = \xi$, when y becomes $K(x,\xi)$ and if

$$\lim_{\varepsilon=0}\int_{\xi-\varepsilon}^{\xi+\varepsilon} Y\,dx = 1,$$

we have

$$(211)\qquad\qquad K(x,\xi) = \sum_{r=1}^{r=\infty} \frac{\varphi_r(x)\,\varphi_r(\xi)}{k_r^2}.$$

This is called Hilbert's bilinear formula. {This formula is of first class importance, since it implies all the properties of normal functions as shown in the next section. On account of these properties we can obtain the developement of any function in a series of normal functions as shown in (229).}

It will be convenient to use the letter λ_r for k_r^2. Thus $\lambda_r = \dfrac{r^2 \pi^2}{l^2}$. Also it is assumed that the normal functions are multiplied by such constants that the integrated squares equal unity.

$$(212) \quad \varphi_r(x) = A_r \sin \frac{r \pi x}{l}, \quad \int_0^l \varphi_r^2 dx = 1 = A_r^2 \int_0^l \sin^2 \frac{r \pi x}{l} dx = \frac{l}{2} A_r^2,$$

$$A_r = \sqrt{\frac{2}{l}}, \quad \varphi_r(x) = \sqrt{\frac{2}{l}} \sin \frac{r \pi x}{l}.$$

Then

$$(213) \qquad K(x, \xi) = \frac{2l}{\pi^2} \sum_{r=1}^{r=\infty} \frac{\sin \frac{r \pi x}{l} \sin \frac{r \pi \xi}{l}}{r^2}.$$

This is the same as (152), and shows the symmetry in x and ξ. $\left(\text{It is to be noticed that a } unit \text{ source makes } b = \dfrac{\xi (l - \xi)}{l}\right)$.

38. Solution of Integral Equation. If we multiply the function $K(x, \xi)$ (211), by a normal function $\varphi_s(\xi)$ and integrate term by term,

$$(214) \quad \int_0^l K(x, \xi) \varphi_s(\xi) d\xi = \sum_{r=1}^{r=\infty} \frac{\varphi_r(x)}{\lambda_r} \int_0^l \varphi_r(\xi) \varphi_s(\xi) d\xi = \frac{\varphi_s(x)}{\lambda_s},$$

and we have

$$(215) \qquad \varphi_s(x) = \lambda_s \int_0^l K(x, \xi) \varphi_s(\xi) d\xi,$$

which is the same as (117).

Thus although the integral equation

$$(202) \qquad \varphi(x) - \lambda \int_0^l K(x, \xi) f(\xi) d\xi = F(x)$$

has no solution when λ is a characteristic number λ_s, (i. e. $\sin \sqrt{\lambda_s} l = 0$), the homogeneous equation (117) with $F(x) = 0$, has as a solution the normal function φ_s. The Green's function $K(x, \xi)$ is called the *kernel* of the integral equation. It will be shown in Chapter IX that every symmetrical kernel has normal functions, defined by the integral equation (117).

The equation (215), as has been previously stated, suffices to define the properties of the normal functions. For suppose that we have two functions belonging to two characteristic numbers λ_r, λ_s,

$$(216) \qquad \begin{aligned} \varphi_r(x) &= \lambda_r \int_0^l K(x, \xi) \varphi_r(\xi) d\xi, \\ \varphi_s(x) &= \lambda_s \int_0^l K(x, \xi) \varphi_s(\xi) d\xi. \end{aligned}$$

Multiplying the first by $\lambda_s \varphi_s(x)$, the second by $\lambda_r \varphi_r'x)$, subtracting and integrating,

$$(217) \quad (\lambda_s - \lambda_r)\int_0^l \varphi_r(x)\varphi_s(x)\,dx = \lambda_r\lambda_s\left[\int_0^l\int_0^l K(x,\,\xi)\varphi_r(\xi)\varphi_s(x)\,d\xi\,dx\right.$$

$$\left.-\int_0^l\int_0^l K(x,\,\xi)\varphi_s(\xi)\varphi_r(x)\,d\xi\,dx\right],$$

which vanishes because the second integral is the same as the first with x and ξ interchanged, and $K(x,\,\xi) = K(\xi,\,x)$. Consequently unless $\lambda_r = \lambda_s$ we must have

$$(144) \qquad\qquad \int_0^l \varphi_r(x)\varphi_s(x)\,dx = 0,$$

or the two normal functions are orthogonal.

Suppose that one of the normal functions belonged to a complex characteristic number $\lambda_r = \mu_r + i\nu_r$, then changing i to $-i$ we should find a number $\lambda_s = \mu_r - i\nu_r$, and hence the two conjugate functions $\varphi_r = \varphi_1 + i\varphi_2$ and $\varphi_s = \varphi_1 - i\varphi_2$, must be orthogonal, that is,

$$(218) \qquad \int_0^l(\varphi_1 + i\varphi_2)(\varphi_1 - i\varphi_2)\,dx = \int_0^l(\varphi_1^2 + \varphi_2^2)\,dx = 0,$$

which is impossible (the integrand being positive). Consequently the characteristic numbers λ are all real. It will be seen that this corresponds exactly to the proof given (62) that the roots of Lagrange's determinant are all real.

Suppose now that the force $Y(x,\,t)$ contains a single harmonic term $\tau f(x)\cos pt$, then

$$(219) \qquad\qquad P_r = \cos pt\int_0^l \tau f(x)\varphi_r(x)\,dx,$$

and as above (158) we have

$$(220) \qquad \psi_r = \frac{F_r\cos pt}{\varrho(p_r^2 - p^2)}, \quad p_r = ak_r, \quad a = \sqrt{\frac{\tau}{\varrho}},\,.$$

$$(221) \qquad y = \cos pt\sum_{r=1}^{r=\infty}\frac{\varphi_r(x)}{k_r^2 - k^2}\int_0^l f(x)\varphi_r(x)\,dx.$$

Let us call the space factor $\psi(x)$, {not to be confused with $\psi_r(t)$}, putting

$$(222) \qquad \psi(x) = \sum_{r=1}^{r=\infty}\frac{\varphi_r(x)}{\lambda_r - \lambda_s}\int_0^l f(x)\varphi_r(x)\,dx.$$

We have to prc·e that this is a solution of the integral equation (201)
If we put, as above,

$$(223) \qquad F(x) = \int_0^l K(x, \xi) f(\xi) d\xi,$$

and use the bilinear formula, (210),

$$(224) \qquad F(x) = \sum_{r=1}^{r=\infty} \frac{\varphi_r(x)}{\lambda_r} \int_0^l \varphi_r(\xi) f(\xi) d\xi.$$

Subtracting this from $\psi(x)$ we have

$$(225) \qquad \psi(x) - F(x) = \sum_{r=1}^{r=\infty} \left[\frac{1}{\lambda_r - \lambda} - \frac{1}{\lambda_r} \right] \varphi_r(x) \int_0^l \varphi_r(\xi) f(\xi) d\xi.$$

$$= \lambda \sum_{r=1}^{r=\infty} \frac{\varphi_r(x)}{\lambda_r (\lambda_r - \lambda)} \int_0^l \varphi_r(\xi) f(\xi) d\xi.$$

But multiplying $\psi(\xi)$ by $K(x, \xi)$ and integrating,

$$(226) \qquad \int_0^l K(x, \xi) \psi(\xi) d\xi = \sum_{r=1}^{r=\infty} \frac{\int_0^l K(x, \xi) \varphi_r(\xi) d\xi \int_0^l \varphi_r(\xi) f(\xi) d\xi}{\lambda_r - \lambda}$$

$$= \sum_{r=1}^{r=\infty} \frac{\varphi_r(x)}{\lambda_r (\lambda_r - \lambda)} \int_0^l \varphi_r(\xi) f(\xi) d\xi,$$

which has been above proved equal to

$$\frac{\psi(x) - F(x)}{\lambda}. \qquad (225)$$

Consequently

$$(227) \qquad \psi(x) = F(x) + \lambda \int_0^l K(x, \xi) \psi(\xi) d\xi,$$

or $\psi(x)$ satisfies our integral equation.

If we develop $F(x)$ in series of normal functions

$$F(x) = A_1 \varphi_1(x) + A_2 \varphi_2(x) + \cdots$$

we find, on multiplying by φ_s and integrating, by (144)

$$(228) \qquad A_s = \int_0^l F(x) \varphi_s(x) dx$$

so that

$$(229) \qquad F(x) = \sum_{r=1}^{r=\infty} \varphi_r(x) \int_0^l F(\xi) \varphi_r(\xi) d\xi.$$

Such a development is said to be in the Fourier manner.

Comparing with the previous expression, (224).

(230)
$$\lambda_r\int_0^l F(\xi)\,\varphi_r(\xi)\,d\xi = \int_0^l \varphi_r(\xi)f(\xi)\,d\xi\,,$$

and inserting the integral on the right in $\psi(x)$, (222)

$$\psi(x) = \sum_{r=1}^{r=\infty} \frac{\lambda_r\varphi_r(x)}{\lambda_r - \lambda}\int_0^l \varphi_r(\xi)F(\xi)\,d\xi$$

(231)
$$= \sum_{r=1}^{r=\infty}\varphi_r(x)\int_0^l \varphi_r(\xi)F(\xi)\,d\xi + \lambda\sum_{r=1}^{r=\infty}\frac{\varphi_r(x)}{\lambda_r - \lambda}\int_0^l \varphi_r(\xi)F(\xi)\,d\xi$$

$$= F(x) + \lambda\sum_{r=0}^{r=\infty}\frac{\varphi_r(x)}{\lambda_r - \lambda}\int_0^l \varphi_r(\xi)F(\xi)\,d\xi\,,$$

and this is Schmidt's solution of the integral equation.[1])

39. Other Boundary Conditions. So far we have considered the example of a string with fixed ends $\varphi = 0$. It will be of interest, both physically and mathematically, to consider other conditions at the ends of the fundamental interval. If the string, instead of being fastened to a fixed point, were attached to a ring without mass sliding without friction on a wire at right angles to the string, since there can be no force parallel to the wire, the string must end at right angles to the wire, or $\frac{d\varphi}{dx} = 0$.

If we consider, on the other hand, u as the longitudinal displacement of air in a pipe, the compression is $s = -\frac{\partial u}{\partial x}$, and at an *open* end of a tube where the pressure must be that of the atmosphere, we have $s = -\frac{\partial u}{\partial x} = 0$. In either of these cases, if the condition is $\frac{\partial u}{\partial x} = 0$ at both ends, the normal functions must be not $\sin kx$ but $\cos kx$. If we put

(232)
$$\varphi = \cos kx\,, \qquad \frac{d\varphi}{dx} = -k\sin kx$$

we must have to determine k, by $\sin kl = 0$, as before (35). Thus the characteristic numbers are the same as before, and the harmonics of an open pipe are the same as of a closed one. On the other hand, suppose the pipe is closed at one end, $x = 0$, and open at the other, $x = l$. Then putting

$$\varphi = A\cos kx + B\sin kx\,,$$

(233)
$$\frac{d\varphi}{dx} = k\{-A\sin kx + B\cos kx\}$$

---

1) E. Schmidt, *Math. Ann.* Bd. 63, p. 454, 1907.

we must have, as before, $A = 0$, but instead of (35),

$$\cos kl = 0,$$

(234)
$$k = \frac{2n+1}{2}\frac{\pi}{l}, \quad n = 0, 1, 2, 3 \ldots$$

and the frequencies are proportional to the *odd* integers, so that a pipe open and closed lacks the even harmonics. It is obvious that for either of these cases, where the normal functions are cosines, we must have a new Green's function.

The source of the sort previously described in which the displacement u is continuous at ξ, but $\frac{\partial u}{\partial x}$, or the compression is discontinuous, would be realized by placing a piston at $x = \xi$, producing a compression in front and a rarefaction behind. In order, however to have an example which will be comparable with the more practical problems of sound in three dimensions, which will be considered in Chapter V, let us consider a source that is symmetrical with reference to the compression on both sides of ξ, and producing a discontinuity in u, the displacement. Such a source will be produced if we introduce or withdraw air from outside at ξ, say through a side tube, or if we move two pistons simultaneously in opposite directions. It will be convenient if we fix our attention, not upon the displacement u, but upon the displacement *potential* φ defined by the equation $u = \frac{d\varphi}{dx}$, which satisfies the same equation as before,

$$\frac{d^2\varphi}{dx^2} + k^2\varphi = 0.$$

The compression is

(235)
$$s = -\frac{du}{dx} = -\frac{d^2\varphi}{dx^2} = k^2\varphi.$$

Consequently we have for an open end $\varphi = 0$, and for a closed one $\frac{d\varphi}{dx} = 0$.

We will define a unit symmetrical source as before (178)

(178)
$$\frac{d\varphi}{dx}\Big|_{\xi-0}^{\xi+0} = -1,$$

which means physically that the velocity u on the left is greater than that on the right by unity, or in other words, a unit volume of air is withdrawn from a tube of unit cross-section in unit time. We will as before assume that φ is continuous at the source.

If the end conditions are $\frac{d\varphi}{dx} = 0$, we can no longer define the Green's function as a solution of the equation

$$\frac{d^2K}{dx^2} = 0,$$

because $\dfrac{dK}{dx}$ being constant would have to be zero throughout, and K would be constant, which would be useless. Let us then put

$$(236) \qquad \frac{d^2 K}{d x^2} = a,$$

representing a constant compression, which may be got by suitably moving the two pistons in opposite directions.

We have then

$$(237) \qquad
\begin{aligned}
K &= \frac{a x^2}{2} + c, & \frac{dK}{dx} &= a x, & x &< \xi, \\
K &= \frac{a}{2}(x - l)^2 + c', & \frac{dK}{dx} &= a(x - l), & x &> \xi,
\end{aligned}$$

and for continuity at $x = \xi$,

$$c' - c = \frac{a}{2}(2 l \xi - l^2),$$

and for the discontinuity of $\dfrac{dK}{dx}$, $a = \dfrac{1}{l}$. Accordingly if we put $c = \dfrac{\xi^2}{2 l} - \xi + b$, we obtain the more symmetrical form,

$$(238) \qquad
\begin{aligned}
K(x, \xi) &= \frac{x^2 + \xi^2}{2 l} - \xi + b, & x &< \xi, \\
K(x, \xi) &= \frac{x^2 + \xi^2}{2 l} - x + b, & x &> \xi.
\end{aligned}$$

The constant b remains undetermined, and we may, if we chose, determine it so as to make

$$(239) \qquad \int_0^l K(x, \xi)\,dx = 0, \quad \text{by taking} \quad b = \frac{l}{3}.$$

The symmetry of the Green's function is proved as before, for if

$$u = K(x, \xi), \quad v = K(x, \eta),$$

we have instead of (183)

$$(240) \qquad \int_0^l \left(u \frac{d^2 v}{d x^2} - v \frac{d^2 u}{d x^2} \right) dx = a \int_0^l u\,dx - a \int_0^l v\,dx,$$

which vanishes by (239). Also any continuous function which satisfies

$$\frac{d^2 \varphi}{d x^2} + k^2 \varphi = 0, \quad \frac{d\varphi}{dx} = 0, \quad \begin{matrix} x = 0, \\ x = l, \end{matrix}$$

must have

$$(241) \qquad k^2 \int_0^l \varphi\,dx = -\int_0^l \frac{d^2 \varphi}{d x^2}\,dx = -\frac{d\varphi}{dx}\Big|_0^l = 0,$$

or its mean value in the fundamental interval vanishes. This is the case for the normal functions

$$\varphi_r = \cos \frac{r \pi x}{l},$$

and only such functions can be developed in a cosine series. If on the other hand φ satisfies (163)

$$(163) \qquad \frac{d^2 \varphi}{dx^2} + k^2 \varphi = - f(x),$$

we have instead of (199)

$$(242) \qquad K \frac{d^2 \varphi}{dx^2} - \varphi \frac{d^2 K}{dx^2} + k^2 K \varphi = - a \varphi - K f$$

and instead of (202) we have the constant term $a \int_0^l \varphi \, dx$ added to F.

If however, the mean value of φ vanishes, the integral equation is the same as before. The bilinear formula is then

$$(243) \qquad K(x, \xi) = \frac{2l}{\pi^2} \sum_{r=1}^{r=\infty} \frac{\cos \dfrac{r \pi x}{l} \cos \dfrac{r \pi \xi}{l}}{r^2},$$

as may be verified by the Fourier development. The Green's function for the equation

$$\frac{d^2 \varphi}{dx^2} + k^2 \varphi = 0,$$

will be in this case

$$(244) \qquad
\begin{aligned}
\Gamma(x, \xi) &= \frac{\cos k x \cos k (l - \xi)}{k \sin k l}, \quad x < \xi, \\
\Gamma(x, \xi) &= \frac{\cos k \xi \cos k (l - x)}{k \sin k l}, \quad x > \xi.
\end{aligned}$$

But we have by no means exhausted the possible, or practical, boundary conditions. Suppose that at the end of a pipe there is a piston without mass, acted upon by a spring so that it has a force exerted on it proportional to the displacement. Then the displacement will be proportional to the pressure, or to the compression, and if it is at the end $x = l$,

$$h u = s = - \frac{\partial u}{\partial x}$$

so that we have the condition

$$(245) \qquad \frac{\partial u}{\partial x} + h u = 0.$$

The same condition answers for a string fastened to a springy or yielding support. The preceding cases are obtained by putting $h = 0$, or $h = \infty$. More generally let the piston or string be attached to a point of any system having a finite number of degrees of freedom, and having energy functions

$$(246) \qquad T = \frac{1}{2}\sum_{i=1}^{i=n}\sum_{j=1}^{j=n} a_{ij} q_i' q_j', \quad W = \frac{1}{2}\sum_{i=1}^{i=n}\sum_{j=1}^{j=n} c_{ij} q_i q_j.$$

Then the force required on the coordinate q_i will be

$$(247) \qquad P_i = \sum_{j=1}^{j=n}(a_{ij} q_j'' + c_{ij} q_j).$$

Now let the piston be connected with a point whose coordinate is q_1, so that $u = q_1$, and the force is proportional to s, so that $P_1 = Cs$, and let there be no other forces impressed on the terminal system. Then

$$(248) \qquad \begin{aligned} &a_{11} q_1'' + a_{12} q_2'' + \cdots + c_{11} q_1 + c_{12} q_2 \cdots = Cs, \\ &a_{21} q_1'' + a_{22} q_2'' + \cdots + c_{21} q_1 + c_{22} q_2 \cdots = 0, \\ &\cdot\ \cdot\ \cdot\ \cdot\ \cdot\ \cdot\ \cdot\ \cdot\ \cdot\ \cdot\ \cdot\ \cdot\ \cdot\ \cdot \\ &\cdot\ \cdot\ \cdot\ \cdot\ \cdot\ \cdot\ \cdot\ \cdot\ \cdot\ \cdot\ \cdot\ \cdot\ \cdot\ \cdot \\ &a_{n1} q_1'' + a_{n2} q_2'' + \cdots + c_{n1} q_1 + c_{n2} q_2 \cdots = 0. \end{aligned}$$

These equations are satisfied by putting

$$u = \varphi \cos pt, \quad s = -\frac{d\varphi}{dx}\cos pt, \quad q_r = A_r \cos pt, \quad \text{giving}$$

$$(249) \qquad \begin{aligned} &A_1(c_{11} - p^2 a_{11}) + A_2(c_{12} - p^2 a_{12}) + \cdots = -C\left(\frac{d\varphi}{dx}\right)_1 \\ &A_1(c_{21} - p^2 a_{21}) + A_2(c_{22} - p^2 a_{22}) + \cdots = 0 \\ &\cdot\ \cdot\ \cdot\ \cdot\ \cdot\ \cdot\ \cdot\ \cdot\ \cdot\ \cdot\ \cdot\ \cdot\ \cdot\ \cdot \end{aligned}$$

Solving for A_1, which is equal to the value of φ at the end in question, we obtain

$$(250) \qquad A_1 = \varphi = -C\frac{d\varphi}{dx}\frac{D(-p^2)}{\Delta(-p^2)},$$

where Δ is the determinant of (60), and D is its first minor, a polynomial in p^2. Writing as before $\lambda = k^2 = \frac{p^2}{a^2}$, we may write the condition

$$(251) \qquad \frac{\varphi'}{\varphi} = f(\lambda),$$

where $f(\lambda)$ is the quotient of two polynomials in λ. Two such end conditions will determine the values of λ. In the present case of the uniform string, if we put

$$(32) \qquad \varphi = A \cos kx + B \sin kx$$

this becomes

$$(252) \qquad A \cos kl + B \sin kl = (A \sin kl - B \cos kl)\, Ck\frac{D_1(-p^2)}{\Delta(-p^2)}.$$

Suppose that at the end $x = 0$ we have a similar condition, with the suffixes 0.

$$(253) \qquad\qquad A = -BCk \frac{D_0(-p^2)}{\Delta_0(-p^2)}.$$

Eliminating A and B between these two equations, we obtain the equation for the frequencies

$$(254) \qquad \tan kl = Ck \frac{D_0 \Delta - D \Delta_0}{C^2 k^2 D D_0 + \Delta \Delta_0} = F(k),$$

which is a transcendental equation in k, $F(k)$ being the quotient of two polynomials in k. We may find the real roots of this equation by drawing two graphs, one of $y = \tan kl$, the other of $y = F(k)$, the points of intersection satisfying the equation (254). We then find $\frac{B}{A}$ from (252), so that there remains only one arbitrary constant multiplying each φ_r.

Now for two normal functions,

$$\frac{d^2 \varphi_r}{dx^2} + \lambda_r \varphi_r = 0, \quad \frac{d^2 \varphi_s}{dx^2} + \lambda_s \varphi_s = 0,$$

we have as in (239)

$$(255) \quad \int_0^l \left(\varphi_r \frac{d^2 \varphi_s}{dx^2} - \varphi_s \frac{d^2 \varphi_r}{dx^2} \right) dx = (\varphi_r \varphi_s' - \varphi_s \varphi_r') \Big|_0^l = (\lambda_r - \lambda_s) \int_0^l \varphi_r \varphi_s \, dx$$

In the two elementary cases which we have treated, we had at both ends either $\varphi = 0$ or $\varphi' = 0$, consequently the normal functions were orthogonal. The same is true if we have the condition (245)

$$(245) \qquad\qquad \varphi' + h\varphi = 0,$$

as we see at once. It is *not* true for the general case

$$(251) \qquad\qquad \varphi_r' = \varphi_r f(\lambda),$$

where we have instead

$$(256) \qquad \int_0^l \varphi_r \varphi_s \, dx = \frac{f(\lambda_s) - f(\lambda_r)}{\lambda_r - \lambda_s} \varphi_r \varphi_s \Big|_0^l.$$

(The functions f may be different at the two ends).

The normal functions φ_r are now *not* orthogonal, and we cannot develop an arbitrary function in terms of them in the Fourier manner. Nevertheless, the fundamental properties of normal coordinates enable us to find the development, as shown by Lord Rayleigh for a particular case.[1]) The reason why the integral above does not vanish is that it is not the whole of the mutual energy of two coordinates, but we must take account of the energy in the terminal apparatus as well.

---

1) *Theory of Sound*, Vol. I, p. 202.

Consider now any normal vibrations in which all quantities are proportional to a normal coordinate ψ_r which varies as $\cos(p_r t - \alpha_r)$. We then have, solving the equations (249) for one end

$$(257) \quad \frac{A_i}{A_1} = \frac{D_i(-p_r^2)}{D_1(-p_r^2)}, \quad A_i = P_i(\lambda_r)\varphi_r(l), \quad q_i = P_i(\lambda_r)\varphi_r(l)\psi_r,$$

where P_i is as before the quotient of two polynomials depending on the constitution of the terminal apparatus.

We may now most conveniently express the energy of the terminal apparatus, if we make use instead of the q's, of normal coordinates for that apparatus. Let us make a transformation

$$\chi_1 = \alpha_{11}q_1 + \alpha_{12}q_2 + \cdots + \alpha_{1n}q_n,$$
$$(258) \qquad \chi_2 = \alpha_{21}q_1 + \alpha_{22}q_2 + \cdots + \alpha_{2n}q_n,$$
$$\cdot \quad \cdot \quad \cdot \quad \cdot \quad \cdot \quad \cdot \quad \cdot \quad \cdot \quad \cdot$$

such that the energy functions for one end are

$$(259) \qquad T = \frac{1}{2}\sum_{r=1}^{r=n} a_r \chi_r'^2, \quad W = \frac{1}{2}\sum_{r=1}^{r=n} c_r \chi_r^2$$

we then obtain for any coordinate χ,

$$\chi_i = Q_i(\lambda_r)\varphi_r(l)\psi_r.$$

Let us now suppose that the general motion of the whole system is expressed in normal coordinates ψ_r,

$$u = \sum_{r=1}^{r=\infty}\varphi_r\psi_r, \quad \frac{\partial u}{\partial t} = \sum_{r=1}^{r=\infty}\varphi_r\psi_r',$$

$$(260) \qquad \chi_i = \sum_{r=1}^{r=\infty} Q_i(\lambda_r)\varphi_r(l)\psi_r,$$

$$\chi_i' = \sum_{r=1}^{r=\infty} Q_i(\lambda_r)\varphi_r(l)\psi_r'.$$

We now have for the whole kinetic energy, including the terminal apparatus,

$$(261) \quad T = \frac{1}{2}\int_0^l \varrho\left(\sum_{r=1}^{r=\infty}\varphi_r\psi_r'\right)^2 dx + \frac{1}{2}\sum_{i=1}^{i=n} a_i\left(\sum_{r=1}^{r=\infty} Q_i(\lambda_r)\varphi_r\psi_r'\right)^2\bigg|_{0,l}.$$

By the property of normal functions, the mutual energy term in $\psi_r'\psi_s'$ must vanish, giving

$$(262) \qquad \varrho\int_0^l \varphi_r\varphi_s dx + \sum_{i=1}^{i=n} a_i Q_i(\lambda_r)Q_i(\lambda_s)\varphi_r\varphi_s\bigg|_{0,l} = 0,$$

where it is understood that the arrangements at the two ends may be totally different, and that we take the sum for both. The equation (262) is a more convenient form to deal with than (256), although in both cases the term not the integral is evidently a symmetric function of λ_r and λ_s.

If now y is the initial value of u, and χ_i^0 the initial value of χ_i, and c_r the initial value of ψ_r, to find the coefficients in the development

$$(263) \qquad y = \sum_{s=1}^{s=\infty} c_s \varphi_s, \qquad \chi_1^0 = \sum_{s=1}^{s=\infty} c_s Q_i(\lambda_s)\, \varphi_s(l),$$

let us form the expression for the mutual energy of this motion and a normal vibration ψ_r,

$$(264) \qquad \begin{aligned} & \varrho \int_0^l y \varphi_r\, dx + \sum_{i=1}^{i=n} a_i \chi_i^0 Q_i(\lambda_r) \varphi_r \big|_{0,l} \\ & = \sum_{s=1}^{s=\infty} c_s \left\{ \varrho \int_0^l \varphi_r \varphi_s\, dx + \sum_{i=1}^{i=n} a_i Q_i(\lambda_r) Q_i(\lambda_s)\, \varphi_r \varphi_s \big|_{0,l} \right\}. \end{aligned}$$

But by equation (262) all the terms vanish except the one for which $r = s$, consequently we have

$$(265) \qquad c_s = \frac{\varrho \int_0^l y \varphi_s\, dx + \sum_{i=1}^{i=n} a_i \chi_i^0 Q_i(\lambda_s) \varphi_s \big|_{0,l}}{\varrho \int_0^l \varphi_s^2\, dx + \sum_{i=1}^{i=n} a_i (Q_i(\lambda_s))^2 \varphi_s^2 \big|_{0,l}}.$$

This development of course includes the Fourier development as a particular case, and we have not yet proved that the development is possible. In this connection we may quote the words of Lord Rayleigh, written in 1877. "So much stress is often laid on special proofs of Fourier's and Laplace's series, that the student is apt to acquire too contracted a view of the nature of those important results of analysis."

As a practical example, let us consider the case of a phonometer

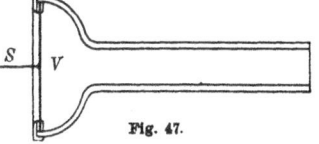

Fig. 47.

used by the author (Fig. 47), in which a pipe communicates at one end with a chamber of volume V, having on the other side a diaphragm, which we may consider as a piston of mass m, area S and stiffness f, so that a displacement ξ requires a force $f\xi$. If u be the displacement of the air at the end of the tube, σ the cross-section of the tube, there enters the chamber the volume σu, while if the diaphragm moves in a distance ξ, there enters the chamber the volume $S\xi$. The compression in the chamber is then the ratio of the volume forced in to V,

$$s = \frac{\sigma u + S\xi}{V},$$

and the pressure $p = es$, which is the same as that at the end of tube $x = l$. This pressure acting on the area S produces the force

$$X = pS = \frac{eS(\sigma u + S\xi)}{V},$$

resisting the entrance of the diaphragm. But the motion of the diaphragm is given by

$$-X = m\frac{d^2\xi}{dt^2} + f\xi.$$

If we neglect the kinetic energy of the air in the chamber, we may put

$$T = \tfrac{1}{2}m\xi'^2, \quad W = \tfrac{1}{2}\frac{e}{V}(\sigma u + S\xi)^2 + \tfrac{1}{2}f\xi^2,$$

for this will give the motion of the diaphragm by

$$\frac{d}{dt}\left(\frac{\partial T}{\partial \xi'}\right) + \frac{\partial W}{\partial \xi} = 0$$

and $-\dfrac{\partial W}{\partial u}$ will give σ times the pressure at the end of the tube.

We may now use as normal coordinates,

$$\chi_1 = \sigma u + S\xi, \quad \chi_2 = \xi,$$

so that with

$$a_1 = 0, \quad a_2 = m, \quad c_1 = \frac{e}{V}, \quad c_2 = f$$

we have

$$T = \tfrac{1}{2}a_1\chi_1'^2, \quad W = \tfrac{1}{2}c_1\chi_1^2 + \tfrac{1}{2}c_2\chi_2^2.$$

The equations (248) are

$$s = -\frac{\partial u}{\partial x} = \frac{\sigma u + S\xi}{V}$$

$$0 = m\xi'' + f\xi + \frac{eS}{V}(\sigma u + S\xi)$$

and putting

$$u = A_1 \cos pt, \quad \xi = A_2 \cos pt,$$

we find

(266)
$$-\frac{d\varphi}{dx} = A_1\frac{\sigma}{V} + A_2\frac{S}{V},$$

$$0 = A_1\frac{eS\sigma}{V} + A_2\left\{f - mp^2 + \frac{eS^2}{V}\right\},$$

(267) $$A_1 = \varphi = \frac{\frac{d\varphi}{dx}\left(mp^2 - f - \frac{eS^2}{V}\right)}{\frac{\sigma}{V}(f - mp^2)}, \quad A_2 = -\frac{d\varphi}{dx}\frac{eS}{(f - mp^2)}.$$

Thus equation (251) is

(268)
$$\frac{\varphi'}{\varphi} = \frac{\sigma}{V} \frac{f - mp^2}{\left(mp^2 - f - \frac{eS^2}{V}\right)}.$$

It is to be observed that the above acoustical problem is exactly like the electrical one of a pair of wires whose ends are directly connected to a condenser K_1, and indirectly to another K_2 through a coil of inductance L (Fig. 48).

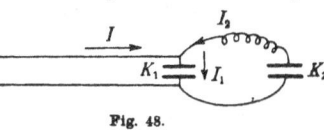

Fig. 48.

The arriving current I is to be put equal to $\sigma \frac{\partial u}{\partial t}$ and we have

$$I = I_1 - I_2$$

where

$$I_1 = \frac{dq_1}{dt}, \quad I_2 = \frac{dq_2}{dt},$$

q_1 and q_2 being the charges of the condensers K_1 and K_2, and

$$T = \tfrac{1}{2} L I_2^2, \quad W = \tfrac{1}{2} K_1 q_1^2 + \tfrac{1}{2} K_2 q_2^2.$$

If the second wire be absent, and both condensers be connected to earth instead of to it, we have the case of a wireless "antenna" or receiving wire. The other end of the antenna makes the current equal to zero, and corresponds to a *closed* pipe $u = 0$. Let us rather consider an open pipe $\frac{\partial u}{\partial x} = 0$, and let this be at the end $x = 0$. We must then put

$$\varphi = \cos kx, \quad \frac{d\varphi}{dx} = -k \sin kx,$$

and (268) gives

(269)
$$\tan kl = \frac{\sigma}{kV} \frac{(ma^2k^2 - f)}{\left(ma^2k^2 - f - \frac{eS^2}{V}\right)} = F(k).$$

The curve $y = F(k)$ is of the fourth order, crossing the k-axis for $k = \pm \sqrt{\frac{f}{ma^2}}$ and having vertical asymptotes at $k = 0$, $k = \pm \sqrt{\frac{f + eS^2/V}{ma^2}}$.

We see that, F being an odd function, we need consider only positive values of k, and that the higher overtones are nearly given by $\tan kl = 0$, that is are those of a pipe open at both ends. If V is infinite, this is true for all periods. If either m or f is infinite, that is if we have an immovable wall, we have $\tan kl = \frac{\sigma}{kV}$, and if $V = 0$ we have $\cos kl = 0$, as for a closed-open pipe.

Let us examine the forced oscillation, supposing that we have at the open end $x = 0$, a sound entering from outside, providing the compression $s = s_0 \cos pt$. The value of $k = ap$ is now prescribed. We must put

$$\varphi = A \cos kx + B \sin kx$$

$$\frac{d\varphi}{dx} = k\,(-A \sin kx + B \cos kx)$$

and the above condition gives $s_0 = -kB$, while (269) gives

$$k\,(-A \sin kl + B \cos kl) = (A \cos kl + B \sin kl)\,\frac{\sigma}{V}\,\frac{f - mp^2}{mp^2 - f - \dfrac{eS^2}{V}}$$

from which we obtain A. We may conveniently put

$$\frac{\sigma}{kV}\,\frac{f - mp^2}{f - mp^2 + \dfrac{eS^2}{V}} = \tan k\alpha$$

when we obtain

$$A = B \operatorname{ctn} k\,(l - \alpha),$$

and with the value found for B

$$\varphi = -\frac{s_0 \cos k\,(x - l + \alpha)}{k \sin k\,(l - \alpha)}, \quad \frac{d\varphi}{dx} = \frac{s_0 \sin k\,(x - l + \alpha)}{\sin k\,(l - \alpha)}.$$

The equations (267) then give for A_2,

$$A_2 = -\frac{s_0\,eS \sin k\alpha}{\sin k\,(l - \alpha)(f - mp^2)}$$

which is the amplitude of the vibration of the diaphragm. Measuring this we find s_0, the condensation of the sound entering the tube, giving the intensity of the sound. By altering the length of the tube, we may by resonance make the instrument as sensitive as desired. (We have omitted the damping, which will in practice set a limit.) The amplitude A_2 will be infinite not only when $\sin k\,(l - \alpha) = 0$, but also when $p^2 = \dfrac{f}{m}$, that is when the terminal apparatus is tuned to resonance.

In all these cases, the integral equation will contain, as we see by comparing with (200), an additional term for each end, so that we have $\varphi' = \varphi f(\lambda)$, $K' = Kf(0)$ at the ends, and

$$(270) \quad \varphi K \{f(0) - f(\lambda)\}\Big|_0^l + \varphi(\xi) - \lambda \int_0^l K(x, \xi)\,\varphi(x)\,dx = f(\xi).$$

Accordingly the normal functions will not be orthogonal, as we have already seen.

40. More general Differential and Integral Equation. In the deduction of the equation of the string, we have assumed the tension τ to be constant. Suppose that it is a function of x, then we have the equation

$$(271) \qquad \varrho\,\frac{\partial^2 y}{\partial t^2} = \frac{\partial}{\partial x}\left(\tau\,\frac{\partial y}{\partial x}\right),$$

and introducing the normal function $y = \varphi(x) \cos pt$,

(272)
$$-\varrho p^2 \varphi = \frac{d}{dx}\left(\tau \frac{d\varphi}{dx}\right)$$

and we have the equation of the normal functions

(273)
$$\frac{d}{dx}\left(\tau(x)\frac{d\varphi}{dx}\right) + \lambda\varphi = 0, \text{ where } \lambda = \varrho p^2.$$

In the case of a heavy chain hanging vertically from one end, we have $\tau = \varrho gx$, where x is measured from the *free* end. We will then put

(274)
$$\frac{d}{dx}\left(x\frac{dy}{dx}\right) + \lambda y = 0, \qquad \lambda = \frac{p^2}{g}.$$

It is easy to show that the general linear equation of order two,

$$P(x)\frac{d^2y}{dx^2} + Q(x)\frac{dy}{dx} + R(x)y = 0,$$

can be put in the form

(275)
$$\frac{d}{dx}\left\{p(x)\frac{dy}{dx}\right\} + q(x)y = 0,$$

for differentiating,

$$p\frac{d^2y}{dx^2} + \frac{dp}{dx}\frac{dy}{dx} + qy = 0,$$

consequently we must have

$$\frac{1}{p}\frac{dp}{dx} = \frac{Q}{P}, \quad \frac{q}{p} = \frac{R}{P},$$

and integrating,

$$p = Ce^{\int \frac{Q}{P}dx}, \quad q = C\frac{R}{p}e^{\int \frac{Q}{P}dx}.$$

Let us call the operator L,

(276)
$$L(y) \equiv \frac{d}{dx}\left(p\frac{dy}{dx}\right) + qy,$$

and let p, p', q be continuous functions in a fundamental interval ab. Let u, v, with their first two derivatives, be continuous in the fundamental interval. Then we have

(277)
$$vL(u) - uL(v) = v\frac{d}{dx}\left(p\frac{du}{dx}\right) - u\frac{d}{dx}\left(p\frac{dv}{dx}\right)$$
$$= \frac{d}{dx}\left(pv\frac{du}{dx} - pu\frac{dv}{dx}\right),$$

and integrating

(278)
$$\int_a^b (vL(u) - uL(v))dx = p\left(v\frac{du}{dx} - u\frac{dv}{dx}\right)\Big|_a^b.$$

This is Green's theorem for one dimension.

Let us define the Green's function for the operator L as the function satisfying the equation $L(u) = 0$, and representing the shape of the string when drawn aside by a unit force at $x = \xi$.

$$p(x)K'(x,\xi)\Big|_{\xi-0}^{\xi+0} = -1,$$

(279)
$$K'(x,\xi)\Big|_{\xi-0}^{\xi+0} = -\frac{1}{p(\xi)}.$$

As before we prove $K(x,\xi) = K(\xi, x)$. We may have various Green's functions for various end conditions, such as $K = 0$, or $K' = 0$ for an end.

In the case of a heavy chain,

(280)
$$p = x, \quad q = 0, \quad L(u) = \frac{d}{dx}\left(x\frac{du}{dx}\right),$$

and if

(281)
$$L(u) = 0, \quad x\frac{du}{dx} = c, \quad \frac{du}{dx} = \frac{c}{x}, \quad u = c\log x + b.$$

In order not to become infinite when $x = 0$ we must have $c = 0$, $x < \xi$, so that we must use the terminal condition $K'(0) = 0$. At the end $x = l$, if $K(l) = 0$, we have

$$c\log l + b = 0, \quad b = -c\log l$$

(282)
$$u = c\log\frac{x}{l}, \quad u' = \frac{c}{x}, \quad u'(\xi) = \frac{c}{\xi} = -\frac{1}{\xi}.$$

Consequently
$$c = -1,$$

and we have

$$K(x,\xi) = -\log\frac{\xi}{l}, \quad x < \xi,$$

(283)
$$K(x,\xi) = -\log\frac{x}{l}, \quad l > x > \xi.$$

For any sources $f(x)$ we obtain as before for the solution of

(284)
$$\frac{d}{dx}\left(p\frac{du}{dx}\right) = -f(x),$$

the integral

(285)
$$u = \int_0^l K(x,\xi)f(\xi)\,d\xi = -\int_0^x f(\xi)\log\frac{x}{l}\,d\xi$$

$$-\int_x^l \log\frac{\xi}{l}f(\xi)\,d(\xi).$$

Consider now the operator,

(286) $$\Lambda \equiv L + \lambda \equiv \frac{d}{dx}\left(p\,\frac{d(\ )}{dx}\right) + (q + \lambda)(\ ),$$

and let Γ be the Green's function for Λ. Then as before we have for the forced oscillation

(287) $$\Lambda\varphi = -f(x)$$

with the solution

(288) $$\varphi(x) = \int_0^l \Gamma(x, \xi) f(\xi)\, d\xi.$$

If $\varphi(x)$ is given, and $f(x)$ is be found, equation (288) is called an integral equation of the *first* kind, and $f(x)$ is found by applying the operation Λ to φ, as shown by (287), always supposing the kernel of the integral equation is the Green's function for Λ.

As before since

$$L(K) = 0,$$

(289) $$L(\varphi) + \lambda\varphi = -f(x),$$

gives as before

$$-\varphi L(K) + KL(\varphi) + \lambda K\varphi = -Kf(x)$$

(290) $$p(K\varphi' - \varphi K')\Big|_{\xi+0}^{\xi-0} + \lambda\int_0^l K\varphi\, dx = -\int_0^l Kf(x)\, dx,$$

and the discontinuity of pK' again gives

(291) $$-\varphi(\xi) + \lambda\int_0^l K(x, \xi)\varphi(x)dx = -\int_0^l K(x, \xi)f(x)dx = -F(\xi)$$

the solution of which we have above. Changing x and ξ,

$$\varphi(x) - \lambda\int_0^l K(x, \xi)\varphi(\xi)d\xi = F(x).$$

Let us now put in Green's theorem

$$u = K(x, \xi), \quad v = \Gamma(x, \eta).$$

Then since

(292) $$\begin{aligned} L(K) &= 0, \quad L(\Gamma) + \lambda\Gamma = 0, \\ \Gamma L(K) &- KL(\Gamma) = \lambda K\Gamma, \end{aligned}$$

and integrating,

(293) $$p(\Gamma K' - K\Gamma')\Big|_{\xi+0}^{\xi-0} + p(\Gamma K' - K\Gamma')\Big|_{\eta+0}^{\eta-0} = \lambda\int_0^l K\Gamma dx,$$

(294) $$\Gamma(\xi, \eta) - K(\eta, \xi) = \lambda\int_0^l K(x, \xi)\Gamma(x, \eta)dx.$$

This equation is symmetrical in K and Γ and is called the *resolvent*. Consequently the function $\Gamma(\xi, \eta)$ satisfies the integral equation with kernel K,

$$(295) \qquad \Gamma(\xi, \eta) - \lambda \int_0^l K(x, \xi)\, \Gamma(x, \eta)\, dx = K(\xi, \eta),$$

whatever the value of η. But in like manner the function $K(\xi, \eta)$ satisfies the integral equation

$$(296) \qquad K(\eta, \xi) + \lambda \int_0^l \Gamma(\eta, x)\, K(x, \xi)\, dx = \Gamma(\eta, \xi),$$

with kernel Γ. These two kernels have the property that either is a *solving* function for the other, that is if

$$(202) \qquad F(x) = \varphi(x) - \lambda \int_0^l K(x, \xi)\, \varphi(\xi)\, d\xi,$$

then we have the solution,

$$(297) \qquad \varphi(x) = F(x) + \lambda \int_0^l \Gamma(x, \xi)\, F(\xi)\, d\xi.$$

This we find by inserting the latter in (202),

$$\varphi(\xi) = F(\xi) + \lambda \int_0^l \Gamma(\xi, \eta)\, F(\eta)\, d\eta,$$

$$(298) \qquad F(x) = F(x) + \lambda \int_0^l \Gamma(x, \xi)\, F(\xi)\, d\xi$$

$$- \lambda \int_0^l K(x, \xi) \left[F(\xi) + \lambda \int_0^l \Gamma(\xi, \eta)\, F(\eta)\, d\eta \right] d\xi,$$

which is an identity, as we see by integrating the resolvent, {equation 294}, after multiplying by $F(\xi)$,

$$(299) \qquad \int_0^l \Gamma(x, \xi)\, F(\xi)\, d\xi - \int_0^l K(x, \xi)\, F(\xi)\, d\xi$$

$$- \lambda \int_0^l F(\xi)\, d\xi \int_0^l K(x, \xi)\, \Gamma(x, \eta)\, dx = 0,$$

which by the symmetry of the kernels is the equivalent of the above. Comparing with Schmidt's solution we find

$$(300) \qquad \Gamma(x, \xi) = \sum_{r=1}^{r=\infty} \frac{\varphi_r(x)\, \varphi_r(\xi)}{\lambda_r - \lambda}.$$

The solving function Γ for the kernel K of (179) is the φ of (197).

Putting (300) and (211) in the equation of the resolvent (294) we find that it is identically satisfied. Now Γ is, by definition (292), a function of λ, and by (300) we see that while it has the same normal functions as K, its characteristic numbers are those of K each diminished by λ.

Returning to the equation for the heavy chain, we had

$$(274) \qquad \frac{d}{dx}\left(x\frac{dy}{dx}\right) + \lambda y = 0, \quad \lambda = \frac{p^2}{g}.$$

We will change the variable by putting $x = z^2$. Then

$$2z\frac{dz}{dx} = 1, \quad \frac{d}{dx} = \frac{dz}{dx}\frac{d}{dz} = \frac{1}{2z}\frac{d}{dz}$$

$$\frac{d}{dx}\left(x\frac{dy}{dx}\right) = \frac{1}{2z}\frac{d}{dz}\left(\frac{z^2}{2z}\frac{dy}{dz}\right) = \frac{1}{4z}\frac{d}{dz}\left(z\frac{dy}{dz}\right),$$

and our equation becomes

$$(301) \qquad \frac{d^2y}{dz^2} + \frac{1}{z}\frac{dy}{dz} + c^2 y = 0, \quad c^2 = 4\lambda,$$

and if we put $cz = \zeta$

$$(302) \qquad \frac{d^2y}{d\zeta^2} + \frac{1}{\zeta}\frac{dy}{d\zeta} + y = 0.$$

This is Bessel's equation, (Chapter VI).

We thus have the solution

$$(303) \qquad y = J_0(c\zeta) = J_0(2\sqrt{\lambda x}).$$

As the point $x = l$ is to be a node, we must have

$$(304) \qquad J_0(2\sqrt{\lambda l}) = 0,$$

which is the transcendental equation for the characteristic numbers λ. Calling a root λ_r we have for the normal functions,

$$(305) \qquad \varphi_r(x) = aJ_0(2\sqrt{\lambda_r x})$$

and the general solution for our equation for the chain is

$$(306) \qquad y = \sum_{r=1}^{r=\infty} A_r \cos(p_r t - \alpha_r)J_0(2\sqrt{\lambda_r x}),$$

where the $\lambda_r's$ are determined by (304). The proof of the orthogonality of the different normal functions has been given for the general case, but may be repeated for this as follows. We have by (274)

$$x\frac{d^2\varphi_r}{dx^2} + \frac{d\varphi_r}{dx} + \lambda_r\varphi_r = 0,$$

$$x\frac{d^2\varphi_s}{dx^2} + \frac{d\varphi_s}{dx} + \lambda_s\varphi_s = 0.$$

Multiply the first by φ_s, the second by φ_r and subtract, giving

$$(307) \qquad \frac{d}{dx}[x(\varphi_s\varphi_r' - \varphi_r\varphi_s')] + (\lambda_r - \lambda_s)\varphi_r\varphi_s = 0.$$

Integrating from 0 to l,

$$(308) \qquad x(\varphi_s\varphi_r' - \varphi_r\varphi_s')\ \Big|_0^l = (\lambda_s - \lambda_r)\int_0^l \varphi_r\varphi_s\,dx.$$

At 0, x vanishes, and at l, φ_r and φ_s vanish. Accordingly if $\lambda_r \neq \lambda_s$, the integral vanishes, and the functions are orthogonal.

The above equation (308) is true for any values of λ, whether or not characteristic values. Let us make λ_r and λ_s approach equality. Put

$$(309) \qquad \lambda_s = \lambda_r + \varepsilon, \quad x(\varphi_{\lambda_r+\varepsilon}\varphi'_{\lambda_r} - \varphi_{\lambda_r}\varphi'_{\lambda_r+\varepsilon})\ \Big|_0^l = \varepsilon\int_0^l \varphi_r\varphi_s\,dx.$$

Now we have

$$(310) \qquad \varphi_{\lambda_r+\varepsilon} = \varphi_r + \varepsilon\frac{\partial\varphi_r}{\partial\lambda}, \quad \varphi'_{\lambda_r+\varepsilon} = \varphi'_r + \varepsilon\frac{\partial\varphi'_r}{\partial\lambda},$$

and the equation (308) becomes, since $\varphi_r' = \dfrac{\partial\varphi_r}{\partial x}$,

$$(311) \qquad x\left(\frac{\partial\varphi_r}{\partial x}\frac{\partial\varphi_r}{\partial\lambda} - \varphi_r\frac{\partial^2\varphi_r}{\partial\lambda\,\partial x}\right)\Big|_0^l = \int_0^l \varphi_r^2\,dx.$$

At the lower limit, x vanishes, and at the upper φ_r vanishes if a normal function. Consequently the integrated square is equal merely to the product,

$$l\left(\frac{\partial\varphi_r}{\partial x}\frac{\partial\varphi_r}{\partial\lambda}\right)_{x=l}$$

But introducing the value

$$(305) \qquad \varphi_r = a J_0\left(2\sqrt{\lambda_r\,x}\right),$$

which is a symmetrical function of x and λ, we have

$$(312) \qquad \begin{aligned} \frac{\partial\varphi_r}{\partial x} &= a J_0'\left(2\sqrt{\lambda_r x}\right)\sqrt{\frac{\lambda_r}{x}}, \\[2mm] \frac{\partial\varphi_r}{\partial\lambda} &= a J_0'\left(2\sqrt{\lambda_r x}\right)\sqrt{\frac{x}{\lambda_r}}, \end{aligned}$$

and finally

$$(313) \qquad \int_0^l \varphi_r^2\,dx = la^2 J_0'^2\left(2\sqrt{\lambda_r\,l}\right).$$

If the functions are to be normalized this must equal unity, so that

$$a = \frac{1}{\sqrt{l}\,J_0'\left(2\sqrt{\lambda_r l}\right)}$$

and the normal function is

(314)
$$\varphi_r = \frac{J_0\left(2\sqrt{\lambda_r x}\right)}{\sqrt{l}\, J_0'\left(2\sqrt{\lambda_r l}\right)}.$$

The bilinear formula is accordingly

(315)
$$K(x, \xi) = \frac{1}{2}\sum_{r=1}^{r=\infty} \frac{J_0\left(2\sqrt{\lambda_r x}\right) J_0\left(2\sqrt{\lambda_r \xi}\right)}{\lambda_r J_0'^2\left(2\sqrt{\lambda_r l}\right)}.$$

As a third example let us take the case of a massive chain revolving about an axis at right angles to its length with angular velocity ω. If x be measured from the axis, the centrifugal force on an element of mass $\varrho\, dx$ is $\omega^2 x \varrho\, dx$, and accordingly the tension due to the centrifugal force at x, due to the chain from x to l, is

(316)
$$\tau = \varrho\, \omega^2 \int_x^l x\, dx = \tfrac{1}{2}\, \varrho\, \omega^2 (l^2 - x^2).$$

Accordingly the differential equation for the vibrations of the chain is

(317)
$$\varrho\, \frac{\partial^2 y}{\partial t^2} = \tfrac{1}{2}\, \varrho\, \omega^2 \frac{\partial}{\partial x}\left\{(l^2 - x^2)\frac{\partial y}{\partial x}\right\},$$

or, putting

$$t' = \frac{\omega t}{\sqrt{2}}, \quad x' = \frac{x}{l}$$

(318)
$$\frac{\partial^2 y}{\partial t'^2} = \frac{\partial}{\partial x'}\left\{(1 - x'^2)\frac{\partial y}{\partial x'}\right\}.$$

We shall suppose the units so taken, and drop the accents. Putting

$$y = \varphi(x)\cos pt,$$

we have

(319)
$$\frac{d}{dx}\left\{(1 - x^2)\frac{d\varphi}{dx}\right\} + \lambda\varphi = 0, \quad \lambda = p^2.$$

This is Legendre's equation, and if λ has the proper values, it has as a solution the Legendre Polynomial $P_n(x)$ (Chapter VI).

The Green's function for

(320)
$$L(u) \equiv \frac{d}{dx}\left\{(1 - x^2)\frac{du}{dx}\right\}$$

is given by integrating

(321)
$$\frac{d}{dx}\left\{(1 - x^2)\frac{dK}{dx}\right\} = 0,$$

$$(1 - x^2)\frac{dK}{dx} = c, \quad \frac{dK}{dx} = \frac{c}{1 - x^2},$$

$$K = \frac{c}{2}\log\frac{1 + x}{1 - x} + b. \quad \text{If } K(0) = 0,\ b = 0.$$

Consequently, for

(322)

$$x < \xi, \ K = \frac{c}{2} \log \frac{1+x}{1-x},$$

$$x > \xi, \ K = \frac{c}{2} \log \frac{1+\xi}{1-\xi},$$

and to make

(323)

$$K'\ \Big|_{\xi-0}^{\xi+0} = -\frac{1}{1-\xi^2},$$

we must have

$$\frac{c}{1-\xi^2} = \frac{1}{1-\xi^2},$$

so that

(324)

$$K = \frac{1}{2} \log \frac{1+x}{1-x}, \quad x < \xi,$$

$$K = \frac{1}{2} \log \frac{1+\xi}{1-\xi}, \quad x > \xi.$$

The solution of the equation

$$\varLambda(u) = 0,$$

is a Legendre's polynomial if we take $\lambda_r = r(r+1)$ and if it is to vanish for $x = 0$, r must be odd.

41. Transverse Vibrations of a Bar. We shall consider as a final example an equation of the fourth order, that for the transverse vibrations of a bar, (147) Chapter I.

(325)
$$\varrho S \frac{\partial^2 u}{\partial t^2} + EI \frac{\partial^4 u}{\partial x^4} = 0.$$

We have no solution of this equation of the form of d'Alembert's solution for the string, but we may consider the propagation of a simple harmonic wave. If we put

$$u = \cos p\left(t - \frac{x}{v}\right)$$

we find

$$\frac{\partial^2 u}{\partial t^2} = -p^2 u, \quad \frac{\partial^4 u}{\partial x^4} = \frac{p^4}{v^4} u,$$

so that we must take

$$-\varrho S p^2 + EI \frac{p^4}{v^4} = 0,$$

(326)
$$v = \sqrt[4]{\frac{EI}{\varrho S}} \sqrt{p}.$$

The velocity of propagation is not constant, as in the case of the string, but varies as the square root of the frequency. Accordingly a disturbance represented by a Fourier's series has all the terms propagated with different velocities, so that the form is at once distorted.

We may however treat the case of standing waves. To find the normal vibrations, put $u = \varphi \cos pt$, giving

(327) $$\frac{d^4\varphi}{dx^4} - k^4\varphi = 0, \quad k^4 = \frac{\varrho S}{EI}p^2 = \lambda.$$

The general solution of (327) is

(328) $$\varphi = A\cos kx + B\sin kx + C\cosh kx + D\sinh kx.$$

The usual terminal conditions are either, if the end of the bar is free, that it has neither shearing force T nor moment G applied, by Chapter I, (140), (144)

(329) $$\frac{d^2\varphi}{dx^2} = 0, \quad \frac{d^3\varphi}{dx^3} = 0,$$

or if the end is clamped, so that its tangent is invariable

(330) $$\varphi = 0, \quad \frac{d\varphi}{dx} = 0.$$

Suppose first the bar is free at both ends. Then since

$$\frac{d^2\varphi}{dx^2} = k^2\left\{ - A\cos kx - B\sin kx + C\cosh kx + D\sinh kx \right\}$$

the first condition for $x = 0$ makes $C = A$.

Since

$$\frac{d^3\varphi}{dx^3} = k^3\left\{ A\sin kx - B\cos kx + A\sinh kx + D\cosh kx \right\}$$

the second condition makes $D = B$.

For $x = l$, the conditions are

(331)
$$A(\cosh kl - \cos kl) + B(\sinh kl - \sin kl) = 0,$$
$$A(\sinh kl + \sin kl) + B(\cosh kl - \cos kl) = 0,$$

or eliminating A and B

(332) $$(\cosh kl - \cos kl)^2 = \sinh^2 kl - \sin^2 kl.$$

But since we have the identical relation

$$\cosh^2 kl - \sinh^2 kl = 1,$$

this reduces to

(333) $$\cos kl \cosh kl = 1.$$

This is the equation for the frequencies, whose roots in k give us the normal functions

(334) $$\varphi_r = A_r(\cos k_r x + \cosh k_r x) + B_r(\sin k_r x + \sinh k_r x).$$

The equation (333) may be easily solved with the help of a table of trigonometric and hyperbolic cosines, or by means of the graphical construction of the curves

(335) $$y = \cos x, \quad y = \frac{1}{\cosh x}$$

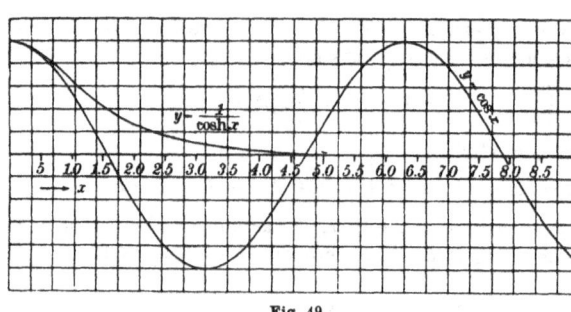

Fig. 49.

whose intersections give by their abscissas the values of $k_r l$. Inspection of Fig. 49 shows that the values approach the roots of the equation $\cos kl = 0$. The frequencies $\frac{p_r}{2\pi}$ are proportional to k_r^2. The values of the roots of (333) are given by Lord Rayleigh,

$$x_1 = 4.7300408,$$
$$x_2 = 7.8532046,$$
$$x_3 = 10.9956078,$$
$$x_4 = 14.1371655,$$
$$x_5 = 17.2787596,$$

after which $x_r = \frac{(2r+1)\pi}{2}$ with an accuracy of seven decimal places.

If the bar is clamped at both ends we obtain the same frequencies, being thus reminded of the open and closed pipe. But if the bar is free at one end and clamped at the other, $x = l$, we have as before

$$(336) \quad \begin{aligned} \varphi &= A(\cos kx + \cosh kx) + B(\sin kx + \sinh kx), \\ \frac{d\varphi}{dx} &= k\{A(-\sin kx + \sinh kx) + B(\cos kx + \cosh kx)\} \end{aligned}$$

while instead of the conditions (329) we have (330),

$$(337) \quad \begin{aligned} A(\cos kl + \cosh kl) + B(\sin kl + \sinh kl) &= 0, \\ A(-\sin kl + \sinh kl) + B(\cos kl + \cosh kl) &= 0, \end{aligned}$$

and eliminating A and B,

$$(338) \quad (\cos kl + \cosh kl)^2 = \sinh^2 kl - \sin^2 kl,$$

or as before

$$(339) \quad \cos kl \cosh kl = -1.$$

By consulting the figure, we see that all roots after the two lowest are nearly the same as before. Lord Rayleigh gives

$$x_1 = 1.875104,$$
$$x_2 = 4.694098,$$
$$x_3 = 7.854757,$$
$$x_4 = 10.995541,$$
$$x_5 = 14.137168,$$
$$x_6 = 17.278759.$$

Any two normal functions may be proved orthogonal by means of the differential equation (327), for since

$$(327) \qquad \frac{d^4\varphi_r}{dx^4} = \lambda_r \varphi_r, \quad \frac{d^4\varphi_s}{dx^4} = \lambda_s \varphi_s,$$

we have

$$(340) \qquad (\lambda_r - \lambda_s)\int_0^l \varphi_r \varphi_s \, dx = \int_0^l \left(\varphi_s \frac{d^4\varphi_r}{dx^4} - \varphi_r \frac{d^4\varphi_s}{dx^4} \right) dx$$

$$= \left\{ \varphi_s \frac{d^3\varphi_r}{dx^3} - \varphi_r \frac{d^3\varphi_s}{dx^3} + \frac{d\varphi_r}{dx}\frac{d^2\varphi_s}{dx^2} - \frac{d\varphi_s}{dx}\frac{d^2\varphi_r}{dx^2} \right\} \Big|_0^l.$$

Now whether the ends be clamped or free, every term vanishes at the limits, so that if $\lambda_r \neq \lambda_s$ the functions are orthogonal. We may use the same formula to find the values of the integrated squares for normalization.

As in (309) *et seq.* let λ_r approach λ_s. Then we have

$$(341) \qquad 4k^3\int_0^l \varphi_r^2 dx = \left\{ \varphi \frac{\partial^4\varphi}{\partial k \partial x^3} - \frac{\partial\varphi}{\partial k}\frac{\partial^3\varphi}{\partial x^3} + \frac{\partial^2\varphi}{\partial k \partial x}\frac{\partial^2\varphi}{\partial x^2} - \frac{\partial^3\varphi}{\partial k \partial x^2}\frac{\partial\varphi}{\partial x} \right\} \Big|_0^l.$$

Now since φ is a function of kx, denoting derivatives by accents,

$$\frac{\partial\varphi}{\partial x} = k\varphi', \quad \frac{\partial\varphi}{\partial k} = x\varphi', \quad \frac{\partial^2\varphi}{\partial x^2} = k^2\varphi'', \quad \frac{\partial^3\varphi}{\partial x^3} = k^3\varphi''',$$

$$\frac{\partial^2\varphi}{\partial k \partial x} = \varphi' + kx\varphi'', \qquad \frac{\partial^3\varphi}{\partial k \partial x^2} = 2k\varphi'' + k^2 x\varphi''',$$

$$\frac{\partial^4\varphi}{\partial k \partial x^3} = 3k^2\varphi''' + k^3 x\varphi^{iv} = 3k^2\varphi''' + k^3 x\varphi,$$

$$4k^3\int_0^l \varphi_r^2 dx =$$

$$= \{ 3k^2\varphi\varphi''' + k^3 x\varphi^2 - k^3 x\varphi'\varphi''' + k^2\varphi''\varphi' + k^3 x\varphi''^2 - 2k^2\varphi'\varphi'' - k^3 x\varphi'\varphi''' \} \Big|_0^l.$$

Now the terms in x vanish at one end, and whether the ends be free or clamped, both $\varphi\varphi'''$ and $\varphi'\varphi''$ vanish, accordingly for all conditions

$$(342) \qquad \int_0^l \varphi_r^2 dx = \frac{l}{4} (\varphi^2 - \varphi'\varphi''' + \varphi''^2)_{x=l},$$

which simplifies on specifying the end conditions.

Let us now treat the forced vibrations, putting

$$(343) \qquad \varrho S\frac{\partial^2 u}{\partial t^2} + EI\frac{\partial^4 u}{\partial x^4} = EIf(x)\cos pt,$$

which gives as before

$$(344) \qquad \frac{d^4\varphi}{dx^4} - \lambda\varphi = f(x).$$

Let us first consider the case of equilibrium $\lambda = 0$,

$$(345) \qquad \frac{d^4\varphi}{dx^4} = f(x),$$

and consider a point-load by making $f(x)$ the jag-function. Consider the ends clamped and a unit load at $x = \xi$. We thus define the Green's function $K(x, \xi)$ as a solution of

$$(346) \qquad \frac{d^4K}{dx^4} = 0, \quad K(0, \xi) = \frac{\partial K}{\partial x}(0, \xi) = 0,$$

with a discontinuity in the *third* derivative defined by

$$(347) \qquad \frac{d^3K}{dx^3}\Big|_{\xi-0}^{\xi+0} = -1.$$

We have then from (344) and (346)

$$(348) \qquad K\frac{d^4\varphi}{dx^4} - \varphi\frac{d^4\varphi}{dx^4} - \lambda K\varphi = Kf(x)$$

and integrating

$$(349)\int_0^l\left(K\frac{d^4\varphi}{dx^4} - \varphi\frac{d^4\varphi}{dx^4}\right)dx - \lambda\int_0^l K(x, \xi)\,\varphi(x)dx = \int_0^l K(x, \xi)f(x)dx,$$

and since the first integral is the derivative of

$$K\frac{d^3\varphi}{dx^3} - \varphi\frac{d^3K}{dx^3} + \frac{d\varphi}{dx}\frac{d^2K}{dx^2} - \frac{dK}{dx}\frac{d^2\varphi}{dx^2},$$

of which every term vanishes at the limits, we have only to consider the discontinuity in d^3K/dx^3 at $x = \xi$, so that finally we obtain

$$(350) \qquad \varphi(\xi) - \lambda\int_0^l K(x, \xi)\varphi(x)dx = \int_0^l K(x, \xi)f(x)dx = F(\xi),$$

the integral equation of the usual sort. By means of it we may treat the question of development in normal functions of the sort in question.

CHAPTER IV.

FOURIER'S SERIES AND INTEGRAL. CAUCHY'S METHOD.
INITIAL DATA.

42. Convergence of Fourier's Series. In the preceding chapter we have determined the coefficients in the trigonometric series by passing to the limit from the case of systems of a finite number of degrees of freedom. Although this process seems satisfactory to the intuition of the physicist, it by no means satisfies the mathematician, who demands proof of the possibility of thus passing to the limit. In all the cases of developments in normal functions, we have had to assume that the series, on being multiplied by the function to be developed, can be integrated term by term. The question of the possibility of this will be taken up in Chapter IX, in the meantime we shall give Dirichlet's[1] celebrated treatment of the question for Fourier's series, which has had a most extraordinary influence on all modern mathematics.

Suppose that we consider the series

$$(1) \qquad S(x) = \tfrac{1}{2} A_0 + \sum_{1}^{\infty} (A_r \cos rx + B_r \sin rx)$$

where the A's and B's have the values

$$(2) \qquad A_r = \frac{1}{\pi} \int_{-\pi}^{\pi} f(\alpha) \cos r\alpha\, d\alpha, \quad B_r = \frac{1}{\pi} \int_{-\pi}^{\pi} f(\alpha) \sin r\alpha\, d\alpha.$$

Let us introduce these values into the series, and write for the first m terms, putting the functions of x under the integral sign

$$(3) \qquad S_m(x) = \frac{1}{\pi} \int_{-\pi}^{\pi} \left[\frac{1}{2} + \sum_{1}^{m} (\cos rx \cos r\alpha + \sin rx \sin r\alpha) \right] f(\alpha)\, d\alpha$$

$$= \frac{1}{\pi} \int_{-\pi}^{\pi} \left[\frac{1}{2} + \sum_{1}^{m} \cos r\,(\alpha - x) \right] f(\alpha)\, d\alpha.$$

We require the limit approached by S_m as m increases without limit. Now by Chapter III (123),

$$(4) \qquad \frac{1}{2} + \sum_{1}^{m} \cos r\,(\alpha - x) = \frac{\sin (2m+1)\dfrac{\alpha - x}{2}}{2 \sin \dfrac{\alpha - x}{2}}$$

so that we have

$$(5) \qquad S_m(x) = \frac{1}{\pi} \int_{-\pi}^{\pi} \frac{\sin (2m+1)\dfrac{\alpha - x}{2}}{2 \sin \dfrac{\alpha - x}{2}} f(\alpha)\, d\alpha$$

or putting $\dfrac{(\alpha - x)}{2} = \beta$,

$$(6) \qquad S_m(x) = \frac{1}{\pi} \int_{-\frac{(\pi + x)}{2}}^{\frac{(\pi - x)}{2}} \frac{\sin (2m+1)\beta}{\sin \beta} f(x + 2\beta)\, d\beta.$$

---

1) *Crelle's Journal*, Bd. IV., Sur la convergence des séries trigonometriques, qui servent à représenter une fonction arbitraire entre des limites données.

If we suppose that x lies within the fundamental interval,

$$-\pi < x < \pi,$$

the limits in (6) lie between $-\pi$ and π.

We shall make the limit of S_m depend on the following, known as Dirichlet's integral,

$$(7) \qquad J = \int_0^b \varphi(x) \frac{\sin kx}{\sin x} \, dx, \qquad 0 < b < \frac{\pi}{2}.$$

In our case k is an odd integer which is to increase indefinitely. All that we have to say will apply equally well if we have x instead of $\sin x$ in the denominator,

$$(8) \qquad J' = \int_0^b \varphi(x) \frac{\sin kx}{x} \, dx.$$

Consider first the integral

$$(9) \quad \int_0^{\frac{\pi}{2}} \frac{\sin(2m+1)x}{\sin x} \, dx = 2 \int_0^{\frac{\pi}{2}} \left(\frac{1}{2} + \cos 2x + \cos 4x + \cdots + \cos 2mx \right) dx = \frac{\pi}{2},$$

since the integral of every term but the first vanishes. Suppose that in J the function $\varphi(x)$ is finite, continuous, positive, and never *increases* with x, it is then said to be monotonically decreasing. The integrand vanishes when $x = \frac{r\pi}{k}$, where r is any integer, let us accordingly divide the range of integration into parts

$$0, \quad \frac{\pi}{k}, \quad \frac{2\pi}{k}, \quad \cdots \quad \frac{p\pi}{k}, \quad b,$$

$$(10) \qquad J = \int_0^{\frac{\pi}{k}} \varphi(x) \frac{\sin kx}{\sin x} \, dx + \int_{\frac{\pi}{k}}^{\frac{2\pi}{k}} + \int_{\frac{2\pi}{k}}^{\frac{3\pi}{k}} + \cdots + \int_{\frac{p\pi}{k}}^{b}.$$

The terms are alternately positive and negative, and decrease in absolute value. For $\sin x$ (or x in J') is continually increasing, and $\varphi(x)$ is not increasing, whereas $\sin kx$ repeats its values in successive intervals with alternating signs. Thus we may write,

$$(11) \qquad J = u_0 - u_1 + u_2 - \cdots$$

where

$$(12) \qquad u_r = (-1)^r \int_{\frac{r\pi}{k}}^{(r+1)\frac{\pi}{k}} \varphi(x) \frac{\sin kx}{\sin x} \, dx$$

In like manner, calling K the integral with $\varphi(x)$ replaced by 1. and $b = \frac{\pi}{2}$

$$(13) \qquad K = \varrho_0 - \varrho_1 + \varrho_2 - \cdots = \frac{\pi}{2},$$

$$\varrho_r = (-1)^r \int_{\frac{r\pi}{k}}^{(r+1)\frac{\pi}{k}} \frac{\sin kx}{\sin x}\, dx.$$

Now in a series where the terms alternate in sign, and each is less in absolute value than the preceding, if we stop at any term the error committed is of the same sign as the last term. Therefore

$$(14) \qquad u_0 - u_1 + \cdots - u_{2r-1} < J < u_0 - u_1 + \cdots + u_{2r},$$

$$(15) \qquad \varrho_0 - \varrho_1 + \cdots - \varrho_{2r-1} < \frac{\pi}{2} < \varrho_0 - \varrho_1 + \cdots + \varrho_{2r}.$$

Now in u_r replacing $\varphi(x)$ by the greatest and least values it has in the rth interval, makes

$$(16) \qquad \varphi\left(\frac{r\pi}{k}\right)\varrho_r > u_r > \varphi\left(\frac{(r+1)\pi}{k}\right)\varrho_r.$$

In (14) replace on the left the even terms by smaller and the odd by larger, on the right the opposite, so that the inequality is emphasized.

$$(17) \quad \varphi\left(\frac{\pi}{k}\right)(\varrho_0 - \varrho_1) + \varphi\left(\frac{3\pi}{k}\right)(\varrho_2 - \varrho_3) \cdots + \varphi\left(\frac{(2r+1)\pi}{k}\right)(\varrho_{2r-2} - \varrho_{2r-1}) < J,$$

$$(18) \quad J < \varphi(0)\varrho_0 - \varphi\left(\frac{2\pi}{k}\right)(\varrho_1 - \varrho_2) \cdots - \varphi\left(\frac{2r\pi}{k}\right)(\varrho_{2r-1} - \varrho_{2r}).$$

Since $\varphi\left(\frac{2r\pi}{k}\right)$ is the smallest φ, we have *a fortiori*,

$$J < \varphi(0)\varrho_0 - \varphi\left(\frac{2r\pi}{k}\right)(\varrho_1 - \varrho_2 + \varrho_3 \cdots + \varrho_{2r-1} - \varrho_{2r})$$

or, adding and subtracting $\varphi\left(\frac{2r\pi}{k}\right)\varrho_0$

$$(19) \quad J < \left[\varphi(0) - \varphi\left(\frac{2r\pi}{k}\right)\right]\varrho_0 + \varphi\left(\frac{2r\pi}{k}\right)(\varrho_0 - \varrho_1 + \cdots + \varrho_{2r}).$$

In like manner (17) gives, adding and subtracting $\varphi\left(\frac{2r\pi}{k}\right)\varrho_{2r}$,

$$(20) \qquad J > \varphi\left(\frac{2r\pi}{k}\right)(\varrho_0 - \varrho_1 + \cdots + \varrho_{2r}) - \varphi\left(\frac{2r\pi}{k}\right)\varrho_{2r}.$$

But since the two sums $\varrho_0 \cdots \varrho_{2r}$, $\varrho_0 \cdots \varrho_{2r-1}$ are respectively greater and less than $\frac{\pi}{2}$

$$(21) \qquad J > \varphi\left(\frac{2r\pi}{k}\right)\frac{\pi}{2} - \varphi\left(\frac{2r\pi}{k}\right)\varrho_{2r}$$

(also subtracting and adding $\varphi\left(\frac{2r\pi}{k}\right)\varrho_{2r}$ from (19)

$$(22) \qquad J < \left[\varphi(0) - \varphi\left(\frac{2r\pi}{k}\right)\right]\varrho_0 + \varphi\left(\frac{2r\pi}{k}\right)\frac{\pi}{2} + \varphi\left(\frac{2r\pi}{k}\right)\varrho_{2r}.$$

Now if k increase without limit, we may calculate the limit approached by ϱ_{2r},

$$(23) \qquad \varrho_r = (-1)^r \int_{\frac{r\pi}{x}}^{(r+1)\frac{\pi}{k}} \frac{\sin kx}{\sin x}\,dx < \frac{(-1)^r}{\sin\frac{r\pi}{k}} \int_{\frac{r\pi}{k}}^{(r+1)\frac{\pi}{k}} \sin kx\,dx < \frac{(-1)^r}{\sin\frac{r\pi}{k}}\frac{2}{k}$$

$$|\varrho_r| < \frac{\frac{r\pi}{k}}{\sin\frac{r\pi}{k}}\frac{2}{r\pi}.$$

Now let k increase without limit, and at the same time let r increase without limit, but so that $\lim\limits_{k=\infty}\frac{r}{k} = 0$ (e. g. let r be the greatest integer in $\sqrt{k}$). Then ϱ_r and also ϱ_{2r} tend toward 0. Moreover ϱ_0 is finite, since

$$\varrho_0 - \varrho_1 < \frac{\pi}{2},$$

$$\varrho_0 < \frac{\pi}{2} + \varrho_1 < \frac{\pi}{2} + \frac{\frac{\pi}{k}}{\sin\frac{\pi}{k}}\frac{2}{\pi}.$$

Thus J remains between two quantities tending both toward the same limit $\frac{\pi}{2}\varphi(0)$.

Accordingly

$$(24) \qquad \lim_{k=\infty}\int_0^b \varphi(x)\frac{\sin kx}{\sin x}\,dx = \frac{\pi}{2}\varphi(+0)$$

for no matter what value of b, if $0 < b < \frac{\pi}{2}$.

We may now remove the restrictions we have made as to the nature of $\varphi(x)$. We have supposed it positive. If it is not, let us add to it a positive constant C large enough to make $\varphi(x) + C$ positive in the field of integration. Then the theorem is true for this function, and being true for a constant, it is also true for $\varphi(x)$.

Secondly, we have supposed φ is monotonically decreasing. If it is monotonically increasing, apply the theorem to $-\varphi(x)$ which is monotonically decreasing.

Further consider the integral

$$\int_a^b \varphi(x)\frac{\sin kx}{\sin x}\,dx, \quad 0 < a < b \leqq \frac{\pi}{2}.$$

If φ is monotonic from a to b the limit for $k = \infty$ is zero, for taking a function $\varphi_1(x)$ which has the constant value $\varphi(a)$ from 0 to a and $\varphi_1 = \varphi$ from a to b, we have as before

$$\lim_{k=\infty}\int_0^a \varphi_1(x)\frac{\sin kx}{\sin x}\,dx = \lim_{k=\infty}\int_0^b \varphi_1(x)\frac{\sin kx}{\sin x}\,dx = \frac{\pi}{2}\varphi_1(+0),$$

consequently taking the difference,

$$(25) \qquad\qquad \lim_{k=\infty}\int_a^b \varphi(x)\frac{\sin kx}{\sin x}\,dx = 0.$$

Now we may remove the restriction of monotonic variation. For if the function has a finite number of maxima and minima from a to b it is monotonic between any two adjacent ones, and the integral accordingly vanishes, except for the first integral, which equals $\frac{\pi}{2}\varphi(+0)$ (such functions as $\sin\frac{1}{x}$, $x\sin\frac{1}{x}$ are excluded).

We have then proved that

$$(24) \qquad\qquad \lim_{k=\infty}\int_0^b \varphi(x)\frac{\sin kx}{\sin x}\,dx = \frac{\pi}{2}\varphi(+0)$$

if φ is *continuous* from 0 to b, and has only a *finite* number of maxima and minima.

Suppose now

$$\frac{\pi}{2} < b < \pi.$$

Put

$$J = \int_0^b = \int_0^{\frac{\pi}{2}} + \int_{\frac{\pi}{2}}^b = J_1 + J_2,$$

and in the second integral J_2 put $x = \pi - x'$.

$$J_2 = -\int_{\frac{\pi}{2}}^{\pi-b}\frac{\sin k(\pi - x')}{\sin(\pi - x')}\varphi(\pi - x')\,dx' = \int_{\pi-b}^{\frac{\pi}{2}}\frac{\sin kx'}{\sin x'}\varphi(\pi - x')\,dx'$$

whose limit is zero. But if $b = \pi$,

$$\lim_{k=\infty} J_2 = \lim_{k=\infty} \int_0^{\frac{\pi}{2}} \frac{\sin kx'}{\sin x'} \varphi(\pi - x') dx' = \varphi(\pi - 0)$$

and we have

(26) $$\lim_{k=\infty} \int_0^\pi \frac{\sin kx}{\sin x} \varphi(x) dx = \frac{\pi}{2}[\varphi(+0) + \varphi(\pi - 0)].$$

It may be that $\varphi(x)$, supposed continuous *from* 0 to b, is discontinuous at 0, that is $\lim_{\varepsilon=0} \varphi(\varepsilon)$ is not equal to $\varphi(0)$ or to $\lim_{\varepsilon=0} \varphi(-\varepsilon)$. Then we must use instead of $\varphi(0)$ the value $\lim_{\varepsilon=0} \varphi(\varepsilon)$, or as we write $\varphi(+0)$. Also if $b = \pi$ we must use in case of a discontinuity there

$$\lim_{\varepsilon=0} \varphi(\pi - \varepsilon) = \varphi(\pi - 0).$$

It is also evident that $\varphi(x)$ may have a finite number of discontinuities of the nature of sudden jumps.

The conditions that we have found, which are *sufficient* (though not necessary) for the behavior of the limit, are known as Dirichlet's conditions, viz.:

The function remains finite in the interval, is in general continuous, but may have a *finite* number of sudden discontinuities, and a finite number of maxima and minima.

We may now find the limit approached by the Fourier's series. We had

(6) $$S_m(x) = \frac{1}{\pi} \int_{-\frac{(\pi+x)}{2}}^{\frac{(\pi-x)}{2}} \frac{\sin(2m+1)\beta}{\sin \beta} f(x + 2\beta) d\beta.$$

Supposing that $f(x)$ satisfies Dirichlet's conditions from $-\pi$ to π. Suppose x is neither $-\pi$ nor π.

(27) $$S_m(x) = \frac{1}{\pi} \int_{-(\pi-x)}^{0} + \frac{1}{\pi} \int_0^{\frac{(\pi-x)}{2}}.$$

The second integral has as limit for $m = \infty$ the value $\frac{\pi}{2} f(x + 0)$. In the first put $\beta' = -\beta$, and inverting the limits of integration we have

(28) $$\lim_{m=\infty} \int_0^{\frac{(\pi+x)}{2}} \frac{\sin(2m+1)\beta'}{\sin \beta'} f(x - 2\beta') d\beta' = \frac{\pi}{2} f(x - 0).$$

Consequently

(29) $$\lim_{m=\infty} S_m(x) = \tfrac{1}{2}[f(x+0) + f(x-0)]$$

and for every point where $f(x)$ is continuous

(30) $$\lim_{m=\infty} S_m(x) = f(x).$$

If $x = -\pi$, by (26),

(31) $$\lim_{m=\infty} S_m(-\pi) = \frac{1}{\pi} \int_0^\pi \frac{\sin(2m+1)\beta}{\sin\beta} f(-\pi + 2\beta) d\beta$$

$$= \frac{\pi}{2}[f(-\pi+0) + f(\pi-0)].$$

If $x = \pi$,

(32) $$\lim_{m=\infty} S_m(\pi) = \frac{1}{\pi} \int_{-\pi}^0 \frac{\sin(2m+1)\beta}{\sin\beta} f(\pi+2\beta) d\beta$$

$$= \int_0^\pi \frac{\sin(2m+1)\beta'}{\sin\beta'} f(\pi-2\beta') d\beta' = \frac{\pi}{2}[f(\pi-0) + f(-\pi+0)].$$

Thus whether the function is discontinuous or not the limit approached by $S_m(x)$ is the mean of values on both sides of the point.

It is interesting to form an idea of the magnitude of the coefficients. We have

(33) $$\frac{d}{dx}(f(x)\cos rx) = f'(x)\cos rx - r\sin rx f(x),$$

$$\frac{d}{dx}(f(x)\sin rx) = f'(x)\sin rx + r\cos rx f(x).$$

If we integrate these equations, and f becomes discontinuous for $x = o$, we have on the left

(34) $$[f(a-0) - f(a+0)]\cos ra,$$

$$[f(a-0) - f(a+0)]\sin ra,$$

and as r increases without limit these oscillate, but remain between finite limits. Also since

$$\int_{-\pi}^\pi f'(x) dx$$

is finite, so are the integrals

$$\int_{-\pi}^\pi f'(x) \cos rx \, dx \quad \text{and} \quad \int_{-\pi}^\pi f'(x) \sin rx \, dx.$$

But the integrals of the last terms in (33), are $-\pi r B_r$ and $\pi r A_r$. Consequently when $f(x)$ is *finite* in the interval, $r A_r$ and $r B_r$ remain finite as r increases. This decrease like $\frac{1}{r}$ is not sufficient to make the series converge — the convergence is therefore generally due to the changes of sign of the terms.

If $f(x)$ is *continuous* in the whole interval, the bracketed terms (34) vanish, and (if $f(\pi) = f(-\pi)$)

$$r A_r = -\frac{1}{\pi}\int_{-\pi}^{\pi} f'(x) \cos rx \, dx \,,$$

(35)

$$r B_r = \frac{1}{\pi}\int_{-\pi}^{\pi} f'(x) \sin rx \, dx \,.$$

If then $f(x)$ is continuous and $f'(x)$ finite, we can replace f by f' above and thus prove that under these circumstances $r^2 A_r$, $r^2 B_r$ are finite, so that the series is *absolutely* convergent.

In case $f(x) = f(-x)$,

$$\int_{-\pi}^{0} f(x) \cos rx \, dx = \int_{0}^{\pi} f(x) \cos rx \, dx,$$

$$\int_{-\pi}^{0} f(x) \sin rx \, dx = -\int_{0}^{\pi} f(x) \sin rx \, dx,$$

and thus

(36)
$$A_r = \frac{2}{\pi}\int_{0}^{\pi} f(x) \cos rx \, dx, \quad B_r = 0 \,.$$

Accordingly an even function is developable in a *cosine* series. It is evident that a cosine series is always an even function.

If $f(x) = -f(-x)$ we have in like manner,

(37)
$$B_r = \frac{2}{\pi}\int_{0}^{\pi} f(x) \sin rx \, dx, \quad A_r = 0 \,,$$

and the odd function is developed in a *sine* series. Conversely a sine series is always an odd function.

The coefficients as written in (2) give the development,

$$f(x) = \frac{1}{2} A_0 + \sum_{1}^{\infty} (A_r \cos rx + B_r \sin rx),$$

valid between $x = -\pi$ and $x = \pi$.

We will now consider a few simple examples of Fourier's series. Let us develop a constant, say unity, in a sine series valid from 0 to π. We have

$$(38)\quad B_r = \frac{2}{\pi}\int_0^\pi \sin r x\, dx = \frac{2}{\pi r}(1 - \cos r\pi) = \frac{4}{\pi r},\ r\ \text{odd, or } 0,\ r\ \text{even}.$$

$$1 = \frac{4}{\pi}\left(\frac{\sin x}{1} + \frac{\sin 3x}{3} + \frac{\sin 5x}{5} + \cdots\right).$$

Since the series represents an odd function, for $-\pi < x < 0$ it represents -1. Consequently the function is discontinuous at $x = 0$. The periodicity of the function represented by the series is shown in Fig. 50. If on the contrary we attempt to develop unity into a cosine series we have

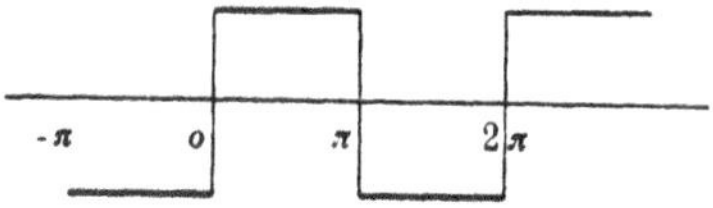

Fig. 50.

$$A_0 = \frac{2}{\pi}\int_0^\pi dx = 2,\quad A_r = \frac{2}{\pi}\int_0^\pi \cos r x\, dx = \frac{2}{\pi r}(\sin \pi - \sin 0) = 0,$$

so that all terms vanish except the constant one.

The function x may be represented by either a sine or a cosine series. In the first case

$$(39)\quad\begin{aligned} B_r &= \frac{2}{\pi}\int_0^\pi x \sin r x\, dx = \frac{2}{\pi}\left\{-\frac{x\cos rx}{r}\Big|_0^\pi + \frac{1}{r}\int_0^\pi \cos r x\, dx\right\} \\ &= -\frac{2}{r}\cos r\pi = (-1)^{r+1}\frac{2}{r} \end{aligned}$$

$$x = 2\left\{\frac{\sin x}{1} - \frac{\sin 2x}{2} + \frac{\sin 3x}{3} - \frac{\sin 4x}{4} + \cdots\right\}.$$

For the value $x = \pi$ the series vanishes, so that the function has a discontinuity, but the series represents the mean of the values on the two sides of the discontinuity, Fig. 51. For the cosine development we find

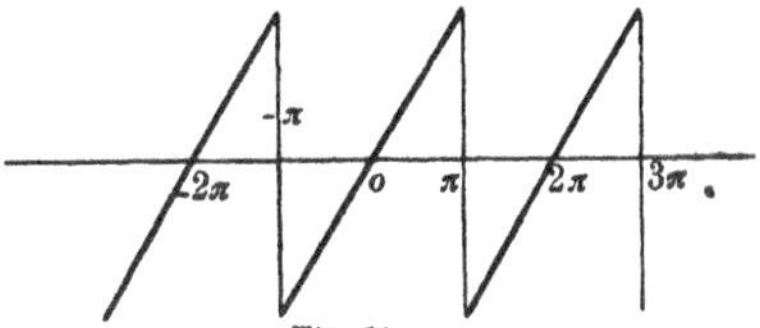

Fig. 51.

$$A_0 = \frac{2}{\pi}\int_0^\pi x\, dx = \pi,$$

$$(40)\quad A_r = \frac{2}{\pi}\int_0^\pi x \cos r x\, dx = \frac{2}{\pi}\left\{\frac{x\sin rx}{r}\Big|_0^\pi - \frac{1}{r}\int_0^\pi \sin r x\, dx\right\} = -\frac{4}{\pi r^2},\ r\ \text{odd},$$

$$x = \frac{\pi}{2} - \frac{4}{\pi}\left\{\frac{\cos x}{1^2} + \frac{\cos 3x}{3^2} + \frac{\cos 5x}{5^2} + \cdots\right\}.$$

This represents for $-\pi < x < 0$ the function $-x$, so that there is in

this case no discontinuity of the function, but the derivative is discontinuous at the points $n\pi$, and the series of derivatives is the series (38),

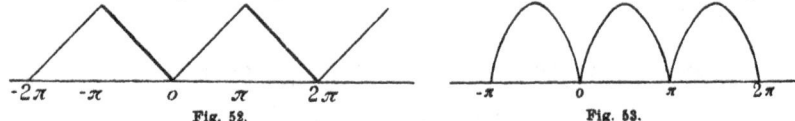

Fig. 52.   Fig. 53.

which is discontinuous in the proper manner, Fig. 52. (It is to be noted that this is not always the case).

We may develop the function $\sin x$ in a cosine series, obtaining

$$A_r = \frac{2}{\pi} \int_0^\pi \sin x \cos r x\, dx = \frac{1}{\pi} \int_0^\pi \{ \sin (r+1)x - \sin (r-1)x \}\, dx$$

$$= -\frac{1}{\pi} \left\{ \frac{\cos(r+1)\pi - 1}{r+1} - \frac{\cos(r-1)\pi - 1}{r-1} \right\},$$

(41)

$$A_0 = \frac{2}{\pi} \int_0^\pi \sin x\, dx = \frac{4}{\pi}$$

$$\sin x = \frac{4}{\pi} \left\{ \frac{1}{2} - \frac{\cos 2x}{1 \cdot 3} - \frac{\cos 4x}{3 \cdot 5} - \frac{\cos 6x}{5 \cdot 6} - \cdots \right\} \quad 0 < x < \pi.$$

Accordingly between $-\pi < x < 0$, the same series represents $-\sin x$, Fig. 53.

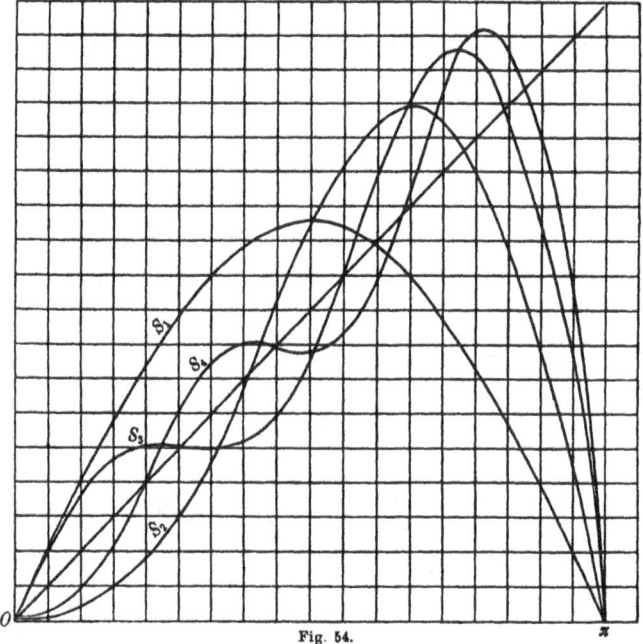

Fig. 54.

In Figs. 54, 55 the curves are drawn which represent the successive approximations S_1, S_2, S_3, and we see how the successive terms remove the curvature and cause continual approximation to the function

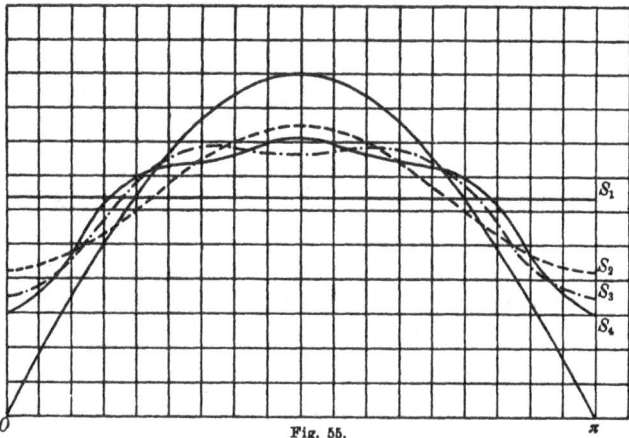

Fig. 55.

represented by the series. We also see plainly how the approximation becomes worse as we approach a point of discontinuity, that is, the nearer we come to the point of discontinuity the more terms must be taken to secure the same degree of accuracy. This is an instance of lack of uniform convergence (see Appendix).

43. Fourier's Integral. We have already shown in Chapter III, (147) that if we integrate from $x = -l$ to $x = l$ we obtain a series

$$(42) \qquad f(x) = \frac{1}{2} A_0 + \sum_1^\infty \left\{ A_r \cos \frac{r \pi x}{l} + B_r \sin \frac{r \pi x}{l} \right\},$$

in which

$$(43) \qquad A_r = \frac{1}{l} \int_{-l}^{l} f(\alpha) \cos \frac{r \pi \alpha}{l} d\alpha, \quad B_r = \frac{1}{l} \int_{-l}^{l} f(\alpha) \sin \frac{r \pi \alpha}{l} d\alpha,$$

convergent in the range $-l < x < l$, if $f(x)$ sastisfies Dirichlet's conditions throughout that range.

We have now to inquire what takes place when the field of integration increases indefinitely, $l = \infty$. We notice that, since the absolute values of the sine and cosine are less than unity,

$$|A_r| \quad \text{and} \quad |B_r| < \frac{1}{l} \int_{-l}^{l} |f(\alpha)| d\alpha$$

and accordingly if $f(\alpha)$ is absolutely integrable in the infinite range, that is if the integral

$$\int_{-\infty}^{\infty} |f(\alpha)| \, d\alpha$$

is convergent, in the limit both A_r and B_r vanish on account of the factor $\frac{1}{l}$. Nevertheless, if as before we insert the values of the coefficients, and take the first m terms for any given value of l, we have

$$(44) \quad S(l, m; x) = \frac{1}{l} \int_{-l}^{l} f(\alpha) \left[-\frac{1}{2} + \sum_{r=0}^{r=m} \cos \frac{r\pi}{l}(\alpha - x) \right] d\alpha .$$

We have now to consider the double limit, both for $m = \infty$, and for $l = \infty$, and as we find in the Appendix, it will probably make a difference in which order we pass to the limit. It is obvious that, if we keep m constant, and put $l = \infty$, since each term in the sum becomes unity, the integrand becomes $(m - \frac{1}{2})f(\alpha)$, and on account of the assumed absolute integrability of $|f(\alpha)|$, the integral is finite, so that $\lim_{l=\infty} S(l, m; x) = 0$. Let us accordingly put

$$\frac{\pi}{l} = \delta, \quad r\delta = \lambda_r, \quad m\delta = \frac{m}{l}\pi = p,$$

and making p the variable instead of m let us write

$$(45) \quad S(l, p; x) = \int_{-l}^{l} f(\alpha) \left[-\frac{1}{2l} + \frac{1}{\pi} \sum_{r=0}^{r=\frac{pl}{\pi}} \delta \cos \lambda_r(\alpha - x) \right] d\alpha .$$

Now by the definition of a definite integral we have

$$\int_{0}^{p} \cos \lambda(\alpha - x) d\lambda = \lim_{l=\infty} \sum_{r=0}^{r=\frac{pl}{\pi}} \delta \cos \lambda_r(\alpha - x),$$

and this means, reverting to the definition of a limit, that if ε be a positive number arbitrarily given, however small, we can find corresponding to it a number l', such that whenever $l > l'$

$$\sum_{r=0}^{r=\frac{pl}{\pi}} \delta \cos \lambda_r(\alpha - x) = \int_{0}^{p} \cos \lambda(\alpha - x) d\lambda + \varepsilon', \quad |\varepsilon'| < \varepsilon.$$

Accordingly we have

$$S(l, p; x) = \int_{-l}^{l} f(\alpha) \left[-\frac{1}{2l} + \frac{1}{\pi} \int_{0}^{p} \cos \lambda(\alpha - x) d\lambda + \frac{\varepsilon'}{\pi} \right] d\alpha,$$

and passing to the limit,

$$(46)\ \lim_{l=\infty} S(l,p;x) = \frac{1}{\pi} \int_{-\infty}^{\infty} f(\alpha)\, d\alpha \int_{0}^{p} \cos \lambda(\alpha - x)\, d\lambda = \frac{1}{\pi} \int_{-\infty}^{\infty} f(\alpha)\, \frac{\sin p(\alpha - x)}{\alpha - x}\, d\alpha.$$

Now putting $\alpha - x = \beta$ this is

$$(47)\qquad\qquad \frac{1}{\pi} \int_{-\infty}^{\infty} f(\beta + x)\, \frac{\sin p\beta}{\beta}\, d\beta$$

which may be divided into the sum of two Dirichlet's integrals of the form (8), and we have therefore,

$$(48)\quad \frac{1}{2}\left[f(x + 0) + f(x - 0)\right] = \lim_{p=\infty} \frac{1}{\pi} \int_{-\infty}^{\infty} f(\alpha)\, d\alpha \int_{0}^{p} \cos \lambda(\alpha - x)\, d\lambda,$$

provided $f(x)$ satisfies Dirichlet's conditions in the infinite range. Accordingly we see that we must make l and m increase so that $\lim \frac{m}{l} = \infty$, and it is obvious why l must not increase first.

If we should now assume that

$$(49)\quad \lim_{p=\infty} \frac{1}{\pi} \int_{-\infty}^{\infty} f(\alpha)\, d\alpha \int_{0}^{p} \cos \lambda(\alpha - x)\, d\lambda = \frac{1}{\pi} \int_{-\infty}^{\infty} f(\alpha)\, d\alpha \int_{0}^{\infty} \cos \lambda(\alpha - x)\, d\lambda,$$

we should obtain Fourier's celebrated double integral, in the form given by him[1]), and repeated by many authors since. But unfortunately the integral

$$\int_{0}^{\infty} \cos \lambda(\alpha - x)\, d\lambda$$

has no sense, for the integral

$$\int_{0}^{p} \cos \lambda(\alpha - x)\, d\lambda = \frac{\sin p(\alpha - x)}{\alpha - x}$$

does not approach any limit as p increases, but merely oscillates. The equation (49) is then incorrect, but we find the correct limit as follows

From the definition of the infinite integral we have

$$(50)\quad \int_{-\infty}^{\infty} f(\alpha)\, d\alpha \int_{0}^{p} \cos \lambda(\alpha - x)\, d\lambda = \lim_{\substack{a=-\infty \\ b=\infty}} \int_{a}^{b} f(\alpha)\, d\alpha \int_{0}^{p} \cos \lambda(\alpha - x)\, d\lambda,$$

1) Fourier, *Théorie de la Chaleur*, p. 408.

and in the iterated integral between finite limits there is no reason why we may not change the order of integration, obtaining

$$\int_0^p d\lambda \int_a^b f(\alpha) \cos \lambda(\alpha - x)d\alpha.$$

The integral with respect to α may now be divided up into three,

(51) $$\int_0^p d\lambda \int_{-\infty}^\infty f(\alpha) \cos \lambda(\alpha - x)d\alpha - \int_0^p d\lambda \int_{-\infty}^a f(\alpha) \cos \lambda(\alpha - x)d\alpha$$

$$-\int_0^p d\lambda \int_b^\infty f(\alpha) \cos \lambda(\alpha - x)d\alpha.$$

Now if the integral,

$$\int_{-\infty}^\infty |f(\alpha)| \, d\alpha$$

is finite and determinate, the two last terms tend toward zero, for their absolute values are less than

$$\int_0^p d\lambda \int_{-\infty}^a |f(\alpha)| \, d\alpha = p \int_{-\infty}^a |f(\alpha)| \, d\alpha \text{ and } \int_0^p d\lambda \int_b^\infty |f(\alpha)| \, d\alpha = p \int_b^\infty |f(\alpha)| \, d\alpha.$$

Consequently we shall have, in the limit for $a = -\infty$, $b = \infty$,

(52) $$\frac{1}{2}[f(x + 0) + f(x - 0)] = \lim_{p = \infty} \frac{1}{\pi} \int_0^p d\lambda \int_{-\infty}^\infty f(\alpha) \cos \lambda(\alpha - x)d\alpha$$

$$= \frac{1}{\pi} \int_0^\infty d\lambda \int_{-\infty}^\infty f(\alpha) \cos \lambda(\alpha - x)d\alpha,$$

and this is the proper form of Fourier's integral theorem.

If we have an even function of λ we may write,

(53) $$f(x) = \frac{1}{2\pi} \int_{-\infty}^\infty d\lambda \int_{-\infty}^\infty f(\alpha) \cos \lambda(\alpha - x)d\alpha.$$

With regard to this theorem Kronecker[1]) says: "This so-called Fourier double-integral made at its discovery a tremendous impression on the mathematical world. It was shown for the first time how an almost arbitrary function, satisfying only the limitations mentioned, fits itself into mathematical forms. The formula (52) maintains its correctness, as was shown by P. du Bois-Reymond, for various fluctuating functions, inserted instead of the cosine." One of these cases will appear in Chapter VIII.

---

1) Kronecker, Einfache und Vielfache Integrale, S. 81.

The difference between the mode of representation in Fourier's series and integral can be made more vivid by means of an application to optics. The general term $\dfrac{\cos}{\sin}\dfrac{r\pi x}{l}$ represents a simple wave of wavelength $\dfrac{2l}{r}$, and we may accordingly say that for the interval $2l$ the arbitrary function can be represented as the sum of an infinite number of such waves of wave-lengths proportional to the reciprocals of the integers, having a *spectrum* of an infinite number of discrete wave-lengths. But if in the integral (52) we expand the cosine and write

$$(54) \qquad \varphi(\lambda) = \frac{1}{\pi}\int_{-\infty}^{\infty} f(\alpha)\cos\lambda\alpha\,d\alpha, \quad \psi(\lambda) = \frac{1}{\pi}\int_{-\infty}^{\infty} f(\alpha)\sin\lambda\alpha\,d\alpha$$

we may write the integral,

$$(55) \qquad f(x) = \int_{0}^{\infty}\{\varphi(\lambda)\cos\lambda x + \psi(\lambda)\sin\lambda x\}\,d\lambda$$

and as the wave-length of the terms $\cos\lambda x$ and $\sin\lambda x$ is $\dfrac{2\pi}{\lambda}$ we see that the function is represented as the resultant of waves of all possible wave-lengths from zero to infinity, forming a *continuous spectrum*. It is by such an integral that white light, containing all colors, must be represented.

As an example let us undertake to find the spectrum of a damped vibration, which begins at a definite instant $t = 0$, and continues indefinitely. We then have

$$f(t) = 0, \quad t < 0; \quad f(t) = e^{-at}\sin bt, \quad t > 0$$

so that

$$(56) \qquad f(t) = \frac{1}{\pi}\int_{0}^{\infty}d\lambda\int_{0}^{\infty}e^{-a\alpha}\sin b\alpha\cos\lambda(\alpha - t)\,d\alpha$$

$$(57) \quad \varphi(\lambda) = \frac{1}{\pi}\int_{0}^{\infty}e^{-a\alpha}\sin b\alpha\cos\lambda\alpha\,d\alpha = \frac{1}{2\pi}\left[\frac{b+\lambda}{(b+\lambda)^2+a^2} + \frac{(b-\lambda)}{(b-\lambda)^2+a^2}\right]$$

$$(58) \quad \psi(\lambda) = \frac{1}{\pi}\int_{0}^{\infty}e^{-a\alpha}\sin b\alpha\sin\lambda\alpha\,d\alpha = \frac{1}{2\pi}\left[\frac{a}{(b-\lambda)^2+a^2} - \frac{a}{(b+\lambda)^2+a^2}\right].$$

Thus we obtain the spectrum of waves, each of which has been in existence for an infinite time, the amplitude of the waves belonging to the spectrum interval from λ to $\lambda + d\lambda$ being

$$[\{\varphi(\lambda)\}^2 + \{\psi(\lambda)\}^2]^{\frac{1}{2}}\,d\lambda.$$

This amplitude, as a function of λ, is shown for several values of $\dfrac{a}{b}$ in in Fig. 56, and we see that the smaller the damping a, the more concentrated the spectrum about the value $\lambda = b$. If a force whose magnitude is represented by such a damped oscillation acts upon an undamped

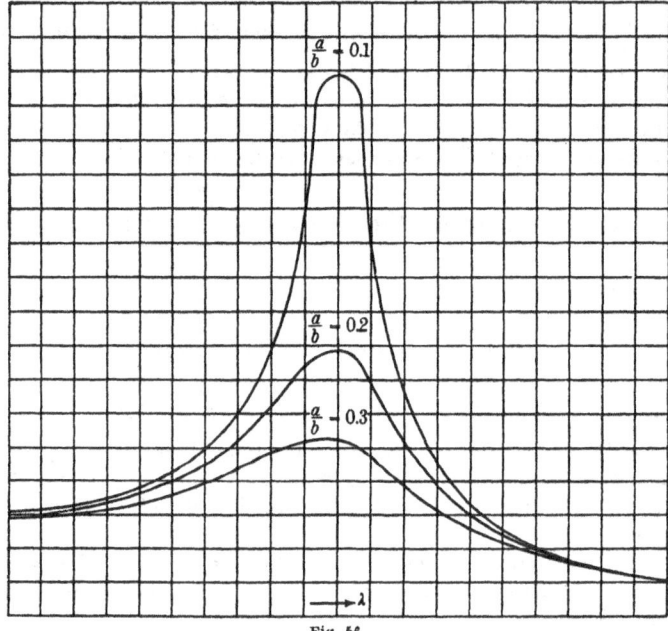

Fig. 56.

resonating system, each component will produce a response, and we accordingly see that even if there is not exact tuning, some response will be produced, and there will be more latitude the more the damping of the force, that is the wider its spectrum. This fact has an important application in wireless telegraphy, where in order to interfere little with foreign receivers, the sending oscillation must be but slightly damped.

If in Fourier's integral (52) we replace the cosine by the sine, the odd function of λ makes the integrals vanish, accordingly since

$$e^{i\lambda(\alpha-x)} = \cos \lambda (\alpha - x) + i \sin \lambda (\alpha - x)$$

we have

(59)
$$f(x) = \frac{1}{2\pi} \int_{-\infty}^{\infty} d\lambda \int_{-\infty}^{\infty} f(\alpha) \, e^{i\lambda(\alpha-x)} d\alpha .$$

An interesting form of Fourier's theorem is arrived at by putting

(60)
$$g(y) = \frac{1}{\sqrt{2\pi}} \int_{-\infty}^{\infty} f(x) \, e^{ixy} dx .$$

We then have

(61) $$f(x) = \frac{1}{\sqrt{2\pi}} \int_{-\infty}^{\infty} g(y) e^{-ixy} dy$$

showing the reciprocal properties of the functions f and g. Either of (60), (61) is an integral equation of the *first* kind, like (288) Chapter III, of which the other is a solution.

If f and g are even functions,

(62) $$f(x) = \sqrt{\frac{2}{\pi}} \int_{0}^{\infty} g(y) \cos xy\, dy, \quad g(y) = \sqrt{\frac{2}{\pi}} \int_{0}^{\infty} f(x) \cos xy\, dx,$$

whereas if they are odd,

(63) $$f(x) = \sqrt{\frac{2}{\pi}} \int_{0}^{\infty} g(y) \sin xy\, dy, \quad g(y) = \sqrt{\frac{2}{\pi}} \int_{0}^{\infty} f(x) \sin xy\, dx.$$

Both Fourier's integral and series may be extended to functions of any number of variables. Let us represent $f(x, y)$ by a Fourier integral. For a given value of y,

(64) $$f(x, y) = \frac{1}{\pi} \int_{0}^{\infty} d\lambda \int_{-\infty}^{\infty} f(\alpha, y) \cos \lambda (\alpha - x)\, d\alpha,$$

but in like manner

(65) $$f(\alpha, y) = \frac{1}{\pi} \int_{0}^{\infty} d\mu \int_{-\infty}^{\infty} f(\alpha, \beta) \cos \mu (\beta - y)\, d\beta,$$

and inserting in (64),

(66) $$f(x, y) = \frac{1}{\pi^2} \int_{0}^{\infty} \int_{0}^{\infty} d\lambda\, d\mu \int_{-\infty}^{\infty} \int_{-\infty}^{\infty} f(\alpha, \beta) \cos \lambda (\alpha - x) \cos \mu (\beta - y)\, d\alpha\, d\beta.$$

In like manner the theorem can be extended to any number of variables. For this purpose the form with the exponential is convenient. We have

(67) $$f(x, y, z \cdots)$$

$$= \frac{1}{(2\pi)^n} \int_{-\infty}^{\infty} \int \cdots \int \int e^{i[\lambda(\alpha - x) + \mu(\beta - y) + \nu(\gamma - z) \cdots]} f(\alpha, \beta, \gamma \cdots)\, d\lambda\, d\mu\, d\nu \cdots d\alpha\, d\beta\, d\gamma \cdots,$$

where it is to be understood that the integration with respect to $\alpha, \beta, \gamma \cdots$ takes place before that with respect to $\lambda, \mu, \nu \cdots$. We may write either $\alpha - x \cdots$ or $x - \alpha \cdots$

44. Cauchy's Method of Integration. Wave-equation. By means of Fourier's multiple double-integral Cauchy[1]) gave a general method of integrating linear partial differential equations with constant coefficients.

Before explaining the general method let us take as a simple example the equation of the string,

$$\frac{\partial^2 u}{\partial t^2} = a^2 \frac{\partial^2 u}{\partial x^2}$$

with

$$t = 0 \begin{cases} u = F(x), \\ \dfrac{\partial u}{\partial t} = G(x). \end{cases}$$

Let us put $u = e^{pt+qx}$, giving

$$p^2 = a^2 q^2, \quad p = \pm aq,$$

so that

$$u = A e^{aqt+qx} + B e^{-aqt+qx}$$

is a solution, where A, B, q, are independent of x. Put $q = -i\lambda$ and let A and B be functions of α and λ, then, introducing the factor $e^{i\lambda\alpha}$,

$$u = A(\alpha) e^{i\lambda(\alpha-x-at)} + B(\alpha) e^{i\lambda(\alpha-x+at)}$$

is a solution whatever α and λ. Also if we multiply by $d\lambda d\alpha$ and if we add any number of such solutions, for different and continuously varying α and λ, that is the definite integral,

$$(68) \qquad u = \frac{1}{\pi} \int_{-\infty}^{\infty} \int_{-\infty}^{\infty} \{ A(\alpha) e^{i\lambda(\alpha-x-at)} + B(\alpha) e^{i\lambda(\alpha-x+at)} \} \, d\lambda \, d\alpha$$

is a solution.

If

$$A(\alpha) + B(\alpha) = F(\alpha).$$

when $t = 0$ this reduces to $F(x)$.

Differentiating by t,

$$(69) \quad \frac{\partial u}{\partial t} = \frac{1}{\pi} \int_{-\infty}^{\infty} \int_{-\infty}^{\infty} -i\lambda a \{ A(\alpha) e^{i\lambda(\alpha-x-at)} - B(\alpha) e^{i\lambda(\alpha-x+at)} \} \, d\lambda \, d\alpha,$$

and when $t = 0$ this reduces to $G(x)$ if

$$-i\lambda a (A(\alpha) - B(\alpha)) = G(\alpha).$$

---

1) Cauchy. Jour. *de l'Ec.* Polytechnique, cah. 19, 1823; p. 571, Mémoire sur l'intégration des équations, linéaires aux différences partielles et à coefficiens constans.

From these two equations we get

$$A(\alpha) = \tfrac{1}{2}\Big[F(\alpha) + \frac{iG(\alpha)}{a\lambda}\Big],$$

$$B(\alpha) = \tfrac{1}{2}\Big[F(\alpha) - \frac{iG(\alpha)}{a\lambda}\Big],$$

(70) $\quad u = \dfrac{1}{2\pi}\displaystyle\int_{-\infty}^{\infty}\int_{-\infty}^{\infty} F(\alpha)\{e^{i\lambda\{\alpha-(x+at)\}} + e^{i\lambda\{\alpha-(x-at)\}}\}\,d\lambda\,d\alpha$

$\qquad + \dfrac{i}{2\pi a}\displaystyle\int_{-\infty}^{\infty}\int_{-\infty}^{\infty}\dfrac{G(\alpha)}{\lambda}\{e^{i\lambda\{\alpha-(x+at)\}} - e^{i\lambda\{\alpha-(x-at)\}}\}\,d\lambda\,d\alpha.$

The first integral represents

$$\tfrac{1}{2}[F(x+at) + F(x-at)],$$

while in the second,

$$i\int_{-\infty}^{\infty}e^{i\lambda(\alpha-x)}\{e^{-i\lambda at} - e^{i\lambda at}\}\frac{d\lambda}{\lambda} = \int_{-\infty}^{\infty}\cos\lambda(\alpha-x)\frac{\sin\lambda at}{\lambda}\,d\lambda$$

$$= \frac{1}{2}\int_{-\infty}^{\infty}\frac{\sin\lambda(\alpha-x+at) - \sin\lambda(\alpha-x-at)}{\lambda}\,d\lambda$$

$$= \int_{0}^{\infty}\frac{\sin\lambda(\alpha-x+at) - \sin\lambda(\alpha-x-at)}{\lambda}\,d\lambda.$$

The last integral is the difference of two, each of which is of the form of the integral

$$\int_{0}^{\infty}\frac{\sin by}{y}\,dy,$$

which is equal to $\frac{\pi}{2}$ if $b > 0$, to $-\frac{\pi}{2}$ if $b < 0$, and to 0 if $b = 0$. Accordingly our integral, being the difference of two equal terms, is zero when $\alpha - x + at$ and $\alpha - x - at$ are of the same sign, that is either

$$\alpha < x - at \quad \text{or} \quad \alpha > x + at$$

but when

$$x - at < \alpha < x + at$$

the two terms are of opposite sign, and their difference equals π. Consequently in the last integral in (70), the integral is zero except between $x - at$ and $x + at$, so that we have

$$\frac{1}{2a}\int_{x-at}^{x+at}G(\alpha)\,d\alpha$$

which was given by d'Alembert's solution, (27) Chapter III.

We will now consider Cauchy's method of integration for any number of independent variables.

Suppose that for the value $t = 0$,

$$(71) \qquad \begin{aligned} u &= F(x, y, z, \ldots), \\ \frac{\partial u}{\partial t} &= G(x, y, z, \ldots). \end{aligned}$$

We are to find a solution in the form

$$(72)\; u = \frac{1}{(2\pi)^n} \int\int \cdots \int\int_{-\infty}^{\infty} U(t, \alpha, \beta, \gamma \ldots) e^{i[\lambda(x-\alpha) + \mu(y-\beta) + \cdots]} d\alpha\, d\beta \cdots d\lambda\, d\mu \cdots$$

which shall satisfy the initial conditions. Now we have

$$(73) \quad \frac{\partial u}{\partial t} = \frac{1}{(2\pi)^n} \int\int \cdots \int\int_{-\infty}^{\infty} \frac{\partial U}{\partial t} e^{i[\lambda(x-\alpha) + \mu(y-\beta)\cdots]} d\alpha\, d\beta \cdots d\lambda\, d\mu \cdots,$$

also

$$(74) \quad \frac{\partial u}{\partial x} = \frac{1}{(2\pi)^n} \int\int \cdots \int\int_{-\infty}^{\infty} i\lambda\, U e^{i[\lambda(x-\alpha) + \mu(y-\beta) + \cdots]} d\alpha\, d\beta \cdots d\lambda\, d\mu \cdots,$$

consequently if we determine U as a function of the single variable t to satisfy the ordinary differential equation obtained by replacing

$$\frac{\partial^k u}{\partial x^k} \text{ by } (i\lambda)^k U, \quad \frac{\partial^l u}{\partial y^l} \text{ by } (i\mu)^l U, \text{ etc.,}$$

the integral (72) will be a solution. The arbitrary constants are to made such functions of $\alpha, \beta, \gamma \ldots$ that for $t = 0$, U shall reduce to $F(\alpha, \beta, \gamma)$ and $\dfrac{dU}{dt}$ to $G(\alpha, \beta, \gamma \ldots)$. This was the method adopted for the last equation.

Let us apply this to the equation of sound for three dimensions, (115) Chapter I.

$$(75) \qquad \frac{\partial^2 u}{\partial t^2} = a^2 \left[\frac{\partial^2 u}{\partial x^2} + \frac{\partial^2 u}{\partial y^2} + \frac{\partial^2 u}{\partial z^2} \right],$$

reducing to the previous equation when u is independent of y and z.

We are accordingly to put

$$(76) \qquad \frac{d^2 U}{dt^2} = -a^2(\lambda^2 + \mu^2 + \nu^2) U = -a^2 \varrho^2 U,$$

if we put

$$\lambda^2 + \mu^2 + \nu^2 = \varrho^2.$$

The solution is
$$U = A \cos a\varrho t + B \sin a\varrho t,$$

where A and B are to be functions of α, β, γ.

Since for $t = 0$, $U = F(\alpha, \beta, \gamma)$ we have $A = F(\alpha, \beta, \gamma)$. Also since

$$\frac{dU}{dt} = -a\varrho A \sin a\varrho t + a\varrho B \cos a\varrho t$$

putting $t = 0$
$$a\varrho B = G(\alpha, \beta, \gamma),$$

so that
$$U = F(\alpha, \beta, \gamma) \cos a\varrho t + G(\alpha, \beta, \gamma) \frac{\sin a\varrho t}{a\varrho},$$

giving the solution,

(77) $u =$

$$\frac{1}{8\pi^3} \int\int\int\int\int\int_{-\infty}^{\infty} F(\alpha,\beta,\gamma) \cos a\varrho t \, e^{i[\lambda(\alpha-x)+\mu(\beta-y)+\nu(\gamma-z)]} \, d\alpha \, d\beta \, d\gamma \, d\lambda \, d\mu \, d\nu$$

$$+ \frac{1}{8\pi^3} \int\int\int\int\int\int_{-\infty}^{\infty} G(\alpha,\beta,\gamma) \frac{\sin a\varrho t}{a\varrho} \, e^{i[\lambda(\alpha-x)+\mu(\beta-y)+\nu(\gamma-z)]} d\alpha \, d\beta \, d\gamma \, d\lambda \, d\mu \, d\nu.$$

We may call these integrals u_1, u_2, so that for

(78) $t = 0$ $\begin{cases} u_1 = F(x, y, z). & \dfrac{\partial u_1}{\partial t} = 0, \\[2mm] u_2 = 0, & \dfrac{\partial u_2}{\partial t} = G(x, y, z). \end{cases}$

Consider first the integral u_2, and let us introduce polar coordinates r, θ, φ for the point α, β, γ with respect to an origin at x, y, z.

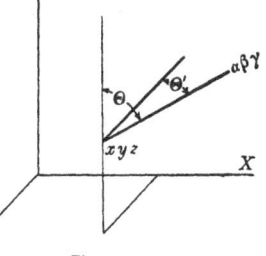

Fig. 57.

(79) $\begin{aligned} \alpha - x &= r\cos\varphi \sin\theta, \\ \beta - y &= r\sin\varphi \sin\theta, \\ \gamma - z &= r\cos\theta. \end{aligned}$

Put also,

(80) $\begin{aligned} \lambda &= \varrho \cos\varphi' \sin\theta', \\ \mu &= \varrho \sin\varphi' \sin\theta', \\ \nu &= \varrho \cos\theta' \end{aligned}$

agreeing with (76). Then the scalar product

(81) $\lambda(\alpha - x) + \mu(\beta - y) + \nu(\gamma - z) = \varrho r \cos(\varrho r),$

and if we take for the polar axis from which to measure θ', the line joining x, y, z to α, β, γ, (81) is $\varrho r \cos\theta'$, Fig. 57.

Then

(82) $u_2 = \dfrac{1}{8\pi^3} \int_0^{\infty}\int_0^{\infty}\int_0^{\pi}\int_0^{\pi}\int_0^{2\pi}\int_0^{2\pi} G(x + r\cos\varphi\sin\theta, \ldots)$

$$\frac{\sin a\varrho t}{a\varrho} e^{i r\varrho \cos\theta'} r^2 \varrho^2 \sin\theta \sin\theta' \, d\varrho \, dr \, d\theta \, d\theta' \, d\varphi \, d\varphi'.$$

The integration with respect to φ' and θ' may be performed, for φ' enters only as $d\varphi'$, and the integration merely multiplies by 2π. On the other hand θ' enters in

$$(83) \qquad \int_0^\pi e^{ir\varrho\cos\theta'}\sin\theta'\,d\theta' = -\frac{e^{ir\varrho\cos\theta'}}{ir\varrho}\Big|_0^\pi$$

$$= \frac{1}{ir\varrho}(e^{ir\varrho} - e^{-ir\varrho}) = \frac{2\sin r\varrho}{r\varrho},$$

and we have, dividing out $r\varrho^2$,

$$u_2 = \frac{1}{2\pi^2 a}\int_0^\infty\int_0^\infty\int_0^\pi\int_0^{2\pi} G(x+r\cos\varphi\sin\theta,\ldots)\sin a\varrho t\sin\varrho r\cdot r\sin\theta\,d\varrho\,dr\,d\theta\,d\varphi$$

$$(84)$$

$$= \frac{1}{4\pi^2 a}\int_0^\infty\int_0^\infty\int_0^\pi\int_0^{2\pi} G(\ldots)\{\cos\varrho\,(r-at)-\cos\varrho\,(r+at)\}\,r\sin\theta\,d\varrho\,dr\,d\theta\,d\varphi.$$

The integral

$$(85) \qquad \frac{1}{\pi}\int_0^\infty d\varrho\int_0^\infty r\,G(x+r\cos\varphi\sin\theta\ldots)\cos\varrho\,(r-at)\,dr$$

is in the form of a Fourier double-integral, (52), if we put

$$\varrho = \lambda,\ r = \alpha,\ f(\alpha) = rG\,(\ldots r)\ \ \alpha > 0,$$
$$f(\alpha) = 0 \qquad\qquad \alpha \leqq 0,$$

and represents $atG(x+at\cos\varphi\sin\theta,\ldots)$ (if G is continuous) when $t > 0$. The second term represents $-atG(-at)$ which is zero if $t > 0$, but which represents the solution extended back to epochs in which $t < 0$.

Accordingly our integral becomes

$$(86)\,u_2 = \frac{t}{4\pi}\int_0^\pi\int_0^{2\pi} G(x+at\cos\varphi\sin\theta, y+at\sin\varphi\sin\theta, z+at\cos\theta)\sin\theta\,d\theta\,d\varphi.$$

Now in u_1 we have a similar integral, with the exception that instead of $\frac{\sin a\varrho t}{a\varrho}$ we have $\cos a\varrho t$ which is its derivative with respect to t. Accordingly

$$(87) \qquad u_1 = \frac{\partial}{\partial t}\left[\frac{t}{4\pi}\int_0^\pi\int_0^{2\pi} F(x+at\cos\varphi\sin\theta,\ldots)\sin\theta\,d\theta\,d\varphi\right],$$

and the complete solution is $u = u_1 + u_2$. This is an example of Stokes's rule, Chapter III, after equation (27), in that the terms from $u_{t=0}$ are derived from those in $\left(\frac{\partial u}{\partial t}\right)_{t=0}$ by differentiation.

The above integral was given by Poisson in 1820.[1])

In these integrals we have functions of a point situated on a sphere of radius at with center at the point x, y, z, whose element of surface is $dS = (at)^2 \sin \theta\, d\theta\, d\varphi$. The mean value of a function on a certain surface is

$$\overline{F} = \frac{1}{S} \int\int F\, dS,$$

and on a sphere,

(88)
$$\overline{F} = \frac{1}{4\pi r^2} \int_0^\pi \int_0^{2\pi} F r^2 \sin\theta\, d\theta\, d\varphi.$$

Consequently putting $r = at$, we find

(89)
$$u_2 = t\,\overline{G}(at),$$

or the part u_2 is obtained at a point P and at a time t by taking the mean of the values for $t = 0$ of $\frac{\partial u}{\partial t}$ at points lying on a sphere of radius at with P as center, and multiplying by t. The part u_1 is found by taking the mean of the initial values of u on the same sphere, multiplying by t, and taking the rate of change as t varies.

(90)
$$u_1 = \frac{\partial}{\partial t} \{ t\,\overline{F}(at) \}.$$

Suppose the functions F and G are zero except in a limited region V and suppose P is outside this region, Fig. 58. Then when t is small the sphere of radius at lies in the zero region, and nothing is contributed to the integral. This continues until $at = r_1$ where r_1 is the distance from P to the nearest point of V. Then the values of F and G begin to come into play, but gradually, since at first the part of S having non-zero values is small. After $at > r_2$, where r_2 is the distance to the most remote point of V, the sphere is again in the zero region, and there is no disturbance there-

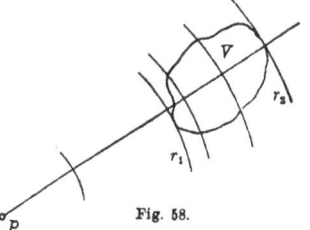

Fig. 58.

after. Thus we see that disturbances are propagated with the velocity a in waves, and that after the wave has passed nothing remains. If the region V is very small, and in it G is very large, while the region covers an area S,

(91)
$$u_2 = \frac{r}{a} \frac{GS}{4\pi r^2} = \frac{GS}{4\pi a r} = \frac{\left(\frac{\partial u}{\partial t}\right)_{t=0} S}{4\pi a^2 t}$$

or u_2 varies inversely as the distance. The function u is the velocity-potential or compression, not the displacement of the air.

---

1) *Nouveaux Mémoires de l'Académie des Sciences*, t. III.

Suppose that initially there is no motion, and that in the whole region V the compression s is of the same sign, say positive. It might be supposed that at all times the compression would then be positive, but this is not the case. Let u now stand for the velocity-potential φ Since the compression is connected with this by the equation (112′) Chapter I, we have $a^2 s = -\dfrac{\partial \varphi}{\partial t}$, and the initial values of $\dfrac{\partial \varphi}{\partial t}$ will be everywhere negative. Accordingly φ, which is represented by the term u_2, will be negative when it is not zero, that is between the times $t = \dfrac{r_1}{a}$ and $t = \dfrac{r_2}{a}$. It will then sink from its zero value to a minimum and then rise again to zero. While φ is falling s will be positive, and while it is rising s will be negative, so that the mean value of the compression $\bar{s}$, where

$$(t_1 - t_0)\bar{s} = \int_{t_0}^{t_1} s\, dt = -\frac{1}{a^2}\int_{t_0}^{t_1}\frac{\partial \varphi}{\partial t}\, dt = -\frac{1}{a^2}(\varphi_1 - \varphi_0) = 0,$$

will be zero.

It might be expected that since in one and three dimensions we have wave-propagation, the same would be true in two dimensions, and that the integral could be obtained by replacing the spheres by circles. This is however not the case, as we shall see in § 49.

45. Equation of Heat Conduction. The equation of Fourier (Chapter I, (83)) $\dfrac{\partial u}{\partial t} = a^2 \Delta u$, where $a^2 = \dfrac{k}{c\varrho}$ differs from that of wave-motion only in having the first derivative by t instead of the second. This makes a most important difference, as we shall see. In fact we can give arbitrarily only the values of u for $t = 0$, because the values of $\dfrac{\partial u}{\partial t}$ are determined by the equation, and are not arbitrary.

Let us solve under the condition,

$$t = 0, \quad u = F(x, y, z).$$

According to our rule we must put

$$\frac{dU}{dt} = -a^2(\lambda^2 + \mu^2 + \nu^2)\, U$$

$$U = A(\alpha, \beta, \gamma)\, e^{-a^2(\lambda^2 + \mu^2 + \nu^2)t}$$

and for

$$t = 0, \quad U = A(\alpha, \beta, \gamma) = F(\alpha, \beta, \gamma).$$

Accordingly,

$$(92) \quad u = \frac{1}{(2\pi)^3}\int\int\int_{-\infty}^{\infty}\int\int\int F(\alpha, \beta, \gamma)\, e^{-a^2(\lambda^2 + \mu^2 + \nu^2)t + i[\lambda(\alpha - x) + \mu(\beta - y) + \nu(\gamma - z)]}$$
$$d\alpha\, d\beta\, d\gamma\, d\lambda\, d\mu\, d\nu.$$

Let us write this in trigonometric form,

$$(93) \quad u = \frac{1}{(2\pi)^3} \int\int\int \int\int\int_{-\infty}^{\infty} F(\alpha,\beta,\gamma) e^{-a^2(\lambda^2+\mu^2+\nu^2)t} \cos\lambda(\alpha-x)$$
$$\cos\mu(\beta-y)\cos\nu(\gamma-z)\,d\alpha\,d\beta\,d\gamma\,d\lambda\,d\mu\,d\nu.$$

In this case also we may perform the integration with respect to λ, μ, ν. Again we need certain definite integrals. In the Appendix, we find

$$(94) \quad \int_0^{\infty} e^{-ax^2}\cos bx\,dx = \frac{1}{2}\sqrt{\frac{\pi}{a}}e^{-\frac{b^2}{4a}}.$$

Put

$$x = \lambda, \quad a = a^2 t, \quad b = \alpha - x,$$

and we have

$$(95) \quad \int_0^{\infty} e^{-a^2\lambda^2 t}\cos\lambda(\alpha-x)\,d\lambda = \frac{1}{2}\sqrt{\frac{\pi}{a^2 t}}e^{-\frac{(x-\alpha)^2}{4a^2 t}}$$

Accordingly,

$$(96) \quad u = \frac{1}{8\pi^{\frac{3}{2}}a^3} \int\int\int_{-\infty}^{\infty} \frac{F(\alpha,\beta,\gamma)}{t^{\frac{3}{2}}} e^{-\frac{(x-\alpha)^2+(y-\beta)^2+(z-\gamma)^2}{4a^2 t}} d\alpha\,d\beta\,d\gamma.$$

If we put

$$(97) \quad \frac{\alpha-x}{2a\sqrt{t}} = \xi, \quad \frac{\beta-y}{2a\sqrt{t}} = \eta, \quad \frac{\gamma-z}{2a\sqrt{t}} = \zeta,$$

this becomes

$$(98) \quad u = \frac{1}{\pi^{\frac{3}{2}}} \int\int\int_{-\infty}^{\infty} F(x+2a\sqrt{t}\cdot\xi,\ y+2a\sqrt{t}\cdot\eta,\ z+2a\sqrt{t}\cdot\zeta)$$
$$\cdot e^{-(\xi^2+\eta^2+\zeta^2)}d\xi\,d\eta\,d\zeta.$$

This integral was given by Laplace. For $t = 0$,

$$(99) \quad u_{t=0} = \frac{1}{\pi^{\frac{3}{2}}} \int\int\int_{-\infty}^{\infty} F(x,y,z)e^{-(\xi^2+\eta^2+\zeta^2)}d\xi\,d\eta\,d\zeta$$
$$= \frac{1}{\pi^{\frac{3}{2}}}F(x,y,z)\left[\int_{-\infty}^{\infty}e^{-\xi^2}d\xi\right]^3 = F(x,y,z),$$

as we see by the value of Laplace's integral, Appendix.

The integral,

$$u = \frac{1}{8\pi^{\frac{3}{2}}a^3} \int\int\int_{-\infty}^{\infty} \frac{F'(\alpha, \beta, \gamma)}{t^{\frac{3}{2}}} e^{-\frac{r^2}{4a^2 t}} d\alpha d\beta d\gamma$$

(100)

$$= \frac{4\pi}{8\pi^{\frac{3}{2}}a^3} \int_0^{\infty} \frac{e^{-\frac{r^2}{4a^2 t}}}{t^{\frac{3}{2}}} \overline{F}(r) r^2 dr ,$$

where

$$r^2 = (x - \alpha)^2 + (y - \beta)^2 + (z - \gamma)^2,$$

shows the nature of the distribution of temperature. The temperature at any point P is found by drawing about it a sphere of radius r, finding the mean over its surface of the initial temperatures, reducing this

in the ratio $\dfrac{r^2}{2\pi^{\frac{1}{2}}a^3} \dfrac{e^{-\frac{r^2}{4a^2 t}}}{t^{\frac{3}{2}}}$, and integrating from $r = 0$ to $r = \infty$. Thus

the effect of points at a distance falls off very rapidly, but most rapidly at first.

Each element of volume $d\tau = d\alpha d\beta d\gamma$ contributes to the temperature at P an amount proportional to its initial temperature u_0 and its volume $d\tau$, equal to

(101)

$$\frac{u_0 d\tau}{8\pi^{\frac{3}{2}}a^3} \frac{e^{-\frac{r^2}{4a^2 t}}}{t^{\frac{3}{2}}}.$$

Let us investigate the function,

$$y = \frac{e^{-\frac{\alpha^2}{t}}}{t^{\frac{n}{2}}}, \quad n > 1.$$

For $t = 0$, numerator and denominator vanish, but on evaluation we find $y = 0$. Also for $t = \infty$, $y = 0$. We have

$$\frac{dy}{dt} = \frac{e^{-\frac{\alpha^2}{t}}}{t^{\frac{n}{2}}} \left\{ \frac{\alpha^2}{t^2} - \frac{n}{2t} \right\}$$

which vanishes for

$$t = 0, \quad t = \infty, \quad t = \frac{2\alpha^2}{n}.$$

The second derivative,

$$\frac{d^2 y}{dt^2} = \frac{e^{-\frac{\alpha^2}{t}}}{t^{\frac{n}{2}+2}} \left\{ \frac{\alpha^4}{t^2} - (n + 2) \frac{\alpha^2}{t} + \frac{n(n + 2)}{4} \right\}$$

vanishes for $t = 0$, $t = \infty$, and the roots of the quadratic in $\dfrac{\alpha^2}{t} = \xi$,

$$\xi^2 - (n + 2)\xi + \frac{n(n + 2)}{4} = 0,$$

$$\xi = \frac{n + 2 \pm \sqrt{2n + 4}}{2},$$

for which we have points of inflexion. Since $\alpha^2 = \dfrac{r^2}{4a^2}$ we see that the effect of the hot point on a point at a distance r is to make its temperature begin *immediately* to rise at first slowly, then more rapidly, reach-

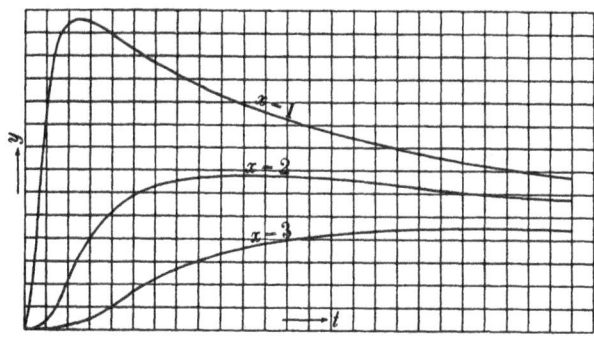

Fig 59.

ing a maximum, and then falling off toward the asymptotic value zero, Fig. 59. Thus there is nothing in the way of wave-propagation, for the influence is felt at once, and lasts forever. The time of reaching the maximum however, depends on the distance.

It is evident in the case of this differential equation that the nature of the solution is the same whether we have one, two or three dimensions, for the only difference is in introducing the factor $t^{\frac{n}{2}}$ in the denominator. The time of the maximum y,

$$(102) \qquad\qquad t = \frac{2\alpha^2}{n} = \frac{r^2}{2na^2},$$

depends on the distance, being proportional to its *square*, and inversely to the number of dimensions in which the heat spreads. The maximum value of y,

$$(103) \qquad\qquad \frac{e^{-\frac{a^2}{t}}}{t^{\frac{n}{2}}} = \frac{e^{-\frac{n}{2}}}{\left(\dfrac{2\alpha^2}{n}\right)^{\frac{n}{2}}} = \frac{e^{-\frac{n}{2}}}{\left(\dfrac{r^2}{2na^2}\right)^{\frac{n}{2}}},$$

varies inversely as r^n.

Considered as a function of r,

(104)
$$\frac{u_0 \, d\tau}{\left(2a\pi^{\frac{1}{2}}\right)^n} \frac{e^{-\frac{r^2}{4a^2 t}}}{t^{\frac{n}{2}}}$$

is zero for $t = 0$, for all values of r, except $r = 0$. However the integral $\iiint u \, d\tau$, which when multiplied by the product of the specific heat and the density, represents the total amount of heat in the body, is finite and constant. In case $n = 1$, we find

(105)
$$\int_{-\infty}^{\infty} u \, dx = \frac{u_0}{2a\pi^{\frac{1}{2}}} \int_{-\infty}^{\infty} \frac{e^{-\frac{x^2}{4a^2 t}}}{t^{\frac{1}{2}}} \, dx,$$

and if we put $\dfrac{x}{2a\sqrt{t}} = \xi$, this reduces to

$$\frac{u_0}{\pi^{\frac{1}{2}}} \int_{-\infty}^{\infty} e^{-\xi^2} d\xi = u_0.$$

The function $\dfrac{e^{-\frac{x^2}{4a^2 t}}}{\sqrt{t}}$, representing the temperature in one dimensional flow from an initially hot point at $x = 0$, is shown in Fig. 60. The initial distribution is represented by the jag-function of § 35.

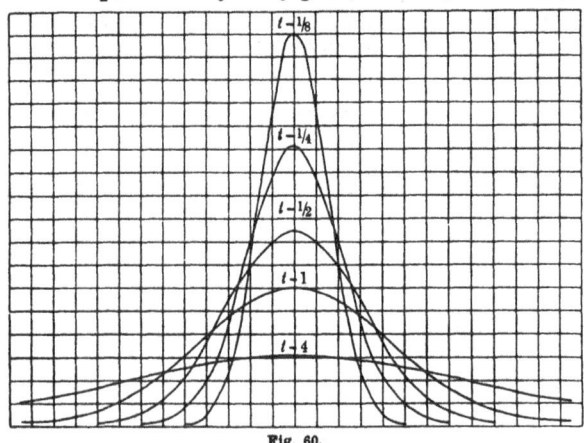

Fig. 60.

Suppose we consider the case of flow in one dimension, in which we have for

$$t = 0, \quad \begin{array}{l} F(x) = 0, \quad x > 0 \\ F(x) = 2, \quad x < 0. \end{array}$$

This will give the same distribution for $x > 0$ as if we considered only the positive half of a rod of infinite length, whose end at $x = 0$ is maintained at constant temperature, or electrically a cable at whose end a constant battery is applied. We then have from (96),

$$(106) \qquad u = \frac{1}{a\sqrt{\pi t}} \int_{-\infty}^{0} e^{-\frac{(x-\alpha)^2}{4a^2 t}} \, d\alpha$$

and if we put

$$\frac{\alpha - x}{2a\sqrt{t}} = \xi$$

this becomes

$$(107) \qquad u = \frac{2}{\sqrt{\pi}} \int_{-\infty}^{-\frac{x}{2a\sqrt{t}}} e^{-\xi^2} d\xi = \frac{2}{\sqrt{\pi}} \int_{\frac{x}{2a\sqrt{t}}}^{\infty} e^{-\xi^2} d\xi.$$

Now the definite integral,

$$(108) \qquad \Phi(x) = \frac{2}{\sqrt{\pi}} \int_{0}^{x} e^{-\xi^2} d\xi,$$

occurs in the theory of probability, and is known as Kramp's integral or the error-function. We have already made use of the result $\Phi(\infty) = 1$. We have then

$$(109) \qquad u = \frac{2}{\sqrt{\pi}} \int_{0}^{\infty} e^{-\xi^2} d\xi - \frac{2}{\sqrt{\pi}} \int_{0}^{\frac{x}{2a\sqrt{t}}} e^{-\xi^2} d\xi = 1 - \Phi\left(\frac{x}{2a\sqrt{t}}\right),$$

as the temperature at the time t at a distance x from the end of an infinitely long rod, originally at temperature zero, whose end since the time $t = 0$ has been maintained at the temperature unity. By means of tables of the error-function, Fig. 61 is easily drawn, giving the tempe-

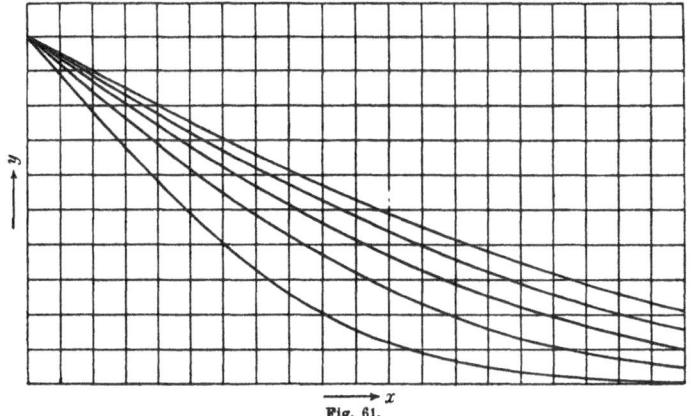

Fig. 61.

rature distribution for times $1, 2, 3, 4, 5$ times a. We see that the discontinuity in u at $x = 0$ is instantly abolished.

The solution (109) enables us to solve the problem of the distribution of temperature in the rod when the temperature of the end, instead of being constant, is made a given function of the time, $f(t)$. Let us first suppose that

$$f(t) = 0, \quad \text{for } \begin{cases} t < \tau, \\ t > \tau + h, \end{cases} \quad f(t) = 1, \quad \text{for } \tau < t < \tau + h.$$

If we call the function of (109) $U(x, t)$, representing the effect at time t of applying a temperature unity to the end of the rod at the time $t = 0$ and thereafter, we may get the result of the present assumption by combining $U(x, t - \tau)$, which represents the effect of heating the end from $t = \tau$ on, and $- U(x, t - \tau - h)$, which represents the effect of *cooling* the end from $\tau + h$ on, the result of both being to reduce the temperature of the end to zero again after $t = \tau + h$. We thus have

$$U(x, t - \tau) - U(x, t - \tau - h).$$

If now the interval h is made shorter, we have for an infinitely short interval $h\left(\dfrac{\partial U}{\partial t}\right)_{t-\tau}$. If the temperature of the end is not unity, we have to multiply this value by the actual temperature. Consequently if during the interval from $t = \tau$ to $\tau + d\tau$ the temperature of the end is $f(\tau)$, the effect at time t at other points will be $\left(\dfrac{\partial U}{\partial t}\right)_{t-\tau} f(\tau) d\tau$, and if we take the sum for all times we shall have

$$(110) \qquad u = \int_{-\infty}^{t} \left(\frac{\partial U}{\partial t}\right)_{t-\tau} f(\tau) d\tau.$$

This result was obtained by Duhamel.[1]) Effecting the differentiation of the definite integral we have

$$(111) \qquad \frac{\partial U}{\partial t} = \Phi'\left(\frac{x}{a\sqrt{t}}\right) \frac{x}{4 a t^{\frac{3}{2}}} = \frac{x e^{-\frac{x^2}{4 a^2 t}}}{2 a \sqrt{\pi} t^{\frac{3}{2}}},$$

so that our solution is

$$(112) \quad u = \frac{x}{2 a \sqrt{\pi}} \int_{-\infty}^{t} f(\tau) \frac{e^{-\frac{x^2}{4 a^2 (t-\tau)}}}{(t-\tau)^{\frac{3}{2}}} d\tau = \frac{x}{2 a \sqrt{\pi}} \int_{0}^{\infty} f(t-\chi) \frac{e^{-\frac{x^2}{4 a^2 \chi}}}{\chi^{\frac{3}{2}}} d\chi.$$

---

[1]) Duhamel, *Journal de l'Ecole Polytechnique*, Tom. 14, Cah. 22, p. 20, 1833.

This may be confirmed by the change of variable from τ to ξ, where

$$(113) \qquad t - \tau = \frac{x^2}{4\,a^2\xi^2} = \chi, \quad d\chi = -\frac{x^2 d\xi}{2\,a^2\xi^3} = -4\,\frac{a}{x}\,\chi^{\frac{3}{2}}d\xi,$$

giving

$$(114) \qquad u = \frac{2}{\sqrt{\pi}}\int_0^\infty f\Big(t - \frac{x^2}{4\,a^2\xi^2}\Big)e^{-\xi^2}d\xi,$$

which on putting $x = 0$ becomes

$$(115) \qquad \frac{2}{\sqrt{\pi}}f(t)\int_0^\infty e^{-\xi^2}d\xi = f(t).$$

Suppose that the temperature of the end varies harmonically, $f(t) = \sin pt$. Then

$$(116) \qquad u = \frac{2}{\sqrt{\pi}}\int_0^\infty \sin p\Big(t - \frac{x^2}{4\,a^2\xi^2}\Big)e^{-\xi^2}d\xi,$$

to evaluate which we expand the sine, obtaining

$$(117) \qquad u = \frac{2}{\sqrt{\pi}}\left\{\sin pt\int_0^\infty \cos\frac{px^2}{4\,a^2\xi^2}\,e^{-\xi^2}d\xi - \cos pt\int_0^\infty \sin\frac{px^2}{4\,a^2\xi^2}\,e^{-\xi^2}d\xi\right\}.$$

The definite integrals are those given in Appendix, if we use the substitution

$$\frac{px^2}{4\,a^2} = \frac{\alpha^2}{2}, \quad \alpha = \frac{x}{a}\sqrt{\frac{p}{2}},$$

giving

$$(118) \qquad \int_0^\infty {\cos\atop\sin}\frac{px^2}{4\,a^2\xi^2}\,e^{-\xi^2}d\xi = \frac{\sqrt{\pi}}{2}e^{-\frac{x}{a}\sqrt{\frac{p}{2}}}{\cos\atop\sin}\frac{x}{a}\sqrt{\frac{p}{2}},$$

so that we have

$$(119) \qquad u = e^{-\frac{x}{a}\sqrt{\frac{p}{2}}}\sin\Big(pt - \frac{x}{a}\sqrt{\frac{p}{2}}\Big).$$

This result will be found directly.

46. Equation of Telegraphy. The equation of telegraphy, § 15 (162)

$$(120) \qquad KL\frac{\partial^2 u}{\partial t^2} + (KR + LS)\frac{\partial u}{\partial t} + RSu = \frac{\partial^2 u}{\partial x^2}$$

when we neglect L, the self-inductance, and S, the leakage, reduces to Fourier's equation, with $a^2 = \frac{1}{KR}$. The equation was treated by Lord

Kelvin[1]), in 1855, for a submarine cable, and from it he deduced his so-called KR law; the time necessary to produce the maximum electrical effect at a distance x is proportional to KRx^2.

Let us now consider the telegraphic equation in its generality, writing it

$$(121) \qquad a\frac{\partial^2 u}{\partial t^2} + 2b\frac{\partial u}{\partial t} + cu = \frac{\partial^2 u}{\partial x^2}.$$

We must, in the Fourier integral, use

$$(122) \qquad a\frac{d^2 U}{dt^2} + 2b\frac{dU}{dt} + cU = -\lambda^2 U$$

and if we put $U = e^{pt}$, we have

$$(123) \qquad ap^2 + 2bp + c + \lambda^2 = 0, \quad p = -\frac{b}{a} \pm \frac{\sqrt{b^2 - a(c + \lambda^2)}}{a},$$

It will be convenient to remove the factor $e^{-\frac{b}{a}t}$ by putting $u = e^{-\frac{b}{a}t}u'$. We find

$$(124) \qquad \frac{\partial u}{\partial t} = e^{-\frac{b}{a}t}\left(\frac{\partial u'}{\partial t} - \frac{b}{a}u'\right), \quad \frac{\partial^2 u}{\partial x^2} = e^{-\frac{b}{a}t}\frac{\partial^2 u'}{\partial x^2},$$
$$\frac{\partial^2 u}{\partial t^2} = e^{-\frac{b}{a}t}\left(\frac{\partial^2 u'}{\partial t^2} - 2\frac{b}{a}\frac{\partial u'}{\partial t} + \frac{b^2}{a^2}u'\right),$$

$$a\left(\frac{\partial^2 u'}{\partial t^2} - 2\frac{b}{a}\frac{\partial u'}{\partial t} + \frac{b^2}{a^2}u'\right) + 2b\left(\frac{\partial u'}{\partial t} - \frac{b}{a}u'\right) + cu' = \frac{\partial^2 u'}{\partial x^2},$$

$$(125) \qquad a\frac{\partial^2 u'}{\partial t^2} + \frac{ac - b^2}{a}u' = \frac{\partial^2 u'}{\partial x^2}.$$

Accordingly the term in the first derivative has been removed. If we now divide by minus the coefficient of u', and change the units by putting

$$(126) \qquad t = \frac{a}{\sqrt{b^2 - ac}}t', \quad x = \sqrt{\frac{a}{b^2 - ac}}x',$$

the equation simplifies into

$$(127) \qquad \frac{\partial^2 u'}{\partial t'^2} = \frac{\partial^2 u'}{\partial x'^2} + u'.$$

(It is be noticed that if there is neither resistance nor leakage $b = c = 0$, we cannot make this simplification, so that the results we are now to get depend on the presence of at least one of these properties.)

We shall now drop the accents, and treat the equation,

$$(128) \qquad \frac{\partial^2 u}{\partial t^2} = \frac{\partial^2 u}{\partial x^2} + u,$$

---

1) On the Theory of the Electric Telegraph, Proc. Roy. Soc., May, 1855, Math. and Phys. Papers, Vol. III, p. 61.

for the solution of which we have

$$\frac{d^2 U}{dt^2} = (1 - \lambda^2)\, U,$$

(129) $$U = A \cos \sqrt{\lambda^2 - 1} \cdot t + B \sin \sqrt{\lambda^2 - 1} \cdot t,$$

$$\frac{dU}{dt} = \sqrt{\lambda^2 - 1} \left(- A \sin\sqrt{\lambda^2 - 1} \cdot t + B \cos\sqrt{\lambda^2 - 1} \cdot t \right),$$

and if for

(130) $$t = 0, \quad u = f(x), \quad \frac{\partial u}{\partial t} = g(x),$$

we must put

$$A = f(\alpha), \quad B = \frac{g(\alpha)}{\sqrt{\lambda^2 - 1}},$$

so that we have the solution,

(131)
$$u = \frac{1}{2\pi} \int_{-\infty}^{\infty}\!\!\int f(\alpha) \cos\sqrt{\lambda^2 - 1} \cdot t\, e^{i\lambda(x - \alpha)} d\alpha\, d\lambda$$

$$+ \frac{1}{2\pi} \int_{-\infty}^{\infty}\!\!\int g(\alpha) \frac{\sin\sqrt{\lambda^2 - 1} \cdot t}{\sqrt{\lambda^2 - 1}} e^{i\lambda(x - \alpha)} d\alpha\, d\lambda.$$

Again we may, though less easily than before, perform the integration with respect to λ. Consider the definite integral,

(132)
$$J(x) = \frac{1}{2\pi} \int_{-\pi}^{\pi} e^{ix \cos \omega} d\omega = \frac{1}{2\pi} \int_{-\pi}^{\pi} \sum_{n=0}^{n=\infty} \left[\frac{i^n x^n \cos^n \omega}{n!} \right] d\omega$$

$$= \frac{1}{2\pi} \sum_{n=0}^{n=\infty} \frac{i^n x^n}{n!} \int_{-\pi}^{\pi} \cos^n \omega\, d\omega,$$

Now we have

(133) $$\int_{0}^{\frac{\pi}{2}} \cos^n \omega\, d\omega = \frac{1 \cdot 3 \cdot 5 \cdots n - 1}{2 \cdot 4 \cdot 6 \cdots n} \frac{\pi}{2}$$

and the same integral from $\frac{\pi}{2}$ to π has the same value if n is *even*, the negative if n is odd. Also

$$\int_{0}^{\pi} \cos^n \omega\, d\omega = \int_{-\pi}^{0} \cos^n \omega\, d\omega,$$

so that

(134) $$\int_{-\pi}^{\pi} \cos^{2s} \omega\, d\omega = \frac{1 \cdot 3 \cdot 5 \cdots 2s - 1}{2 \cdot 4 \cdot 6 \cdots 2s} \cdot 2\pi = \frac{(2s)!}{2^{2s}(s!)^2} \cdot 2\pi,$$

and our integral is, putting $n = 2s$,

(135) $$J(x) = \sum_{s=0}^{s=\infty} \frac{(-1)^s}{(s!)^2}\left(\frac{x}{2}\right)^{2s} = 1 - \frac{x^2}{2^2} + \frac{x^4}{2^2 4^2} - \frac{x^6}{2^2 \cdot 4^2 \cdot 6^2} + \cdots.$$

This series defines the Bessel's function of order zero, (Chap. VII).

If we call

$$I(x) = J(ix) = \frac{1}{2\pi}\int_{-\pi}^{\pi} e^{-x\cos\omega} d\omega$$

(131)

$$= \sum_{s=0}^{s=\infty} \frac{1}{(s!)^2}\left(\frac{x}{2}\right)^{2s} = 1 + \frac{x^2}{2^2} + \frac{x^4}{2^2 \cdot 4^2} + \frac{x^6}{2^2 \cdot 4^2 \cdot 6^2} + \cdots$$

this function increases rapidly with x, and has *no* real roots.

Consider now the definite integral,

(137)

$$\int_0^{\pi} J(r\sin\varphi\sin\theta)e^{ir\cos\varphi\cos\theta}\sin\theta\, d\theta$$

$$= \frac{1}{2\pi}\int_0^{\pi}\int_{-\pi}^{\pi} e^{ir(\cos\varphi\cos\theta + \sin\varphi\sin\theta\cos\omega)}\sin\theta\, d\theta\, d\omega.$$

Let us put

Fig. 62.

$$\cos\Theta = \cos\theta\cos\varphi + \sin\theta\sin\varphi\cos\omega,$$
$$\sin\theta\, d\theta\, d\omega = dS,$$

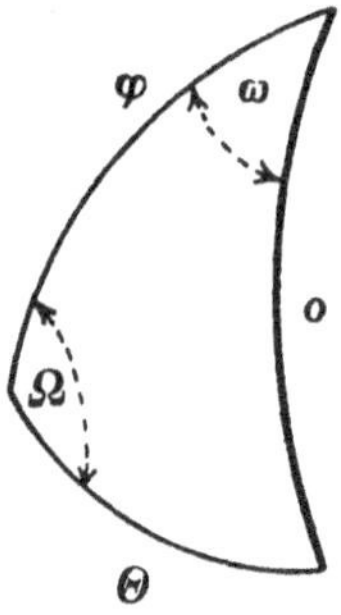

and the integral (137) becomes

(138) $$\frac{1}{2\pi}\int_0^{\pi}\int_{-\pi}^{\pi} e^{ir\cos\Theta} dS$$

and Θ will be one side of the spherical triangle of which the others are θ, φ, including the angle ω, Fig. 62. Let Ω be the angle included between φ, Θ and since our double integral is over the surface of the sphere of unit radius, let us take the pole at Ω instead of ω, when the element of area will be

$$dS = \sin\Theta\, d\Theta\, d\Omega,$$

and our integral is

(139) $$\frac{1}{2\pi}\int_0^{\pi}\int_{-\pi}^{\pi} e^{ir\cos\Theta}\sin\Theta\, d\Theta\, d\Omega = \frac{e^{ir} - e^{-ir}}{ir} = \frac{2\sin r}{r}.$$

Consequently

(140) $$\frac{\sin r}{r} = \frac{1}{2}\int_0^{\pi} J(r\sin\varphi\sin\theta)e^{ir\cos\varphi\cos\theta}\sin\theta\, d\theta.$$

Now put

$$r\cos\varphi = -\lambda t, \quad r\sin\varphi = it, \quad r^2 = t^2(\lambda^2 - 1),$$

$$\cos\theta = \frac{\beta}{t}, \quad -\sin\theta\, d\theta = \frac{d\beta}{t},$$

$$\frac{\sin t\sqrt{\lambda^2 - 1}}{t\sqrt{\lambda^2 - 1}} = \frac{1}{2}\int_{-t}^{t} J\left(it\sqrt{1 - \frac{\beta^2}{t^2}}\right)e^{-i\lambda\beta}\frac{d\beta}{t},$$

$$(141) \qquad \frac{\sin t\sqrt{\lambda^2 - 1}}{\sqrt{\lambda^2 - 1}} = \frac{1}{2}\int_{-t}^{t} I(\sqrt{t^2 - \beta^2})e^{-i\lambda\beta}\, d\beta.$$

Inserting in our second integral of (131),

$$(142) \qquad u_2 = \frac{1}{4\pi}\int_{-\infty}^{\infty}d\lambda\int_{-\infty}^{\infty}d\alpha\, g(\alpha)\int_{-t}^{t} I(\sqrt{t^2 - \beta^2})e^{i\lambda(x - \alpha - \beta)}\, d\beta.$$

If we change the order of integration, so that, after having integrated with respect to β we integrate with respect to λ before α, we may put the result in the form of a Fourier integral. For if in the Fourier integral,

$$(143) \qquad \frac{1}{2\pi}\int_{-\infty}^{\infty}d\lambda\int_{-\infty}^{\infty}\varphi(\beta)e^{i\lambda(x - \beta)}d\beta = \varphi(x),$$

we make

$$\varphi(x) = 0 \qquad\qquad \text{when} \quad |x| > |t|,$$

$$\varphi(x) = I(\sqrt{t^2 - x^2}) \quad \text{when} \quad |x| < |t|,$$

then the integration with respect to β goes only from $-t$ to t and our integral (142) becomes, replacing all the accents (even on α),

$$(144)\ u_2' = \frac{1}{2}\int_{-\infty}^{\infty}g(\alpha')\varphi(x' - \alpha')d\alpha' = \frac{1}{2}\int_{x' - t'}^{x' + t'}g(\alpha')I(\sqrt{t'^2 - (x' - \alpha')^2})d\alpha'.$$

From this we obtain by Stokes's rule the other term

$$(145) \qquad \begin{aligned} u_1' &= \frac{1}{2}\frac{\partial}{\partial t'}\int_{x' - t'}^{x' + t'}f(\alpha')I(\sqrt{t'^2 - (x' - \alpha')^2})d\alpha \\ &= \frac{1}{2}[f(x' + t') + f(x' - t')] + \frac{1}{2}\int_{x' - t'}^{x' + t'}f(\alpha')\frac{\partial}{\partial t'}I(\sqrt{t'^2 - (x' - \alpha')^2})d\alpha. \end{aligned}$$

since $I(0) = 1$.

We may now change the variables back to x, t, u, and introduce the factor $e^{-\frac{b}{a}t}$. If the initial conditions are now

$$t = 0, \quad u = F(x), \quad \frac{\partial u}{\partial t} = G(x),$$

we must put, according to (124), (130),

$$F(x) = f(x'), \quad G(x) = g(x') - \frac{b}{a} f(x'), \quad x' \pm t' = \sqrt{\frac{b^2 - ac}{a}} \left(x \pm \frac{t}{\sqrt{a}} \right).$$

If the relation between α and α' is the same as between x and x', we at once obtain as the definite solution,

$$(146) \quad \begin{aligned} u = \frac{e^{-\frac{b}{a}t}}{2} & \left\{ F\left(x + \frac{t}{\sqrt{a}} \right) + F\left(x - \frac{t}{\sqrt{a}} \right) \right. \\ & + \sqrt{a} \int_{x-\frac{t}{\sqrt{a}}}^{x+\frac{t}{\sqrt{a}}} F(\alpha) \frac{\partial}{\partial t} I \sqrt{\frac{(b^2 - ac)}{a} \left\{ \frac{t^2}{a} - (x - \alpha)^2 \right\}} \, d\alpha \\ & + \sqrt{a} \int_{x-\frac{t}{\sqrt{a}}}^{x+\frac{t}{\sqrt{a}}} \left(G(\alpha) + \frac{b}{a} F(\alpha) \right) I \sqrt{\frac{(b^2 - ac)}{a} \left\{ \frac{t^2}{a} - (x - \alpha)^2 \right\}} \, d\alpha \right\}. \end{aligned}$$

The terms outside the integral signs show wave-propagation with velocity $\frac{1}{\sqrt{a}}$, the same as if the terms in b, c were absent. Thus the telegraphic equation is so far similar to the pure wave-equation, but the wave is damped according to the exponential factor. There is however an entirely new phenomenon, for the integral terms show, even after the wave has passed on, an effect coming from all points where originally F and G are not zero, and within a distance $\frac{t}{\sqrt{a}}$ from the point in question, that may be covered in the time t with the wave-velocity. This effect persists through all time, although dying away, and constitutes a residue or *tail*, like that left by a drop of dirty mercury rolling over a table. As an example, let u stand for the potential V, which is connected with the current by the equation (154), Chapter 1. Let us suppose that there is no leakage $c = 0$, and that there is initially nowhere current in the line, and that the potential is zero except from $x = x_1$ to $x = x_2$. Then we have

for all x, $G(x) = 0$; $F(x) = 0$, except for $x_1 < x < x_2$.

Accordingly for

$$x - \frac{t}{\sqrt{a}} > x_2, \quad \text{or} \quad x + \frac{t}{\sqrt{a}} < x_1,$$

all the functions have the value zero, so that u is zero. This at once distinguishes the solution from that in the case of heat or of the cable, for instead of being felt instantaneously the disturbance does not arrive until the time $t = \sqrt{a}\,(x - x_2)$, on the right, or $t = \sqrt{a}\,(x_1 - x)$ on the left (The velocity is infinite when $a = 0$, as we have seen). When

$$t > \sqrt{a}\,(x - x_1) \quad \text{or} \quad \sqrt{a}\,(x_2 - x)$$

we have

$$F\left(x + \frac{t}{\sqrt{a}}\right) = F\left(x - \frac{t}{\sqrt{a}}\right) = 0,$$

and

$$(147) \quad u = \frac{e^{-\frac{b}{a}t}}{2} \int_{x_1}^{x_2} F(\alpha)\left(\frac{b}{a} + \frac{\partial}{\partial t}\right) I \sqrt{\frac{(b^2 - ac)}{a}\left\{\frac{t^2}{a} - (x - \alpha)^2\right\}}\, d\alpha.$$

This integral never vanishes, and represents the tail of the wave. In this respect the telegraphic equation resembles the Fourier equation.

In order to give a concrete idea of the solution, we give in Fig. 63 the result of permanently putting the end of the line at a constant potential, at times 1, 2, 3, 4, 5, times $\frac{a}{b}$, with the dotted lines showing the same with no inductance, as in Fig. 61.

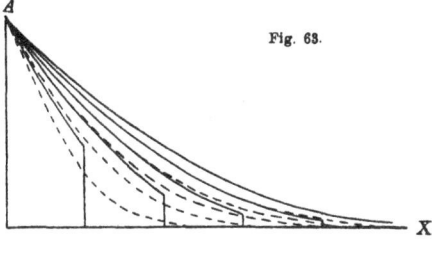

Fig. 63.

In Fig. 64 we show the propagation of a potential which was constant from A to B, the tail being shown after the waves have separated.

47. Periodic Waves in Various Cases.
Having now seen how the equations of heat conduction and of telegraphy differ from that of sound, we may look at them from a point of view which may bring out some similarities, while still showing the characteristic differences. Suppose we inquire under what circumstances a solution which is a simple harmonic function of the time can be propagated according to the equation

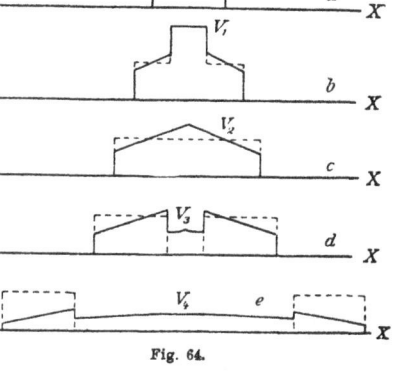

Fig. 64.

$$(121) \quad a\frac{\partial^2 u}{\partial t^2} + 2b\frac{\partial u}{\partial t} + cu = \frac{\partial^2 u}{\partial x^2},$$

whether or not a, b, or c are zero. Let us proceed as in Chapter III, where we have put $u = \varphi \cos pt$, only in this case, it will be more convenient to put $u = \varphi(x)e^{ipt}$, and take the real part of the resulting complex solution. We thus obtain

$$(148) \qquad (c - ap^2 + 2bpi)\varphi = \frac{d^2\varphi}{dx^2}$$

or

$$(149) \qquad \frac{d^2\varphi}{dx^2} + k^2\varphi = 0,$$

if we put

$$(150) \qquad ap^2 - c - 2bpi = k^2.$$

Now, this is the same equation that we had for the normal functions of the string or pipe, but in this case k^2, and therefore k, is complex. Suppose we put

$$k = \alpha - i\beta,$$

then we have, squaring, and equating real parts

$$(151) \qquad \alpha^2 - \beta^2 = ap^2 - c, \quad \alpha\beta = bp,$$

solving which for α and β gives the real positive values

$$(152) \qquad \begin{aligned} \alpha &= \sqrt{\tfrac{1}{2}\left(\sqrt{4b^2p^2 + (ap^2 - c)^2} + ap^2 - c\right)}, \\ \beta &= \sqrt{\tfrac{1}{2}\left(\sqrt{4b^2p^2 + (ap^2 - c)^2} - (ap^2 - c)\right)}. \end{aligned}$$

The solution of (149) is either

$$\cos kx, \quad \sin kx, \quad \text{or} \quad e^{ikx}, \ e^{-ikx}.$$

The two first are appropriate for standing, the two latter for running waves. Let us first consider the latter. We then have

$$u = e^{i(pt - kx)} = e^{-\beta x}e^{i(pt - \alpha x)},$$

of which the real part is

$$e^{-\beta x}\cos(pt - \alpha x)$$

which represents a damped wave progressing to the right, of period $T = \frac{2\pi}{p}$, wave-length $\lambda = \frac{2\pi}{\alpha}$, and velocity $v = \frac{p}{\alpha} = \frac{\beta}{b}$. In advancing a constant distance, the amplitude of the wave falls off in a constant ratio, the ratio of decrease for one wave-length being $e^{-\frac{2\pi\beta}{\alpha}}$. We may call the distance in travelling which the amplitude decreases in the ratio e^{-1} the relaxation-distance $\frac{1}{\beta}$. We see that in general the velocity, wave-length, and relaxation-distance all depend on the frequency, or the waves suffer dispersion, which explains how the general disturbance composed of waves of all frequencies must be distorted in propagation, as the waves separate. This explains the difficulty of long-distance telephony, or of submarine telegraphy or telephony on lines of any length

In order to get an idea of the dependence of these quantities on the frequency, let us eliminate p from equations (151), obtaining

$$\alpha^2 - \beta^2 = \frac{a}{b^2}\,\alpha^2\beta^2 - c, \quad \text{or} \quad \frac{1}{\beta^2} - \frac{1}{\alpha^2} = \frac{a}{b^2} - \frac{c}{\alpha^2\beta^2}.$$

Taking α and β as coordinates, this represents a quartic curve with a horizontal asymptote $\beta = \dfrac{b}{\sqrt{a}}$, and cutting the β-axis at a distance $\sqrt{c}$ from the origin, Fig. 65. We may then plot the frequency from the

equation $p = \dfrac{\alpha\beta}{b}$.
The velocity
$v = \dfrac{\beta}{b}$ is proportional to the
ordinate β, and
the relaxation-
distance to its
reciprocal. It
accordingly is
plain that as p
increases α in-
creases, and
that if $\sqrt{c} < \dfrac{b}{\sqrt{a}}$
the velocity in-

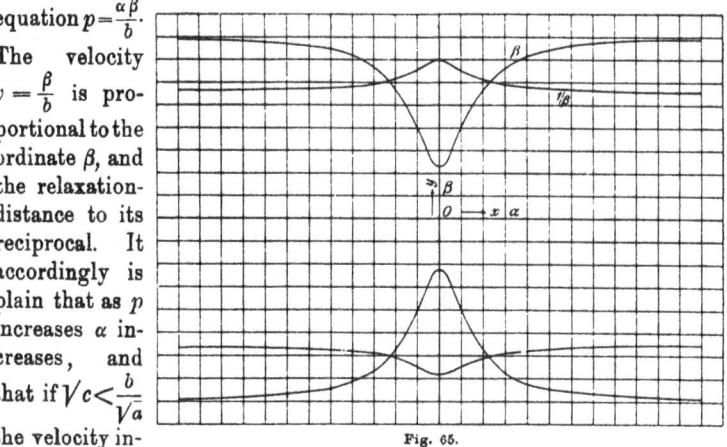

Fig. 65.

creases with p, but if the leakage is so great that $\sqrt{c} > \dfrac{b}{\sqrt{a}}$, the velocity decreases as p increases, in either case approaching the limiting velocity $\dfrac{1}{\sqrt{a}}$ for infinite frequency. This we have found above to be the velocity of a discontinuity, forming the front of a wave.

In the case of vanishing of the discriminant $b^2 - ac$ our locus becomes a straight line $\beta = \sqrt{c}$, the velocity is independent of the frequency $v = \dfrac{1}{\sqrt{a}}$, and there is no dispersion. This is called by Heaviside the distortionless case, and includes the one already dealt with when $b = c = 0$. There is still damping, but as it is the same for all wavelengths all disturbances are propagated without change of shape. If $c = 0$ the locus is a hyperbola.

In so far periodic waves behave alike whether in heat conduction, in the cable, or in the inductive telegraph line. In the two former, with no side loss, $a = c = 0$, $\alpha = \beta = \sqrt{bp}$ and the velocity $v = \sqrt{\dfrac{p}{b}}$ is proportional to the square root of the frequency, as in the case of waves in a bar, § 41. In this case the loss of amplitude per wave-length is $e^{-2\pi} = .001867 = \frac{1}{536}$. This is called by Lord Kelvin the case of pure diffusion, for liquids and gases diffuse according to this law. An interesting case is that of the propagation of the annual wave of temperature below the surface of the earth (considered plane) in which the wave-length is found to be about 15 meters, so that in this small depth the seasonal changes of temperature are practically obliterated. The appearance of either a or c, unless $ap^2 = c$, tends to make α and β variable.

It is to be noted that equation (150) is simply one relation between p and k, and that we may take either one real. Here we have taken p real, making k complex, but in § 44 and the applications we have taken k real, which has made p complex, that is has introduced a damping exponential in the time, which we have removed by a change of variable. It is evidently indifferent for a progressive wave, whether we consider the damping according to time or space, since the wave traverses equal spaces in equal times. Cauchy's method of the Fourier integral then consists in resolving the initial circumstances into a continuous spectrum of waves, considering the propagation of all, and recombining.

48. Standing Waves. Normal Functions. In case we determine instead of p, the frequency, the value of k by means of certain boundary conditions, we obtain normal functions as in Chapter III, but we do not obtain undamped standing waves. Equation (150) now determines the values of p, which will in general be *complex*, unless $b = 0$, thus denoting waves dying away *in situ*. In the case of heat conduction, $a = 0$, we have

$$p\,i = -\frac{k^2 + c}{2\,b}$$

and we have the solutions

(153) $$u = e^{-\frac{k^2 + c}{2b}\,t}\{A\cos kx + B\sin kx\}$$

and there is no oscillation of temperature, but every term dies away, the terms of shorter wave-length dying away most rapidly. Thus all space irregularities of temperature tend to be smoothed out. If there is no sidewise loss, as in the case of an infinite plane slab, $c = 0$, and the rate of damping is proportional to k^2, that is varies inversely as the square of the wave-length. By the combination of such solutions as (153) in series we may represent a large variety of initial and boundary conditions.

We shall take up a few of the more obvious problems. We have already considered in Chapter III the case of steady flow in one dimension, and the Green's function, due to a source of heat. If an infinite slab have its faces maintained at constant temperatures u_0, u_1, at $x = 0$, $x = l$, or a bar without sidewise loss have its ends thus maintained, we have the solution of $\dfrac{d^2 u}{d x^2} = 0$,

$$(154) \qquad u = u_0 + (u_1 - u_0) \frac{x}{l}.$$

If the bar radiates, $c \neq 0$, the solution of

$$(155) \qquad \frac{d^2 u}{d x^2} - cu = 0$$

is

$$(156) \qquad u = \frac{(u_1 \sinh \sqrt{c} \cdot x + u_0 \sinh \sqrt{c}\,(l - x))}{\sinh \sqrt{c} \cdot l}$$

If the flow is not steady, and the conditions at the boundary are the same, we may add these solutions respectively to the solution for any initial distribution of temperature with $u = 0$ at both boundaries. But this solution demands values of k roots of $\sin kl = 0$, and we have the sine series

$$(157) \qquad u = e^{-\frac{ct}{2b}} \sum_{r=0}^{r=\infty} B_r e^{-\frac{r^2 \pi^2}{2 b l^2}t} \sin \frac{r \pi x}{l}$$

where the B's are determined by the distribution of u as a function of x when $t = 0$. Making use of the development of a constant, (38) and of x, (39) in sines, we find the solution, for $c = 0$, which keeps the boundaries at u_0, u_1, and makes initially $u = f(x)$, to be

$$(158) \quad u = u_0 + (u_1 - u_0) \frac{x}{l} - \frac{2}{\pi} \sum_{1}^{\infty} \frac{1}{r} \{u_0 - (-1)^r u_1\} e^{-\frac{r^2 \pi^2}{2 b l^2}t} \sin \frac{r \pi x}{l}$$

$$+ \frac{2}{l} \sum_{1}^{\infty} e^{-\frac{r^2 \pi^2}{2 b l^2}t} \sin \frac{r \pi x}{l} \int_0^l f(\alpha) \sin \frac{r \pi \alpha}{l} \, d\alpha.$$

From this solution we can find the solution when the boundary temperatures u_0, u_1 are arbitrary functions of the time as we did in (110) for the infinite medium. For the effect of the temperatures u_0 and u_1 maintained from $t = 0$ until t has been to cause rise of temperature

$$(159) \qquad \frac{\partial u}{\partial t} dt = dt \frac{\pi}{b l^2} \sum_{1}^{\infty} r \{u_0 - (-1)^r u_1\} e^{-\frac{r^2 \pi^2}{2 b l^2}t} \sin \frac{r \pi x}{l},$$

in the time from t to $t + dt$. Accordingly the effect of temperature u_0 at a time $t = \tau$ to $\tau + d\tau$ is found at a later time t by replacing t by $t - \tau$ in (159) and thus the whole effect is

$$(160) \quad u = \frac{\pi}{b l^2} \sum_{1}^{\infty} r \sin \frac{r \pi x}{l} \int_0^t \{u_0(\tau) - (-1)^r u_1(\tau)\} e^{-\frac{r^2 \pi^2 (t - \tau)}{2 b l^2}} \, d\tau.$$

As a practical example consider the potential in a cable put to earth at the distant end $u_1 = 0$, when the key is closed for a time a at the sending end, so that $u_0 =$ const from $0 < \tau < a$, and otherwise is zero.

(161)
$$u = \frac{\pi u_0}{b l^2} \sum_1^\infty r \sin \frac{r\pi x}{l} \int_0^a e^{-\frac{r^2 \pi^2}{2 b l^2}(t-\tau)} d\tau, \qquad t > a$$

$$= \frac{2 u_0}{\pi} \sum_1^\infty \frac{1}{r} \left\{ e^{\frac{r^2 \pi^2}{2 b l^2}a} - 1 \right\} e^{-\frac{r^2 \pi^2}{2 b l^2}t} \sin \frac{r\pi x}{l},$$

which applies to the time *after* the charging key has been released, or if $t < a$ the limit a must be replaced by t, giving

(162)
$$u = \frac{2 u_0}{\pi} \sum_1^\infty \frac{1}{r} \left\{ 1 - e^{-\frac{r^2 \pi^2}{2 b l^2}t} \right\} \sin \frac{r\pi x}{l}$$

$$= u_0 \left\{ \frac{l-x}{l} - \frac{2}{\pi} \sum_1^\infty \frac{1}{r} e^{-\frac{r^2 \pi^2}{2 b l^2}t} \sin \frac{r\pi x}{l} \right\}$$

which, being independant of a, is the same as if the battery were permanently connected. The value (161) tends toward 0, while (162) tends toward $\frac{u_0(l-x)}{l}$ as t increases.

The current in this case is found by differentiation by x

(163)
$$I = -\frac{1}{R} \frac{\partial u}{\partial x} = \frac{u_0}{R l} \left\{ 1 + 2 \sum_1^\infty e^{-\frac{r^2 \pi^2}{2 b l^2}t} \cos \frac{r\pi x}{l} \right\}.$$

At the distant end $x = l$, this becomes, putting $e^{-\frac{\pi^2}{2 b l^2}t} = v$,

(164)
$$I_{x=l} = \frac{u_0}{R l} \left\{ 1 - 2 (v - v^4 + v^9 - \cdots) \right\} = f(t),$$

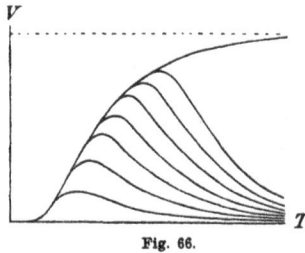

Fig. 66.

a very rapidly convergent series when $v < 1$, from which Lord Kelvin calculated the curves in Fig. 66 given in his famous paper "On the Theory of the Electric Telegraph", appearing in 1855. In the figure the scale is such that the side of one square represents a time a for which $v = \frac{3}{4}$ i.e., $a = \frac{2 b l^2}{\pi^2} \log_e \frac{4}{3}$.

If instead of being held down indefinitely the key is held down for a time a, $2a$, $3a$, etc. the current is found from the formulae

$$f(t) - f(t - a),$$
$$f(t) - f(t - 2a),$$
$$\cdots \cdots \cdots \cdots$$

the effect of which is shown in the curves, 1, 2, 3, etc.

If instead of having the temperatures given at the boundaries, the slab is impenetrable to heat, or the cable is insulated, the condition is $\frac{d\varphi}{dx} = 0$, and we must use a cosine series. Instead of this problem, let us consider the more interesting one in which the boundary faces of the slab radiate heat, assuming the law of Newton that the heat current radiated is proportional to the excess of temperature of the surface of the slab over that of the surrounding medium. Since the heat current arriving from the inside is $- k \frac{\partial u}{\partial x}$, putting this equal to that radiated at the right-hand boundary, we have

$$(165) \qquad - k \frac{\partial u}{\partial x} = b (u - u_0)$$

where u_0 is the temperature of the external medium, and b is called the thermal emissivity of the solid. Since u_0 is constant, we may on the left write $\frac{\partial}{\partial x}(u - u_0)$, and using the single letter u for this excess, and h for $\frac{b}{k}$, we have the condition

$$(166) \qquad \frac{\partial u}{\partial x} + hu = 0,$$

which we have treated in § 39. At the left-hand face the current *toward* the face has the opposite sign, so that we have for the space factor.

$$(167) \qquad \frac{d\varphi}{dx} - h\varphi = 0, \quad x = 0. \quad \frac{d\varphi}{dx} + h\varphi = 0, \quad x = l.$$

Inserting in the solution

$$(168) \qquad \begin{aligned} \varphi &= A \cos kx + B \sin kx, \\ \frac{d\varphi}{dx} &= k \left\{ - A \sin kx + B \cos kx \right\}, \end{aligned}$$

we have

$$(169) \qquad \begin{aligned} Bk - hA &= 0, \\ A(h \cos kl - k \sin kl) + B(h \sin kl + k \cos kl) &= 0. \end{aligned}$$

Eliminating A and B we have the equation for k,

$$(170) \qquad \tan kl = \frac{2hk}{k^2 - h^2}.$$

If we put $kl = x$ we may write this

$$(171) \qquad \frac{2}{\tan x} = \frac{x}{hl} - \frac{hl}{x},$$

and the values of x are found from the intersections of the curves

$$(172) \qquad y = \frac{2}{\tan x}, \quad y = \frac{x}{hl} - \frac{hl}{x},$$

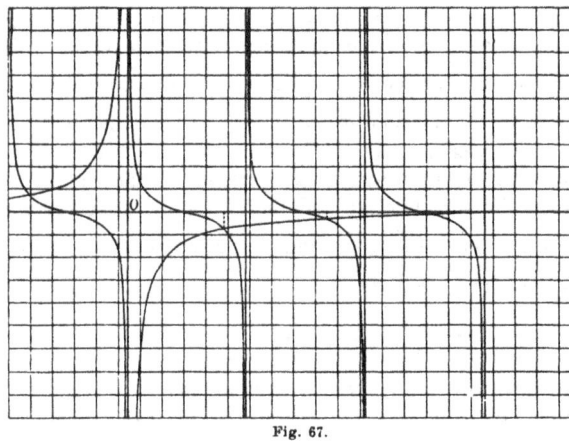

Fig. 67.

of which the second is an hyperbola(Fig. 67). The value $k = 0$ is to be excluded, because, dividing (170) by kl, and passing to the limit $k = 0$, we should have $2 = -hl$, which is impossible, as h is positive. Using the first of equations (169), we have the normal function,

$$(173) \qquad \varphi_r = k_r \cos k_r x + h \sin k_r x.$$

We have shown in § 39 that for the condition (166) the normal functions are orthogonal, it accordingly remains only to normalize them. From the differential equation by integration by parts we find

$$(174) \qquad k_r^2 \int_0^l \varphi_r^2 dx = -\int_0^l \varphi_r \frac{d^2 \varphi_r}{dx^2} dx = -\left. \varphi_r \frac{d\varphi_r}{dx} \right|_0^l + \int_0^l \left(\frac{d\varphi_r}{dx}\right)^2 dx.$$

But from the value of φ_r, (173),

$$(175) \qquad k_r^2 \varphi_r^2 + \left(\frac{d\varphi_r}{dx}\right)^2 = k_r^2(k_r^2 + h^2),$$

and integrating,

$$(176) \qquad k_r^2 \int_0^l \varphi_r^2 dx + \int_0^l \left(\frac{d\varphi_r}{dx}\right)^2 dx = k_r^2(k_r^2 + h^2)l.$$

Adding to (174) we obtain

$$(177) \qquad 2 k_r^2 \int_0^l \varphi_r^2 dx = -\left. \varphi_r \frac{d\varphi_r}{dx} \right|_0^l + k_r^2(k_r^2 + h^2)l.$$

But at either limit, since

$$(167) \qquad \frac{d\varphi_r}{dx} = \pm h\varphi_r$$

we find by (173)

$$\varphi_r^2 = k_r^2, \quad -\varphi_r \frac{d\varphi_r}{dx} = \pm hk_r^2.$$

Thus we finally obtain for the integrated square,

$$(178) \qquad \int_0^l \varphi_r^2 \, dx = (k_r^2 + h^2)\frac{l}{2} + h \,,$$

using which in the development of the initial value $u = f(x)$, we obtain

$$(179) \qquad \sum_{r=1}^{r=\infty} e^{-\frac{k_r^2 t}{2b}} \frac{k_r \cos kx + h \sin k_r x}{h + (k_r^2 + h^2) l/2} \int_0^l (k_r \cos k_r \alpha + h \sin k_r \alpha) f(\alpha) \, d\alpha.$$

49. Wave Equation in two Dimensions. We shall conclude this chapter with the treatment of the wave-equation in two dimensions, which has already been mentioned as presenting a notable difference from the cases of one and three dimensions, the reasons for which will be gone into in Chapter VI. The equation

$$(180) \qquad \frac{\partial^2 u}{\partial t^2} = a^2 \left\{ \frac{\partial^2 u}{\partial x^2} + \frac{\partial^2 u}{\partial y^2} \right\}$$

leads to

$$(181) \qquad \frac{d^2 U}{dt^2} = -a^2(\lambda^2 + \mu^2) U = -a^2 \varrho^2 U \,,$$

and as in the case of three dimensions,

$$(182) \qquad U = F(\alpha, \beta) \cos \varrho t + G(\alpha, \beta) \frac{\sin a\varrho t}{a\varrho} \,,$$

$$(183) \quad u = \frac{1}{4\pi^2} \int\int\int\int_{-\infty}^{\infty} \left\{ F(\alpha, \beta) \cos a\varrho t + G(\alpha, \beta) \frac{\sin a\varrho t}{a\varrho} \right\} e^{i[\lambda(x-\alpha) + \mu(y-\beta)]} \, d\alpha \, d\beta \, d\lambda \, d\mu.$$

We have now to put in the plane,

$$(184) \qquad \begin{aligned} \alpha - x &= r \cos \theta & \lambda &= \varrho \cos \varphi \\ \beta - y &= r \sin \theta & \mu &= \varrho \sin \varphi \\ \lambda(\alpha - x) + \mu(\beta - y) &= r\varrho \cos \theta' \,, \end{aligned}$$

where the angle θ' is counted from the direction of r. Thus the second term is

$$(185) \qquad u_2 = \frac{1}{4\pi^2} \int_0^\infty \int_0^\infty \int_0^{2\pi} \int_0^{2\pi} G(\alpha, \beta) \frac{\sin a\varrho t}{a\varrho} e^{ir\varrho \cos \theta'} r \, dr \, d\varrho \, d\theta \, d\theta'.$$

Now we have by (132),

$$(132) \qquad \frac{1}{2\pi} \int_0^{2\pi} e^{ir\varrho \cos \theta'} \, d\theta' = J(r\varrho)$$

so that

(186) $\qquad u_2 = \dfrac{1}{2\pi a} \displaystyle\int_0^\infty \int_0^\infty \int_0^{2\pi} G(\alpha, \beta) \sin a\varrho t J_0(r\varrho) r\, dr\, d\varrho\, d\theta.$

But we have the definite integral, Appendix

$$\int_0^\infty \sin ax J_0(bx)\, dx = \frac{1}{\sqrt{a^2 - b^2}}, \quad b^2 < a^2 \text{ or } = 0, \quad b^2 > a^2$$

using which we obtain

(187)
$$u_2 = \frac{1}{2\pi a} \int_0^\infty \int_0^{2\pi} G(\alpha, \beta) r\, dr\, d\theta \int_0^\infty \sin a t\varrho\, J_0(r\varrho)\, d\varrho$$

$$= \frac{1}{2\pi a} \int_0^{at} \int_0^{2\pi} \frac{G(\alpha, \beta) r\, dr\, d\theta}{\sqrt{a^2 t^2 - r^2}},$$

(188) $\qquad u = \dfrac{1}{2\pi a}\dfrac{\partial}{\partial t} \displaystyle\int_0^{at} \int_0^{2\pi} \dfrac{F(\alpha, \beta) r\, dr\, d\theta}{\sqrt{a^2 t^2 - r^2}} + \dfrac{1}{2\pi a} \int_0^{at} \int_0^{2\pi} \dfrac{G(\alpha, \beta) r\, dr\, d\theta}{\sqrt{a^2 t^2 - r^2}}.$

This formula, known as that of Poisson and Parseval, shows that at a time t the disturbance at a point depends on the initial values F, G, not only at all points on the circumference of a circle of radius at, but also at all points nearer, that is the effect persists, or the wave leaves a tail behind, as in the case of the damped wave, but quite different from the pure wave-propagation in one and three dimensions.

CHAPTER V.

METHODS OF GREEN. POTENTIALS. BOUNDARY PROBLEMS.

50. Divergence Theorem. The methods next to be described are dependent on the theorem of George Green contained in his celebrated Essay on the Application of Mathematical Analysis to the theories of Electricity and Magnetism, Nottingham, 1828. The theorem has to do with integrations of certain volume integrals. If W is a function of a point x, y, z, which, with its derivative in any direction, is uniform and continuous in a region of space τ bounded by a closed surface S, the volume integral

(1) $\qquad\qquad J = \displaystyle\int \int \int \frac{\partial W}{\partial x}\, dx\, dy\, dz$

throughout τ is finite, for $\dfrac{\partial W}{\partial x}$ is finite.

We have

$$J = \iint dy\, dz \int \frac{\partial W}{\partial x}\, dx .$$

Suppose we perform the summation by taking first all the elements

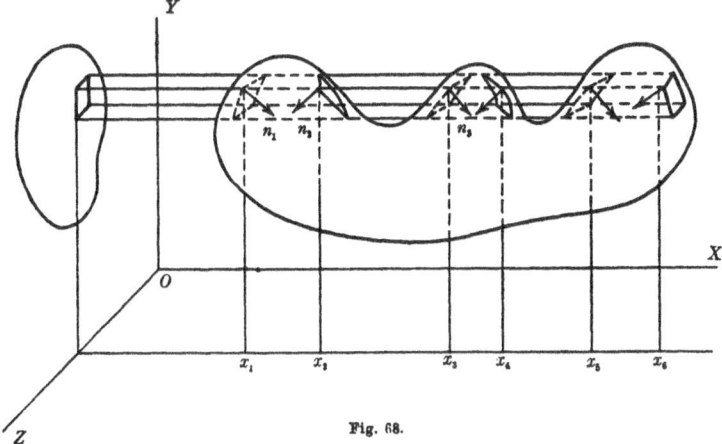

Fig. 68.

of volume $d\tau = dx\, dy\, dz$ contained in a prism parallel to the X-axis, standing on the base $dy\, dz$, and cutting the surface S at

$$x_1, x_2, \ldots x_{2n}, \quad \text{(Fig. 68)}.$$

Then

(2) $$\int \frac{\partial W}{\partial x}\, dx = W_2 - W_1 + W_4 - W_3 \cdots + W_{2n} - W_{2n-1},$$

where

$$W_r = W(x_r, y, z),$$

so that

(3) $$J = \iint dy\, dz\, (W_2 - W_1 \cdots - W_{2n-1}).$$

Now if $dS_1, dS_2, \ldots$ denote the areas cut out of S by the prism where it cuts in to or out from τ, all these areas have the same orthogonal projection $dy\, dz$. If $n_1, n_2, \ldots$ are the normals drawn *inwards* to τ from S, wherever the prism cuts in we have

(4) $$dS_r \cos(n_r x) = dy\, dz ,$$

and where it cuts out,

(5) $$- dS_r \cos(n_r x) = dy\, dz .$$

Thus

(6) $$J = - \iint \sum_{r=1}^{r=2n} W_r \cos(n_r x)\, dS_r .$$

If we now take all the prisms necessary to cover the volume τ every element of S will also appear, and the sum will take in all elements of the surface. Accordingly J becomes the surface-integral

$$(7) \qquad J = -\iint W \cos(nx) dS.$$

In like manner replacing x by y and z,

$$(8) \qquad \begin{aligned} \iiint \frac{\partial W}{\partial x} d\tau &= -\iint W \cos(nx) dS, \\ \iiint \frac{\partial W}{\partial y} d\tau &= -\iint W \cos(ny) dS, \\ \iiint \frac{\partial W}{\partial z} d\tau &= -\iint W \cos(nz) dS. \end{aligned}$$

Let us apply these equalities to three separate functions of x, y, z, X, Y, Z. Adding the three results,

$$(9) \qquad \begin{aligned} &\iiint \left\{ \frac{\partial X}{\partial x} + \frac{\partial Y}{\partial y} + \frac{\partial Z}{\partial z} \right\} d\tau \\ &= -\iint \left\{ X \cos(nx) + Y \cos(ny) + Z \cos(nz) \right\} dS. \end{aligned}$$

This theorem is best interpreted when X, Y, Z, are the components of a *vector* function $\mathbf{A}$,

$$X = \mathbf{A}_x, \quad Y = \mathbf{A}_y, \quad Z = \mathbf{A}_z.$$

Thus we see that the integrand in the surface integral is the negative projection of $\mathbf{A}$ on the internal normal, $\mathbf{A}_n$, or the integral is the outward *flux* of the vector $\mathbf{A}$.

$$(10) \qquad \iiint \left\{ \frac{\partial \mathbf{A}_x}{\partial x} + \frac{\partial \mathbf{A}_y}{\partial y} + \frac{\partial \mathbf{A}_z}{\partial z} \right\} d\tau = -\iint \mathbf{A}_n dS.$$

If the vector $\mathbf{A}$ at all points of the surface S, points outward, $-\mathbf{A}_n > 0$, and the integral on the left is positive. If we apply the theorem to any region however small, and this is still true, the integrand

$$\frac{\partial \mathbf{A}_x}{\partial x} + \frac{\partial \mathbf{A}_y}{\partial y} + \frac{\partial \mathbf{A}_z}{\partial z}$$

must be *positive*. This is therefore called *divergence* of the vector $\mathbf{A}$. It is, of course, a scalar quantity. The proof just given is the proof of the Divergence Theorem promised in Chapter I, § 8.

51. Newtonian Fields. Gauss's Theorem. Vectors whose divergence vanishes identically in a region have zero flux through *any* closed surface in that region, and are called solenoidal. Such a vector we have met in the steady state of temperature of a conducting body. Another is found in the field of force acting according to the law of

Newton, by which two small bodies of masses m, m' attract or repel each other with a force in the direction of the line joining them, of magnitude

$$\mathbf{F} = \gamma \frac{m m'}{r^2},$$

where r is their distance apart. If we make $m' = 1$, the vector $\mathbf{F}$ then obtained is called the field-strength at the point P, where the unit mass is situated, due to the mass m at Q. If the surface S is a sphere with m at the center, the flux

$$(11) \qquad \iint \mathbf{F}_n dS = \gamma m \iint \frac{dS}{r^2},$$

since the normal is in the direction of the radius. If we divide up the sphere into elements dS in any manner, and draw cones with apex at m and bases dS, and if the area cut by such a cone from a sphere of radius 1 is $d\omega$, $d\omega$ is called the solid angle subtended by the cone, and

$$dS = r^2 d\omega$$

Consequently r^2 cancels out and

$$(12) \qquad \iint \mathbf{F}_n dS = \gamma m \iint d\omega = 4\pi\gamma m,$$

which is independent of the size of the sphere. If we consider the force to be repulsive, as in the case of electricity and magnetism, we take γ positive, and $\mathbf{F}$ is in the direction of r drawn away from m, so that if the normal n is drawn inwards, the flux is negative, $-4\pi\gamma m$. If S is *any* surface, Fig. 69, and $d\Sigma$ is the projection of dS on a sphere with center m, $d\Sigma = \pm dS \cos(nr)$, r being drawn from m, and the upper sign being used for r cutting in, the lower for r cutting out. As before,

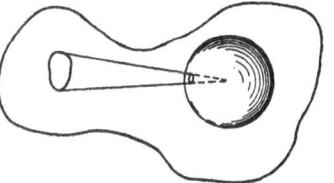

Fig. 69.

$$d\Sigma = r^2 d\omega$$

$$(13) \quad \mathbf{F}_n dS = \mathbf{F}\cos(\mathbf{F}n)dS = \mathbf{F}\cos(rn)dS = \pm \mathbf{F}d\Sigma = \pm \gamma m d\omega.$$

If m is *outside* of S, any radius from m cuts the surface an even number of times, Fig. 70, so that for every element $d\omega$ where the radius cuts in there is an equal but opposite one $-d\omega$ for cutting out, and thus the elements of the integral cancel in pairs, and the flux is zero. Thus the field $\mathbf{F}$ is a solenoidal vector in any region not containing m. If m is inside, any radius cuts the surface an odd

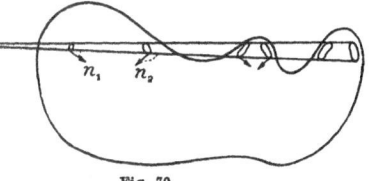

Fig. 70.

number of times, and there remains one uncancelled element $d\omega$ for cutting out. The sum of all these is the same as for a sphere, 4π, so that for *any* surface the flux is $-4\pi\gamma m$. These theorems are due to Gauss.

If the point m is *on* the surface at a non-singular point, every radius from it strikes the surface an odd number of times, and the solid angle is that on *one* side of the tangent plane, or 2π. Accordingly Gauss's integral

$$(14) \qquad \Omega = \int\int \frac{\cos(rn)\,dS}{r^2}$$

has two discontinuities, as the point Q passes to or from the surface on the inside or outside. We shall write

$$(15) \qquad \Omega_e = 0, \quad \Omega_s = -2\pi, \quad \Omega_i = -4\pi.$$

Let us show analytically that div $\mathbf{F} = 0$. Take m for origin, then

$$r^2 = x^2 + y^2 + z^2, \quad r\frac{\partial r}{\partial x} = x,$$

$$\frac{\partial r}{\partial x} = \frac{x}{r} = \cos(rx), \quad \frac{\partial r}{\partial y} = \frac{y}{r} = \cos(ry), \quad \frac{\partial r}{\partial z} = \frac{z}{r} = \cos(rz),$$

$$\mathbf{F}_x = \gamma\frac{m}{r^2}\cos(rx) = \gamma m\frac{x}{r^3}, \qquad \mathbf{F}_y = \gamma\frac{m}{r^2}\cos(ry) = \gamma m\frac{y}{r^3},$$

$$\mathbf{F}_z = \gamma\frac{m}{r^2}\cos(rz) = \gamma m\frac{z}{r^3},$$

$$\frac{\partial \mathbf{F}_x}{\partial x} = \gamma m\left\{\frac{1}{r^3} - \frac{3x^2}{r^5}\right\}, \qquad \frac{\partial \mathbf{F}_y}{\partial y} = \gamma m\left\{\frac{1}{r^3} - \frac{3y^2}{r^5}\right\},$$

$$\frac{\partial \mathbf{F}_z}{\partial z} = \gamma m\left\{\frac{1}{r^3} - \frac{3z^2}{r^5}\right\},$$

$$\operatorname{div}\mathbf{F} = \frac{3}{r^3} - \frac{3(x^2 + y^2 + z^2)}{r^5} = 0.$$

Where $r = 0$, this is $\infty - \infty$, and does not vanish, but is indeterminate.

Let us apply the divergence theorem (10) to a lamellar vector, $\mathbf{A} = \operatorname{grad} V$,

$$\mathbf{A}_x = \frac{\partial V}{\partial x}, \quad \mathbf{A}_y = \frac{\partial V}{\partial y}, \quad \mathbf{A}_z = \frac{\partial V}{\partial z},$$

where V and its derivatives are continuous uniform functions. Then (10) becomes

$$(16) \qquad \int\int\int \Delta V\,d\tau = -\int\int (\operatorname{grad} V)_n\,dS = -\int\int \frac{\partial V}{\partial n}\,dS.$$

This is called the gradient theorem, and shows the relation between the gradient, or first (vector) differential parameter, and the Laplacian, or second, (scalar) differential parameter.

52. Green's Theorem. Now apply the divergence theorem to a *compound* lamellar vector, (Chapter I, 65),

$$\mathbf{A} = U \operatorname{grad} V,$$

$$\mathbf{A}_x = U \frac{\partial V}{\partial x}, \quad \mathbf{A}_y = U \frac{\partial V}{\partial y}, \quad \mathbf{A}_z = U \frac{\partial V}{\partial z},$$

where U and V are both continuous uniform functions with first derivatives also such.

Forming the divergence,

$$\operatorname{div} \mathbf{A} = U \left\{ \frac{\partial^2 V}{\partial x^2} + \frac{\partial^2 V}{\partial y^2} + \frac{\partial^2 V}{\partial z^2} \right\} + \frac{\partial U}{\partial x} \frac{\partial V}{\partial x} + \frac{\partial U}{\partial y} \frac{\partial V}{\partial y} + \frac{\partial U}{\partial z} \frac{\partial V}{\partial z}$$

$$= U \Delta V + \Delta(U, V).$$

The last term is the scalar product of the gradients of U and V and is called the mixed or mutual differential parameter (see § 5).

The theorem now reads

$$(17) \qquad \iiint \{ U \Delta V + \Delta(U, V) \} \, d\tau = - \iint U (\operatorname{grad} V)_n \, dS =$$

$$= - \iint U \frac{\partial V}{\partial n} \, dS.$$

Transposing we have

$$(18) \qquad \iiint \Delta(U, V) d\tau = - \iint U \frac{\partial V}{\partial n} \, dS - \iiint U \Delta V d\tau.$$

This result is *Green's Theorem*, (A). Since the left hand member is a symmetrical function of U and V, we may interchange them, writing

$$(19) \qquad \iiint \Delta(U, V) d\tau = - \iint V \frac{\partial U}{\partial n} \, dS - \iiint V \Delta U d\tau.$$

Now writing the right-hand members equal, and transposing,

$$(20) \qquad \iiint (U \Delta V - V \Delta U) d\tau = \iint \left(V \frac{\partial U}{\partial n} - U \frac{\partial V}{\partial n} \right) dS.$$

This we shall speak of as Green's Theorem, (B).

Thus we have the volume integral converted entirely into a surface integral.

53. Newtonian Potential. Let us return to the theory of Newtonian fields. Let the coordinates of the repelling point Q be a, b, c, and of the point of observation P, x, y, z, as before. Then

$$r^2 = (x - a)^2 + (y - b)^2 + (z - c)^2, \qquad r \frac{\partial r}{\partial x} = x - a,$$

and the direction cosines of r drawn from Q to P, are

$$\cos(rx) = \frac{x-a}{r} = \frac{\partial r}{\partial x}, \quad \cos(ry) = \frac{y-b}{r} = \frac{\partial r}{\partial y},$$

(21)

$$\cos(rz) = \frac{z-c}{r} = \frac{\partial r}{\partial z}.$$

The components of the field are

$$\mathbf{F}_x = \gamma \frac{m}{r^2}\cos(rx) = \gamma \frac{m}{r^2}\frac{\partial r}{\partial x}, \quad \mathbf{F}_y = \gamma \frac{m}{r^2}\cos(ry) = \gamma \frac{m}{r^2}\frac{\partial r}{\partial y},$$

(22)

$$\mathbf{F}_z = \gamma \frac{m}{r^2}\cos(rz) = \gamma \frac{m}{r^2}\frac{\partial r}{\partial z},$$

and if we call

$$V = \frac{m}{r},$$

(23)

$$\mathbf{F}_x = -\gamma \frac{\partial V}{\partial x}, \quad \mathbf{F}_y = -\gamma \frac{\partial V}{\partial y}, \quad \mathbf{F}_z = -\gamma \frac{\partial V}{\partial z}, \qquad \text{or}$$

$$\mathbf{F} = -\gamma \operatorname{grad} V.$$

Thus the field is a lamellar vector. The function V is called the *potential function* of the field. By differentiating as before, we find

(24) $\operatorname{div}\mathbf{F} = -\gamma \Delta V = 0$,

or the field is solenoidal as well as lamellar, and the potential function satisfies Laplace's equation.

A function which is finite and continuous, (i. e. holomorphic) in a certain region, and satisfies Laplace's equation, is said to be *harmonic* in that region.

A function that is harmonic in a certain region can have at no point of the region a point of maximum or minimum. For if P be, say, a point of maximum, draw a small sphere enclosing it, and at each of its points V must be less than at P, consequently the inward normal derivative must be positive everywhere on this sphere, and the flux

$$\iint \frac{\partial V}{\partial n}\, dS$$

will be positive. But by the gradient theorem this is equal to the volume integral of $-\Delta V$, which is zero. If not zero, for the above reason $-\Delta V$ is called the *concentration* of V, because proportional to the excess of the value of V at P over its values at surrounding points.

If we have a number of repelling points $m_1, m_2 \ldots$ and call

$$V_1 = \frac{m_1}{r_1}, \quad V_2 = \frac{m_2}{r_2}, \quad \ldots$$

where $r_1, r_2 \ldots$ are the distances of P from the various points Q, the various actions have the components,

$$\mathbf{F}_{1x} = -\gamma \frac{\partial V_1}{\partial x}, \quad \mathbf{F}_{2x} = -\gamma \frac{\partial V_2}{\partial x},$$

and for the resultant field,

$$(25) \qquad \mathbf{F}_x = \mathbf{F}_{1x} + \mathbf{F}_{2x} + \cdots = -\gamma \frac{\partial}{\partial x}(V_1 + V_2 + \cdots),$$

and if we call

$$(26) \qquad V = \frac{m_1}{r_1} + \frac{m_2}{r_2} + \cdots$$

we have the resultant field derived from V, the total potential, as before. The advantage of the function V is that from this one scalar function we obtain the whole vector field as its vector differential parameter or gradient.

Since every term $\frac{m_s}{r_s}$ in V satisfies Laplace's equation, (except for $r_s = 0$), so does V their sum. Thus the potential function is harmonic, except at the points $Q_1, Q_2 \cdots$

If instead of discontinuously distributed points we have a continuous distribution of gravitating matter, instead of a sum we have the definite integral

$$(27) \qquad V = \int\int\int \frac{dm}{r},$$

where r is the distance from the point of observation P to the point Q at which the mass dm is situated.

If $dm = \rho\, d\tau$, ρ is called the density of the acting matter. If the coordinates of Q are a, b, c, we have in rectangular coordinates,

$$dm = \rho(a, b, c)\, da\, db\, dc,$$

$$r^2 = (x - a)^2 + (y - b)^2 + (z - c)^2,$$

$$(28) \quad V_P = V(x, y, z) = \int\int\int \frac{\rho(a, b, c)\, da\, db\, dc}{\{(x - a)^2 + (y - b)^2 + (z - c)^2\}^{\frac{1}{2}}}$$

It is evident that for each value of x, y, z, V has a single finite definite value. As P moves away from the distribution, every $\frac{1}{r}$ becomes less, and where P is at an infinite distance V vanishes like $\frac{M}{r}$ where $M = \int\int\int dm$.

The function V has derivatives according to x, y, z, and the field is given by $-\gamma$ times its gradient. For instance (since ρ does not depend on x, y, z)

$$(29) \quad \frac{\partial V}{\partial x} = \frac{\partial}{\partial x}\int\int\int \frac{\rho\, d\tau}{r} = \int\int\int \rho\, \frac{\partial}{\partial x}\left(\frac{1}{r}\right)d\tau = -\int\int\int \frac{\rho}{r^2}\frac{\partial r}{\partial x}\, d\tau$$

$$= -\int\int\int \frac{\rho}{r^2}\frac{x - a}{r}\, d\tau.$$

$$(30) \qquad \frac{\partial^2 V}{\partial x^2} = \int\int\int \rho\left\{\frac{3(x - a)^2 - r^2}{r^5}\right\}d\tau.$$

The derivatives are all finite, as long as r is not zero, that is, as long as P is *outside* of the distribution of matter, and Laplace's equation is satisfied. Thus the potential is a harmonic function outside of gravitating matter, and has neither maximum nor minimum. The field is a solenoidal, as well as a lamellar vector, like the temperature in a steady flow of heat. What is the analogue of a *source* of heat? This will be answered in the next section.

54. Poisson's Equation. We have found (Gauss's theorem) that the flux through a closed surface S due to a point m is zero if the point is outside S, and $-4\pi\gamma m$ if it is within, no matter where. In like manner for a number of points each contributes a flux proportional to its mass, so that for all we have

$$(31) \qquad \iint F_n \, dS = -\gamma \iint \frac{\partial V}{\partial n} \, dS = -4\pi\gamma \Sigma m.$$

In like manner for a continuous distribution, instead of Σm we have $\iiint dm$, the field of integration being all the volume within S.

$$(32) \qquad \iint \frac{\partial V}{\partial n} \, dS = 4\pi \iiint dm = 4\pi \iiint \varrho \, d\tau.$$

Now by the application of the gradient theorem (16)

$$(33) \qquad \iint \frac{\partial V}{\partial n} \, dS = -\iiint \Delta V d\tau = 4\pi \iiint \varrho \, d\tau.$$

Take for the surface S any surface inside the distribution. The mass considered is only that *within* S, so that both integrals are over the same volume. Transposing

$$(34) \qquad \iiint (\Delta V + 4\pi\varrho) \, d\tau = 0,$$

and since this is true for any volume, the integrand must vanish everywhere, so that

$$(35) \qquad \Delta V + 4\pi\varrho = 0.$$

This is known as Poisson's equation. Accordingly the *concentration* of the potential at any point is 4π times the density at that point, and any point of repelling matter, producing divergence of the field, is the analogue of a source of heat. We may also say that the integral (28) is the expression of the action at a distance, while the differential equation (35) expresses the action of neighboring regions.

In the theory of electrostatics, electrical quantities act upon each other according to the Newtonian Law, accordingly Poisson's equation is the relation between the density of electrification and the electric potential promised in Chapter I, § 10.

We have thus found a solution of the equation

$$\frac{\partial^2 V}{\partial x^2} + \frac{\partial^2 V}{\partial y^2} + \frac{\partial^2 V}{\partial z^2} + 4\pi\varrho(x,y,z) = 0$$

to be

$$V = \iiint \frac{\varrho(a,b,c)\,da\,db\,dc}{\{(x-a)^2 + (y-b)^2 + (z-c)^2\}^{\frac{1}{2}}}$$

We have seen that at all points outside of matter V and its derivatives are finite, and the finiteness of those of any order implies that those of order one less are continuous. Thus the gradient of V is a continuous vector-function, and its lines have no singularities. This is always supposing ϱ to be a *finite* function of the point Q.

We may use Poisson's equation to calculate the potential due to certain symmetrical distributions. Suppose that V depends only on the distance from a fixed point. Let us transform Laplace's operator for this case, for any number of dimensions. Let $V = V(r)$ where

$$r^2 = (x_1 - a_1)^2 + (x_2 - a_2)^2 + \cdots + (x_n - a_n)^2.$$

Then we have

$$\frac{\partial r}{\partial x_s} = \frac{x_s - a_s}{r}, \qquad \frac{\partial^2 r}{\partial x_s^2} = \frac{1}{r} - \frac{(x_s - a_s)^2}{r^3}$$

$$\frac{\partial V}{\partial x_s} = \frac{dV}{dr}\frac{\partial r}{\partial x_s}, \qquad \frac{\partial^2 V}{\partial x_s^2} = \frac{d^2 V}{dr^2}\left(\frac{\partial r}{\partial x_s}\right)^2 + \frac{dV}{dr}\frac{\partial^2 r}{\partial x_s^2}$$

and accordingly

$$(36) \qquad \Delta V = \sum_{s=1}^{s=n}\frac{\partial^2 V}{\partial x_s^2} = \frac{d^2 V}{dr^2}\sum_{s=1}^{s=n}\left(\frac{\partial r}{\partial x_s}\right)^2 + \frac{dV}{dr}\sum_{s=1}^{s=n}\frac{\partial^2 r}{\partial x_s^2}$$

$$= \frac{d^2 V}{dr^2} + \frac{n-1}{r}\frac{dV}{dr} = \frac{1}{r^{n-1}}\frac{d}{dr}\left(r^{n-1}\frac{dV}{dr}\right).$$

Suppose first $n = 3$, and let ϱ depend only on r. Such a distribution we shall call a spherical distribution, and outside of it we have

$$(37) \qquad \Delta V = \frac{1}{r^2}\frac{d}{dr}\left(r^2\frac{dV}{dr}\right) = 0.$$

Integrating,

$$\frac{d}{dr}\left(r^2\frac{dV}{dr}\right) = 0, \quad r^2\frac{dV}{dr} = c, \quad \frac{dV}{dr} = \frac{c}{r^2}.$$

Integrating again,

$$\frac{dV}{dr} = \frac{c}{r^2}, \quad V = -\frac{c}{r} + c',$$

and since V vanishes for $r = \infty$, $c' = 0$.

The constant $-c$ must be equal to M, and we see that the spherical distribution acts as if concentrated at its center and $\dfrac{dV}{dr} = -\dfrac{M}{r^2}$. Now consider internal points, for which

(38) $$\Delta V = \frac{1}{r^2} \frac{d}{dr} \left(r^2 \frac{dV}{dr} \right) = -4 \pi \varrho(r).$$

Integrating $\dfrac{d}{dr} \left(r^2 \dfrac{dV}{dr} \right) = -4 \pi r^2 \varrho$ we have

$$r^2 \frac{dV}{dr} = -4 \pi \int_0^r r^2 \varrho \, dr$$

Let us suppose that the sphere is homogeneous, that is ϱ is constant. Then

$$r^2 \frac{dV}{dr} = -\frac{4\pi}{3} \varrho r^3,$$

so that the field,

$$\frac{dV}{dr} = -\frac{4\pi}{3} \varrho r,$$

is proportional to the distance from the center. At the boundary of the sphere $r = R$, this is continuous with the external value,

$$-\frac{M}{r^2} = -\frac{4\pi}{3} \frac{R^3 \varrho}{R^2}.$$

Integrating again,

$$V = C - \frac{2\pi}{3} \varrho r^2$$

and if this is to be continuous with the external value at the surface, we must have $C = 2\pi\varrho R^2$.

These results enable us to examine the potential (28) and its derivatives at points in the substance of any acting distribution. When the point P is within, there is one element for which the point Q coincides with P, $r = 0$, and one element of the integrands in (27), (28), (29), (30) becomes infinite. In order to see whether this makes the integral cease to be finite, we pursue the usual course of excluding from the region of integration a small region about the point P, where the discontinuity of the integrand occurs, and finding the limit as the region excluded grows smaller. In this case let us exclude a small sphere about P. In the remainder of the volume, r is not zero, and the integrals are all finite The integrals over the small sphere are not greater in absolute value than if we gave ϱ the constant value ϱ_m representing its greatest value in the small sphere. But as we have just seen, at any point of a sphere, V and $\dfrac{dV}{dr}$ are finite and *vanish* with the radius of the sphere. Consequently V and its first derivatives are finite at internal points of any distribution, for which ϱ is finite.

Suppose however, that in a certain region ϱ becomes infinite in such a way that as the volume diminishes $\lim \iiint \varrho \, d\tau = m$ remains finite. Then applying Poisson's equation and the gradient theorem

$$(39) \qquad \iiint \Delta V d\tau = -4\pi \iiint \varrho \, d\tau = -4\pi m$$

$$\iint \frac{\partial V}{\partial n} dS = 4\pi m,$$

and we have the flux from a point mass, from which we started.

Suppose the singularity of ϱ is along an attracting *line* or limit of a cylinder of cross-section ω, so that $d\tau = d\omega \, dl$ and $dS = dl \, ds$, where dl is the element of length and ds is the element of the circumference. Then applying to an element of length dl,

$$(40) \qquad \iiint \Delta V d\tau = -\iint \frac{\partial V}{\partial n} ds \, dl = -4\pi \iint \varrho \, d\omega \, dl,$$

the integrals over the ends vanishing to a higher order, and dividing by dl, and putting

$$\bar{\omega} = 2 \lim \int \varrho \, d\omega,$$

we have the integral around the belt of the cylinder,

$$(41) \qquad \int \frac{\partial V}{\partial n} ds = 2\pi \bar{\omega}.$$

As an example, consider the singular line to be straight and of infinite length, and take it for the Z-axis. Then V does not depend on z, and

$$\frac{\partial^2 V}{\partial x^2} + \frac{\partial^2 V}{\partial y^2} = -4\pi \varrho.$$

Using polar coordinates, consider *any* cylindrically symmetrical distribution with $\varrho = \varrho(r)$ in two dimensions. Then from (36) we have

$$(42) \qquad \frac{1}{r} \frac{d}{dr} \left(r \frac{dV}{dr} \right) = 0,$$

outside the acting matter.

Integrating

$$\frac{d}{dr} \left(r \frac{dV}{dr} \right) = 0, \quad r \frac{dV}{dr} = C.$$

Integrating

$$\frac{dV}{dr} = \frac{C}{r}, \quad V = C \log r + C'.$$

The field is inversely proportional to the *first* power of the distance from the center, and the potential is proportional to the logarithm, and is *infinite* at an infinite distance. Inside the distribution

$$\frac{1}{r} \frac{d}{dr} \left(r \frac{dV}{dr} \right) = -4\pi \varrho,$$

$$r \frac{dV}{dr} = -4\pi \int_0^r r \varrho \, dr,$$

and if again, we put ϱ constant,

$$\frac{dV}{dr} = -2\pi\varrho r, \quad V = C'' - \pi\varrho r^2.$$

If again the field is to be continuous at the boundary $r = R$, we must have $C = -2\pi\varrho R^2 = -2\varrho\omega$ and we have the flux

$$(43) \qquad \int \frac{\partial V}{\partial n}\, ds = 2\pi \int \varrho R\, ds = 4\pi^2 R^2\varrho = 4\pi\varrho\omega,$$

and we may now let ϱ become infinite as the cross-section ω diminishes, if $\lim 2\omega\varrho = \varpi$ when (43) agrees with (41).

55. Logarithmic Potential. From the element

$$V = -\varpi \log r = \varpi \log \frac{1}{r},$$

which may be interpreted either as above, or as the potential of a field due to a point of mass ϖ acting on points situated in a *plane* with a force inversely proportional to the *first* power of the distance, we may form the *logarithmic potential* due to a continuous distribution in the plane, such that the mass on the area dS is $d\varpi = \varrho\,dS$ (ϱ being now used in a new sense, but still a density),

$$(44) \qquad V_P = V(x, y) = \iint \varrho\,(a, b) \log \frac{1}{r}\, dS,$$

where

$$r^2 = (x - a)^2 + (y - b)^2.$$

For this logarithmic potential we find a series of theorems analogous to those already deduced for the Newtonian potential, by employment of the theorems of Gauss and Green for the plane. For Gauss's theorem we consider the flux of a vector $\mathbf{F} = \frac{1}{r}$ through a closed *contour*, (use Fig. 69 in the plane)

$$(45) \qquad \Omega = \int \mathbf{F}_n\, ds = \int \frac{\cos(nr)\, ds}{r} = \int \frac{d\Sigma}{r} = \int d\varphi$$

where $d\Sigma = r\,d\varphi$ and φ is the plane angle subtended by the arc ds. Consequently the integral is 2π for an inside point, and

$$(46) \qquad \Omega_e = 0, \quad \Omega_s = \pi, \quad \Omega_i = 2\pi.$$

Consequently for two dimensions 2π takes the place occupied by 4π in three dimensional problems. The gradient theorem and Green's theorem become for the plane

$$(47) \qquad \iint \Delta V\, dS = \iint \left\{ \frac{\partial^2 V}{\partial x^2} + \frac{\partial^2 V}{\partial y^2} \right\} dx\, dy = -\int \frac{\partial V}{\partial n}\, ds.$$

$$(48) \qquad \iint \left\{ \frac{\partial U}{\partial x} \frac{\partial V}{\partial y} + \frac{\partial U}{\partial y} \frac{\partial V}{\partial x} \right\} dS = -\int U \frac{\partial V}{\partial n}\, ds - \iint U \Delta V\, dS,$$

by means of which and Gauss's theorem we prove Poisson's equation

$$(49) \qquad \Delta V = \frac{\partial^2 V}{\partial x^2} + \frac{\partial^2 V}{\partial y^2} = -2\pi\varrho,$$

(ϱ now has a new signification).

56. Surface Distributions. Returning now to three dimensions, suppose the density ϱ becomes infinite on a certain surface S, in such a way that in a thin sheet of thickness n there is distributed matter of density ϱ so that ϱ increases indefinitely while n decreases indefinitely and that the product remains finite. If dS is the element of surface of the final distribution, we have $d\tau = n\,dS$, and if $\lim\limits_{n=0}(\varrho\,d\tau) = \lim\limits_{n=0}(\varrho n)dS$ $= \sigma\,dS$, σ is called the surface density. We have then

$$(50) \qquad V = \int\int \frac{\sigma\,dS}{r}.$$

As the point P approaches the surface S, one of the elements of the integrand becomes infinite, and the question arises whether the integral ceases to be finite. This we treat as before, by removing a small region in the form of a circular plate, and determining the limit of its attraction. Before doing so let us however consider the derivative as P approaches the surface. Applying Poisson's equation and the gradient theorem to a tube composed of normals to the surface S and bounded by two caps drawn close to S on the sides 1 and 2, say S_1 and S_2, the integral over the narrow ribbon connecting the caps vanishes as the caps approach each other, and

$$\int\int\int \Delta V d\tau = \int\int \frac{\partial V}{\partial n_1}\,dS_1 + \int\int \frac{\partial V}{\partial n_2}\,dS_2 = -4\pi\int\int\int \varrho n\,dS,$$

and in the limit

$(51)\ dS_1 = dS_2 = dS,\ \lim(\varrho n) = \sigma,\ \int\int \left\{\dfrac{\partial V}{\partial n_1} + \dfrac{\partial V}{\partial n_2} + 4\pi\sigma\right\}dS = 0.$

This being true for any portion of the surface S we have

$$(52) \qquad \frac{\partial V}{\partial n_1} + \frac{\partial V}{\partial n_2} = -4\pi\sigma,$$

which is Poisson's surface equation. If we consider the derivative in the *same* direction n on the sides 1 and 2, if n is in the direction 12,

$$\frac{\partial V}{\partial n_1} = -\left(\frac{\partial V}{\partial n}\right)_1, \quad \frac{\partial V}{\partial n_2} = \left(\frac{\partial V}{\partial n}\right)_2$$

the suffixes denoting the sides of S, and

$$(53) \qquad \left(\frac{\partial V}{\partial n}\right)_2 - \left(\frac{\partial V}{\partial n}\right)_1 = -4\pi\sigma = \frac{\partial V}{\partial n}\Big|_{-0}^{+0}$$

or the derivative in the direction of the *normal* experiences a *discontinuity* of magnitude 4π times the surface density in crossing a surface distribution. This theorem was discovered by Coulomb for an electrified surface.

As an illustration let us consider a distribution on an infinite plane, which we will consider as the limit of a slab of finite thickness R normal to the X-axis. As the slab is infinite, its action will be independent of y and z, and we have V a one-dimensional potential function, depending on x alone. We thus have ΔV reducing to $\dfrac{d^2 V}{dx^2}$, in agreement with (36) for $n = 1$. Accordingly outside the slab, integrating

$$\frac{d^2 V}{dx^2} = 0, \quad \frac{dV}{dx} = C, \quad V = Cx + C',$$

the field, which is normal to the slab, is constant, and the potential is a linear function of x. Inside the slab,

(54)
$$\frac{d^2 V}{dx^2} = -4\pi\varrho$$

and if, as in the case of three and two dimensions, ϱ is constant, integrating

$$\frac{dV}{dx} = -4\pi\int_0^x \varrho\, dx = -4\pi\varrho x, \quad V = -2\pi\varrho x^2 + C'',$$

and at the surface, $x = R$, for continuity with the external values, we must have $C = -4\pi\varrho R$. Suppose the slab extends from $x = -R$ to $x = R$. At the left-hand face $x = -R$, $\dfrac{dV}{dx} = 4\pi\varrho R$, so that *outside* on the left we must take $C = 4\pi\varrho R$. The constant C' is indeterminate, and does not affect the result, which is, that there is a difference of amount $4\pi\varrho \cdot 2R$ in the derivative $\dfrac{dV}{dx}$ on the two sides of the slab.

If now the thickness $2R$ diminishes and ϱ increases, so that $\lim (2R\varrho) = \sigma$ we find the discontinuity $4\pi\sigma$. We thus see that a distri-

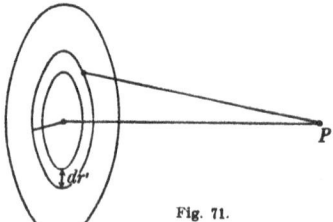

bution, though infinite in amount, produces a finite field, as in the case of the infinite columnar distribution, although the potential of one is a linear, the other a logarithmic function of the distance.

Fig. 71.

In order to emphasize this example, and to get rid of the infinite mass let us calculate the potential of a circular plate of *finite* radius R, at a point P on the normal to its center at a distance x (Fig. 71). Let us divide the area into circular rings of radius r', $r' + dr'$, so that $dS = 2\pi r'dr'$, $r = \sqrt{x^2 + r'^2}$ and

(55) $$V = \int\int \frac{\sigma\, dS}{r} = 2\pi\sigma \int_0^R \frac{r'dr'}{\sqrt{x^2 + r'^2}} = 2\pi\sigma \left\{ \sqrt{x^2 + R^2} - x \right\}.$$

We find the field,

$$\frac{dV}{dx} = 2\pi\sigma \left\{ \frac{x}{\sqrt{x^2 + R^2}} - 1 \right\},$$

so that both the potential and the field vanish at ∞, as in Fig. 72. The field being evidently symmetrical on both sides of the disc, we see the discontinuity in the slope $\frac{dV}{dx}$. As R in-creases, although V becomes infinite $\frac{dV}{dx}$

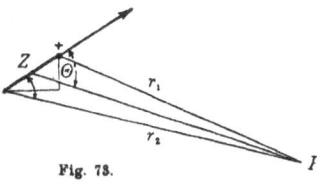

Fig. 72.

approaches the finite constant value already found, of magnitude $-2\pi\sigma$.

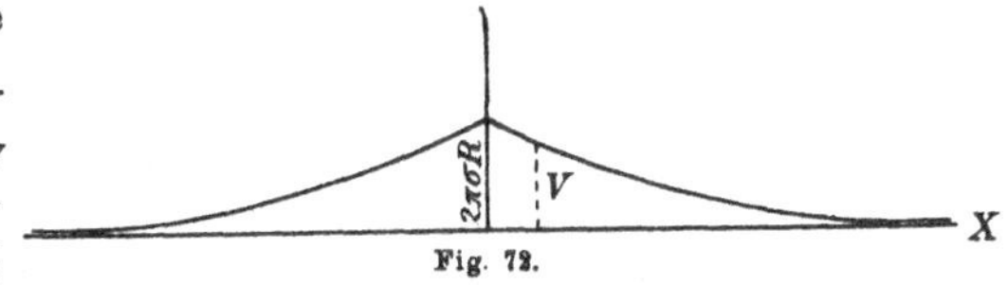

As a final example consider a spherical distribution of constant surface-density σ. Outside we have seen that

$$V = \frac{M}{r} = \frac{4\pi R^2\sigma}{r}, \quad \frac{dV}{dr} = -\frac{4\pi R^2\sigma}{r^2}$$

and at the surface

$$\left(\frac{dV}{dr}\right)_s = -4\pi\sigma.$$

Inside we have

$$\Delta V = 0, \quad V = \frac{C}{r} + C',$$

and since V would be infinite at the center unless $C = 0$, this must be the case, so that inside V is constant, and $\frac{dV}{dr} = 0$. Again the discontinuity in $\frac{dV}{dn}$ is $4\pi\sigma$.

57. Double Surface Layers. Let us now consider the potential due to a *doublet*, composed of two points of masses m and $-m$ at a distance apart l where m increases and l diminishes without limit so that the *moment* of the doublet approaches a finite limit,

$$\lim_{l=0} (ml) = M.$$

If P is at distances r_1 and r_2 from the positive and negative points (Fig. 73),

$$V = \frac{m}{r_1} - \frac{m}{r_2} = m\frac{r_2 - r_1}{r_1 r_2}.$$

If θ is the angle made by the line from $-m$ to m with the line drawn to P.

$$r_2 - r_1 = l\cos\theta,$$
$$V = \frac{ml\cos\theta}{r_1 r_2},.$$

Fig. 73.

and in the limit,

(56) $$V = \frac{M}{r^2} \cos \theta.$$

If the direction from the negative to the positive end of the doublet is s,

$$dr = - ds \cos \theta, \quad \frac{\partial r}{\partial s} = - \cos \theta$$

(57) $$V = M \frac{\partial \frac{1}{r}}{\partial s} = - \frac{M}{r^2} \frac{\partial r}{\partial s} = \frac{M \cos \theta}{r^2},$$

so that V depends on the *direction* of QP as well as on its magnitude, being positive on the side nearer the positive end of the doublet. Such a field is nearly that of a short *magnet*, which can be approximately represented by two equal and opposite poles.

If we now form on a surface S a *double layer*, composed of doublets with axes normal to the surface such that the doublets form two surface layers of density σ and $-\sigma$ separated by the normal distance n, an element dS has the moment $dM = n\sigma dS$, and produces the potential at P

$$dW = n\sigma dS \frac{\partial \frac{1}{r}}{\partial n}.$$

The moment per unit of area $\lim_{n=0} (\sigma n) = \varkappa$ is called the *strength* of the double-layer. Thus we have for the potential due to a double-layer

(58) $$W = \int\int \varkappa \frac{\partial \frac{1}{r}}{\partial n} dS,$$

n being drawn on the positive side.

It is noticeable that if $\varkappa$ is constant,

(59) $$W = \varkappa \int\int \frac{\partial \frac{1}{r}}{\partial n} dS = - \varkappa \int\int \frac{\cos(rn) dS}{r^2} = \varkappa \Omega,$$

where Ω is the solid angle subtended at P by the double layer, taken positive on the positive side, negative on the negative side of the layer. If S is closed $W = 0$ outside, $W = 4\pi\varkappa$ on the inside. W is accordingly *discontinuous* in crossing the double-layer. This is to be contrasted with the behavior at a *single* surface-layer, where V is continuous, $\frac{\partial V}{\partial n}$ is discontinuous.

Consider the normal derivatives on both sides of the surfaces $-\sigma, \sigma$. In crossing the first $\frac{\partial W}{\partial n}$ *increases* by $4\pi\sigma$, in crossing the second it decreases by the same amount. So that in the limit it has the *same* value on both sides of the double-layer, or is *not* discontinuous, Fig. 74. But whereas W does not change suddenly on crossing either single layer, and $\dfrac{\partial W}{\partial n}$ does, the change in W between points $1'\,2'$ is

$$\int_{1'}^{2'}\frac{\partial W}{\partial n}\,dn = W_{2'} - W_{1'} = W_2 - W_1 = \int\left\{\left(\frac{\partial W}{\partial n}\right)_1 + 4\pi\sigma\right\}dn,$$

and as $1'$ and $2'$ come together, the first term vanishes, but since $\lim(\sigma dn) = \varkappa$ we have

(60) $$W_2 - W_1 = 4\pi\varkappa.$$

Consequently the above-mentioned phenomenon is general, and the *potential* of a double-layer has a discontinuity of 4π times the strength at the point of crossing the surface, while the normal derivative is continuous.

58. Surface Discontinuities. We may prove these theorems otherwise by the usual process of removing a small portion of the re-

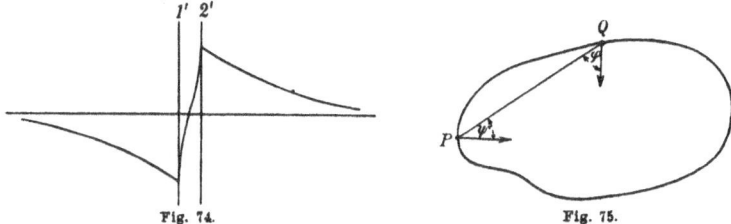

Fig. 74.   Fig. 75.

gion of integration where the integrand becomes infinite, in this case a small region surrounding the point P on the surface, small enough to be treated as a plane circular plate of constant strength $\varkappa$, producing the potential W', Fig. 75. The potential and derivative due to the remainder of the surface are continuous, as we pass through the hole, so that any discontinuities must arise from the double distribution on the small disk. This is equal to $\varkappa$ times the solid angle subtended by the disc, and is evidently, from symmetry equal and of opposite signs at symmetrical points on the axis on opposite sides. As these points approach the disc, the derivative of W accordingly approaches the same value on both sides. As for W' itself, the solid angle subtended at P approaches the value 2π as P approaches the disc, so that on one side $W' = 2\pi\varkappa$. on the other $-2\pi\varkappa$, and the difference is $4\pi\varkappa$.

For points on the surface S, as we have seen in (46) that Ω_s is the mean of the limits approached on the two sides, we find the same thing true for the potential W.

We may accordingly write, denoting the inside and outside limits by the letters i and e,

$$
(61) \qquad
\begin{aligned}
W_i &= \ \ 2\pi\varkappa + \int\int \varkappa\,\frac{\partial\frac{1}{r}}{\partial n}\,dS, \\[2mm]
W_e &= -2\pi\varkappa + \int\int \varkappa\,\frac{\partial\frac{1}{r}}{\partial n}\,dS,
\end{aligned}
$$

subtracting the second from the first of which gives (60), and adding

$$
(62) \qquad \tfrac{1}{2}(W_i + W_e) = \int\int \varkappa\,\frac{\partial\frac{1}{r}}{\partial n}\,dS.
$$

In the same way we may consider the limits of the normal derivatives of the potential of the single layer V. Differentiating the integral

$$
V_p = \int\int \frac{\sigma_q\,dS_q}{r_{pq}}
$$

in the direction of n_p, when P is on the surface S,

$$
(63) \qquad \frac{\partial V_p}{\partial n_p} = \int\int \sigma_q\,\frac{\partial}{\partial n_p}\left(\frac{1}{r_{pq}}\right)dS_q = \int\int \frac{\sigma_q\cos\psi}{r^2}\,dS_q,
$$

where ψ is the angle between the normal at P and the line PQ. Comparing this with the potential of a double layer of the same strength

$$
W_p = \int\int \sigma_q\,\frac{\partial}{\partial n_q}\left(\frac{1}{r_{pq}}\right)dS_q = \int\int \frac{\sigma_q\cos\varphi}{r^2}\,dS_q,
$$

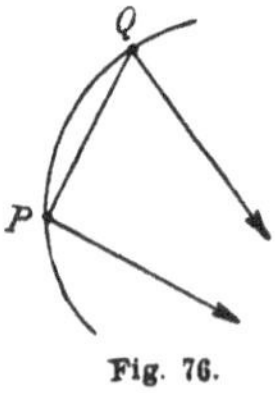

Fig. 76.

we find that they differ only in the latter having the angle φ between the normal at Q and QP. When however we consider the small element containing the infinite element, that in which Q approaches P, Fig. 76, we find that as the two normals approach each other the ratio of $\dfrac{\cos\psi}{\cos\varphi}$ approaches -1, so that we may write for the small element,

$$
\frac{\partial V'}{\partial n_p} = -\int\int \frac{\sigma_q\cos\varphi}{r^2}\,dS;
$$

which is the potential of a double layer, and accordingly the two limits approached on the inside and on the outside differ from the value on S by $2\pi\sigma$. Accordingly

$$
(64) \qquad
\begin{aligned}
\left(\frac{\partial V}{\partial n}\right)_i &= -2\pi\sigma + \int\int \frac{\sigma\cos\psi}{r^2}\,dS, \\[2mm]
\left(\frac{\partial V}{\partial n}\right)_e &= \ \ 2\pi\sigma + \int\int \frac{\sigma\cos\psi}{r^2}\,dS,
\end{aligned}
$$

taking the difference of which gives (52), and adding

$$
(65) \qquad \tfrac{1}{2}\left\{\left(\frac{\partial V}{\partial n}\right)_e + \left(\frac{\partial V}{\partial n}\right)_i\right\} = \int\int \frac{\sigma\cos\psi}{r^2}\,dS
$$

While the formulae (52) and (60) are classic, the additional ones (64) and (61) seem to be quite recent, appearing first in a paper by Fredholm in 1900.

All that has been said of surface distributions both single and double may be extended to the logarithmic potential for distributions on a contour, substituting $\log \frac{1}{r}$ for $\frac{1}{r}$ and 2π for 4π. In particular formulae (61), (64) become

(61′)
$$W_i = \quad \pi\varkappa + \int \varkappa \frac{\partial \log \frac{1}{r}}{\partial n}\, ds,$$

$$W_e = -\,\pi\varkappa + \int \varkappa \frac{\partial \log \frac{1}{r}}{\partial n}\, ds,$$

(64′)
$$\left(\frac{\partial V}{\partial n}\right)_i = -\,\pi\sigma + \int \frac{\sigma \cos \psi}{r}\, ds,$$

$$\left(\frac{\partial V}{\partial n}\right)_e = \quad \pi\sigma + \int \frac{\sigma \cos \psi}{r}\, ds.$$

59. Green's formula. We shall next prove that any potential-function which is discontinuous at certain surfaces or whose normal derivative is discontinuous, can be made up of single and double surface layers.

We have seen that the solution of a partial differential equation always contains arbitrary functions, and that these may be determined by giving, for a particular value of one variable, the values of the dependent variable and one or more of its derivatives as functions of the remaining variables. For instance, in Chapter III, for the equation

$$\frac{\partial^2 u}{\partial t^2} = a^2 \frac{\partial^2 u}{\partial^2 x},$$

we can take for

$$t = 0,\ u = f(x),\ \frac{\partial u}{\partial t} = g(x).$$

We might then expect to apply the method of Cauchy to Laplace's equation

$$\frac{\partial^2 V}{\partial x^2} + \frac{\partial^2 V}{\partial y^2} + \frac{\partial^2 V}{\partial^2 x} = 0,$$

and require that for

$$x = a,\ V = f(y, z),\ \frac{\partial V}{\partial x} = g(y, z).$$

Let us see whether we can determine V so that at any *closed* surface, instead of at the plane $x = a$, we may give V and $\frac{\partial V}{\partial n}$ as arbitrary functions of a point on the surface.

We have seen that the function $U = \frac{1}{r}$ satisfies Laplace's equation except for $r = 0$. Also we have

$$\frac{\partial U}{\partial n} = -\frac{1}{r^2}\frac{\partial r}{\partial n} = -\frac{\cos(rn)}{r^2}.$$

Let us apply Green's theorem to two functions, one $U = \frac{1}{r}$, the other V the potential function due to some distribution of matter. The theorem (B) (20) applies to a space in which both U, V are finite, with their derivatives. Accordingly the point P from which r is measured, must not be in the volume of integration. If P lies inside the closed surface S, let us draw a small sphere Σ of radius ε about P as a center, and exclude the space inside Σ from the integration. We must accordingly extend the surface integral over the surface Σ as well as S. In the space inside S and outside Σ, $\Delta U = \Delta\left(\frac{1}{r}\right) = 0$, so that

$$(66) \qquad \iint_{S+\Sigma}\left(V\frac{\partial\frac{1}{r}}{\partial n} - \frac{1}{r}\frac{\partial V}{\partial n}\right)dS = \iiint \frac{1}{r}\Delta V d\tau.$$

On the surface of the sphere Σ, $\frac{\partial}{\partial n} = \frac{\partial}{\partial r}$, the normals being always drawn *into* the volume τ,

$$(67) \qquad \iint_{\Sigma} V\frac{\partial\frac{1}{r}}{\partial n} d\Sigma = -\iint \frac{V}{r^2} d\Sigma = -\iint V_\Sigma d\omega,$$

while

$$-\iint_{\Sigma}\frac{1}{r}\frac{\partial V}{\partial r} d\Sigma = -\varepsilon\iint\frac{\partial V}{\partial r}d\omega.$$

As we make ε approach zero as a limit, the last integral vanishes, since $\frac{\partial V}{\partial r}$ is finite, while the first has $\lim_{\varepsilon=0} V_\Sigma = V_P$, and becomes

$$(68) \qquad -V_P\iint d\omega = -4\pi V_P.$$

Consequently we have

$$-4\pi V_P + \iint\left(V\frac{\partial\frac{1}{r}}{\partial n} - \frac{1}{r}\frac{\partial V}{\partial n}\right)dS = \iiint \frac{1}{r}\Delta V d\tau.$$

Transposing, and dividing by 4π,

$$(69) \quad V_P = \frac{1}{4\pi}\iint\left(V\frac{\partial\frac{1}{r}}{\partial n} - \frac{1}{r}\frac{\partial V}{\partial n}\right)dS - \frac{1}{4\pi}\iiint \frac{1}{r}\Delta V d\tau.$$

This important formula is due to Green. It shows that a function continuous in a certain space inside a closed surface S is determined at every internal point P if we know

1. The concentration $-\Delta V$ at every point of τ.
2. The value of V at every point of the surface S.
3. The gradient resolved along the normal to S at every point of S.

In particular if V is harmonic in the space τ, it is determined by the values of V and $\dfrac{\partial V}{\partial n}$ at the surface, thus solving the problem proposed for Laplace's equation.

If V is not harmonic, by Poisson's equation,

$$\Delta V = -4\pi\varrho,$$

and the volume integral

$$-\frac{1}{4\pi}\int\int\int \frac{\Delta V}{r}\,d\tau = \int\int\int \varrho\frac{d\tau}{r}$$

represents the potential due to all the matter lying in the volume of integration, that is *within* S. Accordingly the potential of the matter outside of S is represented at all points within S by the surface integral

$$\frac{1}{4\pi}\int\int V\frac{\partial \frac{1}{r}}{\partial n}\,dS - \frac{1}{4\pi}\int\int \frac{1}{r}\frac{\partial V}{\partial n}\,dS.$$

Suppose we distribute on the surface S a surface layer of density

$$\sigma = -\frac{1}{4\pi}\frac{\partial V}{\partial n}.$$

Then the second integral is $\int\int \dfrac{\sigma\,ds}{r}$ and represents the potential due to this surface layer. Let us also distribute a double layer of variable moment $\varkappa = \dfrac{V_s}{4\pi}$. Then the first integral represents the potential due to this double-layer. We thus get the theorem:

The effect of matter lying outside of any closed surface S may be replaced at all interior points by the superposition of a single and a double layer on S. (If the surface should be equipotential, $V_s = \text{const.}$, and the effect of the double-layer vanishes by Gauss's theorem.

If the point P lies *outside* of S, we integrate over all space outside of S, and inside of the infinite sphere, and assuming that for $r = \infty$, $|V| \leqq \dfrac{M}{r}$, $\left|\dfrac{\partial V}{\partial x}\right| \leqq \dfrac{M}{r^2}$, we easily show that the integrals over the infinite sphere vanish. The theorem is then that the potential of *inside* matter is replaceable at outside points by surface distributions. The normal n in the formula is then drawn *out* from S. Since if we reverse the sign of all densities, we reverse the sign of the action, we may by the proper surface distribution, *destroy* on one side of a closed surface, the action of all matter lying on the other. The potential in the region in question will be constant, say zero. Thus in crossing the surface, V will be discontinuous, changing from V to zero, but as we have seen this is produced by a double layer of moment $\varkappa = \frac{V_s}{4\pi}$ directed toward the side having the value V. Also the normal derivative is zero on one side, and on the other $\frac{\partial V}{\partial n}$. But this discontinuity is produced by a layer $\sigma = -\frac{1}{4\pi}\frac{\partial V}{\partial n}$. It is worth while calling attention to the fact, emphasized by Larmor[1]), that the surface distributions that replace the outside masses are not unique, but depend on the field internal to the surface S due to *both* the outside and inside masses, and as the inside masses may be varied indefinitely, so may the surface distribution representing the *same* field due to the outside masses.

60. Dirichlet's Problem. The formula of Green for a function satisfying Laplace's equation,

$$(70) \qquad V_P = \frac{1}{4\pi} \iint \left\{ V \frac{\partial \frac{1}{r}}{\partial n} - \frac{1}{r}\frac{\partial V}{\partial n} \right\} dS$$

affords us an *analytical continuation* of the surface values V_S into the space τ, provided we know *both* V and $\frac{\partial V}{\partial n}$ on S.

But it cannot be assumed that *both* V and $\frac{\partial V}{\partial n}$ may be arbitrarily given. In fact we can show that if V alone is given at the surface, the continuation within is completely determined, accordingly the gradient is determined at every point, and with it $\frac{\partial V}{\partial n}$

For if not, let V_1 and V_2 be two analytical continuations of the surface values. Then the difference $U = V_1 - V_2$ is harmonic, since V_1 and V_2 are, and $U_S = 0$, since on the surface V_1 and V_2 are equal to V_S.

Now by Green's theorem A), putting $U = V$,

$$(71) \qquad \iiint \left\{ \left(\frac{\partial U}{\partial x}\right)^2 + \left(\frac{\partial U}{\partial y}\right)^2 + \left(\frac{\partial U}{\partial z}\right)^2 \right\} d\tau$$
$$= -\iint U\frac{\partial U}{\partial n} dS - \iiint U\Delta U d\tau.$$

Now within S, $\Delta U = 0$, and on S, $U = 0$, hence both integrals on the right vanish. But the integrand on the left is everywhere positive (un-

---

less zero), being a sum of real squares. Thus a positive integral is proved equal to zero, which is absurd. Therefore we must have at *every* point inside S,

$$\frac{\partial U}{\partial x} = \frac{\partial U}{\partial y} = \frac{\partial U}{\partial z} = 0,$$

U is constant, and since it is zero on S, it must be zero throughout. Thus $V_1 = V_2$, and the continuation is unique for a harmonic function.

This theorem is due to Dirichlet. By similar reasoning we establish the same result for the outside of S, only assuming in addition that V vanishes at infinity. The question whether such a continuation always exists we shall leave open. It was answered in the affirmative by Dirichlet, and the principle is known by his name. His demonstration is faulty, but has been corrected by Hilbert. The problem of finding the harmonic continuation for given surface values is known as Dirichlet's Problem. It is also called the first boundary problem. The second boundary problem is known by the name of Neumann, and differs from it in that the values, not of V, but of $\frac{\partial V}{\partial n}$ are prescribed at the boundary surface. The same proof will establish the uniqueness of the solution, the surface integral in (71) being now made to vanish by the vanishing of the other factor $\frac{\partial V}{\partial n}$. Neumann's problem arises in hydrodynamics, in the problem of finding the flow of an incompressible fluid inside of a closed surface when the normal component of the velocity, that is, the normal derivative of the velocity potential is given at every point of the boundary. On account of the gradient theorem, the given boundary values must satisfy the condition $\iint \frac{\partial V}{\partial n} dS = 0$.

In addition to these two boundary problems, we may consider a third arising in the theory of conduction of heat. In this case neither V nor $\frac{\partial V}{\partial n}$ is given but the linear combination of the two

$$\frac{\partial V}{\partial n} + hV, \quad h > 0.$$

We shall not treat these problems until we have generalized Green's formula (70) for the wave-equation.

61. Wave-equation. The equation of wave-motion, § 12, 15).

(72) $$\frac{\partial^2 \varphi}{\partial t^2} = a^2 \Delta \varphi,$$

may be looked at from two points of view, already distinguished. We may consider the initial data, as we have done in Chapter IV, or the boundary data, as in the present chapter. Let us first consider once more the initial data, by a different method.

Let us apply the gradient-theorem to the wave-equation. Integrate throughout the volume of a sphere of radius r.

$$(73) \quad \frac{1}{a^2}\iiint \frac{\partial^2 \varphi}{\partial t^2}\,d\tau - \iiint \Delta\varphi\,d\tau = -\iint \frac{\partial \varphi}{\partial n}\,dS = \iint \frac{\partial \varphi}{\partial r} r^2\,d\omega$$
$$= r^2 \frac{\partial}{\partial r}\iint \varphi\,d\omega$$

since we may differentiate outside the integral, in which r is constant. Introducing polar coordinates on the left, and differentiating by t outside the integral,

$$(74) \quad \frac{1}{a^2}\frac{\partial^2}{\partial t^2}\iiint \varphi\,d\tau = \frac{1}{a^2}\frac{\partial^2}{\partial t^2}\iiint \varphi r^2\,dr\,d\omega = \frac{1}{a^2}\frac{\partial^2}{\partial t^2}\iint d\omega \int_0^r \varphi r^2\,dr.$$

Differentiating this and the right-hand member of (73) by r,

$$(75) \quad \frac{r^2}{a^2}\frac{\partial^2}{\partial t^2}\iint \varphi\,d\omega = \frac{\partial}{\partial r}\left\{r^2 \frac{\partial}{\partial r}\iint \varphi\,d\omega\right\}.$$

We now have on both sides the integral $\iint \varphi\,d\omega$, which is 4π times the mean value $\bar{\varphi}$ of φ' on a sphere of radius r, so that

$$(75') \quad r^2 \frac{\partial^2 \bar{\varphi}}{\partial t^2} = a^2 \frac{\partial}{\partial r}\left(r^2 \frac{\partial \bar{\varphi}}{\partial r}\right)$$

or dividing by r we may write

$$(76) \quad \frac{\partial^2(r\bar{\varphi})}{\partial t^2} = a^2 \frac{\partial^2(r\bar{\varphi})}{\partial r^2}.$$

But the solution of this we know to be

$$(77) \quad r\bar{\varphi} = f_1(at + r) + f_2(at - r)$$

where f_1 and f_2 are arbitrary.

For $r = 0$,
$$0 = f_1(at) + f_2(at)$$

Consequently for all values of the variable

$$(77') \quad \begin{aligned} f_2(\xi) &= -f_1(\xi)\\ r\bar{\varphi} &= f(at+r) - f(at-r). \end{aligned}$$

Differentiating by r,

$$(78) \quad \bar{\varphi} + r\frac{\partial \bar{\varphi}}{\partial r} = f'(at+r) + f'(at-r)$$

and putting $r = 0$,

$$(79) \quad \bar{\varphi}_{r=0} = 2f'(at).$$

Now $\bar{\varphi}$ is the mean value of φ on a sphere of radius r, and for $r = 0$ this is the value at the point P itself. Accordingly

$$(80) \quad \varphi_P = 2f'(at).$$

Differentiating $r\overline{\varphi}$ by r and t

(78)
$$\frac{\partial}{\partial r}(r\overline{\varphi}) = f'(at+r) + f'(at-r),$$

(81)
$$\frac{1}{a}\frac{\partial}{\partial t}(r\overline{\varphi}) = f'(at+r) - f'(at-r),$$

and adding

(82)
$$\frac{\partial}{\partial r}(r\overline{\varphi}) + \frac{1}{a}\frac{\partial}{\partial t}(r\overline{\varphi}) = 2f'(at+r),$$

and putting $t = 0$,

(83)
$$\left[\frac{\partial}{\partial r}(r\overline{\varphi}) + \frac{1}{a}\frac{\partial}{\partial t}(r\overline{\varphi})\right]_{t=0} = 2f'(r).$$

Inserting the value of

(84)
$$\overline{\varphi} = \frac{1}{4\pi}\int\int\varphi\,d\omega$$
$$2f'(r) = \left[\frac{\partial}{\partial r}\left(\frac{r}{4\pi}\int\int\varphi_r d\omega\right) + \frac{r}{4\pi a}\int\int\frac{\partial \varphi_r}{\partial t}d\omega\right]_{t=0}$$

Now if we suppose that when $t = 0$, $\varphi = F(x, y, z)$, $\frac{\partial \varphi}{\partial t} = G(x, y, z)$,

(85)
$$2f'(r) = \frac{1}{4\pi}\left[\frac{\partial}{\partial r}\left(r\int\int F_r d\omega\right) + \frac{r}{a}\int\int G_r d\omega\right].$$

But by (80) when $r = at$ the left-hand side is equal to φ_P. Accordingly

(86)
$$\varphi_P = \frac{1}{4\pi}\left[\frac{\partial}{\partial(at)}\left(at\int\int F_{r=at}d\omega\right) + t\int\int G_{r=at}d\omega\right],$$

which is Poisson's solution, as already found by Fourier's method (86) (87) Chapter IV. This method is in principle due to Liouville.

62. Wave-Potential. Huygens's Principle. Let us seek now the analogue of the simple potential $\frac{1}{r}$ satisfying Laplace's equation. Let us suppose that φ depends only on r and t. Then our equation becomes

$$\frac{\partial^2 \varphi}{\partial t^2} = a^2\left\{\frac{\partial^2 \varphi}{\partial r^2} + \frac{2}{r}\frac{\partial \varphi}{\partial r}\right\} = \frac{a^2}{r^2}\frac{\partial}{\partial r}\left(r^2\frac{\partial \varphi}{\partial r}\right)$$

which we have already (77') shown to have a solution,

(87)
$$\varphi = \frac{f\left(t - \dfrac{r}{a}\right)}{r}.$$

We may call a point, emitting such waves symmetrical in all directions, a *source* of waves. A source may be represented in practice by a small sphere dilating and contracting. The air alternately ejected and drawn in at the mouth of an organ-pipe is perhaps the commonest source of this nature. Such a function φ we may call a simple wave-potential, corresponding to the Newtonian potential $\frac{1}{r}$ It may also be called a *retarded* potential, being equal to the Newtonian potential due to a point of mass $f(t)$, but arriving at a distance r with a retardation due to propagation with velocity a.

Corresponding to the Newtonian potential of a distributed mass, $V = \int\int\int \frac{\varrho\,d\tau}{r}$ we may take the wave-potential,

$$\text{(88)} \qquad \varphi = \int\int\int \frac{f\left(t - \frac{r}{a}\right)}{r}\,d\tau .$$

where r has the same meaning as before. We now need equations analogous to Poisson's.

Let us find the flux of normal gradient.

$$\varphi = \frac{f\left(t - \frac{r}{a}\right)}{r}, \quad \frac{\partial\varphi}{\partial n} = -\left(\frac{f}{r^2} + \frac{f'}{ar}\right)\frac{\partial r}{\partial n},$$

$$\text{(89)} \qquad \int\int \frac{\partial\varphi}{\partial n}\,dS = -\int\int\left(\frac{f}{r^2} + \frac{f'}{ar}\right)\cos(rn)\,dS$$

$$= -\int\int\left(f + \frac{rf'}{a}\right)d\omega .$$

This flux is not the same for all surfaces, because the fluid is compressible, but if we apply the treatment to an infinitely small sphere about the source, and n is the *outward* normal,

$$\int\int \frac{\partial\varphi}{\partial n}\,dS = -\int\int f(t - 0)\,d\omega = -4\pi f(t).$$

Thus $f(t)$ is the analogue of m, and may be called the *strength* of the source. At the source the wave equation is not satisfied, since $r = 0$.

If now we consider a continuous distribution of sources, $\varrho(t)$ being the strength per unit volume, and the total strength $\int\int\int \varrho(t)\,d\tau$, every elementary source produces its own flux, and we have an equation analogous to that of Poisson. It is not so easily found, however, for the flux of our point-source gradient through one surface is not the same as through another. We have found the flux through a surface *close* to the source We may apply this to a surface distribution with surface density of source $\sigma(t)$, and we find

$$\text{(91)} \qquad \frac{\partial\varphi}{\partial n_1} + \frac{\partial\varphi}{\partial n_2} = -4\pi\sigma(t),$$

for the discontinuity of normal gradient at the surface layer.

Consider now a doublet of strengths $f(t)$, $-f'(t)$ and let the moment $\lim\limits_{l=0} lf(t) = F(t)$. Then the potential is as before (57),

$$(91')\qquad \frac{\partial}{\partial s}\left\{\frac{F\left(t-\dfrac{r}{a}\right)}{r}\right\} = -\frac{1}{r^2}\left(F + \frac{rF''}{a}\right)\frac{\partial r}{\partial s},$$

but in differentiating F is differentiated. For small values of r, we may neglect the second term, and put $F(t)\dfrac{\partial\dfrac{1}{r}}{\partial n}$, corresponding to the single source $\dfrac{f(t)}{r}$ and we have the exact analogues of the gravitation potential. (Such a doublet may be approximated in practice by the to and fro motion of a small object, compressing the air in front, and rarefying it behind. A tuning-fork, on the other hand, is more like two *opposite* doublets.)

Thus if we have a sheet of doublets, they produce a discontinuity in φ, of magnitude $4\pi\varkappa(t) = 4\pi F(t)$.so that the largest value of φ is on the side *towards* which point the positive ends of the doublets.

We may accordingly produce, on the inside of a surface S, the same potential φ as would be produced by sources inside and out, by replacing the outside sources by a surface layer of density $\sigma(t) = -\dfrac{1}{4\pi}\left(\dfrac{\partial\varphi}{\partial n}\right)_t$, ($n$ pointing in) and by a double layer of strength $\varkappa(t) = \dfrac{\varphi_t}{4\pi}$, which produces at a point P the effect

$$(92)\qquad \varphi_P = \iint \frac{\sigma\left(t-\dfrac{r}{a}\right)}{r}\,dS + \iint \frac{\delta}{\delta n}\left\{\frac{\varkappa\left(t-\dfrac{r}{a}\right)}{r}\right\}dS,$$

where in $\dfrac{\delta}{\delta n}$, $\varkappa$ is not supposed to vary except as a function of $t-\dfrac{r}{a}$

That is, if $\varkappa = \varkappa(\xi,\eta,\zeta,l)$ where $l = \left(t-\dfrac{r}{a}\right)$ and l depends on ξ,η,ζ, we must distinguish between the derivatives by ξ appearing explicitly and implicitly, which we may do as follows. Put brackets on all derivatives in which ξ,η,ζ,l are considered independent variables,

$$(93)\qquad \frac{\partial\varphi}{\partial\xi} = \left[\frac{\partial\varphi}{\partial\xi}\right] + \left[\frac{\partial\varphi}{\partial l}\right]\frac{\partial l}{\partial\xi},$$

Then by

$$\frac{\delta}{\delta n}\left(\frac{\varphi\left(t-\dfrac{r}{a}\right)}{r}\right)$$

we mean

$$(94)\qquad \varphi\left(t-\frac{r}{a}\right)\frac{\partial\dfrac{1}{r}}{\partial n} + \frac{1}{r}\left[\frac{\partial\varphi}{\partial l}\right]\frac{\partial l}{\partial n}\quad\text{where}$$

$$l = t - \frac{r}{a},\quad \left[\frac{\partial\varphi}{\partial l}\right] = \frac{\partial\varphi}{\partial t} = -a\frac{\partial\varphi}{\partial r}$$

and we ignore $\left[\dfrac{\partial \varphi}{\partial \xi}\right]$. With this understanding, we have

$$(95) \quad \varphi_P = -\frac{1}{4\pi}\iint \frac{1}{r}\left(\frac{\partial \varphi}{\partial n}\right)_{t-\frac{r}{a}} dS + \frac{1}{4\pi}\iint \frac{\delta}{\delta n}\left\{\frac{\varphi\left(t-\frac{r}{a}\right)}{r}\right\} dS,$$

or more explicitly still, if we put

$$(96) \quad G(t) = \frac{\sigma(t)}{r} + \frac{\partial}{\partial n}\left(\frac{\varkappa(t)}{r}\right) = \left\{-\frac{1}{r}\frac{\partial \varphi}{\partial n} + \varphi\frac{\partial \frac{1}{r}}{\partial n} - \frac{1}{ar}\frac{\partial \varphi}{\partial t}\frac{\partial r}{\partial n}\right\},$$

$$(97) \quad 4\pi\varphi_P = \iint G\left(t-\frac{r}{a}\right) dS.$$

This is the analytical statement of Huygens's Principle, given by Kirchhoff[1]), of great importance in optics, which states that if φ and $\dfrac{\partial \varphi}{\partial t}$ be known for all points of a closed surface at any time, the values of φ can be calculated for all points inside at all times. If the surface is a sphere of radius $r = at$ $\left(\text{internal normal } \dfrac{\partial}{\partial n} = -\dfrac{\partial}{\partial r}\right)$,

$$G(t) = \frac{1}{r}\frac{\partial \varphi}{\partial r} + \frac{\varphi}{r^2} + \frac{1}{ar}\frac{\partial \varphi}{\partial t} - \frac{1}{r^2}\frac{\partial(r\varphi)}{\partial r} + \frac{1}{ar}\frac{\partial \varphi}{\partial t},$$

$$\iint G\left(t-\frac{r}{a}\right) dS = \iint \left(\frac{\partial(r\varphi)}{\partial r}\right)_{t-\frac{r}{a}} d\omega + \frac{r}{a}\iint \left(\frac{\partial \varphi}{\partial t}\right)_{t-\frac{r}{a}} d\omega,$$

or putting $r = at$,

$$(98) \quad \varphi_P = \frac{1}{4\pi}\left\{\frac{\partial}{\partial(at)}at\iint \varphi_{t=0}\,d\omega + t\iint \left(\frac{\partial \varphi}{\partial t}\right)_{t=0}d\omega\right\}$$

which is Poisson's solution.

63. Lorenz's Equation. Let us now seek the equation satisfied by the volume retarded potential,

$$(99) \quad \varphi_P = \varphi(x, y, z, t) = \iiint \frac{\varrho\left(\xi, \eta, \zeta, t-\frac{r}{a}\right) d\tau}{r}.$$

We know that the integrand and therefore the integral, satisfies

$$(72) \quad \Box \varphi = \Delta \varphi - \frac{1}{a^2}\frac{\partial^2 \varphi}{\partial t^2} = 0,$$

for each point P outside the region of sources. Inside the region, let us remove a small sphere Σ about P, and call the potential due to this φ'.

---

1) Kirchhoff, Zur Theorie der Lichtstrahlen, Sitz. d. k. Akad. d. Wissensch. zu Berlin vom 22. Juni 1882, Wied. Ann. Bd. 18, p. 663, Ges. Abh., Nachtrag p. 22.

The potential of the rest satisfies the above equation, and it remains to find only that satisfied by φ'.

Let us write

$$(100) \quad \varphi' = \iiint \frac{\varrho\left(\xi, \eta, \zeta, t - \frac{r}{a}\right) - \varrho(\xi, \eta, \zeta, t)}{r} \, d\tau + \iiint \frac{\varrho(\xi, \eta, \zeta, t)}{r} \, d\tau.$$

In the first integral, when r vanishes, the integrand remains finite, so that its Δ is finite as well as its time derivatives, so that the integrals of both vanish in the limit. The second integral is an ordinary potential, and its Laplacian is consequently $-4\pi\varrho(t)$ while its second time derivative $\iint \frac{\partial^2 \varrho}{\partial t^2} \frac{d\tau}{r}$ vanishes with the radius of the sphere. Accordingly there appears in $\square\varphi$ only the new term $-4\pi\varrho(t)$ and we have

$$(101) \qquad \square\varphi = \triangle\varphi - \frac{1}{a^2} \frac{\partial^2 \varphi}{\partial t^2} = -4\pi\varrho(t).$$

(It may be noted that the method that we have just adopted of removing the infinite element is in effect what we did above in dealing with the surface potentials.) The equation (101) was dealt with successively by L. Lorenz, Lord Rayleigh, H. A. Lorentz, Beltrami, and Poincaré, and has in the last twenty years become of great importance in connection with the theory of electricity, for both the scalar and vector potentials, (Chap. I, § 17), are propagated in accordance with it. On account of its importance, and for the sake of variety in the proof of Poisson's and similar equations, we shall give a treatment by Beltrami.

64. Beltrami's Theorem. Consider the integral

$$(102) \qquad U(x, y, z) = \iiint f(\xi, \eta, \zeta, r) \, d\tau,$$

where

$$r^2 = (x - \xi)^2 + (y - \eta)^2 + (z - \zeta)^2, \quad \frac{\partial r}{\partial x} = -\frac{\partial r}{\partial \xi},$$

and on account of the explicit and implicit dependence on ξ let us write as before (93),

$$(103) \qquad \frac{\partial f}{\partial \xi} = \left[\frac{\partial f}{\partial \xi}\right] + \left[\frac{\partial f}{\partial r}\right]\frac{\partial r}{\partial \xi}.$$

Now as in the Newtonian potential, $\left(\text{where } f = \frac{\varrho(\xi, \eta, \zeta)}{r}\right)$ (8),

$$(104) \qquad \frac{\partial U}{\partial x} = \iiint \left[\frac{\partial f}{\partial r}\right]\frac{\partial r}{\partial x} \, d\tau = -\iiint \left[\frac{\partial f}{\partial r}\right]\frac{\partial r}{\partial \xi} \, d\tau$$

$$= \iiint \left(\left[\frac{\partial f}{\partial \xi}\right] - \frac{\partial f}{\partial \xi}\right) d\xi \, d\eta \, d\zeta = \iiint \left[\frac{\partial f}{\partial \xi}\right] d\tau + \iint f \cos(nx) \, dS.$$

Differentiating again, the only change being due to r,

$$(105) \quad \frac{\partial^2 U}{\partial x^2} = \iiint \left[\frac{\partial}{\partial r} \left[\frac{\partial f}{\partial \xi} \right] \right] \frac{\partial r}{\partial x} \, d\tau + \iint \left[\frac{\partial f}{\partial r} \right] \frac{\partial r}{\partial x} \cos(nx) \, dS.$$

Now we may interchange the order of the differentiation in brackets by ξ and r, so that, replacing $\frac{\partial r}{\partial x}$ by $- \frac{\partial r}{\partial \xi}$,

$$(106) \quad \frac{\partial^2 U}{\partial x^2} = - \iiint \left[\frac{\partial}{\partial \xi} \left[\frac{\partial f}{\partial r} \right] \right] \frac{\partial r}{\partial \xi} \, d\tau - \iint \left[\frac{\partial f}{\partial r} \right] \frac{\partial r}{\partial \xi} \cos(nx) \, dS,$$

$$\frac{\partial r}{\partial \xi} = \cos(rx).$$

Now since for the total change in the direction of r at the end ξ, η, ζ, putting for brevity $\psi = \left[\frac{\partial f}{\partial r} \right]$,

$$(107) \quad \frac{\partial \psi}{\partial r} = \left[\frac{\partial \psi}{\partial r} \right] + \left[\frac{\partial \psi}{\partial \xi} \right] \cos(rx) + \left[\frac{\partial \psi}{\partial \eta} \right] \cos(ry) + \left[\frac{\partial \psi}{\partial \zeta} \right] \cos(rz)$$

Accordingly, writing formulae similar to (106) for y and z and adding, since

$$\cos(rx) \cos(nx) + \cos(ry) \cos(ny) + \cos(rz) \cos(nz) = \cos(rn)$$

$$(108) \quad \Delta U = \iiint \left(\left[\frac{\partial \psi}{\partial r} \right] - \frac{\partial \psi}{\partial r} \right) d\tau - \iint \psi \cos(rn) \, dS$$

Now we have

$$(109) \quad \frac{\partial \psi}{\partial r} = \frac{1}{r^2} \frac{\partial(r^2 \psi)}{\partial r} - \frac{2\psi}{r} = \frac{1}{r^2} \frac{\partial}{\partial r} \left(r^2 \left[\frac{\partial f}{\partial r} \right] \right) - \frac{2}{r} \left[\frac{\partial f}{\partial r} \right],$$

$$\left[\frac{\partial \psi}{\partial r} \right] = \left[\frac{\partial^2 f}{\partial r^2} \right], \quad \frac{1}{r} \left[\frac{\partial^2 (rf)}{\partial r^2} \right] = \left[\frac{\partial^2 f}{\partial r^2} \right] + \frac{2}{r} \left[\frac{\partial f}{\partial r} \right]$$

so that we obtain

$$(110) \quad \left[\frac{\partial \psi}{\partial r} \right] - \frac{\partial \psi}{\partial r} = \frac{1}{r} \left[\frac{\partial^2 (rf)}{\partial r^2} \right] - \frac{1}{r^2} \frac{\partial}{\partial r} \left(r^2 \left[\frac{\partial f}{\partial r} \right] \right)$$

$$(111) \quad \Delta U = \iiint \left[\frac{\partial^2 (rf)}{\partial r^2} \right] \frac{d\tau}{r} - \iiint \frac{\partial}{\partial r} \left(r^2 \left[\frac{\partial f}{\partial r} \right] \right) \frac{d\tau}{r^2}$$
$$- \iint \left[\frac{\partial f}{\partial r} \right] \cos(rn) \, dS.$$

But the second volume integral may be otherwise written

$$(112) \quad \iiint \frac{\partial}{\partial r} \left(r^2 \left[\frac{\partial f}{\partial r} \right] \right) \frac{d\tau}{r^2} = \iint d\omega \int_0^r \frac{\partial}{\partial r} \left(r^2 \left[\frac{\partial f}{\partial r} \right] \right) dr$$
$$= \iint r^2 \left[\frac{\partial f}{\partial r} \right] d\omega - 4\pi \left(r^2 \left[\frac{\partial f}{\partial r} \right] \right)_{r=0}$$

The first term is equal to the surface integral in (111), which disappears in subtracting, leaving

$$(113) \qquad \Delta U = \int \int \int \left[\frac{\partial^2 (rf)}{\partial r^2} \right] \frac{d\tau}{r} + 4\pi \lim_{r=0} \left(r^2 \left[\frac{\partial f}{\partial r} \right] \right).$$

This is Beltrami's generalization of Poisson's equation.

If we put $rf = \varrho$ and suppose it is holomorphic and $\frac{\partial \varrho}{\partial r}$ finite,

$$U = \int \int \int \varrho \, \frac{d\tau}{r}$$

and we have

$$\left[\frac{\partial f}{\partial r} \right] = -\frac{\varrho}{r^2}, \quad \left[\frac{\partial^2 (rf)}{\partial r^2} \right] = 0, \quad r^2 \left[\frac{\partial f}{\partial r} \right] = -\varrho,$$

and accordingly

$$\Delta U = -4\pi\varrho,$$

furnishing a new proof of Poisson's equation, if ϱ is a continuous function.

Suppose

$$\varrho(\xi, \eta, \zeta, r) = \varrho\left(\xi, \eta, \zeta, t - \frac{r}{a}\right).$$

Then as in (94)

$$\left[\frac{\partial \varrho}{\partial r} \right] = -\frac{1}{a} \frac{\partial \varrho}{\partial t}, \quad \left[\frac{\partial^2 \varrho}{\partial r^2} \right] = \frac{1}{a^2} \frac{\partial^2 \varrho}{\partial t^2}$$

$$(114) \quad \Delta U = -4\pi\varrho + \frac{1}{a^2} \int \int \int \frac{\partial^2 \varrho}{\partial t^2} \frac{d\tau}{r} = -4\pi\varrho + \frac{1}{a^2} \frac{\partial^2 U}{\partial t^2}$$

which is the equation satisfied by the *retarded potential* of Lorentz,

$$(115) \qquad U = \int \int \int \frac{\varrho\left(\xi, \eta, \zeta, t - \frac{r}{a}\right)}{r} \, d\tau.$$

65. Helmholtz's Equation. Let us now treat the Lorenz equation in another manner.

$$(101) \qquad \Box \varphi + 4\pi\varrho(x, y, z, t) = 0.$$

We may develop ϱ as a function of t by Fourier's theorem,

$$(116) \qquad \varrho(t) = \frac{1}{2\pi} \int_{-\infty}^{\infty} \int \varrho(\alpha) e^{i\lambda(t-\alpha)} d\alpha \, d\lambda$$

and if in like manner we put for φ,

$$(117) \qquad \varphi(t) = \frac{1}{2\pi} \int_{-\infty}^{\infty} \int \varphi(\alpha) e^{i\lambda(t-\alpha)} d\alpha \, d\lambda,$$

$$\frac{\partial^2 \varphi}{\partial t^2} = -\frac{1}{2\pi} \int_{-\infty}^{\infty} \int \varphi(\alpha) \lambda^2 e^{i\lambda(t-\alpha)} d\alpha \, d\lambda,$$

and inserting in the differential equation,

$$(118) \qquad \int\!\!\int_{-\infty}^{\infty} \left\{ \frac{\lambda^2}{a^2}\varphi(\alpha) + \triangle\varphi(\alpha) + 4\pi\varrho(\alpha) \right\} e^{i\lambda(t-\alpha)} d\alpha \, d\lambda = 0.$$

The functions φ, ϱ are both functions of x, y, z. The equation will be satisfied if for all values of α,

$$(119) \qquad \triangle\varphi + k^2\varphi + 4\pi\varrho = 0, \quad k = \frac{\lambda}{a}.$$

This equation may be called Helmholtz's equation, being the first of such equations to be treated after Laplace's, and it is a special case of Lorenz's obtained by putting φ and ϱ equal to a simple harmonic function of the time multiplied by a function of the point, $\varphi = U\cos pt$ as in Chapter III for one dimension. If $k = 0$ this reduces to Poisson's equation. Let us consider the case where

$$(120) \qquad \triangle U + k^2 U = 0.$$

This equation is the subject of an excellent treatise by Pockels (Teubner).

Let us seek an elementary solution, a function of r alone. We then have

$$(121) \qquad \frac{d^2U}{dr^2} + \frac{2}{r}\frac{dU}{dr} + k^2 U = 0,$$

a linear differential equation with variable coefficients. Multiplying by r we may write

$$\frac{d^2(r\,U)}{dr^2} + k^2 r\,U = 0$$

giving

$$rU = Ae^{ikr} + Be^{-ikr} \quad \text{or} \quad A\cos kr + B\sin kr.$$

We have the particular solutions of (121),

$$\frac{\cos kr}{r}, \quad \frac{\sin kr}{r}, \quad \frac{e^{ikr}}{r}, \quad \frac{e^{-ikr}}{r}.$$

Let us consider as the analogue of $\frac{1}{r}$ the function $\frac{e^{\pm ikr}}{r}$ which is infinite like it for $r = 0$, and vanishes at ∞.

With this elementary solution we can form the integral of Helmholtz

$$(122) \qquad U = \int\!\!\int\!\!\int \frac{\varrho(\xi, \eta, \zeta)e^{\pm ikr}}{r} d\tau,$$

which satisfies $\triangle U + k^2 U = 0$ at outside points.

In order to find ΔU we have the same difficulty for inside points as before. We may apply Beltrami's theorem with

$$f = \frac{\varrho e^{\pm ikr}}{r}, \quad rf = \varrho e^{\pm ikr}, \quad \left[\frac{\partial^2 (rf)}{\partial r^2}\right] = -k^2 \varrho e^{\pm ikr},$$

$$r^2 \left[\frac{\partial f}{\partial r}\right] = \varrho e^{\pm ikr} \{\pm ikr - 1\}, \quad \lim_{r=0} r^2 \left[\frac{\partial f}{\partial r}\right] = -\varrho$$

$$\Delta U = -k^2 U - 4\pi \varrho.$$

Accordingly we have as the general solution of Helmholtz's equation (119) the integral (122).

Putting again $k = \frac{\lambda}{a}$, we have the solution of (119)

$$(123) \qquad \varphi(\alpha) = \iiint \frac{\varrho(\xi, \eta, \zeta, \alpha) e^{\pm \frac{i\lambda r}{a}}}{r} \, d\tau,$$

inserting which in (117),

$$(124) \quad \varphi(t) = \frac{1}{2\pi} \int\!\!\int_{-\infty}^{\infty} d\alpha \, d\lambda \iiint \frac{\varrho(\xi, \eta, \zeta, \alpha) e^{i\left[\lambda\left(t \pm \frac{r}{a} - \alpha\right)\right]}}{r} \, d\tau$$

and since

$$\frac{1}{2\pi} \int\!\!\int_{-\infty}^{\infty} \varrho(\alpha) e^{i\left[\lambda\left(t \pm \frac{r}{a} - \alpha\right)\right]} d\alpha \, d\lambda = \varrho\left(t \pm \frac{r}{a}\right),$$

we have

$$(125) \qquad \varphi(x, y, z, t) = \iiint \frac{\varrho\left(\xi, \eta, \zeta, t \pm \frac{r}{a}\right) d\tau}{r}.$$

It was somewhat in this manner that the solution of Lorenz's equation was deduced from Helmholtz's solution by Lord Rayleigh.[1]

All that has been said of the wave equation may of course be applied to Helmholtz's equation as a particular case, simpler, as not involving the time. In particular we have corresponding to equation (95)

$$(126) \quad U_P = -\frac{1}{4\pi} \int\!\!\int \frac{\partial U}{\partial n} \frac{e^{-ikr}}{r} \, dS + \frac{1}{4\pi} \int\!\!\int U \frac{\partial}{\partial n}\left(\frac{e^{-ikr}}{r}\right) dS$$

which may be at once proved by applying Green's theorem to U and $\frac{e^{-ikr}}{r}$ instead of $\frac{1}{r}$ as in (69). It is to be noticed that in (126) we may replace e^{-ikr} by $\cos kr$, while if we replace it by $\sin kr$, the left-hand side will be zero instead of U, since $\frac{\sin kr}{r}$ has no singularity for $r = 0$

---

1) Theory of Sound, Vol. II. p. 104.

In the case of two dimensions for the symmetrical case, Helmholtz's equation becomes, instead of (121)

$$(127) \qquad \frac{\partial^2 u}{\partial r^2} + \frac{1}{r}\frac{\partial u}{\partial r} + k^2 u = 0.$$

But this is Bessel's equation (Chap. VII), satisfied by the two Bessel's functions $J_0(kr)$ and $Y_0(kr)$, the first of which is finite for $r = 0$ like $\frac{\sin kr}{r}$, the second is infinite like $\log r$. We can therefore form the potential analogous to the logarithmic potential

$$(128) \qquad U = -\iint \varrho\, Y_0(kr)\, dS$$

which satisfies the equation

$$(129) \qquad \frac{\partial^2 U}{\partial x^2} + \frac{\partial^2 U}{\partial y^2} + k^2 U = -2\pi\varrho.$$

The functions $J_0(kr)$ and $Y_0(kr)$ are oscillatory, like $\frac{\cos kr}{r}$ and $\frac{\sin k r}{r}$ Corresponding to equation (126) we have

$$(130) \qquad U_P = \frac{1}{2\pi}\int \left\{ \frac{\partial U}{\partial n} Y_0(kr) - U\frac{\partial Y_0(kr)}{\partial n} \right\} dS.$$

It is to be expressly noted that our treatment of the wave-equation is not to be extended to two dimensions, as we have seen in Chapter IV, for there is no analogue of the elementary solution $\frac{f(r-at)}{r}$, nor can the solution just found $Y_0(kr)$ when multiplied by e^{ipt} be reduced to any simple function of $r - at$, as is the case for the function in three dimensions $\frac{e^{ipt}e^{-ikr}}{r} = \frac{e^{ik(at-r)}}{r}$, denoting a progressive wave.

66. Green's Function. Let us now define the Green's function for three dimensional regions. For one dimension we defined it (§ 35) as a solution of the equation $\frac{d^2 y}{dx^2} = 0$, vanishing at the boundaries of a certain fundamental region, and having one discontinuity in the first derivative. In like manner for a three dimensional fundamental region bounded by a surface S let us define the Green's function G corresponding to a point Q of discontinuity with coordinates ξ, η, ζ as a solution of Laplace's equation, vanishing at the surface S, and having a singularity at Q like $\frac{1}{4\pi r_{pq}}$,

$$(131) \qquad \Delta G = 0, \quad G_S = 0, \quad G = \frac{1}{4\pi r_{pq}} + \omega,$$

where ω is a harmonic function in the whole of the fundamental region. We may employ the notion of the jag-function, as in § 35, as a function

f which is zero everywhere except in an infinitely small region, in which it becomes infinite in such a way that $\iiint f d\tau = 1$. Then correspon-ding to equations (176), (177), § 35, we have, for a sphere of radius ε

$$\Delta U + k^2 U = -f,$$

$$\iiint \Delta U d\tau = -\iint \frac{\partial U}{\partial n} dS = -k^2 \iiint U d\tau - \iiint f d\tau$$

$$\lim_{\varepsilon=0} \iint \frac{\partial U}{\partial n} dS = \lim_{\varepsilon=0} \iiint f d\tau = 1.$$

For $k = 0$ this agrees with (131). We shall from now on adopt this definition of a unit source, agreeing with (178) § 35, and making $4\pi m$, $4\pi\varrho$, $4\pi\sigma$ the strengths and densities of sources in the sections just preceding this. That is instead of following the analogy of mass or electric charge, we now follow that of flow of heat or water, unit source corresponding to unit outflow per unit time.

In order to show the pole, jag or point of discontinuity, we may write $G(x,y,z;\xi,\eta,\zeta)$ or briefly, G_{pq}. The function G may be inter-preted physically as the steady state of temperature due to a source of heat of strength unity at Q while the surface S is maintained at temperature zero, or as the potential due to a charge $\frac{1}{4\pi}$ of electricity, while the sur-face S is made conducting and maintained at potential zero. The func-tion ω is the potential due to the charge induced by Q on the surface, which is called the Green's layer.

As for one dimension, let us first show the symmetry of the func-tion with respect to its two points. Consider two functions G with dif-ferent poles Q, Q' such that

$$(132) \qquad G = G_{pq} = \frac{1}{4\pi r_{pq}} + \omega, \quad G' = G_{pq'} = \frac{1}{4\pi r_{pq'}} + \omega'.$$

Now apply Green's theorem to the two functions G, G', integrating over the space inside S, but excluding the points Q and Q' by two small spheres enclosing them. Then since both G's are harmonic in the space considered, we have

$$(133) \qquad \iint \left(G \frac{\partial G'}{\partial n} - G' \frac{\partial G}{\partial n} \right) dS = 0.$$

On the surface S, both G and G' vanish, so that we have only the inte-grals over the small spheres. But on the sphere about Q, $\iint \frac{\partial G'}{\partial n} dS = 0$, since G' is harmonic inside, while since G behaves like $\frac{1}{4\pi r}$, as in (67), $-\iint \frac{\partial G}{\partial n} dS = 1$, consequently the limit gives us $G'(q) = G_{qq'}$

On the sphere about Q', $\iint \frac{\partial G}{\partial n} dS = 0$, while $\iint \frac{\partial G'}{\partial n} dS = -1$, so that the limit is $-G(q') = -G_{q'q}$. Accordingly we have

(134) $$G_{qq'} = G_{q'q}.$$

Physically this means that the temperature at q_1 when a source is placed at q_2 is the same as that at q_2 when an equal source is at q_1, the temperature of the surface S being always zero.

Let us now apply Green's theorem to a function satisfying Poisson's equation

(135) $$\Delta V = -f(x, y, z).$$

We have

(136) $$\iiint (G\Delta V - V\Delta G)d\tau = -\iiint Gf d\tau = -\iint \left(V\frac{\partial G}{\partial n} - G\frac{\partial V}{\partial n}\right) dS$$

Let G have the pole P, and let the surface integral be extended to the small sphere about P. Consequently, we have as in (67), and since $G_S = 0$,

(137) $$-V_P + \iint V\frac{\partial G}{\partial n} dS = -\iiint Gf d\tau.$$

First let us consider $f = 0$. Then V is harmonic inside S, and we have

(138) $$V_P = \iint V\frac{\partial G}{\partial n} dS,$$

which gives us a solution of Dirichlet's problem, if we can find the Green's function for every position of the pole P. (Fig. 77.)

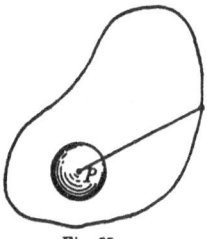

Fig. 77.

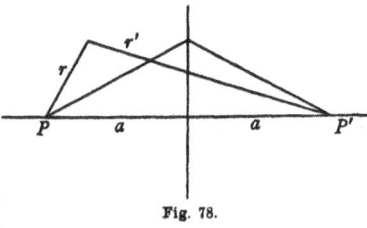

Fig. 78.

67. Examples of Green's Function for a Plane and Sphere.

Let us consider that portion of space which lies on one side of a given plane. Let P be the given pole at a distance a from the plane and let P' be its geometrical image in the plane, Fig. 78. Let the distances of any point on the same side of the plane as P from P and P' be r and r' respectively. Now for every point on the side towards P the function $-\frac{1}{r'}$ is harmonic, and for points on the plane, where $r = r'$, it assumes the value $-\frac{1}{r}$, and is accordingly the function $4 \pi \omega$ of equation (131). We have then

$$(139) \qquad G = \frac{1}{4\pi}\left(\frac{1}{r} - \frac{1}{r'}\right),$$

$$\left[\frac{\partial G}{\partial n}\right]_s = \frac{1}{4\pi}\left(-\frac{\cos(n r)}{r^2} + \frac{\cos(n r')}{r^2}\right) = \frac{1}{2\pi}\frac{\cos \varphi}{r^2}$$

where φ is the acute angle included between the radius r and the normal to the plane. Consequently the equation

$$(140) \qquad V_P = \iint V \frac{\partial G}{\partial n}\, dS = \frac{1}{2\pi}\iint \frac{V \cos \varphi}{r^2}\, dS = \frac{a}{2\pi}\iint \frac{V\, dS}{r^3},$$

solves Dirichlet's problem for the side of the plane on which P is situated. If we suppose a unit positive charge of electricity placed at P with an equal negative charged placed at P', the potential of the plane will be zero, and accordingly, if we consider it made electrically conducting and connected to the earth, the charge at P' may be removed and the electricity that will be induced on this conducting plane by the charge at P, will produce the same effect on the side of P as the charge at P'. The charge at P' is

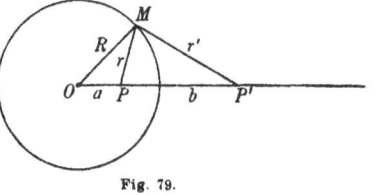

Fig. 79.

called the electrical image of that at P, because exactly as in optics, the plane mirror produces on the side of P exactly the same effect as a real charge at P' would do in the absence of the mirror.

The same method may be extended without much greater complication to the case of the sphere. Let P be the given pole at a distance a from the center of the sphere of radius R, Fig. 79. Take the point P' lying on the same radius as P at a distance b from the center such that $ab = R^2$. Then P and P' are said to be inverse points with respect to the sphere. If M be any point on the surface of the sphere, the triangles OMP and OMP' are similar, for they have a common angle at O and the sides including it are proportional, for by hypothesis $\frac{a}{R} = \frac{R}{b}$. Accordingly if r and r' are the distances from P and P' respectively, we have for points on the surface

$$(141) \qquad \frac{r'}{r} = \frac{R}{a}, \qquad \frac{1}{r} = \frac{R}{a}\frac{1}{r'}.$$

Therefore since $\frac{1}{r'}$ is harmonic in the space containing P, we have the Green's function for that space

$$(142) \qquad G = \frac{1}{4\pi}\left(\frac{1}{r} - \frac{R}{a}\frac{1}{r'}\right),$$

$$(143)\quad \frac{\partial G}{\partial n} = \frac{1}{4\pi}\left(-\frac{1}{r^2}\frac{\partial r}{\partial n} + \frac{R}{a}\frac{1}{r'^2}\frac{\partial r'}{\partial n}\right) = \frac{1}{4\pi}\left(-\frac{\cos(nr)}{r^2} + \frac{R}{a}\frac{\cos(nr')}{r'^2}\right).$$

The density of the Green's layer at potential zero induced on the sphere made conducting by unit charge at P is $\dfrac{\partial G}{\partial n}$.

Now in the triangles OMP and OMP' we have

$$(144)\qquad a^2 = R^2 + r^2 - 2\,Rr\cos(nr),$$

$$b^2 = R^2 + r'^2 - 2\,Rr'\cos(nr')$$

so that

$$(145)\qquad \frac{\partial G}{\partial n} = \frac{1}{4\pi}\left\{\frac{a^2-(R^2+r^2)}{2\,Rr^2} - \frac{R}{a}\frac{b^2-(R^2+r'^2)}{2\,Rr'^2}\right\},$$

which by (141) and $ab = R^2$ gives $\dfrac{\partial G}{\partial n} = \dfrac{a^2 - R^2}{4\pi Rr^2}$

$$(146)\qquad V_P = \frac{a^2-R^2}{4\pi R}\iint \frac{V}{r^3}\,dS.$$

If we introduce polar coordinates θ, φ, θ being the angle measured from the radius OP, we have $dS = R^2\sin\theta\,d\theta\,d\varphi$ and the formula becomes

$$(147)\qquad V_P = \frac{R(R^2-a^2)}{4\pi}\int_0^{\pi}\int_0^{2\pi} \frac{V(\theta,\varphi)\sin\theta\,d\theta\,d\varphi}{\{a^2+R^2-2\,aR\cos\theta\}^{3/2}}.$$

This formula was given by Poisson.

For the logarithmic potential the formula is

$$(148)\qquad V_P = \frac{R^2-a^2}{2\pi}\int_0^{2\pi} \frac{V(\theta)\,d\theta}{R^2+a^2-2\,Ra\cos\theta}.$$

68. Forced Vibrations. Integral Equations. Let us now consider the case of equation (135) where f is not zero, but V vanishes on S. We then have, by (137)

$$(149)\qquad V_P = \iiint f_q G_{pq}\,d\tau_q,$$

and the function V is represented sourcewise. (Compare (189) § 35.) In fact, since G_{pq} vanishes at every point P on the surface, the integral V must also vanish on the surface. Since G_{pq} is the temperature due to a source unity, the integral gives the temperature due to density f.

Suppose now we deal with the forced vibration for the wave-equation

$$(150)\qquad \frac{\partial^2\varphi}{\partial t^2} = a^2\{\Delta\varphi + \Phi(x,y,z,t)\},$$

putting $\Phi = f\cos pt$, $\varphi = U\cos pt$, so that

$$(151) \qquad \Delta U + k^2 U = -f,$$

which is the three-dimensional analogue of (163), § 36.

Applying the same process to U and the Green's function

$$(152) \quad \iiint \{G\Delta U - U\Delta G\}d\tau + k^2\iiint GUd\tau = -\iiint fG d\tau,$$

and as before, (69), (133) and (137),

$$(153) \qquad -U_p + k^2\iiint U_q G_{pq} d\tau_q = -\iiint f_q G_{pq} d\tau_q.$$

Calling

$$(154) \qquad F_p = \iiint f_q G_{pq} d\tau_q, \quad k^2 = \lambda,$$

we have accordingly

$$(155) \qquad U_p - \lambda\iiint U_q G_{pq} d\tau_q = F_p$$

which is an integral equation for U, with the symmetrical kernel G_{pq}, corresponding to (202) § 35, and F being represented source-wise.

If F is zero, the homogeneous integral equation

$$(156) \qquad U_p = \lambda\iiint U_q G_{pq} d\tau_q$$

is satisfied only for values of λ corresponding to the normal functions for the volume in question, giving the series of vibrations possible for the air in the volume.

The normal functions satisfy the equations

$$(157) \qquad \Delta u_r + \lambda_r u_r = 0,$$

vanish on the surface S, and we will suppose them normalized so that

$$(158) \qquad \iiint u_r^2 d\tau = 1.$$

If we apply Green's theorem to two normal functions u_r, u_s as in Chapter III,

$$\Delta u_r + \lambda_r u_r = 0, \quad \Delta u_s + \lambda_s u_s = 0,$$

we obtain

$$(159) \quad \iiint (u_s\Delta u_r - u_r\Delta u_s)d\tau + (\lambda_r - \lambda_s)\iiint u_r u_s d\tau = 0,$$

and since u_r and u_s both vanish on the surface

$$(160) \qquad (\lambda_r - \lambda_s)\iiint u_r u_s d\tau = 0,$$

and if the λ's are unequal, the functions are orthogonal. The same thing may be proved from the integral equation, as in Chapter III, (217). Also the characteristic numbers λ_r are real. Consequently we can develop a continuous function in a series of normal functions,

$$(161) \qquad U = \sum_{r=1}^{r=\infty} c_r u_r,$$

and by multiplying by u_s and integrating we find as in Chapter III (228),

$$(162) \qquad c_r = \iiint U u_r d\tau.$$

In particular if U is the kernel G_{pq}, since, by (156),

$$(163) \qquad \iiint G_{pq} u_{rp} d\tau_p = \frac{u_{rq}}{\lambda_r},$$

we find $c_r = \frac{u_{rq}}{\lambda_r}$ and we have the bilinear formula,

$$(164) \qquad G_{pq} = \sum_{r=1}^{r=\infty} \frac{u_{rp} u_{rq}}{\lambda_r}.$$

If this is uniformly convergent, we may apply the method of §§ 36, 38, which is in abbreviated form as follows. If f and U, a solution of the equation (151), are developable in series of normal functions, let us put

$$f = \sum a_r u_r, \quad F = \sum b_r u_r, \quad U = \sum c_r u_r,$$

$$(165) \quad a_r = \iiint f u_r d\tau, \quad b_r = \iiint F u_r d\tau, \quad c_r = \iiint U u_r d\tau$$

and inserting in the differential equation (151),

$$(166) \qquad \sum c_r (\Delta u_r + \lambda u_r) + \sum a_r u_r = 0$$

which is by (157)

$$\sum \{c_r (\lambda - \lambda_r) + a_r\} u_r = 0.$$

Accordingly $c_r = \frac{a_r}{\lambda_r - \lambda}$ and we have the solution

$$(167) \qquad U = \sum \frac{a_r u_r}{\lambda_r - \lambda} = \sum \frac{u_r}{\lambda_r - \lambda} \iiint f u_r d\tau$$

as in § 38, (222). The work of proving this to be a solution of the integral equation and converting it into Schmidt's solution is the same as there given. For by the definition of F, the development of f, and the equation (156)

$$(168) \quad F = \iiint f G d\tau = \sum a_r \iiint u_r G d\tau = \sum \frac{a_r u_r}{\lambda_r} = \sum b_r u_r,$$

$$
(169) \quad U - F - \sum a_r u_r \left\{ \frac{1}{\lambda_r - \lambda} - \frac{1}{\lambda_r} \right\} = \lambda \sum \frac{a_r u_r}{\lambda_r (\lambda_r - \lambda)}
$$

$$
= \lambda \sum \frac{b_r u_r}{\lambda_r - \lambda} = \lambda \sum \frac{u_r}{\lambda_r - \lambda} \iiint F u_r \, d\tau,
$$

$$
(170) \quad U = F + \lambda \sum \frac{u_r}{\lambda_r - \lambda} \iiint F u_r \, d\tau,
$$

which is Schmidt's solution.

As a matter of fact, although what we have said above is mathematically correct, the interpretation of the boundary condition $U = 0$ or $u_r = 0$ is physically not easy for air-vibrations, for if U represents the velocity-potential, it is not U but $\frac{\partial U}{\partial n}$ that vanishes on the surface. In the case of one dimension, if u is the displacement, $s = -\frac{\partial u}{\partial x}$ is the compression, and this vanishes at the open end of a pipe. Such a condition is not to be achieved all over a surface bounding a three-dimensional region, although it may be for the edge of a two-dimensional region such as the space between two parallel planes near together, in which the air vibrates parallel to them, and the space is open at the edge.

If we do have the boundary condition $\frac{\partial U}{\partial n} = 0$, we cannot use the Green's function as just defined, nor can we define it by

$$
\Delta G = 0, \quad G = \frac{1}{4\pi r} + \omega, \quad \left(\frac{\partial G}{\partial n} \right)_s = 0,
$$

for since

$$
\Delta \omega = 0, \quad \iint \frac{\partial \omega}{\partial n} \, dS = - \iiint \Delta \omega \, d\tau = 0,
$$

$$
\iint \frac{\partial G}{\partial n} \, dS = \frac{1}{4\pi} \iint \frac{\partial \frac{1}{r}}{\partial n} \, dS = -1,
$$

which is not zero. In fact, if no heat flows out, a source emitting heat cannot maintain a steady temperature, for the temperature will rise. If however, we define G by the equations

$$
(171) \quad \Delta G = a, \quad G = \frac{1}{4\pi r} + \omega, \quad \left(\frac{\partial G}{\partial n} \right)_s = 0,
$$

we may satisfy all conditions. To find a, integrate ΔG throughout the volume, excluding the sphere about the pole,

$$
\iiint \Delta G \, d\tau = - \int_{S, \Sigma} \iint \frac{\partial G}{\partial n} \, dS = a \iiint d\tau = a\tau,
$$

and since the surface integral is equal to unity on the small sphere and to zero on S, we must take a equal to the reciprocal of the whole volume. That is the source-density a must be negative and just sufficient to absorb all the heat emitted by the unit source at the pole, and letting none escape at the boundary. We still have an arbitrary constant in G, which we will determine presently.

Any function satisfying the equation

$$\Delta U + \lambda U = 0$$

with the boundary condition $\dfrac{\partial U}{\partial n} = 0$ must give

$$(172) \qquad \lambda \iiint U d\tau = -\iiint \Delta U d\tau = \iint \frac{\partial U}{\partial n} dS = 0,$$

so that

$$(173) \qquad\qquad \iiint U d\tau = 0$$

or the mean value of U through the volume must vanish. This must be the case for every normal function.

We shall now prove the symmetry of the new Green's function, for as in (133)

$$(174) \qquad \iint \left(G \frac{\partial G'}{\partial n} - G' \frac{\partial G}{\partial n} \right) dS = \iiint (G' \Delta G - G \Delta G') d\tau$$
$$= a \iiint G' d\tau - a \iiint G d\tau.$$

This will be zero, if we define the arbitrary constant in G so that it shall satisfy equation (173). Now since

$$\Delta G = a, \quad \Delta U + \lambda U = -f,$$

we have instead of (152),

$$(175) \iiint (G \Delta U - U \Delta G) d\tau + \lambda \iiint G U d\tau = -\iiint G f d\tau - a \iiint U d\tau,$$

and as before,

$$- U_P + \iint \left(U \frac{\partial G}{\partial n} - G \frac{\partial U}{\partial n} \right) dS + \lambda \iiint G U d\tau = -F - \frac{1}{\tau} \iiint U d\tau.$$

The last term is the mean value $\bar{U}$ of U in τ and since $\dfrac{\partial U}{\partial n}$ and $\dfrac{\partial G}{\partial n}$ vanish on the surface, we obtain (155) as before, except that the first term is $U_P - \bar{U}$. But we now have, instead of (173),

$$\lambda \iiint U d\tau = -\iiint f d\tau$$

and consequently $\bar{U}$ will vanish if $\bar{f}$ does. But only such a function can be developed in normal functions of the present sort. Also a function can be represented source-wise if it fulfills (173). This method may be employed for one dimension, and gives the kernel for the normal functions for the open pipe.

When we come to the solving function Γ, the Green's function for the operator Λ, where

(176) $$\Lambda\Gamma \equiv L\Gamma + \lambda\Gamma = 0, \quad LG \equiv \Delta G = 0,$$

we have no difficulty in putting $\dfrac{\partial\Gamma}{\partial n} = 0$ on the surface. We also have the obvious physical interpretation for Γ as the velocity-potential due to a unit source of sound at its pole, with the effect of the rigid walls. The symmetry of Γ expresses the important acoustical and optical theorem of reciprocity that if the source and the receiving point be interchanged, the effect is the same.

Now, as in § 39 (252), we apply Green's theorem to two functions

$$u = G_{pq_1}, \qquad U = \Gamma_{pq_2},$$

(177) $$\iiint(\Gamma\Delta G - G\Delta\Gamma)d\tau = \lambda\iiint G\,\Gamma d\tau,$$

and excluding the small sphere about the poles $q_1, q_2,$

(178) $$\Gamma_{q_1 q_2} - G_{q_2 q_1} = \lambda\iiint G_{pq_1}\Gamma_{pq_2}d\tau_p$$

which is the resolvent, corresponding to (294), Chapter III, and as before (300) the solving kernel is

(179) $$\Gamma_{pq} = \sum {}^l\frac{u_{r\,p}\,u_{r\,q}}{\lambda_r - \lambda}.$$

Example. Rectangular Membrane.

Let us now consider as an example the forced vibrations of a rectangular region, and for simplicity we will take the case of two dimensions, say a membrane, which at its edges, satisfies the condition $u = 0$, where u stands for the transverse displacement of the membrane. For the Green's function G we have the simple representation by the shape of the membrane when pushed aside normally by a sharp point, so as in the neighborhood to assume a conical shape. For Γ we have a similar representation when the pushing point is vibrating.

Although we have deduced the equation for a membrane in Chapter I, § 12, we shall now give a somewhat more elegant deduction depending on Green's theorem.

If the plane of the membrane is that of x, y, the normal displacement u, the tension, or force exerted across a line of unit length τ, the surface density ρ, if we consider the forces on any small area we find for the resolved force perpendicular to the xy plane on an element of boundary ds, $\tau ds\dfrac{\partial u}{\partial n}$ where n is the outward normal to the boundary. But in the plane

$$\int\frac{\partial u}{\partial n}\,ds = \iint\Delta u\,dS,$$

so that for the equation of motion we have

$$\varrho\, dS \frac{\partial^2 u}{\partial t^2} = \tau \Delta u\, dS, \quad \frac{\partial^2 u}{\partial t^2} = c^2 \left[\frac{\partial^2 u}{\partial x^2} + \frac{\partial^2 u}{\partial y^2} \right], \quad c^2 = \frac{\tau}{\varrho}$$

for the transverse vibration of a membrane. For the standing waves, putting $u = U \cos pt$, we have

$$(180) \qquad \frac{\partial^2 U}{\partial x^2} + \frac{\partial^2 U}{\partial y^2} + k^2 U = 0.$$

Let us undertake, as usual, to find solutions of the form $X(x)\, Y(y)$. We then obtain

$$(181) \qquad \frac{1}{X} \frac{d^2 X}{dx^2} + \frac{1}{Y} \frac{d^2 Y}{dy^2} + k^2 = 0$$

requiring each term to be constant, say

$$(182) \qquad \frac{d^2 X}{dx^2} + q^2 X = 0, \quad \frac{d^2 Y}{dy^2} + r^2 Y = 0, \quad q^2 + r^2 = k^2,$$

$$(183) \qquad X = A \cos qx + B \sin qx, \quad Y = C \cos ry + D \sin ry.$$

If the rectangle is between $x = 0$, $x = a$, $y = 0$, $y = b$ we must have for $U = 0$ at the boundary,

$$A = C = 0, \quad \sin qa = \sin rb = 0, \quad q = \frac{m\pi}{a}, \quad r = \frac{n\pi}{b},$$

consequently the normal functions and characteristic numbers are given by

$$(184) \qquad \begin{aligned} u_{mn} &= \sin \frac{m\pi x}{a} \sin \frac{n\pi y}{b} \\ \lambda_{mn} &= k_{mn}^2 = \pi^2 \left(\frac{m^2}{a^2} + \frac{n^2}{b^2} \right), \quad p_{mn} = k_{mn} c, \end{aligned}$$

where m and n are integers. The general solution is

$$(185) \quad U = \sum_{m=1}^{m=\infty} \sum_{n=1}^{n=\infty} \sin \frac{m\pi x}{a} \sin \frac{n\pi y}{b} \{ A_{mn} \cos p_{mn} t + B_{mn} \sin p_{mn} t \},$$

and we are led to the problem of the development of a function

$$(186) \qquad f(x, y) = \sum_{m=1}^{m=\infty} \sum_{n=1}^{n=\infty} A_{mn} \sin \frac{m\pi x}{a} \sin \frac{n\pi y}{b}$$

in a double Fourier's series.

Considering a single normal vibration of frequency $\frac{p_{mn}}{2\pi}$ we may take the sum for all the values of m and n for which

$$\frac{m^2}{a^2} + \frac{n^2}{b^2} = \frac{k^2}{\pi^2}$$

is constant.

We may now ask whether the various normal vibrations are harmonic, that is, their frequencies proportional to the series of integers. If we consider the nodal lines of the solution u_{mn}, we find that beside the edges, they are given by $m - 1$ lines parallel to the y-axis, $n - 1$ parallel to the x-axis, and if we put first $m = n = 1$, we find

$$k_{11} = \pi \sqrt{\frac{1}{a^2} + \frac{1}{b^2}},$$

which gives the lowest frequency.

If m and n are equal, we have

$$k_{mm} = \pi m \sqrt{\frac{1}{a^2} + \frac{1}{b^2}} = m k_{11},$$

so that for this case we have a set of harmonics. The question arises whether there are any others. If so let k, k' corresponding to $m. n$ and m', n' be in the ratios of two integers h, h', then

$$\frac{1}{h^2}\left(\frac{m^2}{a^2} + \frac{n^2}{b^2}\right) = \frac{1}{h'^2}\left(\frac{m'^2}{a^2} + \frac{n'^2}{b^2}\right)$$

or otherwise

$$\frac{h'^2 m^2 - h^2 m'^2}{a^2} = \frac{h^2 n'^2 - h'^2 n^2}{b^2},$$

and as the numerators are integers, $\frac{a^2}{b^2}$ must be a rational number. If not then both sides must vanish, or

$$\frac{m^2}{m'^2} = \frac{h^2}{h'^2} = \frac{n^2}{n'^2}, \quad \frac{m}{n} = \frac{m'}{n'}.$$

But then m', n' are equal multiples of m, n and we have the generalization of the case already treated. If m and n are prime to each other, any other pair of prime numbers gives a set of harmonics, which are not harmonic to the first set.

But if a^2 and b^2 are in a rational relation, the different sets may be harmonic to one another. For example, consider a square membrane, $a = b$. Then we may have the equation

$$m^2 + n^2 = \frac{a^2 k^2}{\pi^2},$$

satisfied for several values of m, n giving the same k. We then have more than one normal function to the same characteristic number.

Consider, for instance the two terms for which $m = 1, n = 2$, $m = 2, n = 1$,

(187) $$u = A_{12} \sin\frac{\pi x}{a} \sin\frac{2\pi y}{a} + A_{21} \sin\frac{2\pi x}{a} \sin\frac{\pi y}{a}.$$

The nodal lines are given by

$$\sin\frac{\pi x}{a} \sin\frac{\pi y}{a}\left[A_{12} \cos\frac{\pi y}{a} + A_{21} \cos\frac{\pi x}{a}\right] = 0$$

giving the edges, and a curve whose equation is

$$(188) \qquad A_{12} \cos \frac{\pi y}{a} + A_{21} \cos \frac{\pi x}{a} = 0.$$

Thus the membrane may vibrate in an infinite variety of forms. If we call two normal functions for the same period u and u', the function $Au + A'u'$ has nodal lines passing through the intersections of those of u and u', which gives us some idea of their appearance.

If we normalize the function

$$u_{mn} = A_{mn} \sin \frac{m\pi x}{a} \sin \frac{n\pi y}{b}$$

by making the integrated square equal to unity, we find

$$(189) \qquad A_{mn}^2 \int_0^a\!\!\int_0^b \sin^2 \frac{m\pi x}{a} \sin^2 \frac{n\pi y}{b}\, dx\, dy = 1, \quad A_{mn} = \sqrt{\frac{2}{a}\frac{2}{b}},$$

so that we have the bilinear formula,

$$(190) \quad K(x, y; \xi, \eta) = \frac{2}{a}\frac{2}{b} \sum_{m=1}^{m=\infty} \sum_{n=1}^{n=\infty} \frac{\sin \dfrac{m\pi x}{a} \sin \dfrac{n\pi y}{b} \sin \dfrac{m\pi \xi}{a} \sin \dfrac{n\pi \eta}{b}}{\pi^2 \left(\dfrac{m^2}{a^2} + \dfrac{n^2}{b^2}\right)}.$$

The solving function becomes

$$(191) \quad \Gamma(x, y; \xi, \eta) = \frac{2}{a}\frac{2}{b} \sum_{m=1}^{m=\infty} \sum_{n=1}^{n=\infty} \frac{\sin \dfrac{m\pi x}{a} \sin \dfrac{n\pi y}{b} \sin \dfrac{m\pi \xi}{a} \sin \dfrac{n\pi \eta}{b}}{\pi^2 \left(\dfrac{m^2}{a^2} + \dfrac{n^2}{b^2}\right) - \lambda},$$

corresponding to (300), Chapter III, for one dimension. In the same way we may treat the three-dimensional problem of the vibration of air in a rectangular chamber, leading to series in three variables.

69. Infinite Region. The case in which the region considered is infinite presents some difficulties and has been treated in detail by Sommerfeld.[1])

We shall deal only briefly with the matter. We have seen that there is a difference between Laplace's equation and the equation

$$\Delta u + k^2 u = 0,$$

in that, if the solution of the former vanishes at the boundary of a region and has no singularities within, it vanishes identically, while under similar circumstances a solution of the latter does not vanish, but

---

1) A. Sommerfeld: Die Greensche Funktion der Schwingungsgleichung. Jahresbericht der Deutschen Mathematiker-Vereinigung, Bd. 21, p. 309, 1913. It is to be noted that the Green's function denoted by Sommerfeld by G is our second Green's function Γ.

is a normal function for the space in question, provided that k^2 is one of the characteristic numbers. Likewise for the solution of the equations with second member, for Poisson's equation $\Delta u = -f$, the solution is determined by its sources by the formula (69)

$$u = \frac{1}{4\pi} \iiint \frac{f\,d\tau}{r}$$

while in the case of Helmholtz's equation this is not the case, for if k^2 is a characteristic number the vibration becomes infinite. If the space is external to a given surface and extends to infinity the case is the same for Poisson's equation, but for Helmholtz's equation the question arises as to what is meant by normal vibrations and characteristic numbers.

We may look at the matter from two points of view, apparently contradictory. Let us first consider the case of one dimension. We have seen that for a string of length l the characteristic numbers are given by $k_r = \frac{r\pi}{l}$. (It is sometimes said that these condense at infinity, meaning by that that the reciprocals, which are here proportional to the wave-lengths, condense at zero, that is, in any region however small including zero there are as many of them as we please.) The difference between two consecutive numbers is $\frac{\pi}{l}$, and the spectrum is discontinuous. If now we let l increase without limit, the interval between consecutive characteristic numbers decreases without limit, or the numbers become everywhere dense, or in close order, and we approach a continuous spectrum.[1])

Consequently it seems as if every number were a possible characteristic number. We might then expect the Green's function K, represented by the bilinear formula (211) Chapter III, to be represented by an integral instead of a sum, in this case by a Fourier's integral. Referring to the form of the Green's function, (116) § 32, we find that in the limit for $l = \infty$ it becomes $K(x, \xi) = x$ when $x < \xi$, $K(x, \xi) = \xi$, when $x > \xi$. But such a function cannot be expressed as a Fourier integral, since a constant is not finitely integrable in the infinite region. Nevertheless there is no doubt that a standing wave would be represented by $\sin kx$, no matter what the value of k, if the string is unlimited.

Let us however look at the matter from another point of view. From the above treatment of the wave-potential in three dimensions we have seen that the effect of sources is propagated in the form of retarded potentials to infinity, and that there is no sign of resonance phe-

---

1) A set of points in close order is not continuous.

nomena, no matter what may be the period of harmonic vibrations. It therefore appears that there are no characteristic numbers at all. What is the discrepancy between these two modes of consideration? It is due to a very simple physical reason. For the equation

$$\frac{d^2u}{dx^2} + k^2u = 0$$

we have the solutions

$$\cos kx, \quad \sin kx, \quad e^{-ikx}, \quad e^{ikx},$$

which are all finite, but indeterminate at infinity. From the first two by multiplication with $\cos nt$ or $\sin nt$ we form the solutions of the wave equation

$$\cos nt \cos kx, \quad \sin nt \cos kx, \quad \cos nt \sin kx, \quad \sin nt \sin kx,$$

which represent standing waves. On the other hand from the last two multiplying by e^{int} we get

$$e^{i(nt-kx)}, \quad e^{i(nt+kx)},$$

of which the real part represents progressive waves running off to infinity, or running in from infinity respectively, the combination of which gives the standing waves. In the case of three dimensions we have as solutions of the equation

(120) $$\Delta u + k^2u = 0$$

the functions

$$\frac{\cos kr}{r}, \quad \frac{\sin kr}{r}, \quad \frac{e^{-ikr}}{r}, \quad \frac{e^{ikr}}{r}$$

from which we obtain the standing waves

$$\frac{\cos nt \cos kr}{r}, \quad \frac{\sin nt \cos kr}{r}, \quad \frac{\cos nt \sin kr}{r}, \quad \frac{\sin nt \sin kr}{r},$$

the first two having a singularity, that is a source at $r = 0$, and denoting forced vibrations, the last two having no source, and denoting free vibrations. From the other solutions we obtain the progressive waves

$$\frac{e^{i(nt-kr)}}{r}, \quad \frac{e^{i(nt+kr)}}{r}$$

of which the real part of the first represents diverging, of the second converging waves, the combination of both being required to produce standing waves. Now whereas incoming waves from infinity are mathematically equally possible with diverging ones, physically they do not occur, so that we have in practice no standing waves in the infinite region, and no normal functions in the same sense as for finite regions.

By means of an extra condition at infinity, which he calls the "emission condition" namely,

$$\lim_{r=\infty} r\left(\frac{\partial u}{\partial r} + iku\right) = 0,$$

for three dimensions.

$$\lim_{r=\infty} \sqrt{r}\left(\frac{\partial u}{\partial r} + iku\right) = 0,$$

for two dimensions,

$$\lim_{r=\infty} \left(\frac{\partial u}{\partial x} + iku\right) = 0,$$

for one dimension, Sommerfeld is able to show that there are no normal functions, but otherwise to give the same treatment as for finite regions. At the same time he is able to represent the second Green's function as a definite integral, which is, however, taken with respect to a complex variable.

Let us consider the case of a one-dimensional region, infinite in both directions. We have the solutions satisfying the emission condition

$$u = A e^{-ik(x-\xi)}, \; x > \xi, \; u = B e^{ik(x-\xi)}, \; x < \xi,$$

while the condition of a unit source at $x = \xi$ gives

$$u\Big|_{\xi-0}^{\xi+0} = 0, \quad \frac{du}{dx}\Big|_{\xi-0}^{\xi+0} = -1, \quad A = B = \frac{1}{2ik}.$$

Accordingly the Green's function is

(192)
$$\Gamma(x,\xi) = \frac{e^{-ik(x-\xi)}}{2ik} \quad x > \xi$$
$$\Gamma(x,\xi) = \frac{e^{ik(x-\xi)}}{2ik} \quad x < \xi.$$

But we may prove that this Green's function is expressible in the form of the definite integral

(193)
$$\Gamma(x,\xi) = \frac{1}{2\pi}\int \frac{e^{i(x-\xi)z}}{z^2 - k^2} \, dz$$

where the complex variable z describes the path indicated in Fig. 80, that is passes from $-\infty$ to $+\infty$ along the real axis, avoiding the points $-k$ and $+k$, leaving the first to the left, the second to the right. Since the function in the numerator of the integrand has no singular points in the finite region, the integrand has no singular points except the two poles at $z = k$ and $z = -k$, for which the residue is $\dfrac{e^{ik(x-\xi)}}{2k}$ and $\dfrac{-e^{-ik(x-\xi)}}{2k}$ respectively. We may then apply Cauchy's residue theorem

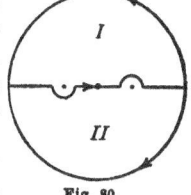

Fig. 80.

(see Appendix 5) to the regions I or II, Fig. 80, each of which contains a single pole. If we put $s = r(\cos\theta + i\sin\theta)$ we find the modulus of the numerator to be $e^{-r\sin\theta\,(x-\xi)}$ which for $r = \infty$ vanishes for the upper half of the plane if $x > \xi$, for the lower if $x < \xi$. But by the residue theorem the integral around the path I is $2\pi i$ times the residue for the pole $s = -k$, that is $\dfrac{-2\pi i e^{-ik(x-\xi)}}{2k}$, so that

$$(194) \qquad \frac{1}{2\pi}\int \frac{e^{i(x-\xi)s}}{k^2 - s^2}\,ds = \frac{-i}{2k}\,e^{-ik(x-\xi)} \quad x > \xi$$

while for the path II the residue is for $s = k$, and the direction of integration opposite so that

$$(195) \qquad \frac{1}{2\pi}\int \frac{e^{i(x-\xi)s}}{k^2 - s^2}\,ds = \frac{-i}{2k}\,e^{ik(x-\xi)} \quad x < \xi.$$

Accordingly the formula (193) is proved, and the bilinear formula appears as a definite integral, in which the emission condition is taken account of by the shape of the path.

To make further evident the analogy with our original Green's function of § 40, we will consider the region infinite on the right, but beginning at $x = 0$, with the condition that the Green's function vanishes at $x = 0$. This can be treated by subtracting the effect of a similar source at $x = -\xi$, so that we have

$$(196) \qquad \Gamma(x, \xi) = \frac{1}{2\pi}\int \frac{e^{i(x-\xi)s} - e^{i(x+\xi)s}}{s^2 - k^2}\,ds,$$

with the same path of integration as before. Writing trigonometric instead of exponential functions, we find, on account of the symmetry of the path, that the sine terms vanish, and we have

$$(197) \qquad \begin{aligned} \Gamma(x, \xi) &= \frac{1}{2\pi}\int \frac{\cos(x-\xi)s - \cos(x+\xi)s}{s^2 - k^2}\,ds, \\ &= \frac{1}{\pi}\int \frac{\sin(sx)\sin s\xi}{s^2 - k^2}\,ds, \end{aligned}$$

which shows the complete analogy with the formula (300'), § 40. Sommerfeld also shows how we may pass to the limit $l = \infty$ in the determination of the characteristic numbers, all of which are found to be complex.

CHAPTER VI.

METHOD OF RIEMANN-VOLTERRA. CHARACTERISTICS.

70. Cauchy's Problem. Characteristics. We have in Chapter III dealt with a particular case of Cauchy's problem for two independent variables, namely for the equation

$$\frac{\partial^2 u}{\partial t^2} - \frac{\partial^2 u}{\partial x^2} = 0,$$

we have shown that u is determined if we give for the line $t = 0$ the values of u and its derivative in the direction of the normal to this line as arbitrary functions of x, $u_{t=0} = F(x)$, $\left(\frac{\partial u}{\partial t}\right)_{t=0} = G(x)$. In Chapter V we have seen in § 60 that for Laplace's equation,

$$\frac{\partial^2 u}{\partial x^2} + \frac{\partial^2 u}{\partial y^2} = 0, \ \text{or} \ \frac{\partial^2 u}{\partial x^2} + \frac{\partial^2 u}{\partial y^2} + \frac{\partial^2 u}{\partial z^2} = 0,$$

it is not possible to give both u and $\frac{\partial u}{\partial n}$ independently, since one alone determines the function u. In this chapter we shall examine into the nature of this fundamental difference in the two problems of Cauchy and Dirichlet, and shall at the same time make use of a great generalization of the methods of Green described in the last chapter, first proposed by Riemann, and developed during the last twenty years by Volterra and others so as to constitute probably the most important method in the theory of partial differential equations. The method is applicable to any number of independent variables, but for the sake of simplicity we shall begin with two, so that geometrically we may deal with the plane. All that is said for this case will be then generalized for any number of dimensions.

Let us consider the equation in two independent variables, linear in the second derivatives,

$$(1) \qquad A\frac{\partial^2 u}{\partial x^2} + 2B\frac{\partial^2 u}{\partial x \partial y} + C\frac{\partial^2 u}{\partial y^2} = F\left(x, y, u, \frac{\partial u}{\partial x}, \ \frac{\partial u}{\partial y}\right),$$

where A, B, C, are functions of x, y, and let us consider the possibility of giving the values of u and $\frac{\partial u}{\partial n}$ along a given curve. Geometrically, as x, y, u are coordinates, this means giving a twisted curve, and a torsion strip along it, giving the tangent plane at each point. If the equation of the projection of this curve, which may be called the *support*, since it supports the data, is

$$(2) \qquad\qquad f(x, y) = 0.$$

the direction cosines of the normal to this projection are proportional to $\frac{\partial f}{\partial x}$, $\frac{\partial f}{\partial y}$, while those of the tangent are $\frac{dx}{ds}$, $\frac{dy}{ds}$, so that if we draw the normal toward the right hand side of the curve as we advance in the direction of positive ds, we have

$$\frac{\partial f}{\partial x} dx + \frac{\partial f}{\partial y} dy = 0,$$

(3)
$$\cos(nx) = \frac{\partial f}{\partial x} \Big/ h = \frac{dy}{ds},$$
$$\cos(ny) = \frac{\partial f}{\partial y} \Big/ h = -\frac{dx}{ds}, \qquad h^2 = \left(\frac{\partial f}{\partial x}\right)^2 + \left(\frac{\partial f}{\partial y}\right)^2$$

If we express the coordinates of a point of the curve in terms of a parameter t, $x = x(t)$, $y = y(t)$, so that (2) is identically satisfied, and also give the boundary values in terms of t, $u = H(t)$, $\frac{\partial u}{\partial n} = G(t)$, we shall have along the curve,

(4)
$$\frac{du}{dt} = \frac{\partial u}{\partial x} \frac{dx}{dt} + \frac{\partial u}{\partial y} \frac{dy}{dt} = H'(t),$$
$$\frac{\partial u}{\partial n} = \frac{\partial u}{\partial x} \cos(nx) + \frac{\partial u}{\partial y} \cos(ny) = \frac{\partial u}{\partial x} \frac{dy}{ds} - \frac{\partial u}{\partial y} \frac{dx}{ds} = G'(t).$$

These two equations, linear in $\frac{\partial u}{\partial x}$ and $\frac{\partial u}{\partial y}$, in which $\frac{dx}{dt}$, $\frac{dy}{dt}$, $\frac{dx}{ds}$, $\frac{dy}{ds}$ are given functions of t, determine $\frac{\partial u}{\partial x}$, $\frac{\partial u}{\partial y}$, since the determinant

(5)
$$\begin{vmatrix} \dfrac{dx}{dt}, & \dfrac{dy}{dt} \\[2mm] \dfrac{dy}{ds}, & -\dfrac{dx}{ds} \end{vmatrix} = -\frac{dx^2 + dy^2}{dt\,ds}$$

cannot vanish. Accordingly, we have the two derivatives given *along the curve*,

$$\frac{\partial u}{\partial x} = p(t), \qquad \frac{\partial u}{\partial y} = q(t).$$

We may then proceed as for u, and differentiating we get

(6)
$$\frac{\partial^2 u}{\partial x^2} \frac{dx}{dt} + \frac{\partial^2 u}{\partial x \partial y} \frac{dy}{dt} = p'(t),$$
$$\frac{\partial^2 u}{\partial x \partial y} \frac{dx}{dt} + \frac{\partial^2 u}{\partial y^2} \frac{dy}{dt} = q'(t),$$

as two linear equations for the *three* derivatives $\frac{\partial^2 u}{\partial y^2}$, $\frac{\partial^2 u}{\partial x \partial y}$, $\frac{\partial^2 u}{\partial y^2}$. But in addition we have at all points the differential equation,

(1)
$$A \frac{\partial^2 u}{\partial x^2} + 2B \frac{\partial^2 u}{\partial x \partial y} + C \frac{\partial^2 u}{\partial y^2} = F,$$

as a third equation, so that we have just enough to determine the second derivatives, unless the determinant

(7)
$$\begin{vmatrix} \frac{dx}{dt}, & \frac{dy}{dt}, & 0 \\ 0, & \frac{dx}{dt}, & \frac{dy}{dt} \\ A, & 2B, & C \end{vmatrix} = C\left(\frac{dx}{dt}\right)^2 - 2B\frac{dx}{dt}\frac{dy}{dt} + A\left(\frac{dy}{dt}\right)^2$$

vanishes. Thus we may proceed in succession, and obtain the values of the derivatives of u of all orders at points on the curve. We may thus form a development of u in a Taylor's series for a region including the points in question, and thus obtain a function, analytic in a certain region, satisfying the given differential equation, and taking the given values on the boundary. It would of course be necessary to examine the field of convergence of the development in order to determine how far the validity of the solution would extend.

If, on the other hand, the equation (t is irrelevant)

(8)
$$C\,dx^2 - 2B\,dx\,dy + A\,dy^2 = 0,$$

holds for the boundary curve, the process is impossible. This is an ordinary differential equation of the first order but of the second degree, which defines two families of curves in the xy-plane. These are called the *characteristics* of the differential equation (1) Accordingly if the given curve is a characteristic, the Cauchy boundary-problem cannot be solved.

Resolving into linear factors equation (8) becomes

(9)
$$\{A\,dy - (B + \sqrt{B^2 - AC})\,dx\}\{A\,dy - (B - \sqrt{B^2 - AC})\,dx\} = 0.$$

If the discriminant $B^2 - AC > 0$ the factors are real, and the equation (1) is hyperbolic. If $B^2 - AC < 0$, the factors are imaginary, and so are the characteristics, the equation is elliptic. If $B^2 - AC = 0$, the two families of characteristics coincide, and the equation is parabolic. This is the same classification that we have used in Chapter III, § 25, for the equation with constant coefficients.

We may obtain a clearer notion of the appropriateness of the name characteristics in a somewhat different manner. If $f(x,y) =$ const. is an integral of the differential equation (8), then by means of the relation

$$\frac{\partial f}{\partial x}dx + \frac{\partial f}{\partial y}dy = 0$$

the equation (8) becomes

(10)
$$A\left(\frac{\partial f}{\partial x}\right)^2 + 2B\frac{\partial f}{\partial x}\frac{\partial f}{\partial y} + C\left(\frac{\partial f}{\partial y}\right)^2 = 0$$

a partial differential equation of the first order, whose characteristics, as defined in § 24, equation (45), are

$$(11) \qquad \frac{dx}{Ap + Bq} = \frac{dy}{Bp + Cq} = \frac{dz}{0}$$

if we write the equation,

$$(10) \qquad Ap^2 + 2Bpq + Cq^2 = 0.$$

But these are easily shown to give the same values for $\frac{dy}{dx}$ by making use of (10) as the equation (8) or (9). Accordingly the characteristics of the equation (1) of the second order are the same as the projections on the $x\,y$-plane of the Cauchy characteristics of the equation (10) of the first order.

By a change of variables suggested by the form of the characteristics the equation (1) may be reduced to a simpler form. If we change from x, y to ξ, η as independent variables we have

$$\frac{\partial u}{\partial x} = \frac{\partial u}{\partial \xi}\frac{\partial \xi}{\partial x} + \frac{\partial u}{\partial \eta}\frac{\partial \eta}{\partial x}, \quad \frac{\partial u}{\partial y} = \frac{\partial u}{\partial \xi}\frac{\partial \xi}{\partial y} + \frac{\partial u}{\partial \eta}\frac{\partial \eta}{\partial y},$$

$$\frac{\partial^2 u}{\partial x^2} = \frac{\partial^2 u}{\partial \xi^2}\left(\frac{\partial \xi}{\partial x}\right)^2 + 2\frac{\partial^2 u}{\partial \xi \partial \eta}\frac{\partial \xi}{\partial x}\frac{\partial \eta}{\partial x} + \frac{\partial^2 u}{\partial \eta^2}\left(\frac{\partial \eta}{\partial x}\right)^2 + \frac{\partial u}{\partial \xi}\frac{\partial^2 \xi}{\partial x^2} + \frac{\partial u}{\partial \eta}\frac{\partial^2 \eta}{\partial x^2},$$

$$(12) \quad \frac{\partial^2 u}{\partial y^2} = \frac{\partial^2 u}{\partial \xi^2}\left(\frac{\partial \xi}{\partial y}\right)^2 + 2\frac{\partial^2 u}{\partial \xi \partial \eta}\frac{\partial \xi}{\partial y}\frac{\partial \eta}{\partial y} + \frac{\partial^2 u}{\partial \eta^2}\left(\frac{\partial \eta}{\partial y}\right)^2 + \frac{\partial u}{\partial \xi}\frac{\partial^2 \xi}{\partial y^2} + \frac{\partial u}{\partial \eta}\frac{\partial^2 \eta}{\partial y^2},$$

$$\frac{\partial^2 u}{\partial x \partial y} = \frac{\partial^2 u}{\partial \xi^2}\frac{\partial \xi}{\partial x}\frac{\partial \xi}{\partial y} + \frac{\partial^2 u}{\partial \xi \partial \eta}\left(\frac{\partial \xi}{\partial x}\frac{\partial \eta}{\partial y} + \frac{\partial \eta}{\partial x}\frac{\partial \xi}{\partial y}\right) + \frac{\partial^2 u}{\partial \eta^2}\frac{\partial \eta}{\partial x}\frac{\partial \eta}{\partial y} + \frac{\partial u}{\partial \xi}\frac{\partial^2 \xi}{\partial x \partial y}$$
$$+ \frac{\partial u}{\partial \eta}\frac{\partial^2 \eta}{\partial x \partial y}$$

substituting which in the equation (1) gives

$$\frac{\partial^2 u}{\partial^2 \xi}\left\{ A\left(\frac{\partial \xi}{\partial x}\right)^2 + 2B\frac{\partial \xi}{\partial x}\frac{\partial \xi}{\partial y} + C\left(\frac{\partial \xi}{\partial y}\right)^2 \right\}$$

$$(13) \qquad + 2\frac{\partial^2 u}{\partial \xi \partial \eta}\left\{ A\frac{\partial \xi}{\partial x}\frac{\partial \eta}{\partial x} + B\left(\frac{\partial \xi}{\partial x}\frac{\partial \eta}{\partial y} + \frac{\partial \eta}{\partial x}\frac{\partial \xi}{\partial y}\right) + C\frac{\partial \xi}{\partial y}\frac{\partial \eta}{\partial y} \right\}$$

$$+ \frac{\partial^2 u}{\partial \eta^2}\left\{ A\left(\frac{\partial \eta}{\partial x}\right)^2 + 2B\frac{\partial \eta}{\partial x}\frac{\partial \eta}{\partial y} + C\left(\frac{\partial \eta}{\partial y}\right)^2 \right\} = F$$

where F now depends on the new variables ξ, η, as well as on u and its first derivatives according to ξ, η.

Let the integrated equations of the characteristics be

$$\varphi(x, y) = \text{const.}, \quad \psi(x, y) = \text{const.}$$

then φ and ψ satisfy equation (10) and if we take $\xi = \varphi$, $\eta = \psi$, the coefficients of $\dfrac{\partial^2 u}{\partial \xi^2}$ and $\dfrac{\partial^2 u}{\partial \eta^2}$ both vanish, and the equation (13) reduces, on division by the coefficient of $\dfrac{\partial^2 u}{\partial \xi \partial \eta}$, to the form

$$(14) \qquad \frac{\partial^2 u}{\partial \xi \partial \eta} = G\left(\xi, \, \eta, \, u, \, \frac{\partial u}{\partial \xi}, \, \frac{\partial u}{\partial \eta}\right).$$

If the equation is hyperbolic, both φ and ψ are real, so that (14) may be taken as the normal form for the hyperbolic equation in two independent variables. If the equation is elliptic, let us take

$$\xi + i\eta = \varphi, \quad \xi - i\eta = \psi,$$

and substituting φ in (10) we find

$$A\left(\frac{\partial \xi}{\partial x}\right)^2 + 2B\frac{\partial \xi}{\partial x}\frac{\partial \xi}{\partial y} + C\left(\frac{\partial \xi}{\partial y}\right)^2$$

$$- \left\{ A\left(\frac{\partial \eta}{\partial x}\right)^2 + 2B\frac{\partial \eta}{\partial x}\frac{\partial \eta}{\partial y} + C\left(\frac{\partial \eta}{\partial y}\right)^2 \right\}$$

$$+ 2i\left\{ A\frac{\partial \xi}{\partial x}\frac{\partial \eta}{\partial x} + B\left(\frac{\partial \xi}{\partial x}\frac{\partial \eta}{\partial y} + \frac{\partial \eta}{\partial x}\frac{\partial \xi}{\partial y}\right) + C\frac{\partial \xi}{\partial y}\frac{\partial \eta}{\partial y} \right\} = 0.$$

Equating the real and imaginary parts both to zero, we find that the coefficient of $\dfrac{\partial^2 u}{\partial \xi \partial \eta}$ in (13) vanishes, while those of $\dfrac{\partial^2 u}{\partial \xi}$, $\dfrac{\partial^2 u}{\partial \eta^2}$ become equal, so that the equation may be reduced to the form

$$(15) \qquad \frac{\partial^2 u}{\partial \xi^2} + \frac{\partial^2 u}{\partial \eta^2} = G\left(\xi, \, \eta, \, u, \, \frac{\partial u}{\partial \xi}, \, \frac{\partial u}{\partial \eta}\right).$$

This is the normal form for the elliptic differential equation.

If the equation is parabolic, take

$$\xi = \varphi, \quad \eta = y.$$

The coefficient of $\dfrac{\partial^2 u}{\partial \xi^2}$ vanishes as before, while that of $2\dfrac{\partial^2 u}{\partial \xi \partial \eta}$ becomes $B\dfrac{\partial \varphi}{\partial x} + C\dfrac{\partial \varphi}{\partial y}$, which, in virtue of the relation $B^2 - AC = 0$ and (10), is found to vanish. Accordingly, since the coefficient of $\dfrac{\partial^2 u}{\partial \eta^2}$ does not vanish, we have

$$(16) \qquad \frac{\partial^2 u}{\partial \eta^2} = G\left(\xi, \, \eta, \, u, \, \frac{\partial u}{\partial \xi}, \, \frac{\partial u}{\partial \eta}\right)$$

as the normal form for the parabolic differential equation. In case the coefficients A, B, C are constant, this method of change of variables gives Euler's transformation which we have used in § 25. We have as types of the elliptic equation Laplace's equation, of the hyperbolic, the wave-equation, and of the parabolic Fourier's equation $a^2\dfrac{\partial^2 u}{\partial x^2} = \dfrac{\partial u}{\partial t}$.

71. Linear Equation. Green's Theorem. Suppose that our equation, like all those of the second order with which we have become acquainted, is linear, so that we may write it

$$(17) \quad L(u) \equiv A\frac{\partial^2 u}{\partial x^2} + 2B\frac{\partial^2 u}{\partial x \partial y} + C\frac{\partial^2 u}{\partial y^2} + D\frac{\partial u}{\partial x} + E\frac{\partial u}{\partial y} + F \cdot u = 0,$$

where all the coefficients A, B, C, D, E, F depend only on x, y. Let us multiply the expression $L(u)$ by a function v and integrate over any region of the plane bounded by a closed contour C. We will integrate by parts in Green's manner, as in § 50 (8), n being the internal normal, and ds the *absolute value* of the element of arc. Integrating the first term according to x,

$$\iint A v \frac{\partial^2 u}{\partial x^2} dx dy = -\int A v \frac{\partial u}{\partial x}\cos(nx) ds - \iint \frac{\partial u}{\partial x}\frac{\partial(Av)}{\partial x} dx dy,$$

and repeating the process for the double integral,

$$\iint \frac{\partial u}{\partial x}\frac{\partial(Av)}{\partial x} dx dy = -\int u\frac{\partial(Av)}{\partial x}\cos(nx) ds - \iint u\frac{\partial^2(Av)}{\partial x^2} dx dy,$$

so that finally, differentiating Av,

$$(18) \quad \iint A v \frac{\partial^2 u}{\partial x^2} dx dy = \int \left[A\left(u\frac{\partial v}{\partial x} - v\frac{\partial u}{\partial x}\right) + uv\frac{\partial A}{\partial x}\right]\cos(nx) ds$$
$$+ \iint u\frac{\partial^2(Av)}{\partial x^2} dx dy.$$

In like manner for the third term,

$$(19) \quad \iint C v \frac{\partial^2 u}{\partial y^2} dx dy = \int \left[C\left(u\frac{\partial v}{\partial y} - v\frac{\partial u}{\partial y}\right) + uv\frac{\partial C}{\partial y}\right]\cos(ny) ds$$
$$+ \iint u\frac{\partial^2(Cv)}{\partial y^2} dx dy.$$

The second term is to be divided into two. Integrating first by x,

$$\iint B v \frac{\partial^2 u}{\partial x \partial y} dx dy = -\int B v \frac{\partial u}{\partial y}\cos(nx) ds - \iint \frac{\partial u}{\partial y}\frac{\partial(Bv)}{\partial x} dx dy,$$

and then by y,

$$\iint \frac{\partial u}{\partial y}\frac{\partial(Bv)}{\partial x} dx dy = -\int u\frac{\partial(Bv)}{\partial x}\cos(ny) ds + \iint u\frac{\partial^2(Bx)}{\partial x \partial y} dx dy$$

we obtain

$$(20) \quad \iint B v \frac{\partial^2 u}{\partial x \partial y} dx dy = \int \left[B\left(u\frac{\partial v}{\partial x}\cos(ny) - v\frac{\partial u}{\partial y}\cos(nx)\right)\right.$$
$$\left. + uv\frac{\partial B}{\partial x}\cos(ny)\right]ds + \iint u\frac{\partial^2(Bv)}{\partial x \partial y} dx dy.$$

On the other hand since the term is symmetrical in x and y, we may write, interchanging them,

$$\iint Bv\frac{\partial^2 u}{\partial x\,\partial y}dx\,dy = \int\Big[B\Big(u\frac{\partial v}{\partial y}\cos(nx)-v\frac{\partial u}{\partial x}\cos(ny)\Big)+uv\frac{\partial B}{\partial y}\cos(nx)\Big]ds$$
$$(21) \qquad\qquad +\iint u\frac{\partial^2(Bv)}{\partial x\,\partial y}\,dx\,dy.$$

For the terms in the first derivatives we integrate only once,

$$(22)\quad \iint Dv\frac{\partial u}{\partial x}dx\,dy = -\int \dot{D}uv\cos(nx)ds-\iint u\frac{\partial(Dv)}{\partial x}dx\,dy,$$
$$\iint Ev\frac{\partial u}{\partial y}dx\,dy = -\int Euv\cos(ny)ds-\iint u\frac{\partial(Ev)}{\partial y}dx\,dy.$$

Making use of these results, transposing all the double integrals to the left, we have finally

$$(23)\quad \iint\big(vL(u)-uM(v)\big)dx\,dy = \int\big(uD_n v-vD_n(u)+uvP_n\big)ds$$

in which M is called the *adjoint* differential operator,

$$M(v) = \frac{\partial^2(Av)}{\partial x^2}+2\frac{\partial^2(Bv)}{\partial x\,\partial y}+\frac{\partial^2(Cv)}{\partial y^2}-\frac{\partial(Dv)}{\partial x}-\frac{\partial(Ev)}{\partial y}+F\cdot v$$
$$(24)\qquad = A\frac{\partial^2 v}{\partial x^2}+2B\frac{\partial^2 v}{\partial x\,\partial y}+C\frac{\partial^2 v}{\partial y^2}+2\frac{\partial v}{\partial x}\Big(\frac{\partial A}{\partial x}+\frac{\partial B}{\partial y}-\frac{1}{2}D\Big)$$
$$+2\frac{\partial v}{\partial y}\Big(\frac{\partial B}{\partial x}+\frac{\partial C}{\partial y}-\frac{1}{2}E\Big)+v\Big(\frac{\partial^2 A}{\partial x^2}+2\frac{\partial^2 B}{\partial x\,\partial y}+\frac{\partial^2 C}{\partial y^2}-\frac{\partial D}{\partial x}-\frac{\partial E}{\partial y}+F\Big),$$

D_n stands for the linear differential operator,

$$(25)\quad D_n u \equiv \Big(A\frac{\partial u}{\partial x}+B\frac{\partial u}{\partial y}\Big)\cos(nx)+\Big(B\frac{\partial u}{\partial x}+C\frac{\partial u}{\partial y}\Big)\cos(ny),$$

and P_n may be called the adjoint function

$$(26)\quad P_n \equiv \Big(\frac{\partial A}{\partial x}+\frac{\partial B}{\partial y}-D\Big)\cos(nx)+\Big(\frac{\partial B}{\partial x}+\frac{\partial C}{\partial y}-E\Big)\cos(ny).$$

It is to be noticed that the operator M depends only on the given operator L, while D_n and P_n depend on the form of the support or contour C as well. In case we have the relations,

$$(27)\qquad \frac{\partial A}{\partial x}+\frac{\partial B}{\partial y}=D, \qquad \frac{\partial B}{\partial x}+\frac{\partial C}{\partial y}=E,$$

we see that the two operators L, M are identical, and that at the same time the adjoint function P_n vanishes. The equation (17) is then said to be self-adjoint. In particular if the coefficients are constant and

$D = E = 0$, this will be the case. In general if (27) are true, the form of the self-adjoint equation will be

(28) $\dfrac{\partial}{\partial x}\left\{A\dfrac{\partial u}{\partial x} + B\dfrac{\partial u}{\partial y}\right\} + \dfrac{\partial}{\partial y}\left\{B\dfrac{\partial u}{\partial x} + C\dfrac{\partial u}{\partial y}\right\} + F\cdot u = 0$.

The nature of the operator D_n is of great importance. It is to be noticed that if we define a direction ν by the equations

$$A\cos(nx) + B\cos(ny) = \varLambda\cos(\nu x),$$
(29) $$B\cos(nx) + C\cos(ny) = \varLambda\cos(\nu y),$$
$$\varLambda^2 = (A^2+B^2)\cos^2(nx) + 2B(A+C)\cos(nx)\cos(ny) + (B^2+C^2)\cos^2(ny).$$

then (25) becomes

(30) $D_n(u) = \varLambda\left(\dfrac{\partial u}{\partial x}\cos(\nu x) + \dfrac{\partial u}{\partial y}\cos(\nu y)\right) = \varLambda\dfrac{\partial u}{\partial \nu}$,

and $\dfrac{\partial u}{\partial \nu}$ is a derivative in the direction ν. This direction has been called by d'Adhémar the *conormal*. Its relation to the normal is easily found from the following geometrical relations. If in the characteristic function $\varPhi$, of (10),

(31) $\varPhi \equiv A\left(\dfrac{\partial f}{\partial x}\right)^2 + 2B\dfrac{\partial f}{\partial x}\dfrac{\partial f}{\partial y} + C\left(\dfrac{\partial f}{\partial y}\right)^2$

we replace $\dfrac{\partial f}{\partial x}, \dfrac{\partial f}{\partial y}$ by $X = \cos(nx)$, $Y = \cos(ny)$, the equation

(32) $\varPhi(X, Y) = AX^2 + 2BXY + CY^2 = 1$,

represents a central conic, and the equations (29), which may be written

(33) $\dfrac{\cos(\nu x)}{\dfrac{\partial \varPhi}{\partial X}} = \dfrac{\cos(\nu y)}{\dfrac{\partial \varPhi}{\partial Y}}$

show that the direction of the conormal ν at any point of the support is related to that of the normal n as the direction of the perpendicular

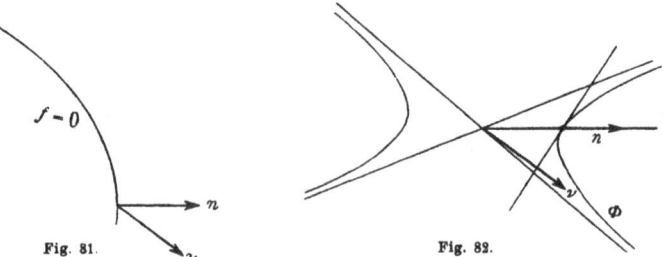

Fig. 81.   Fig. 82.

on the tangent is to that of the radius vector of the quadric $\varPhi = 1$. But this conic has as asymptotes the tangents to the characteristics at the point in question, whose equations are $\Phi(X, Y) = 0$, Fig. 81 and Fig. 82.

We may then rewrite (23) in terms of the conormal,

$$(34) \quad \iint (v L(u) - u M(v)) dx dy = \int \left\{ \Lambda \left(u \frac{\partial v}{\partial \nu} - v \frac{\partial u}{\partial \nu} \right) + u v P_n \right\} ds,$$

and we have the generalization of Green's theorem for two dimensions, corresponding to § 40, (78) for one dimension. As an example, in the case of Laplace's equation,

$$(35) \quad \begin{array}{c} L(u) = M(u) = \Delta u, \; A = C = 1, \; B = D = E = F = 0 \\ \Lambda = 1 \quad \nu = n \end{array}$$

and (34) reduces to the two dimensional case of § 52, (20).

72. Elliptic Equation. If the equation is elliptic, we will write the normal form for the linear equation,

$$(36) \quad L(u) \equiv \Delta u + a \frac{\partial u}{\partial x} + b \frac{\partial u}{\partial y} + cu = 0,$$

when the adjoint equation will be

$$(37) \quad M(v) \equiv \Delta v - \frac{\partial (av)}{\partial x} - \frac{\partial (bv)}{\partial y} + cv = 0.$$

The characteristic function will be the same as for Laplace's equation, and the conormal will coincide with the normal.

In the case of Laplace's equation we have seen that we have a particular solution $\log r$, where r is the distance from a fixed point, also in the case of Helmholtz's equation we have found the particular solution $Y_0(kr)$, which is of the form

$$(38) \quad U \log r + V,$$

where U and V are finite and continuous functions. If we assume that for the elliptic equation in general there is such a fundamental solution, that is one having a logarithmic singularity at a single point ξ, η, and in addition suppose $U(\xi, \eta) = -1$, and use such a solution of the *adjoint* equation for v, we must, as usual in applying Green's Theorem, § 59, exclude the point ξ, η by a small circle. Then since

$$P_n = - \left\{ a \cos(nx) + b \cos(ny) \right\}, \quad \nu = n, \quad \Lambda = 1,$$

we have

$$(39) \quad \begin{aligned} 2\pi u(\xi, \eta) = &- \iint v L(u) dx dy \\ &- \int \left\{ v \frac{\partial u}{\partial n} - u \frac{\partial v}{\partial n} + (a \cos(nx) + b \cos(ny)) uv \right\} ds, \end{aligned}$$

as a generalization of Green's formula § 59, (69).

If then u satisfies the equation $L(u) = 0$, we have u for any point ξ, η inside a closed contour C expressed in terms of a contour integral involving u and $\dfrac{\partial u}{\partial n}$. As these values are not independent of each other, we cannot give both at pleasure. If we define a Green's function as a solution G of the adjoint equation $M(v) = 0$, which has the above properties with regard to ξ, η, is continuous at other points, and vanishes on the contour C, we have

(40) $$u(\xi, \eta) = \frac{1}{2\pi} \int u \frac{\partial G}{\partial n} ds$$

as in the case of Laplace's equation. We may also show that if $H(x, y, \xi', \eta')$ is the Green's function for the adjoint operator M, then we obtain by (34) the relation

(41) $$H(\xi, \eta, \xi', \eta') = G(\xi', \eta', \xi, \eta)$$

corresponding to (134) § 66.

73. Hyperbolic Equation. Let us now consider the hyperbolic equation, which we will write in the normal form

(42) $$L(u) \equiv \frac{\partial^2 u}{\partial x \partial y} + a \frac{\partial u}{\partial x} + b \frac{\partial u}{\partial y} + cu = 0,$$

(43) $$M(v) \equiv \frac{\partial^2 v}{\partial x \partial y} - \frac{\partial(av)}{\partial x} - \frac{\partial(bv)}{\partial y} + cv.$$

For the hyperbolic equation there are no solutions with point-singularities, but we shall see later that there are fundamental solutions which are singular all along the characteristics.

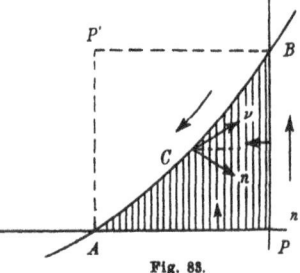

Fig. 83.

For (42) the characteristics are given by $dx\,dy = 0$, which represents straight lines parallel to the axes of coordinates. Let us undertake to determine u by giving its values and those of $\dfrac{\partial u}{\partial v}$ along a supporting curve C that is cut only once by any characteristic. Let us apply Green's theorem (23) to the area bounded by the contour C and two characteristics drawn through the point P whose coordinates are ξ, η, and cutting the contour in A and B, Fig. 83. By (25), (26), (29) we have, since

$$A = C = 0, \quad B = \tfrac{1}{2}, \quad A = \tfrac{1}{2},$$

$$D_n(u) = \frac{1}{2}\left(\frac{\partial u}{\partial y} \cos(nx) + \frac{\partial u}{\partial x} \cos(ny)\right), \quad P_n = -(a\cos(nx) + b\cos(ny)),$$

on AP, $\cos(nx) = 0$, $\cos(ny) = 1$, $ds = dx$,

on PB, $\cos(nx) = -1$, $\cos(ny) = 0$, $ds = dy$.

Inserting in (34) gives

$$\iint (v\,L(u) - u\,M(v))\,dx\,dy$$

$$(44) \quad = \frac{1}{2}\int_A^P\left(u\frac{\partial v}{\partial x} - v\frac{\partial u}{\partial x} - 2buv\right)dx + \frac{1}{2}\int_P^B\left(v\frac{\partial u}{\partial y} - u\frac{\partial v}{\partial y} + 2auv\right)dy$$

$$+ \int_B^A\left[\frac{1}{2}\left(u\frac{\partial v}{\partial \nu} - v\frac{\partial u}{\partial \nu}\right) - (a\cos(nx) + b\cos(ny))uv\right]ds.$$

Now integrating by parts the terms containing derivatives of u in the line integrals on AP and PB, and combining terms, we obtain

$$\iint (v\,L(u) - u\,M(v))\,dx\,dy = -(uv)_P + \frac{1}{2}(uv)_A + \frac{1}{2}(uv)_B$$

$$(45) \quad + \frac{1}{2}\int_A^P u\left(\frac{\partial v}{\partial x} - 2bv\right)dx - \frac{1}{2}\int_P^B u\left(\frac{\partial v}{\partial y} - 2av\right)dy$$

$$+ \int_B^A\left[\frac{1}{2}\left(u\frac{\partial v}{\partial \nu} - v\frac{\partial u}{\partial \nu}\right) - (a\cos(nx) + b\cos(ny))uv\right]ds.$$

According to Riemann we will now define the Green's function for this case as a solution of the adjoint equation $M(v) = 0$ which along the characteristic AP, $y = \eta$, satisfies the equation,

$$(46) \qquad\qquad \frac{\partial G}{\partial x} - 2bG = 0,$$

along the characteristic PB, $x = \xi$, satisfies the equation,

$$(47) \qquad\qquad \frac{\partial G}{\partial y} - 2aG = 0,$$

and at P takes the value $G(\xi, \eta) = 1$. Then if $L(u) = 0$, we have

$$u(\xi, \eta) = \frac{1}{2}[(uG)_A + (uG)_B]$$

$$(48) \quad + \int_B^A\left[\frac{1}{2}\left(u\frac{\partial G}{\partial \nu} - G\frac{\partial u}{\partial \nu}\right) - (a\cos(nx) + b\cos(ny))uv\right]ds$$

giving the value of u at P in terms of the values of u and $\dfrac{\partial u}{\partial \nu}$ along the support.

Along the characteristics we have by integration,

$$\frac{1}{G}\frac{\partial G}{\partial x} = \frac{\partial \log G}{\partial x} = 2b(x, \eta), \quad \log G = 2\int_{x}^{\xi} b(x, \eta)dx,$$

(49)

$$G = e^{2\int_{x}^{\xi} b(x, \eta)dx}$$

and

$$\frac{1}{G}\frac{\partial G}{\partial y} = \frac{\partial (\log G)}{\partial y} = 2a(\xi, y), \quad \log G = 2\int_{\eta}^{y} a(\xi, y)dy$$

(49)

$$G = e^{2\int_{\eta}^{y} a(\xi, y)dy},$$

respectively. Thus the determination of u by the data along the support is made to depend on the determination of G so as to take prescribed values along the characteristics, that is upon a particular Cauchy problem. If $a = b = 0$, these values reduce to constants, namely the value of G at P, that is unity.

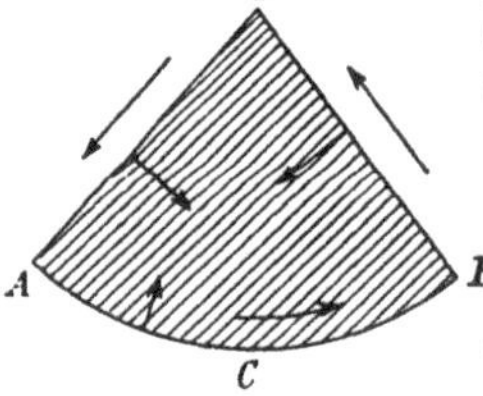

As an example take the equation of the string,

$$\frac{\partial^2 u}{\partial t^2} = a^2 \frac{\partial^2 u}{\partial x^2},$$

which on putting $y = at$ becomes the self-adjoint equation,

Fig. 84.

(50) $L(u) \equiv \dfrac{\partial^2 u}{\partial x^2} - \dfrac{\partial^2 u}{\partial y^2} = 0, \; A = 1, \; B = 0, \; C = -1, \; \varDelta = 1,$

which we will treat without passing to the normal form. The characteristics are given by $dx^2 - dy^2 = 0$, which represents straight lines making equal angles with the coordinate axes. We have, Fig. 84,

on PA, $dx = dy$, $\cos(nx) = \dfrac{1}{\sqrt{2}}$, $\cos(ny) = -\dfrac{1}{\sqrt{2}}$, $ds = -\sqrt{2}\,dx = -\sqrt{2}\,dy$

on PB, $dx = -dy$, $\cos(nx) = -\dfrac{1}{\sqrt{2}}$, $\cos(ny) = -\dfrac{1}{\sqrt{2}}$, $ds = \sqrt{2}\,dy = -\sqrt{2}\,dx$

so that our formula is

$$\int_{P}^{A}\left[v\left(\frac{\partial u}{\partial x}dx + \frac{\partial u}{\partial y}dy\right) - u\left(\frac{\partial v}{\partial x}dx + \frac{\partial v}{\partial y}dy\right)\right]$$

(51)

$$+ \int_{B}^{P}\left[u\left(\frac{\partial v}{\partial x}dx + \frac{\partial v}{\partial y}dy\right) - v\left(\frac{\partial u}{\partial x}dx + \frac{\partial u}{\partial y}dy\right)\right]$$

$$+ \int_{A}^{B}\left(u\frac{\partial v}{\partial \nu} - v\frac{\partial u}{\partial \nu}\right)ds = 0$$

Since $a = b = 0$, the value of G on the characteristics is unity, consequently we may take for G the very simple solution $v = 1$, giving

$$(52) \quad \int_P^A\left(\frac{\partial u}{\partial x}dx + \frac{\partial u}{\partial y}dy\right) - \int_B^P\left(\frac{\partial u}{\partial x}dx + \frac{\partial u}{\partial y}dy\right) - \int_A^B \frac{\partial u}{\partial \nu}ds = 0,$$

or integrating,

$$(53) \qquad 2u_P = u_A + u_B - \int_A^B \frac{\partial u}{\partial \nu}ds.$$

Suppose the curve AB is the line $y = 0$, and that for

$$at = y = 0, \quad u = F(x), \quad -\frac{\partial u}{\partial \nu} = \frac{\partial u}{\partial y} = \frac{1}{a}\frac{\partial u}{\partial t} = \frac{1}{a}G(x).$$

Then if P has the coordinates ξ, η, A has $x = \xi - \eta$, B has $x = \xi + \eta$, and we have

$$(54) \qquad 2u(\xi, \eta) = F(\xi - \eta) + F(\xi + \eta) + \frac{1}{a}\int_{\xi-\eta}^{\xi+\eta} G(x)dx,$$

or in the more familiar notation,

$$(55) \qquad u = \frac{1}{2}\left\{F(x - at) + F(x + at) + \frac{1}{a}\int_{x-at}^{x+at} G(x)dx\right\},$$

which is d'Alembert's solution, Chapter III, (27). The equation (53) or (55) shows how the values of u at A and B arrive at P, or are *propagated* along the characteristics.

The method not only gives us this familiar solution, but obtains a solution whatever the contour AB, that is for any relation between x and y or t, that is if we give u and $\frac{\partial u}{\partial \nu}$ for a point on the string that moves in any prescribed manner. As an example suppose the point is moving with a constant velocity which is in the ratio $\beta < 1$ to the wave-velocity a. Then the contour C is the straight line $x = \beta y$, Fig. 85. It is to be noticed in all these formulae that if, as in the present case, the points PAB succeed each other in the reverse order to that for which the formulae (44), (53) were proved the sign of the integral must be changed. From the equations of the contour and of the characteristics we find the coordinates of A and B,

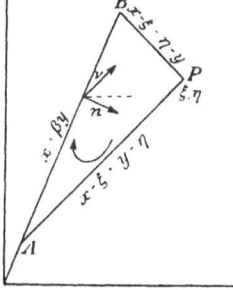

Fig. 85.

for A, $x = \frac{\beta}{1-\beta}(\eta - \xi)$, $y = \frac{\eta - \xi}{1 - \beta}$; for B, $x = \frac{\beta}{1+\beta}(\eta + \xi)$, $y = \frac{\eta + \xi}{1 + \beta}$, for the line AB,

$$\cos{(nx)} = \frac{1}{\sqrt{1+\beta^2}}, \quad \cos{(ny)} = \frac{-\beta}{\sqrt{1+\beta^2}}, \quad ds = \sqrt{1+\beta^2}\,dy,$$

$$\frac{\partial u}{\partial \nu} = \frac{1}{\sqrt{1+\beta^2}}\left(\frac{\partial u}{\partial x} + \beta\frac{\partial u}{\partial y}\right),$$

so that we have the result,

$$(56) \qquad 2u_P = u_A + u_B + \int_A^B \left(\frac{\partial u}{\partial x} + \beta\frac{\partial u}{\partial y}\right)dy.$$

If $\beta = 0$ we have the result for a point at rest, and the formula gives us a result precisely similar to (54), interchanging x, y, but if $\beta \neq 0$, we have the result, supposing that for

$$x = \beta y, \quad u = F(y), \quad \frac{\partial u}{\partial x} = G(y), \quad \text{necessitating} \quad \frac{\partial u}{\partial y} = F'(y) - \beta G(y),$$

$$(57) \quad 2u(\xi, \eta) = F\left(\frac{\eta-\xi}{1-\beta}\right) + F\left(\frac{\eta+\xi}{1+\beta}\right) + \int_{\frac{\eta-\xi}{1-\beta}}^{\frac{\eta+\xi}{1+\beta}} ((1 - \beta^2)G(y) + \beta F'(y))dy$$

or performing the integration of the last term,

$$(58) \quad 2u(\xi, \eta) = (1+\beta)F\left(\frac{\eta+\xi}{1+\beta}\right) + (1-\beta)F\left(\frac{\eta-\xi}{1-\beta}\right) + (1-\beta^2)\int_{\frac{\eta-\xi}{1-\beta}}^{\frac{\eta+\xi}{1+\beta}} G(y)dy,$$

or in the usual notation,

$$(59) \quad 2u = (1+\beta)F\left(\frac{at+x}{1+\beta}\right) + (1-\beta)F\left(\frac{at-x}{1-\beta}\right) + (1-\beta^2)a\int_{\frac{t-x/a}{1-\beta}}^{\frac{t+x/a}{1+\beta}} G(at)dt.$$

Suppose we seek the effect of a traveling source of sound of frequency $\frac{p}{2\pi}$, so that

$$G(at) = \cos pt, \quad F(at) = 0.$$

We then have

$$(60) \qquad u = \frac{a}{2}\frac{(1-\beta^2)}{p}\left\{\sin\frac{p\,(t+x/a)}{1+\beta} - \sin\frac{p\,(t-x/a)}{1-\beta}\right\}$$

that is to say the two waves emitted travel with the velocity a as if the source were not moving, but the waves travelling to the right (in the same direction as the source, have the frequency increased in the ratio $1 : 1 - \beta$, while those traveling backward have the frequency lowered in the ratio $1 : 1 + \beta$. This is known as Doppler's Principle, and by means of it the spectroscope enables us to find the velocities of stars approaching or receding from us by the change in the wave-lengths of their spectrum lines. It is to be noticed that if u is given instead of $\frac{\partial u}{\partial x}$, the approaching wave is increased in the ratio $1 + \beta$ on the forward side, and is diminished in the ratio $1 - \beta$ on the rear side. In a source of sound it is the compression $s = -\frac{\partial u}{\partial x}$ that must be supposed to be unchanged by the motion. The change in the pitch of the bell or whistle of a passing locomotive is easily recognized.

74. Telegraphist's Equation. As a second example let us take the equation of telegraphy, from which we have removed an exponential factor, as in § 46. Writing z for the space-coordinate,

$$(61) \qquad \frac{\partial^2 u}{\partial t^2} = a^2 \frac{\partial^2 u}{\partial z^2} + b^2 u,$$

or changing to the normal form by putting

$$(62) \qquad x = \frac{b}{a}(z + at), \quad y = \frac{b}{a}(z - at).$$

$$(63) \qquad L(u) \equiv \frac{\partial^2 u}{\partial x \partial y} + \frac{u}{4} = 0.$$

This was integrated by du Bois Reymond and by Picard[1]) by Riemann's method.

As before, the Green's function is constant along the characteristics, which are the same as for equation (42). Along the characteristics through P, ξ, η, the function $\varphi = (x - \xi)(y - \eta)$ has the constant value zero. Accordingly Picard attempts to find a function G which depends only on φ. We have then

$$(64) \qquad \begin{aligned} \frac{\partial G}{\partial x} &= \frac{dG}{d\varphi}\frac{\partial \varphi}{\partial x} = \frac{dG}{d\varphi}(y - \eta), \\ \frac{\partial^2 G}{\partial x \partial y} &= \frac{dG}{d\varphi} + \frac{d^2 G}{d\varphi^2}(y - \eta)\frac{\partial \varphi}{\partial y} = \frac{dG}{d\varphi} + \varphi \frac{d^2 G}{d\varphi^2}, \end{aligned}$$

so that the equation (63) becomes,

$$(65) \qquad \varphi \frac{d^2 G}{d\varphi^2} + \frac{dG}{d\varphi} + \frac{G}{4} = 0,$$

which is an ordinary differential equation, linear, with variable coefficients. It may accordingly be solved by a power-series, (see Appendix)

---

1) Picard: Sur l'équation aux dérivées partielles qui se rencontre dans la théorie de la propagation de l'électricité. *Comptes Rendus*, 118, p. 16, 1894. P. du Bois Reymond: Über lineare partielle Differentialgleichungen zweiter Ordnung, Crelle's Journal, 104, p. 241, 1889.

but assumes a more familiar form if we change the independent variable by the substitution,

$$\varphi = \psi^2, \quad d\varphi = 2\psi d\psi,$$

$$\frac{dG}{d\varphi} = \frac{1}{2\psi}\frac{dG}{d\psi}, \quad \frac{d^2G}{d\varphi^2} = \frac{1}{4\psi}\left\{\frac{1}{\psi}\frac{d^2G}{d\psi^2} - \frac{1}{\psi^2}\frac{dG}{d\psi}\right\},$$

inserting which in (65) gives

(66)
$$\frac{d^2G}{d\psi^2} + \frac{1}{\psi}\frac{dG}{d\psi} + G = 0.$$

But this is Bessel's equation (Chap. VII) of which we have the solution $G = J_0(\psi)$, which reduces to unity for $\psi = 0$. Accordingly we have, as in (48)

(67)
$$2u(\xi, \eta) = (uG)_A + (uG)_B + \int_A^B \left(u\frac{\partial G}{\partial \nu} - G\frac{\partial u}{\partial \nu}\right)ds$$

with

$$G = J_0\big(\sqrt{(x-\xi)(y-\eta)}\big).$$

Suppose as before the contour is $t = 0$, that is, $x = y$, Fig. 86. Then

$$- \cos(nx) = \cos(ny) = \frac{1}{\sqrt{2}}, \quad \frac{ds}{\sqrt{2}} = dx = dy, \quad A = C = 0, \quad B = A = \frac{1}{2},$$

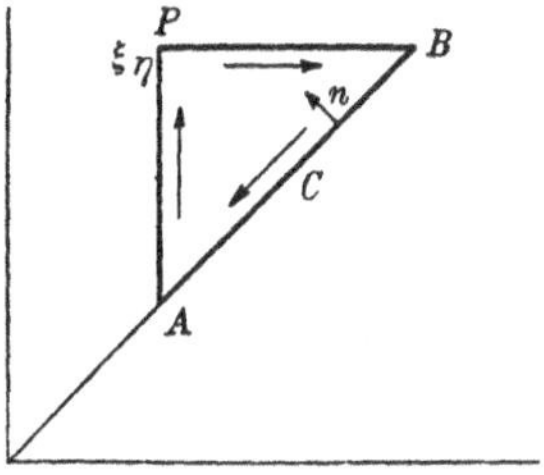

Fig. 86.

$$\frac{\partial u}{\partial \nu} = \frac{1}{\sqrt{2}}\left(\frac{\partial u}{\partial x} - \frac{\partial u}{\partial y}\right), \quad \frac{\partial G}{\partial \nu} = \frac{1}{\sqrt{2}}\left(\frac{\partial G}{\partial x} - \frac{\partial G}{\partial y}\right).$$

Now

$$\frac{\partial G}{\partial x} = \frac{dG}{d\varphi}(y - \eta), \quad \frac{\partial G}{\partial y} = \frac{dG}{d\varphi}(x - \xi),$$

or since on AB, $x = y$,

$$\frac{\partial G}{\partial \nu} = \frac{1}{\sqrt{2}}(\xi - \eta)\frac{\partial G}{\partial \varphi}, \quad G_A = G_B = G_P = 1,$$

(68)
$$2u(\xi, \eta) = u_A + u_B + (\xi - \eta)\int_A^B u\frac{dG}{d\varphi}dx - \int_A^B G\left(\frac{\partial u}{\partial x} - \frac{\partial u}{\partial y}\right)dx.$$

In order to pass from this solution to that in terms of z and t, if we have given for

$$t = 0, \quad u = f(z), \quad \frac{\partial u}{\partial t} = g(z),$$

since

$$x = \frac{b}{a}(z + at), \quad y = \frac{b}{a}(z - at),$$

$$\xi = \frac{b}{a}(\zeta + a\tau), \quad \eta = \frac{b}{a}(\zeta - a\tau),$$

and for A, $x = y = \xi, \quad t = 0, \quad z = \zeta + a\tau,$

B, $x = y = \eta, \quad t = 0, \quad z = \zeta - a\tau$

Also

$$\frac{\partial u}{\partial z} = \frac{b}{a}\left(\frac{\partial u}{\partial x} + \frac{\partial u}{\partial y}\right), \quad \frac{\partial u}{\partial t} = b\left(\frac{\partial u}{\partial x} - \frac{\partial u}{\partial y}\right),$$

$$x - \xi = \frac{b}{a}\left\{z - \zeta + a(t - \tau)\right\},$$

$$y - \eta = \frac{b}{a}\left\{z - \zeta - a(t - \tau)\right\},$$

$$\xi - \eta = 2b\tau,$$

$$\varphi = (x - \xi)(y - \eta) = \frac{b^2}{a^2}\left\{(z - \zeta)^2 - a^2(t - \tau)^2\right\},$$

$$G = J_0(\sqrt{\varphi}) = J_0\left(\frac{b}{a}\sqrt{(z - \zeta)^2 - a^2(t - \tau)^2}\right),$$

$$\frac{dG}{d\varphi} = \frac{J_0'(\sqrt{\varphi})}{2\sqrt{\varphi}},$$

and putting $t = 0$, $dx = \frac{b}{a}dz$, we have finally,

$$u(\zeta, \tau) = \frac{1}{2}\left\{f(\zeta + a\tau) + f(\zeta - a\tau)\right\} + \frac{1}{2a}\int_{\zeta - a\tau}^{\zeta + a\tau} g(z) J_0\left(\frac{b}{a}\sqrt{(z - \zeta)^2 - a^2\tau^2}\right)dz$$

(69)

$$+ \frac{b\tau}{2}\int_{\zeta - a\tau}^{\zeta + a\tau} f(z) \frac{J_0'\left(\frac{b}{a}\sqrt{(z - \zeta)^2 - a^2\tau^2}\right)}{\sqrt{(z - \zeta)^2 - a^2\tau^2}}\,dz,$$

which gives the form already obtained, (146) § 46, if we replace ζ, τ, z, a, by $x, t, \alpha, \frac{1}{\sqrt{a}}$, reducing to (55) if $b = 0$.

75. Lorentz's Equation. We shall now, with Abraham[1]), apply the above method to the general wave-equation with second member of Lorentz, § 17, (195),

$$\Box \varphi = \Delta \varphi - \frac{1}{a^2}\frac{\partial^2 \varphi}{\partial t^2} = -4\pi\varrho(x, y, z, t)$$

(70)

$$t = 0, \quad \begin{cases} \varphi = f(x, y, z), \\ \dfrac{\partial \varphi}{\partial t} = g(x, y, z). \end{cases}$$

Introducing as in the treatment by Green's theorem of the equation with $\varrho = 0$, the mean of φ on a sphere about P,

$$\overline{\varphi} = \frac{1}{4\pi}\int\int \varphi\,d\omega,$$

---

1) Abraham, Acc. dei Lincei, (5) 14, p. 7, 1905. Theorie der Elektrizität, Bd. 2, p. 40.

and putting $u = r\,\overline{\varphi}$ we find by the divergence theorem as in Chapter V, (75),

$$(71) \qquad \frac{1}{a^2}\frac{\partial^2 u}{\partial t^2} - \frac{\partial^2 u}{\partial r^2} = \chi(r, at) = r\int\int \varrho\, d\omega.$$

Putting $r = x$, $at = y$, we have

$$(72) \qquad L(u) = \frac{\partial^2 u}{\partial x^2} - \frac{\partial^2 u}{\partial y^2} = -\chi(x, y),$$

an equation of hyperbolic type, but differing from the preceding by having a second member.

In terms of these variables let the boundary conditions be

$$(73) \qquad u = \frac{r}{4\pi}\int\int f(x, y, z)d\omega = F(r) = F(x), \qquad\qquad y = 0,$$

$$(74) \qquad \frac{\partial u}{\partial y} - \frac{1}{a}\frac{\partial u}{\partial t} = \frac{r}{4\pi a}\int\int g(x, y, z)d\omega = G(r) = G(x), \qquad x > 0.$$

We have now Green's formula, (34),

$$\int\int \{v\,L(u) - u\,L(v)\}\,dxdy = \int\left(u\frac{\partial v}{\partial \nu} - v\frac{\partial u}{\partial \nu}\right)ds$$

where

$$L(u) = -\chi, \quad L(v) = 0,$$

and as before we integrate around the triangle, Fig. 87. Put $v = 1$ and we obtain as before, (53), (54),

$$(75) \qquad 2u_P = u_A + u_B + \int_A^B G(x)\,dx + \int\int \chi\,dxdy.$$

$$2u(\xi, \eta) = F(\xi + \eta) + F(\xi - \eta) + \int_{\xi-\eta}^{\xi+\eta} G(x)\,dx + \int\int \chi\,dxdy.$$

Now F, G, χ are defined only for positive values of x. Let us assume that for negative values they behave like *odd* functions. The surface integral over the triangle ACD then vanishes, and

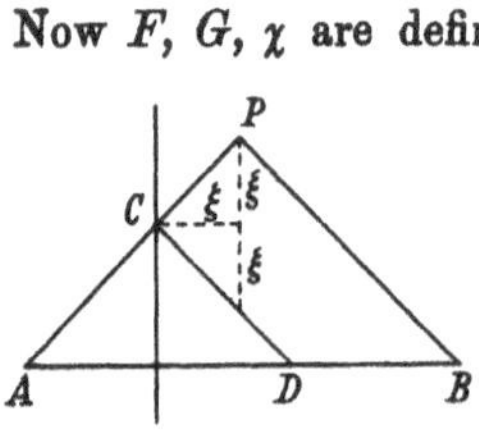

Fig. 87.

$$(76) \qquad \int_{\xi-\eta}^{\xi+\eta} G(x)\,dx = \int_0^{\xi+\eta} - \int_0^{\xi-\eta} = \int_0^{\xi+\eta} - \int_0^{\eta-\xi} = \int_{\eta-\xi}^{\eta+\xi} G(x)\,dx.$$

Dividing by ξ, which represents a value of r,

$$(77) \qquad \overline{\varphi} = \frac{u(\xi, \eta)}{\xi} = \frac{F(\xi+\eta) - F(\eta-\xi)}{2\xi} + \frac{1}{2\xi}\int_{\eta-\xi}^{\eta+\xi} G(x)\,dx + \frac{1}{2\xi}\int\int_{BPCD} \chi\,dxdy.$$

The limit for $\xi = 0$ is the value of φ required. The limit of the surface integral is the limit of an integral over the strip of width $\xi\sqrt{2}$, Fig. 88, in which

$$\chi(x, y) = \chi(\lambda, \eta - \lambda)$$

and the element of area is $2\xi\,d\lambda$, so that in the limit 2ξ disappears, and

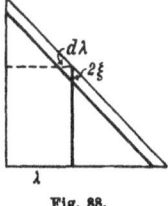

Fig. 88.

$$(78) \quad \varphi_{\xi=0} = F'(\eta) + G(\eta) + \int_0^\eta \chi(\lambda, \eta - \lambda)\,d\lambda ,$$

and replacing $\chi(x, y)$ by its value,

$$(79) \quad \begin{aligned} \chi(r, at) &= r\iint \varrho\,d\omega , \\ \chi(\lambda, \eta - \lambda) &= \lambda\iint \varrho(\lambda, \eta - \lambda)\,d\omega . \end{aligned}$$

Now since, by (73), (74), we have the values of F, G, we obtain,

$$F'(r) = \frac{1}{4\pi}\frac{\partial}{\partial r}\left\{r\iint f(x, y, z)\,d\omega\right\}$$

$$(80) \qquad F'(\eta) = \frac{1}{4\pi}\left[\frac{\partial}{\partial r}\left\{r\iint f(x, y, z)\,d\omega\right\}\right]_{r=\eta=at}$$

$$G(\eta) = \frac{\eta}{4\pi a}\iint g(x, y, z)\,d\omega .$$

The integral in (78),

$$\int_0^\eta \chi(\lambda, \eta - \lambda)\,d\lambda = \int_0^\eta \lambda\,d\lambda \iint \varrho(\lambda, \eta - \lambda)\,d\omega ,$$

becomes a volume integral, in which λ represents the radial coordinate, so that the element of volume is $d\tau = \lambda^2 d\lambda d\omega$, and we may put $\lambda\,d\lambda\,d\omega = \dfrac{d\tau}{\lambda}$,

$$\int_0^\eta \chi(\lambda, \eta - \lambda)\,d\lambda = \iiint_{\lambda=0}^{\lambda=at} \frac{\varrho(\lambda, \eta - \lambda)\,d\tau}{\lambda} ,$$

or replacing by the usual variables,

$$(81)\ \varphi = \frac{1}{4\pi}\frac{\partial}{\partial t}\left\{t\iint \varphi_{t=0}\,d\omega\right\} + \frac{t}{4\pi}\iint\left(\frac{\partial\varphi}{\partial t}\right)_{t=0}d\omega + \iiint \frac{\varrho(r, t - r/a)\,d\tau}{r} ,$$

we find the integral to be the retarded potential of Lorentz, (99) § 63, while the other terms give the familiar Poisson's solution.

76. Equation in n Variables. Let us now consider the equation in any number n of independent variables, $x_1, x_2, \ldots x_n$,

$$(82)\ L(u) \equiv \sum_{r=1}^{r=n}\sum_{s=1}^{s=n} A_{rs}\frac{\partial^2 u}{\partial x_r \partial x_s} + \sum_{r=1}^{r=n} B_r\frac{\partial u}{\partial x_r} + F \cdot u = 0, \quad A_{rs} = A_{sr}$$

and a hypersurface, that is an extension depending on $n - 1$ independent parameters, whose equation is

$$(83) \qquad f(x_1, x_2, \ldots x_n) = 0,$$

which we take for the *support*, that is, on which we give the values of u and $\dfrac{\partial u}{\partial n}$. In conformity with our method of procedure for two independent variables § 70, let us suppose that the support is represented parametrically in terms of $n - 1$ independent parameters $t_1, t_2, \ldots t_{n-1}$, in terms of which u and $\dfrac{\partial u}{\partial n}$ are also given. We shall denote the support values, when expressed in terms of $t_1, \ldots t_{n-1}$, by a bar, so that on the support

$$(84) \qquad \begin{aligned} x_1 &= \bar{x}_1(t_1, t_2, \ldots t_{n-1}), \quad x_n = \bar{x}_n(t_1, t_2, \ldots t_{n-1}), \\ \bar{u} &= F(t_1, t_2, \ldots t_{n-1}), \quad \frac{\overline{\partial u}}{\partial n} = G(t_1, t_2, \ldots t_{n-1}), \end{aligned}$$

the values of $x_1 \ldots x_n$, identically satisfying the equation (83).

Let us make use of the abbreviations,

$$(85) \qquad \frac{\partial f}{\partial x_r} = \pi_r, \quad \frac{\partial x_r}{\partial t_s} = P_{rs}, \quad \frac{\partial u}{\partial x_r} = p_r, \quad \frac{\partial^2 u}{\partial x_r \partial x_s} = p_{rs} = p_{sr}.$$

Then differentiating (83) totally,

$$df = \sum_{r=1}^{r=n} \sum_{s=1}^{s=n-1} \pi_r P_{rs} dt_s = 0,$$

and since the t's are independent, the coefficient of each dt_s vanishes, giving

$$(86) \qquad \begin{aligned} P_{11}\pi_1 + P_{21}\pi_2 + \cdots + P_{n1}\pi_n &= 0, \\ P_{12}\pi_1 + P_{22}\pi_2 + \cdots + P_{n2}\pi_n &= 0, \\ \cdots \cdots \cdots \cdots \cdots \cdots \\ P_{1,n-1}\pi_1 + P_{2,n-1}\pi_2 + \cdots + P_{n,n-1}\pi_n &= 0. \end{aligned}$$

These are only $n - 1$ equations for the n quantities π_s, so that only their ratios are determined. We have

$$(87) \qquad \frac{\pi_1}{D_1} = \frac{\pi_2}{D_2} = \cdots = \frac{\pi_n}{D_n} = \varrho$$

where by D_r we denote the determinant of order $n - 1$ made by omitting the r'th column of the coefficients P_{rs}. It is to be noted that D_r is the Jacobian of all the x's, omitting x_r, with respect to all the $t_1, t_2, \ldots t_{n-1}$.

Now taking the total derivative of $\bar{u}$,

$$(88) \qquad d\bar{u} = dF = \sum_{r=1}^{r=n-1} \frac{\partial F}{\partial t_r} dt_r = \sum_{s=1}^{s=n} \sum_{r=1}^{r=n-1} p_s P_{sr} dt_r,$$

so that we have the $n - 1$ equations,

$$(89)\quad\begin{aligned}
P_{11}p_1 &+ P_{21}p_2 &+ \cdots + P_{n1}p_n &= \frac{\partial F}{\partial t_1},\\
P_{12}p_1 &+ P_{22}p_2 &+ \cdots + P_{n2}p_n &= \frac{\partial F}{\partial t_2},\\
&\qquad\cdot\;\cdot\;\cdot\;\cdot\;\cdot\;\cdot\;\cdot\;\cdot\;\cdot\;\cdot\;\cdot\\
P_{1,\,n-1}p_1 &+ P_{2,\,n-1}p_2 &+ \cdots + P_{n,\,n-1}p_n &= \frac{\partial F}{\partial t_{n-1}},
\end{aligned}$$

to which we adjoin the equation defining $\dfrac{\partial u}{\partial n}$,

$$(90)\qquad \cos(nx_1)p_1 + \cos(nx_2)p_2 + \cdots + \cos(nx_n)p_n = G.$$

But since (83),

$$\frac{\cos(nx_1)}{\dfrac{\partial f}{\partial x_1}} = \frac{\cos(nx_2)}{\dfrac{\partial f}{\partial x_2}} = \cdots = \frac{\cos(nx_n)}{\dfrac{\partial f}{\partial x_n}} = \sigma,$$

by (87) we have $\cos(nx_r) = \varrho\sigma D_r$, so that the determinant of the equations (89) and (90), developing in terms of the last row and its minors, becomes

$$\varrho\sigma(D_1^2 + D_2^2 + \cdots + D_n^2), \quad \text{(compare equation (5))}$$

which does not vanish, and hence the p_r's are completely determined by the values of F and G, the coefficients P_{rs} depending only on the support.

Let us now proceed in like manner to determine the second derivatives. Applying to $\bar{p}_r$ the process that we have just applied to $\bar{u}$, we get

$$(91)\quad\begin{aligned}
P_{11}p_{r1} &+ P_{21}p_{r2} &+ \cdots + P_{n1}p_{rn} &= \frac{\partial \bar{p}_r}{\partial t_1},\\
P_{12}p_{r1} &+ P_{22}p_{r2} &+ \cdots + P_{n2}p_{rn} &= \frac{\partial \bar{p}_r}{\partial t_2},\\
&\qquad\cdot\;\cdot\;\cdot\;\cdot\;\cdot\;\cdot\;\cdot\;\cdot\;\cdot\;\cdot\;\cdot\\
P_{1,\,n-1}p_{r1} &+ P_{2,\,n-1}p_{r2} &+ \cdots + P_{n,\,n-1}p_{rn} &= \frac{\partial \bar{p}_r}{\partial t_{n-1}}.
\end{aligned}$$

These are only $n - 1$ equations to determine the n unknowns $p_{r1}, p_{r2}, \ldots p_{rn}$, so that one of the p's is arbitrary. Notice that it is not the case (as in 86), where the coefficients P_{rs} are the same, but the right-hand members are zero) that the *ratios* are determined, but, as we see by transposing the term in the arbitrary P_{rs} to the right, all the other P_{rs}'s are determined as linear functions of this one. Or what is more symmetrical, we may take as arbitrary any given linear function of the unknowns. Let us therefore adjoin to the equations (91) an equation

$$(92)\qquad a_1p_{r1} + a_2p_{r2} + \cdots + a_np_{rn} = \lambda_r,$$

where $a_1, a_2, \ldots a_n$ are numerical constants taken at pleasure, and λ_r is the arbitrary parameter, in terms of which all the p_{rs}'s are linearly determined, the same values being obtainable no matter how $a_1, a_2, \ldots a_n$ are chosen (that is if we give λ_r all possible values).

If then the a's are so chosen that the determinant,

$$(93) \qquad D = \begin{vmatrix} P_{11}, & P_{21}, & \ldots P_{n1} \\ P_{12}, & P_{22}, & \ldots P_{n2} \\ \cdot & \cdot & \cdot \quad \cdot \quad \cdot \\ P_{1,n-1}, & P_{2,n-1}, & \ldots P_{n,n-1} \\ a_1, & a_2, & \ldots a_n \end{vmatrix}$$

does not vanish, we have

$$(94) \quad Dp_{rs} = \frac{\partial \bar{p}_r}{\partial t_1} D_{s1} + \frac{\partial \bar{p}_r}{\partial t_2} D_{s2} + \cdots + \frac{\partial \bar{p}_r}{\partial t_{n-1}} D_{s,n-1} + \lambda_r D_s, \quad s = 1, 2, \ldots n.$$

where D_{st} is the proper minor of D, and D_s is as in (87).

If we now apply the same process to $\bar{p}_s$ instead of to $\bar{p}_r$ (the a's of course being the same as before), we get

$$(95) \quad Dp_{sr} = \frac{\partial \bar{p}_s}{\partial t_1} D_{r1} + \frac{\partial \bar{p}_s}{\partial t_2} D_{r2} + \cdots + \frac{\partial \bar{p}_s}{\partial t_{n-1}} D_{r,n-1} + \lambda_s D_r, \quad r = 1, 2, \ldots n,$$

where the meaning of the D_{ij}'s is the same as before, but different ones occur. It is to be noticed that all terms except the last in (94), (95), are linear in the a's, and that the values of p_{rs} and p_{sr} are determined as linear functions of the right-hand members of (91), (92), and the corresponding equations with r changed to s.

So far all our work is algebraic, and in (91) we may put on the right any values we please. If we should put in each line zero, we would get a set of solutions, depending on λ_r, of the form

$$(96) \qquad Dp_{rs} = \lambda_r D_s, \quad s = 1, 2, \ldots n.$$

In the same way we should get solutions depending on λ_s, of the form,

$$(97) \qquad Dp_{sr} = \lambda_s D_r, \quad r = 1, 2, \ldots n.$$

Now if we are to have $p_{rs} = p_{sr}$, we must take the λ's so that

$$(98) \qquad \lambda_r D_s = \lambda_s D_r.$$

But this means, comparing with (87), that the λ's are proportional to the π's.

$$(99) \qquad \frac{\lambda_r}{\lambda_s} = \frac{D_r}{D_s} = \frac{\pi_r}{\pi_s}, \quad \frac{\lambda_r}{\pi_r} = \mu, \quad r = 1, 2, \ldots n,$$

so that finally

$$(100) \qquad \lambda_r = \mu \pi_r, \quad \lambda_s = \mu \pi_s, \quad \lambda_r D_s = \lambda_s D_r = \frac{\mu}{\varrho} \pi_r \pi_s$$

that is, instead of n parameters λ_r we have a single one, $\lambda = \frac{\mu}{\varrho}$, and

(101) $$p_{rs} = p_{rs}^{(0)} + \lambda\,\pi_r\pi_s,$$

where $p_{rs}^{(0)}$ is completely determined, standing for all but the last term in (94), (95), divided by D, as we see by putting $\lambda = \lambda_r = \lambda_s = 0$.

If the above reasoning does not seem clear, we may replace it by the following. Let any set of solutions of (91) for any value of the arbitrary parameter be denoted by $p_{rs}^{(0)}$. Then inserting them in (91) and subtracting we find

(102)
$$
\begin{aligned}
P_{11}(p_{r_1} - p_{r_1}^{(0)}) &+ P_{21}(p_{r_2} - p_{r_2}^{(0)}) &+ \cdots = 0, \\
P_{12}(p_{r_1} - p_{r_1}^{(0)}) &+ P_{22}(p_{r_2} - p_{r_2}^{(0)}) &+ \cdots = 0, \\
&\ \cdots \cdots \cdots \cdots \cdots \\
P_{1,n-1}(p_{r_1} - p_{r_1}^{(0)}) &+ P_{2,n-1}(p_{r_2} - p_{r_2}^{(0)}) &+ \cdots = 0.
\end{aligned}
$$

But from these we at once find, as from (86),

(103) $$\frac{p_{rs} - p_{rs}^0}{D_s} = \varrho_r, \quad s = 1, 2, \ldots n.$$

Consequently
$$p_{rs} = p_{rs}^0 + \varrho_r D_s,$$

and since
$$p_{rs} = p_{sr} \text{ and } p_{rs}^{(0)} = p_{sr}^{(0)},$$

we find as before,
$$\varrho_r D_s = \varrho_s D_r = \lambda\,\pi_r\pi_s.$$

Now inserting in the differential equation (82) we obtain,

(104) $$\lambda \sum_{r=1}^{r=n}\sum_{s=1}^{s=n} A_{rs}\pi_r\pi_s + \sum_{r=1}^{r=n}\sum_{s=1}^{s=n} A_{rs}p_{rs}^0 + \sum_{r=1}^{r=n} B_r p_r + Fu = 0,$$

as a final equation for determining λ, so that then all the second derivatives are determined. We then proceed to determine the derivatives of higher order in a similar manner. The process fails, however, if we have

(105) $$\Phi \equiv \sum_{r=1}^{r=n}\sum_{s=1}^{s=n} A_{rs}\pi_r\pi_s = 0,$$

that is

(106) $$\Phi \equiv \sum_{r=1}^{r=n}\sum_{s=1}^{s=n} A_{rs}\frac{\partial f}{\partial x_r}\frac{\partial f}{\partial x_s} = 0.$$

This is a differential equation of the first order, the generalization of (10), which defines the characteristic or exceptional hypersurfaces.

If $f(x_1, x_2, \ldots x_n) = 0$ is a solution of (106), representing a hypersurface, the direction cosines of the normal at any point $\cos(nx_n)$ are proportional to π_r, so that if the π_r's are taken as the running co-ordinates of a point, equation (105) is the equation of a cone, that is at any point $x_1, x_2 \ldots, x_n$, the possible normals lie on a cone. The planes at right angles to these normals envelope a second cone, called the characteristic cone, which is tangent to all the characteristic hypersurfaces at the point. The function Φ is called the characteristic function.

The equations of the Cauchy characteristics of the equation (106), defined in § 24, are

$$(107) \qquad \frac{dx_r}{\dfrac{\partial \Phi}{\partial \pi_r}} = -\frac{d\pi_r}{\dfrac{\partial \Phi}{\partial x_r}} = dt, \quad r = 1, 2, \ldots n,$$

see § 24 (39) and § 70 (10).

These equations are to be integrated with the initial conditions,

$$(108) \qquad \Phi_0 = \sum_{r=1}^{r=n} \sum_{s=1}^{s=n} A_{rs}^{(0)} \pi_r^{(0)} \pi_s^{(0)} = 0,$$

where the affix zero denotes the initial point $x_1^0, \ldots x_n^0$. They repre-

Fig. 89.

sent one-dimensional extensions, or lines, which are called by Hadamard the *bicharacteristics* of the equation (82). All the bicharacteristics passing through the point $x_1^0, \ldots x_n^0$ generate a hypersurface with a conical or singular point $x_1^0 \ldots x_n^0$, Fig. 89, which we will call a conoid, although this term is also used in other senses. If the coefficients A_{rs} are constants, the conoid reduces to the characteristic cone.

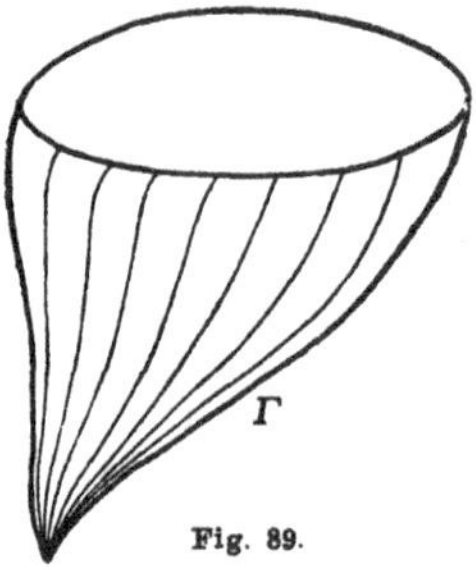

If we consider for a point $x_1^0, \ldots x_n^0$ the function

$$(109) \qquad \Phi = \sum \sum A_{rs}^0 \pi_r \pi_s$$

three cases may arise. First Φ may be a definite positive form[1]) in the π's.

In this case the characteristic cones and conoids are imaginary, and the equation may be called elliptic. Secondly the form Φ may be indefinite. Then the characteristic cones are real, and the equation is hyperbolic. Thirdly, if the determinant

$$\begin{vmatrix} A_{11} & A_{12} & \cdots & A_{1n} \\ A_{21} & A_{22} & \cdot & A_{2n} \\ A_{n1}, & A_{n2} & \cdots & A_{nn} \end{vmatrix}$$

---

1) A definite form is a homogeneous function whose sign is the same for all possible values of its variables, such as $x^2 + 2xy + y^2$. The forms xy and $x^2 - y^2$ are indefinite. By a linear transformation of the variables a form may be transformed so as to contain only terms with squares (see § 29). If the coefficients of these are all of the same sign the form is definite, otherwise it is indefinite. If the determinant vanishes, then the form can be reduced to contain *less* than n squares.

vanishes, the equation is parabolic. This corresponds to the classification for two dimensions.

77. Green's Theorem and Characteristics for n Dimensions. Let us now consider Green's Theorem. To find the adjoint equation, proceed as in (20) to integrate by parts with respect first to x_r and then to x_s the n-fold integral

$$\underset{n}{\int\int \cdots \int} A_{rs} v \frac{\partial^2 u}{\partial x_r \partial x_s} dx_1 dx_2 \ldots dx_n = -\underset{n-1}{\int \cdots \int} A_{rs} v \frac{\partial u}{\partial x_s} \cos(nx_r) dS,$$

$$-\underset{n}{\int\int \cdots \int} \frac{\partial u}{\partial x_s} \frac{\partial (A_{rs} v)}{\partial x_r} dx_1 dx_2 \ldots dx_n,$$

$$(110) \quad = -\underset{n-1}{\int \cdots \int} \left\{ A_{rs} v \frac{\partial u}{\partial x_s} \cos(nx_r) - u \frac{\partial (A_{rs} v)}{\partial x_r} \cos(nx_s) \right\} dS$$

$$+ \underset{n}{\int\int \cdots \int} u \frac{\partial^2 (A_{rs} v)}{\partial x_r \partial x_s} dx_1 dx_2 \ldots dx_n,$$

$$= \underset{n-1}{\int \cdots \int} \left\{ A_{rs} \left(u \frac{\partial v}{\partial x_r} \cos(nx_s) - v \frac{\partial u}{\partial x_s} \cos(nx_r) \right) + u v \frac{\partial A_{rs}}{\partial x_r} \cos(nx_s) \right\} dS$$

$$+ \underset{n}{\int\int \cdots \int} u \frac{\partial^2 (A_{rs} v)}{\partial x_r \partial x_s} dx_1 dx_2 \ldots dx_n.$$

On the other hand since the term is symmetrical in x_r and x_s we may interchange them, and take the half sum of the two expressions, instead of the unsymmetrical expression of (110).

For the terms of first order we integrate but once, obtaining

$$\underset{n}{\int\int \cdots \int} B_r v \frac{\partial u}{\partial x_r} dx_1 dx_2 \cdots dx_n = -\underset{n-1}{\int \cdots \int} u v B_r \cos(nx_r) dS$$

$$(111) \qquad\qquad -\underset{n}{\int\int \cdots \int} u \frac{\partial (B_r v)}{\partial x_r} dx_1 dx_2 \cdots dx_n.$$

Collecting all terms we have finally Green's Theorem,

$$\underset{n}{\int\int \cdots \int} \left\{ v L(u) - u M(v) \right\} dx_1 dx_2 \cdots dx_n$$

$$(112) \qquad\qquad = \underset{n-1}{\int \cdots \int} \left\{ u D_n v - v D_n u + u v P_n \right\} dS,$$

where M is the adjoint differential operator,

$$M(v) \equiv \sum_{r=1}^{r=n} \sum_{s=1}^{s=n} \frac{\partial^2 (A_{rs} v)}{\partial x_r \partial x_s} - \sum_{r=1}^{r=n} \frac{\partial (B_r v)}{\partial x_r} + F \cdot v$$

(113)
$$\equiv \sum_{r=1}^{r=n} \sum_{s=1}^{s=n} A_{rs} \frac{\partial^2 v}{\partial x_r \partial x_s} + 2 \sum_{r=1}^{r=n} \left(\sum_{s=1}^{s=n} \frac{\partial A_{rs}}{\partial x_s} - \frac{1}{2} B_r \right) \frac{\partial v}{\partial x_r}$$

$$+ v \left\{ \sum_{r=1}^{r=n} \sum_{s=1}^{s=n} \frac{\partial^2 A_{rs}}{\partial x_r \partial x_s} - \sum_{r=1}^{r=n} \frac{\partial B_r}{\partial x_r} + F \right\},$$

the adjoint function is

(114)
$$P_n \equiv \sum_{r=1}^{r=n} \left(\sum_{s=1}^{s=n} \frac{\partial A_{rs}}{\partial x_s} - B_r \right) \cos (n x_r),$$

and D_n stands for the differential operator,

(115)
$$D_n u = \sum_{r=1}^{r=n} \sum_{s=1}^{s=n} A_{rs} \cos (n x_s) \frac{\partial u}{\partial x_r}.$$

Now if we write

(116)
$$\sum_{s=1}^{s=n} A_{rs} \cos (n x_s) = \varLambda_r = |\varLambda| \cos (\nu x_r)$$

$\varLambda$ will be a vector whose absolute magnitude is

(117)
$$\varLambda = \sqrt{\sum_{r=1}^{r=n} \left(\sum_{s=1}^{s=n} A_{rs} \cos (n x_s) \right)^2}$$

and we have

(118)
$$D_n u = \varLambda \sum_{r=1}^{r=n} \frac{\partial u}{\partial x_r} \cos (\nu x_r) = \varLambda \frac{\partial u}{\partial \nu}.$$

The direction ν is that of the conormal. Its relation to the normal may be found as before. If we call the direction cosines of the normal, $X_r = \cos (n x_r)$ the equation of the cone of normals to the characteristic hypersurfaces is

$$\Phi = \sum_{r=1}^{r=n} \sum_{s=1}^{s=n} A_{rs} X_r X_s = 0,$$

and the quadric $\Phi = 1$ has normals whose direction cosines are proportional to

(119)
$$\frac{1}{2} \frac{\partial \Phi}{\partial X_r} = \sum_{s=1}^{s=n} A_{rs} X_s = \varLambda (\cos \nu x_r).$$

Accordingly at any point of a hypersurface S the conormal ν is related to the normal n as the perpendicular on the diametral plane conjugate with respect to the quadric $\Phi = 1$ to the radius vector parallel to n. Referring to the equations of the bicharacteristics (107) we also see that, since the direction cosines of the tangent to the bicharacteristic are equal to $\frac{dx_r}{ds}$, which are proportional to $\frac{\partial \Phi}{\partial \pi_r}$ and therefore to $\frac{\partial \Phi}{\partial X_r}$, the conormal is in the direction of the tangent to the bicharacteristic. Accordingly if the hypersurface S is a characteristic hypersurface the conormal is tangent to it.

The formula (112) embraces every case of Green's Theorem that we have heretofore used, and by means of it we shall show that a solution of the equation $L(u) = 0$ is determined if we can find a solution of the adjoint equation $M(v) = 0$, by the surface values of u, v and their conormal derivatives. But if the hypersurface S is characteristic, giving the values of u alone determines the conormal derivative, since the conormal is tangent to the surface at all points.

If we have the conditions,

$$(120) \qquad \sum_{s=1}^{s=n} \frac{\partial A_{rs}}{\partial x_s} = B_r, \qquad\qquad r = 1, 2, \ldots n$$

the adjoint function P_n vanishes, the equation $L(u) = 0$ is self-adjoint, and may be written,

$$(121) \qquad \sum_{r=1}^{r=n} \frac{\partial}{\partial x_r} \sum_{s=1}^{s=n} A_{rs} \frac{\partial u}{\partial x_s} + F \cdot u = 0 .$$

The manner of applying (112) to the solution of Cauchy's problem proposed by Volterra is an extension of Riemann's method explained in § 72. In order to find the value of u at a point P construct the characteristic conoid Γ having its vertex at P, and cutting the support S in the closed $n - 2$-dimensional extension σ, Fig. 90. The figure is drawn as if $n = 3$, when σ is a *curve*, but by an effort of the imagination we may also use it for higher values of n. The formula (112) is now to be applied to the space enclosed by S and the conoid Γ, the

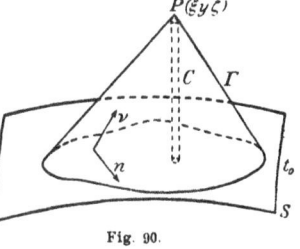

Fig. 90.

surface integrals covering both. On the conoid, as on any characteristic surface we may perform an integration by parts, as in (44). We have seen that the bicharacteristics form a set of curves generating the conoid, and that at any point the conormal is in the direction of the bicharacteristic. Let the length of the arc of any bicharacteristic counted from P be s, which may be expressed in terms of t in (107) by the equation

$$ds^2 = \sum_{r=1}^{r=n} dx_r^2 = dt^2 \sum \left(\frac{\partial \Phi}{\partial \pi_r}\right)^2.$$

Let the coordinates of a point on the conoid Γ be expressed in terms of $n-1$ parameters, of which s is one, the others we will call $t_1, \cdots t_{n-2}$, and let the value of the element of surface of the conoid be

$$dS = A(s, t_1, t_2 \cdots t_{n-2}) ds \, dt_1 \, dt_2 \cdots dt_{n-2}.$$

In the formula (112) we may accordingly write the surface-integral over Γ, by means of (118), with $d\nu = ds$,

$$(122) \quad \underset{n-1}{\int\int \cdots \int} \left\{ A\left(u \frac{\partial v}{\partial s} - v \frac{\partial u}{\partial s}\right) + u v P_n \right\} A \, ds \, dt_1 \cdots dt_{n-2}$$

and we can then perform an integration by parts according to s on the term containing the derivative of u, obtaining

$$(123) \quad \begin{aligned} &-\underset{n-2}{\int\int \cdots \int} A \, A u v \, dt_1 \cdots dt_{n-2} \\ &+\underset{n-1}{\int\int \cdots \int} u \left\{ 2 A \frac{\partial v}{\partial s} + \frac{v}{A} \frac{\partial (A A)}{\partial s} + P_n v \right\} dS \end{aligned}$$

where the $n-2$-fold integral contains the values along σ and at P. The Riemann function is now to be defined as one satisfying the equation $M(v) = 0$, and along the bicharacteristics the differential equation

$$(124) \quad 2 A \frac{\partial v}{\partial s} + v \left(\frac{1}{A} \frac{\partial (A A)}{\partial s} + P_n\right) = 0$$

corresponding to (46), (47) (in which $A = A = 1$). This will be satisfied by $v = 0$. It will be found that the value of u at P does not emerge from the integral as in (48), but that nevertheless its value at P may be found in terms of the values on σ, which have been propagated to P along the bicharacteristics. These lines have then the physical characteristics of rays. The manner in which this comes about will be best shown by examples.

78. Volterra's Method for Wave-Equation. Let us consider the wave-equation in two dimensions, as integrated by Volterra[1]),

$$\frac{1}{a^2} \frac{\partial^2 u}{\partial t^2} = \frac{\partial^2 u}{\partial x^2} + \frac{\partial^2 u}{\partial y^2} + f(x, y, t)$$

---

1) Volterra, Sur les vibrations des corps élastiques isotropes. Acta Mathematica, t. 18, pp. 161—232, 1894.

and put for simplicity $at = z$, then we may use the form

$$(125) \qquad L(u) \equiv \square\, u \equiv \frac{\partial^2 u}{\partial x^2} + \frac{\partial^2 u}{\partial y^2} - \frac{\partial^2 u}{\partial z^2} = - f(x, y, z)$$

and since the coefficients are constant, the equation is self-adjoint, and the characteristic function is

$$(126) \qquad\qquad \Phi \equiv X^2 + Y^2 - Z^2.$$

The characteristic cone is a circular cone of 45^0 opening.

For the conormal we find

$$\cos(\nu x) = \cos(n x), \quad \cos(\nu y) = \cos(n y), \quad \cos(\nu z) = - \cos(n z), \quad \Lambda = 1.$$

The conormal and the normal have the same projections on the xy-plane, opposite projections on the z-axis, Fig. 90. If S is the supporting surface, on which u and $\frac{\partial u}{\partial \nu}$ are given, we may find the value at any point $P(\xi, \eta, \zeta)$ by drawing the characteristic cone Γ with vertex at P cutting S. On this cone the conormal evidently lies *in* the cone. If then we find a solution v of the equation $L(v) = 0$, constant on the cone Γ, we have at the same time $\frac{\partial v}{\partial \nu} = 0$ on the cone, (124) is satisfied and consequently

$$(127) \iiint \{ v L(u) - u L(v) \}\, d\tau = \iint_S \left(u \frac{\partial v}{\partial \nu} - v \frac{\partial u}{\partial \nu} \right) dS - v \int \!\!\int_{\Gamma} \frac{\partial u}{\partial \nu}\, dS.$$

Let us put

$$x' = x - \xi, \quad y' = y - \eta, \quad z' = z - \zeta, \quad r^2 = x'^2 + y'^2,$$

$$(128) \qquad \theta = \frac{z'}{r}, \quad \frac{\partial \theta}{\partial z'} = \frac{1}{r}, \quad \frac{\partial \theta}{\partial r} = - \frac{z'}{r^2},$$

$$\frac{\partial \theta}{\partial x'} = - \frac{z' x'}{r^3} = - \theta \frac{x'}{r^2}, \quad \frac{\partial \theta}{\partial y'} = - \frac{z' y'}{r^3} = - \theta \frac{y'}{r^2}.$$

The equation of the characteristic cone is then $\theta = 1$. Let us seek a solution which vanishes on the cone, and is a function of θ alone. We have

$$\frac{\partial v}{\partial x'} = \frac{dv}{d\theta} \frac{\partial \theta}{\partial x'} = - \theta \frac{dv}{d\theta} \frac{x'}{r^2}, \quad \frac{\partial v}{\partial y'} = \frac{dv}{d\theta} \frac{\partial \theta}{\partial y'} = - \theta \frac{dv}{d\theta} \frac{y'}{r^2}, \quad \frac{\partial v}{\partial z'} = \frac{dv}{d\theta} \frac{1}{r},$$

$$(128) \qquad
\begin{aligned}
\frac{\partial^2 v}{\partial x'^2} &= \theta \frac{d}{d\theta} \left(\theta \frac{dv}{d\theta} \right) \frac{x'^2}{r^4} - \theta \frac{dv}{d\theta} \left(\frac{1}{r^2} - \frac{2 x'^2}{r^4} \right) \\[4pt]
\frac{\partial^2 v}{\partial y'^2} &= \theta \frac{d}{d\theta} \left(\theta \frac{dv}{d\theta} \right) \frac{y'^2}{r^4} - \theta \frac{dv}{d\theta} \left(\frac{1}{r^2} - \frac{2 y'^2}{r^4} \right) \\[4pt]
\frac{\partial^2 v}{\partial z'^2} &= \frac{d^2 v}{d\theta^2} \frac{1}{r^2}
\end{aligned}$$

and consequently

$$(129) \qquad L(v) = \frac{1}{r^3}\left\{(\theta^2-1)\frac{d^2v}{d\theta^2} + \theta\frac{dv}{d\theta}\right\} = 0.$$

From this ordinary differential equation we obtain the solution

$$(130) \qquad v = \log\left(\theta + \sqrt{\theta^2-1}\right) = \log\left\{\frac{z'}{r} + \sqrt{\frac{z'^2}{r^2}-1}\right\}.$$

When $\theta = 1$, $v = 0$, but when $r = 0$, $z' \neq 0$, $v = \infty$. The solution accordingly has the singular *line* $r = 0$, through P parallel to the z-axis. We will accordingly exclude this by a cylinder C of radius ε and apply the volume integration to the volume outside C, and inside of the surface made up of S and Γ. We accordingly must include the surface integral over the cylinder C.

The formula (127) gives, with (125),

$$(131) \qquad \iint\limits_{C+S}\left(u\frac{\partial v}{\partial \nu} - v\frac{\partial u}{\partial \nu}\right)dS = -\iiint vf\, d\tau.$$

On the cylinder C, φ being the angle of azimuth, we have $dS = \varepsilon\, d\varphi\, dz$,

$$(132) \qquad \iint\limits_{C} v\frac{\partial u}{\partial \nu}\, dS = \iint\limits_{C}\frac{\partial u}{\partial \nu} v\varepsilon\, d\varphi\, dz,$$

and since $\lim\limits_{\varepsilon=0}(\varepsilon\log\varepsilon) = 0$ this vanishes in the limit. Also on C

$$(133) \qquad \frac{\partial v}{\partial \nu} = \frac{\partial v}{\partial r} = -\frac{dv}{d\theta}\frac{z'}{r^2} = -\frac{z'}{r^2\sqrt{\theta^2-1}} = -\frac{z'}{r\sqrt{z'^2-r^2}}$$

and u is a function of z alone. Accordingly

$$(134)\lim\limits_{\varepsilon=0}\int\limits_{C}\int u\frac{\partial v}{\partial \nu}\, dS = -\lim\limits_{\varepsilon=0}\iint\frac{z'\, u(\xi,\eta,z)}{\sqrt{z'^2-\varepsilon^2}}\, d\varphi\, dz = -2\pi\int\limits_{z}^{z_S}u(\xi,\eta,z)\, dz$$

where z_S is the value of z at the intersection of the line through P with the support S. We thus obtain finally

$$(135) \qquad -2\pi\int\limits_{z_S}^{\zeta}u(\xi,\eta,z)\, dz = \iint\limits_{S}\left(u\frac{\partial v}{\partial \nu} - v\frac{\partial u}{\partial \nu}\right)dS + \iiint vf\, d\tau$$

a formula which, to be sure, does not give us the value of u at P, but its integral along the line through P parallel to the z-axis. By differentiation by the limit ζ we obtain

$$(136) \qquad 2\pi u(\xi,\eta,\zeta) = -\frac{\partial}{\partial \zeta}\left[\iint\limits_{S}\left(u\frac{\partial v}{\partial \nu} - v\frac{\partial u}{\partial \nu}\right)dS + \iiint vf\, d\tau\right]$$

It remains to show that this result satisfies the conditions. This has been carried out by d'Adhémar[1]), and presents considerable difficulty, since the differentiation involves varying the position of the cone $\varGamma$.

We may compare this result with a familiar one for the elliptic equation of Poisson. If

$$(137) \qquad \Delta u = - f(x, y, z)$$

we have Green's formula, § 59, (69).

$$(138) \quad \varpi u(\xi, \eta, \zeta) = \frac{1}{4\pi} \iint \left\{ u \frac{\partial \frac{1}{r}}{\partial n} - \frac{1}{r} \frac{\partial u}{\partial n} \right\} dS + \frac{1}{4\pi} \iiint \frac{f d\tau}{r}$$

where $\varpi = 1$, inside, $\varpi = 0$ outside, and where

$$r^2 = (x - \xi)^2 + (y - \eta)^2 + (z - \zeta)^2 .$$

Let us integrate both sides according to ζ and then differentiate according to the same variable. The only way in which ζ enters is in $\frac{1}{r}$ and if we integrate we have

$$(139) -\int \frac{d\zeta}{r} = \log \left\{ \frac{z - \zeta}{\sqrt{(\xi - x)^2 + (\eta - y)^2}} + \sqrt{\frac{(z - \zeta)^2}{(\xi - x)^2 + (\eta - y)^2} + 1} \right\} = \varphi$$

so that (139) becomes

$$(140) \; \varpi u(\xi, \eta, \zeta) = - \frac{1}{4\pi} \frac{\partial}{\partial \zeta} \left\{ \iint \left\{ u \frac{\partial \varphi}{\partial n} - \varphi \frac{\partial u}{\partial n} \right\} dS + \iiint f \varphi d\tau \right\} .$$

If we write iz for z the equation (137) changes to

$$(141) \qquad \frac{\partial^2 u}{\partial x^2} + \frac{\partial^2 u}{\partial y^2} - \frac{\partial^2 u}{\partial z^2} = \Box u = - f(x, y, z),$$

and if we write

$$(142) \quad \varphi' = \log \left\{ \frac{z - \zeta}{\sqrt{(\xi - x)^2 + (\eta - y)^2}} + \sqrt{\frac{(z - \zeta)^2}{(\xi - x)^2 + (\eta - y)^2} - 1} \right\}$$

we have

$$(143) \quad u = - \frac{1}{2\pi} \frac{\partial}{\partial \xi} \left\{ \iint \left\{ u \frac{\partial \varphi'}{\partial \nu} - \varphi' \frac{\partial u}{\partial \nu} \right\} dS + \iiint f \varphi' d\tau \right\},$$

which is the formula (136) already deduced. Instead of the normal we have the conormal and instead of the distance r the *codistance*

$$r' = \sqrt{(z - \zeta)^2 - (x - \xi)^2 - (y - \eta)^2} .$$

It is important to notice that in the elliptic equation u being given on the surface gives $\frac{\partial u}{\partial n}$ and the surface of integration is always the same,

---

1) D'Adhémar, Thèse, Journal de Mathématiques, 1904. See also d'Adhémar, Les Equations aux dérivées partielles a caractéristiques réelles.

while in the hyperbolic equation the area of the surface varies with ξ, η, ζ, and u and $\frac{\partial u}{\partial \nu}$ may *both* be given arbitrarily.

If the surface S is a plane, that is if $t = 0$ (initial conditions), $z = 0$, $\zeta > 0$, the conormal is in the direction opposite to the normal,

$$\frac{\partial \varphi'}{\partial \nu} = -\frac{\partial \varphi'}{\partial z} = -\frac{1}{r'},$$

and we have, if $f = 0$,

$$(144) \quad u(\xi, \eta, \zeta) = \frac{1}{2\pi} \frac{\partial}{\partial \zeta} \iint \frac{u}{\sqrt{(\zeta - z)^2 - (\xi - x)^2 - (\eta - y)^2}} dS$$
$$+ \frac{1}{2\pi} \iint \frac{1}{\sqrt{(\zeta - z)^2 - (\xi - x)^2 - (\eta - y)^2}} \frac{\partial u}{\partial z} dS$$

which is the Poisson-Parseval solution, (188) § 49.

Let us now apply the method to the equation for waves in three dimensions,

$$(145) \quad L(u) \equiv \square u \equiv \frac{\partial^2 u}{\partial x^2} + \frac{\partial^2 u}{\partial y^2} + \frac{\partial^2 u}{\partial z^2} - \frac{\partial^2 u}{\partial t^2} = -f(x, y, z, t).$$

Consider x, y, z, t as the coordinates of a point in four-dimensional space. The characteristic cone is

$$(146) \quad \Phi = X^2 + Y^2 + Z^2 - T^2 = 0$$

a cone of 45° opening. Green's Theorem is

$$(147) \quad \iiiint \left\{ v L(u) - u L(v) \right\} dx\,dy\,dz\,dt = \iiint \left\{ u \frac{\partial v}{\partial \nu} - v \frac{\partial u}{\partial \nu} \right\} dS$$

where

$$\cos(\nu x) = \cos(nx), \quad \cos(\nu y) = \cos(ny), \quad \cos(\nu z) = \cos(nz),$$
$$\cos(\nu t) = -\cos(nt).$$

Put as before

$$x' = x - \xi, \quad y' = y - \eta, \quad z' = z - \zeta, \quad t' = t - \tau,$$

$$(148) \quad r^2 = x'^2 + y'^2 + z'^2, \quad \theta = \frac{t'}{r}$$

$$\frac{\partial \theta}{\partial x'} = -\theta \frac{x'}{r^2}, \quad \frac{\partial \theta}{\partial y'} = -\theta \frac{y'}{r^2}, \quad \frac{\partial \theta}{\partial z'} = -\theta \frac{z'}{r^2}, \quad \frac{\partial \theta}{\partial t'} = \frac{1}{r},$$

and let us again seek a solution $v(\theta)$. As in (128'),

$$(149) \quad \begin{aligned} \frac{\partial^2 v}{\partial x'^2} &= \theta \frac{d}{d\theta}\left(\theta \frac{dv}{d\theta}\right)\frac{x'^2}{r^4} - \theta \frac{dv}{d\theta}\left(\frac{1}{r^2} - \frac{2x'^2}{r^4}\right), \\ \frac{\partial^2 v}{\partial y'^2} &= \theta \frac{d}{d\theta}\left(\theta \frac{dv}{d\theta}\right)\frac{y'^2}{r^4} - \theta \frac{dv}{d\theta}\left(\frac{1}{r^2} - \frac{2y'^2}{r^4}\right), \\ \frac{\partial^2 v}{\partial z'^2} &= \theta \frac{d}{d\theta}\left(\theta \frac{dv}{d\theta}\right)\frac{z'^2}{r^4} - \theta \frac{dv}{d\theta}\left(\frac{1}{r^2} - \frac{2z'^2}{r^4}\right), \\ \frac{\partial^2 v}{\partial t^2} &= \frac{1}{r^2}\frac{d^2 v}{d\theta^2}, \end{aligned}$$

but on account of the three space terms,

$$(150) \qquad L(v) = \frac{\theta^2 - 1}{r^2}\frac{d^2 v}{d\theta^2} = 0, \quad \frac{d^2 v}{d\theta^2} = 0, \quad v = a\theta + b\,.$$

If $v = 0$ on the cone Γ, $\theta = 1$,

$$v = \theta - 1 = \frac{t - \tau}{r} - 1$$

having again a singularity for $r = 0$, $t' \doteq 0$.

On the hypercylinder C, $r = \varepsilon$ (representing a small sphere in three dimensional space), $\dfrac{\partial v}{\partial \nu} = \dfrac{\partial v}{\partial r} = -\dfrac{t - \tau}{r^2}$, $dS = r^2 d\omega\, dt$, where $d\omega$ is the element of solid angle, and the hypersurface integral becomes

$$\lim_{\varepsilon = 0}\int\!\!\!\int_C\!\!\!\int \left(u\frac{\partial v}{\partial \nu} - v\frac{\partial u}{\partial \nu}\right)dS$$

$$(151) \qquad = \lim_{\varepsilon = 0}\int\!\!\!\int\!\!\!\int \left\{\frac{u(\tau - t)}{\varepsilon^2} - \left(\frac{t - \tau}{\varepsilon} - 1\right)\frac{\partial u}{\partial r}\right\}\varepsilon^2 d\omega\, dt$$

$$= 4\pi\int_\tau^{t_S}(\tau - t)\,u(\xi, \eta, \zeta, t)\,dt\,.$$

Inserting in (147),

$$(152) \qquad 4\pi\int_\tau^{t_S}(\tau - t)u(\xi, \eta, \zeta, t)\,dt + \int\!\!\!\int\!\!\!\int \left\{u\frac{dv}{d\theta}\frac{\partial \theta}{\partial \nu} - \left(\frac{t - \tau}{r} - 1\right)\frac{\partial u}{\partial \nu}\right\}dS$$

$$= -\int\!\!\!\int\!\!\!\int\!\!\!\int f\left(\frac{t - \tau}{r} - 1\right)dx\,dy\,dz\,dt\,,$$

where

$$\frac{\partial \theta}{\partial \nu} = \frac{\partial \theta}{\partial x}\cos(nx) + \frac{\partial \theta}{\partial y}\cos(ny) + \frac{\partial \theta}{\partial z}\cos(nz) - \frac{\partial \theta}{\partial t}\cos(nt)$$

$$(153) \qquad = -\frac{\theta}{r}\left\{\frac{x'}{r}\cos(nx) + \frac{y'}{r}\cos(ny) + \frac{z'}{r}\cos(nz)\right\} - \frac{\cos nt}{r}$$

$$= \frac{\tau - t}{r^2}\cos(nr) - \frac{\cos(nt)}{r}\,.$$

Differentiating according to τ,

$$(154) \qquad 4\pi\int_\tau^{t_S}u(\xi, \eta, \zeta, t)\,dt = \frac{\partial}{\partial \tau}\int\!\!\!\int\!\!\!\int u\left\{\frac{\tau - t}{r^2}\cos(nr) - \frac{\cos(nt)}{r}\right\}dS$$

$$+ \int\!\!\!\int\!\!\!\int \frac{1}{r}\frac{\partial u}{\partial \nu}\,dS - \int\!\!\!\int\!\!\!\int\!\!\!\int \frac{f}{r}\,dx\,dy\,dz\,dt\,,$$

which does not yet bring u out from under the integral sign. Differentiating again according to τ,

$$
\begin{aligned}
(155) \quad -4\pi u(\xi,\eta,\zeta,\tau) = & \frac{\partial^2}{\partial\tau^2} \iiint u\left\{\frac{\tau-t}{r^2}\cos(nr) - \frac{\cos(nt)}{r}\right\} dS \\
& + \frac{\partial}{\partial\tau} \iiint \frac{1}{r}\frac{\partial u}{\partial\nu}\,dS - \frac{\partial}{\partial\tau}\iiiint \frac{f}{r}\,dx\,dy\,dz\,dt,
\end{aligned}
$$

which is a generalization of Kirchhoff's formula (95) § 62. To obtain Kirchhoff's formula, we put for the hypersurface S the hyperplane $t = 0$, on which x, y, z have any values inside of a closed boundary σ (a surface in three-dimensional space) and a hypercylinder C with ge-

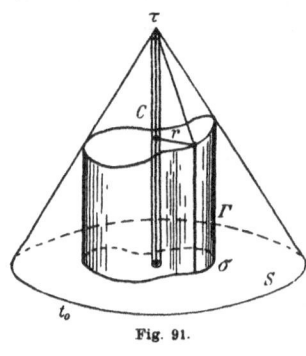

Fig. 91.

nerators parallel to the t-axis, drawn from the points of σ, Fig. 91. On the hyperplane,

$$\cos(nt) = 1,$$
$$\cos(nx) = \cos(ny) = \cos(nz) = 0,$$
$$\cos(\nu t) = -1,$$
$$\cos(\nu x) = \cos(\nu y) = \cos(\nu z) = 0.$$

On the hypercylinder, $dS = dt\,d\sigma$,
$$\cos(nt) = \cos(\nu t) = 0,$$
$$\cos(\nu x) = \cos(nx), \quad \cos(\nu y) = \cos(ny),$$
$$\cos(\nu z) = \cos(nz),$$

and the second hypersurface integral in (155) is

$$
\begin{aligned}
(156) \quad & \frac{\partial}{\partial\tau}\left[-\iiint_{t=0} \frac{1}{r}\frac{\partial u}{\partial t}\,dS + \int_0^{\tau-r} dt \iint \frac{1}{r}\frac{\partial u}{\partial n}\,d\sigma\right] \\
& = \iint \frac{1}{r}\frac{\partial u(x,y,z,\tau-r)}{\partial n}\,d\sigma
\end{aligned}
$$

the first term, over the area insite σ being independent of τ. In like manner

$$
\begin{aligned}
(157) \quad & \frac{\partial^2}{\partial\tau^2}\iiint u\left\{\frac{\tau-t}{r^2}\cos(nr) - \frac{\cos(nt)}{r}\right\} dS \\
& = \frac{\partial^2}{\partial\tau^2}\left[\int_0^{\tau-r}(\tau-t)\,dt\iint\frac{u}{r^2}\cos(nr)\,d\sigma - \iiint_{t=0}\frac{u}{r}\,dS\right].
\end{aligned}
$$

In the last integral nothing depends on τ, so that the derivative vanishes. Now

$$
(158) \quad \frac{\partial}{\partial\tau}\int_0^{\tau-r}(\tau-t)u\,dt = ru(x,y,z,\tau-r) + \int_0^{\tau-r}u(x,y,z,t)\,dt,
$$

and thus

(159)
$$\frac{\partial^2}{\partial \tau^2} \int_0^{\tau-r} (\tau - t) \int\int \frac{u \cos(nr)}{r^2}\, d\sigma = \frac{\partial}{\partial \tau} \int\int \frac{u(x,y,z,\tau-r)\cos(nr)}{r}\, d\sigma$$

$$+ \int\int \frac{u(x,y,z,\tau-r)\cos(nr)}{r^2}\, d\sigma = -\int\int \frac{\delta}{\delta n}\Big(\frac{u(x,y,z,\tau-r)}{r}\Big)\, d\sigma$$

where $\frac{\delta}{\delta n} = \cos(nr)\frac{\partial}{\partial r}$, as in § 62, equations (92) et seq.

In the same way for the quadruple integral

(160)
$$\frac{\partial}{\partial \tau} \int_0^{\tau-r} dt \int\int\int \frac{f}{r}\, dx\, dy\, dz = \int\int\int \frac{f(x,y,z,\tau-r)}{r}\, dx\, dy\, dz.$$

Thus we have finally, by (156), (159) and (160),

(161)
$$4\pi u(\xi, \eta, \zeta, \tau) = \int\int\int \frac{f(x,y,z,\tau-t)}{r}\, dx\, dy\, dz$$

$$- \int\int \frac{1}{r}\frac{\partial u(x,y,z,\tau-r)}{\partial n}\, d\sigma + \int\int \frac{\delta}{\delta n}\frac{(u(x,y,z,\tau-r))}{r}\, d\sigma$$

which is a combination of the formulae of Kirchhoff, (95) § 62, and Lorentz, (99) 63. This result is due to Tedone.[1]

We may in this case easily show that the bicharacteristics correspond to rays. The equation for the characteristic hypersurfaces for (145),

(162)
$$\Big(\frac{\partial f}{\partial x}\Big)^2 + \Big(\frac{\partial f}{\partial y}\Big)^2 + \Big(\frac{\partial f}{\partial z}\Big)^2 - \Big(\frac{\partial f}{\partial t}\Big)^2 = 0$$

has the complete integral, § 23, with four arbitrary constants,

(163)
$$f = lx + my + nz - \sqrt{l^2 + m^2 + n^2}\, t + c.$$

This complete integral represents a plane wave advancing with unr velocity, (since we have chosen our units so as to make $a = 1$) as we see by finding its distance from the origin. Its intersection P_0 with the hyperplane $t = t_0$ represents the position of the plane wave at the time t_0. The general integral is the envelope of a set of planes $f = 0$, forming a characteristic hypersurface, and its intersection σ_0 with the hyperplane $t = t_0$, shows the position of the wave at the instant t_0

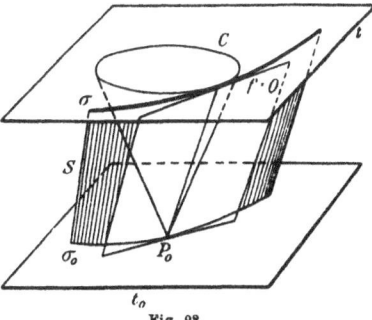

Fig. 92.

Fig. 92. The hyperplane $f = 0$ is tangent to the envelope S along a

1) Tedone, Sulla dimostrazione della formola che rappresenta analiticamente il principio di Huygens. Atti dei Lincei. Rend. ser. 5, v. 5, p. 357, 1896.

bicharacteristic, so that the point of tangency of the plane wave with the general wave moves along the bicharacteristic, and this is the definition of a ray. We may also consider the wave represented by the characteristic surface S as the envelope of all the characteristic conoids with vertices at points on the intersection σ_0 of the characteristic with the hyperplane $t = t_0$. Each conoid represents a wave starting from a particular point. The last construction gives a graphic description of Huygens's principle.

79. Volterra's Method for Equation of Damped Waves in Space.

The equation of telegraphy, or for electric waves in a conducting medium, as deduced in § 17, (190), reduced to the form § 46, (128), and generalized for n dimensions, has been treated by Coulon and Tedone.[1])

$$(164) \qquad \frac{\partial^2 u}{\partial x_1{}^2} + \frac{\partial^2 u}{\partial x_2{}^2} + \cdots + \frac{\partial^2 u}{\partial x_n{}^2} - \frac{\partial^2 u}{\partial t^2} + k^2 u = 0.$$

(We introduce the factor k^2 in order, by putting it equal to zero, to include the case of undamped waves just treated.)

If we put, as in the preceding section

$$r^2 = (x_1 - \xi_1)^2 + (x_2 - \xi_2)^2 + \cdots + (x_n - \xi_n)^2$$

we have by the result of § 54, (36)

$$\frac{\partial^2 u}{\partial r^2} + \frac{n-1}{r}\frac{\partial u}{\partial r} - \frac{\partial^2 u}{\partial t^2} + k^2 u = 0$$

and we may find solutions depending only on r and t. It will be convenient to take out a factor t^λ, by means of the equations

$$u = t^\lambda \varphi, \quad \frac{\partial u}{\partial t} = \lambda t^{\lambda-1}\varphi + t^\lambda \frac{\partial \varphi}{\partial t},$$

$$\frac{\partial^2 u}{\partial t^2} = \lambda(\lambda-1)t^{\lambda-2} + 2\lambda t^{\lambda-1}\frac{\partial \varphi}{\partial t} + t^\lambda \frac{\partial^2 \varphi}{\partial t^2}$$

which reduces our equation to the form

$$(165) \quad \frac{\partial^2 \varphi}{\partial r^2} - \frac{\partial^2 \varphi}{\partial t^2} + \frac{n-1}{r}\frac{\partial \varphi}{\partial r} - \frac{2\lambda}{t}\frac{\partial \varphi}{\partial t} + \left(k^2 - \frac{\lambda(\lambda-1)}{t^2}\right)\varphi = 0.$$

In order to obtain a solution that is constant on the characteristic cone we shall seek a solution not of the form $v(\theta)$, as in the preceding section, but of the form $\varphi = v(\theta)f(\varepsilon)$ where

---

1) Coulon. Sur l'intégration des équations aux dérivées partielles du second ordre. Thèse de doctorat, Paris 1902. Tedone. Sulla integrazione dell' equazione delle onde smorzate col metodo delle caratteristiche. Atti dei Lincei, 1913, p. 757. Ibid. 1914, p. 145.

$$\theta = \frac{t'}{r}, \quad \frac{\partial \theta}{\partial t'} = \frac{1}{r}, \quad \frac{\partial \theta}{\partial r} = -\frac{t'}{r^2} = -\frac{\theta}{r}, \quad \frac{\partial^2 \theta}{\partial t'^2} = 0, \quad \frac{\partial^2 \theta}{\partial r^2} = \frac{2t'}{r^3} = \frac{2\theta}{r^2},$$

$$(166)\ z^2 = k^2(t'^2 - r^2), \quad \frac{\partial z}{\partial t'} = \frac{k^2 t'}{z}, \quad \frac{\partial z}{\partial r} = -\frac{k^2 r}{z},$$

$$\frac{\partial^2 z}{\partial t'^2} = k^2 \left(\frac{1}{z} - \frac{k^2 t'^2}{z^3} \right) = -\frac{k^4 r^2}{z^3} \quad \frac{\partial^2 z}{\partial r^2} = -k^2 \left(\frac{1}{z} + \frac{k^2 r^2}{z^3} \right) = -\frac{k^4 t'^2}{z^3}.$$

Inserting the result of differentiating the product $\varphi = v(\theta)f(z)$ in (165) we may group the terms

$$v \left\{ \frac{\partial^2 f}{\partial r^2} - \frac{\partial^2 f}{\partial t^2} + \frac{n-1}{r} \frac{\partial f}{\partial r} - \frac{2\lambda}{t} \frac{\partial f}{\partial t} + k^2 f \right\}$$

$$(167) \qquad + 2 \left\{ \frac{\partial v}{\partial r} \frac{\partial f}{\partial r} - \frac{\partial v}{\partial t} \frac{\partial f}{\partial t} \right\}$$

$$+ f \left\{ \frac{\partial^2 v}{\partial r^2} - \frac{\partial^2 v}{\partial t^2} + \frac{n-1}{r} \frac{\partial v}{\partial r} - \frac{2\lambda}{t} \frac{\partial v}{\partial t} - \frac{\lambda(\lambda-1)}{t^2} v \right\} = 0.$$

Making use of the values of the derivatives (166) we find that the second line above vanishes, and that if we put the first and third equal to zero the variables are separated, and we obtain the two ordinary differential equations,

$$(168) \qquad \frac{d^2 f}{dz^2} + \frac{(n+2\lambda)}{z} \frac{df}{dz} - f = 0,$$

$$(169) \qquad (\theta^2 - 1) \frac{d^2 v}{d\theta^2} + \left\{ (3-n)\theta - \frac{2\lambda}{\theta} \right\} \frac{dv}{d\theta} - \frac{\lambda(\lambda-1)}{\theta^2} v = 0$$

Of these (168) is simply related to Bessel's equation (see Chapter VII) and has a solution $f(z) = \dfrac{I_{\frac{n-1}{2}+\lambda}(z)}{z^{\frac{n-1}{2}+\lambda}}$, where $I_r(z)$ represents the Bessel function

$$(170) \qquad I_r(z) = \sum_{s=0}^{s=\infty} \frac{\left(\frac{z}{2} \right)^{r+2s}}{s!(r+s)!}.$$

We shall consider only the case $n = 3$, and shall choose first $\lambda = 0$ and second $\lambda = 1$. Thus (169) reduces to $\dfrac{d^2 v}{d\theta^2} = 0$. We get a solution vanishing on the characteristic cone by taking

$$v = \theta - 1 = \frac{t-\tau}{r} - 1,$$

as in the previous section. Thus we obtain the desired solution of (165) and (164)

$$(171) \qquad u_0 = \left(\frac{t-\tau}{r} - 1 \right) \frac{I_1(z)}{z}.$$

Putting $\lambda = 1$, (169) reduces to

$$(\theta^2 - 1) \frac{d^2 v}{d\theta^2} - \frac{2}{\theta} \frac{dv}{d\theta} = 0,$$

of which a solution vanishing for $\theta = 1$ is given by

$$v = \frac{(\theta - 1)^2}{\theta}.$$

Inserting the factor $t' = \theta r$, we obtain for a second particular solution of (164)

(172) $$u_1 = \frac{r}{(t - \tau)}\left(\frac{t - \tau}{r} - 1\right)^2 \frac{I_2(z)}{z^2}.$$

We are now ready to introduce first the function u_0 and second the function u_1, into the formula (147) in which the left hand side will vanish, while the hypersurface integrals on the right are to be extended, as before to the support S, the characteristic cone Γ, and the hypercylinder C, in which we pass to the limit $\varepsilon = 0$. It is to be noted that both u_0 and u_1, while vanishing on the characteristic cone, as before become infinite like $1/r$ on the hypercylinder. Thus in the term with the conormal derivative if we write $u_0 = \psi(r, t)/r$ we have

(173) $$\frac{\partial u_0}{\partial r} = -\frac{1}{r^2}\psi + \frac{1}{r}\frac{\partial \psi}{\partial r}$$

and in passing to the limit all that remains is the first term, in which the negative power of $\varepsilon = kt'$ cancels the power of $t - \tau$. Furthermore since $\tau > t$, we must replace $t - \tau$ by $\tau - t$, which can be done since the equation (164) contains only the square of dt. The hypersurface S is below the vertex of the cone, so that we obtain in place of (152)

(174)
$$\int_{t_s}^{\tau} u(\xi, \eta, \zeta, t) I_1[k(\tau - t)]\, dt = \frac{k}{4\pi} \iiint_S \left(u\frac{\partial u_0}{\partial \nu} - u_0\frac{\partial u}{\partial \nu}\right) dS = \Phi_0,$$

$$\int_{t_s}^{\tau} u(\xi, \eta, \zeta, t) I_2[k(\tau - t)]\, dt = \frac{k}{4\pi} \iiint_S \left(u\frac{\partial u_1}{\partial \nu} - u_1\frac{\partial u}{\partial \nu}\right) dS = \Phi_1.$$

The hypersurface integrals on the right are given functions, containing the support values, but we see that as in (135) and (152), as stated at the end of 78, we do not obtain u, but it appears in an integral equation, of which the kernel is not a constant, as in (135), nor $\tau - t$, as in (152), but a Bessel function. Integral equations of the form

$$\int_{x_0}^{x} k(x, \xi) f(\xi)\, d\xi = \varphi(x)$$

like those of the first kind, § 40, (288), but with variable instead of constant limit, are known by the name of Volterra, and will be briefly treated in Chapter IX. It has been shown by Tedone that the present equation may be solved by differentiation, making use of the formulae

$$\frac{dI_0}{dz} = I_1(z), \quad 2\frac{dI_1}{dz} = I_2(z) + I_0(z),$$

from (170) on § 100, (18), (19). Since we have $I_1(0) = I_0(0) - 0$, in differentiating according to the limit, there remains only the term in which the integrand is differentiated. Accordingly we obtain from the first of (174)

$$(175) \quad \frac{\partial \Phi_0}{\partial \tau} = \int_{t_s}^{\tau} u(\xi, \eta, \zeta, t) \, k \, \frac{\partial I_1}{\partial \tau} \, dt = \frac{k}{2} \int_{t_s}^{\tau} u(I_2 + I_0) \, dt = \frac{k}{2} \, \Phi_1 + \frac{k}{2} \int_{t_s}^{\tau} u I_0 \, dt.$$

Differentiating again, and noticing that $I_0(0) = 1$, we have finally the desired formula,

$$(176) \qquad u = \frac{2}{k} \frac{\partial^2 \Phi_0}{\partial \tau^2} - k \Phi_0 - \frac{\partial \Phi_1}{\partial t}.$$

From this formula it would be possible to show that, as in the case of one dimension, the wave leaves a residue behind it.

80. Hadamard's Method. Fundamental Solutions. We now come to a method which we can only briefly sketch, referring for completeness to a number of remarkable papers by Hadamard[1]) for details.

We have seen the importance for the elliptic equation of Laplace of the fundamental solution $1/r$ for the case of three variables and $\log r$ for the case of two variables, and how the theorem of Green enables us to solve the boundary problem. We have also seen in equations (139) *et seq.* of the previous section how we may pass from the elliptic to the hyperbolic equation, and solve the problem of Cauchy. There is however an important difference. In the elliptic case, the bounding surface is closed, and we cannot give both the function and the normal derivative on the boundary. In the hyperbolic case the supporting boundary is not closed, we can give both the function and the conormal derivative, while the characteristic conoid is here real, and occurs in the boundary integral. Moreover, if we take in the hyperbolic case as the fundamental solution r' or $\log r'$ where r' is the codistance of the two points x, y, z and ξ, η, ζ, we find that this solution becomes infinite on the characteristic cone. This is the reason that it was not used by Volterra, but rather its integral with respect to ζ, so that we arrive at an integral equation. In fact when we come to effect the differentiation in the case of Volterra or of the Poisson-Parseval solution (144) or (188) § 49, we are confronted with a difficulty which we shall now elucidate. Consider the simple integral

$$\int^s \frac{dx}{(b - x)^{1 + \alpha}} = \frac{1}{\alpha (b - x)^\alpha}$$

1) Hadamard. Les problèmes aux limites dans la théorie des équations aux dérivées partielles. Journal de Physique, 6, p. 202, 1907. Théorie des équations aux dérivées partielles linéaires hyperboliques et du problème de Cauchy. Acta Mathematica, 31, p. 333, 1908.

in which $0 < \alpha < 1$. If we consider the definite integral

$$\int_a^b \frac{dx}{(b-x)^{1+\alpha}}$$

the insertion of the limit b gives rise to an infinite term, so that the integral has no meaning. The same thing is true of the integral

(177)
$$\int_a^b \frac{A(x)\,dx}{(b-x)^{1+\alpha}}\;.$$

It is however possible to add to the integral a function

$$\frac{B(x)}{(b-x)^\alpha}$$

which also becomes infinite for $x = b$ so that the sum of the two shall remain finite, that is,

$$\lim_{x=b}\left[\int_a^x \frac{A(x)\,dx}{(b-x)^{1+\alpha}} + \frac{B(x)}{(b-x)^\alpha}\right]$$

is a finite quantity. In fact, if we add and subtract $A(b)$ to and from the numerator in the integral, and $B(b)$ outside the integral, the above expression becomes

(178)
$$\int_a^x \frac{A(x)-A(b)}{(b-x)^{1+\alpha}}\,dx + \frac{B(x)-B(b)}{(b-x)^\alpha} + \frac{A(b)}{\alpha}\left(\frac{1}{(b-x)^\alpha} - \frac{1}{(b-a)^\alpha}\right)$$
$$+ \frac{B(b)}{(b-x)^\alpha}\;.$$

Now the two numerators $A(x) - A(b)$ and $B(x) - B(b)$ contain $x - b$ as a factor, hence the integral remains finite, and the first term after the integral vanishes when we put $x = b$, the only terms that become infinite are then those in $1/(b-x)^\alpha$, and these will disappear if we take

$$\frac{A(b)}{\alpha} + B(b) = 0.$$

Consequently our integral tends toward the limit

(179)
$$\int_a^b \frac{A(x)-A(b)}{(b-x)^{1+\alpha}}\,dx - \frac{A(b)}{\alpha(b-a)^\alpha}$$

which is entirely independent of the function $B(x)$, and is defined solely by the integral (177). It is therefore called by Hadamard the *finite part* of the integral (177) and denoted by the symbol

$$(180) \qquad \overline{\int_a^b \frac{A(x)\,dx}{(b-x)^{1+\alpha}}} \, .$$

The same notion is easily extended to integrals in which the exponent in the denominator is $p + \alpha$, where p is an integer, and to multiple integrals, in which the n-tuple integral is completed by an n-l-tuple integral in an analogous manner.

As an example let us consider the equation for waves in two dimensions, already treated by the method of Volterra.

$$\frac{\partial^2 u}{\partial x^2} + \frac{\partial^2 u}{\partial y^2} - \frac{\partial^2 u}{\partial z^2} = 0.$$

If we write

$$\Gamma = (\xi - x)^2 + (\eta - y)^2 - (\zeta - z)^2,$$

we have already seen that a fundamental solution is given by the co-distance $1/r' = \Gamma^{-\frac{1}{2}}$. Bearing in mind that

$$\frac{\partial}{\partial n} = \cos(nx)\frac{\partial}{\partial x} + \cos(ny)\frac{\partial}{\partial y} - \cos(nz)\frac{\partial}{\partial z}$$

and inserting in the fundamental formula,

$$v = \Gamma^{-\frac{1}{2}}\frac{\partial v}{\partial n} = \Gamma^{-\frac{3}{2}}\{\cos(nx)(\xi - x) + \cos(ny)(\eta - y) + \cos(nz)(\zeta - z)\}$$

we obtain,

$$(181) \qquad 0 = \int\int \left[\frac{1}{\sqrt{\Gamma}}\frac{\partial u}{\partial n} - \frac{u}{\Gamma^{\frac{3}{2}}} \right.$$

$$\left. \{\cos(nx)(\xi - x) + \cos(ny)(\eta - y) + \cos(nz)(\zeta - z)\} \right] dS.$$

Now at the vertex of the characteristic cone $\Gamma^{-\frac{3}{2}}$ is infinite of higher order, and this point is to be excluded by a small sphere. Passing to the limit this adds the term $2\pi u(\xi, \eta, \zeta)$ agreeing with (136). We consequently have

$$2\pi u(\xi, \eta, \zeta) = \int\int \left[\frac{1}{\sqrt{\Gamma}}\frac{\partial u}{\partial n} - \frac{u}{\Gamma^{\frac{3}{2}}}\{\cos(nx)(\xi - x)\cdots\} \right] dS$$

where the second term is an integral like (177), in which $\alpha = 1/2$ and the upper limit is the intersection of the support with the cone $\Gamma = 0$. If the support is the plane $z = 0$, and on it we introduce polar coordinates $x - \xi = r \cos \varphi$, $\eta - y = r \sin \varphi$ we find the finite part of the integral

$$(182) \quad \overline{\int_0^\zeta \int_0^{2\pi} \frac{\zeta u r \, dr \, d\varphi}{(r^2 - \zeta^2)^{\frac{3}{2}}}} = \int_0^\zeta \int_0^{2\pi} \frac{\zeta (u(r) - u(\zeta)) r \, dr \, d\varphi}{(r^2 - \zeta^2)^{\frac{3}{2}}} - \int_0^{2\pi} u(\zeta) \, d\varphi$$

according to (179). Accordingly we have the solution of Cauchy's problem

$$(183) \quad 2\pi u(\xi, \eta, \zeta) = \int_0^\zeta \int_0^{2\pi} \left[\frac{1}{\sqrt{\Gamma}} \frac{\partial u}{\partial z} - \frac{\zeta \{ u(r) - u(\zeta) \}}{(r^2 - \zeta^2)^{\frac{3}{2}}} \right] r \, dr \, d\varphi$$

$$- \int_0^{2\pi} u(\zeta) \, d\varphi.$$

For the explanation of the reason for ignoring the integral on the cone $\Gamma = 0$ we must refer to Hadamard's paper in the Acta Mathematica, in which the general case is dealt with in a very profound manner.

81. Propagation of Discontinuities. Until now we have tacitly assumed that the solutions of our differential equations are continous functions of the variables representing both time and space. Also in the deduction of the equations of wave motion we have assumed in § 12 that the displacements are very small quantities, in order that the differential equations may be linear. But it is often desirable to free ourselves from this limitation, and to deduce the equations in a more careful manner. We shall find that in the propagation of waves of sound in a gas where the motion of the particles is not infinitely small but finite we have phenomena of quite another nature to consider, and especially that we must consider the possibility of discontinuities in our functions. Such phenomena have been investigated by Riemann [1]) and by Hugoniot, and inasmuch as they have been made the subject of a whole treatise by Hadamard (Leçons sur la propagation des ondes, Paris 1903) it is obvious that the merest sketch will here be possible.

---

1) Riemann, Über die Fortpflanzung ebener Luftwellen von endlicher Schwingungsweite, *Abh. Königl. Ges. der Wiss. Göttingen,* Bd. 8, 1860. Hugoniot, Sur la propagation du mouvement dans les corps. *Journal de l'École Polytechnique* 57, 1887, Mémoire sur la propagation du mouvement dans un fluide indefini. *Journal de Mathématiques* 3—4, 1887. It is of interest to know that Hugoniot was an artillery officer, and was led to these researches through his interest in the motion of the powder gases in a gun.

The possibility of the arising of discontinuities may be seen when we consider the simplest case of a gas enclosed by a material surface or envelope to which a certain motion is imparted. Let us take propagation in one dimension in a cylindrical tube, with movable pistons at the points $x = 0$ and $x = l$, in which u denotes the displacement of the air, and under the conditions previously supposed we have the equation

$$\frac{\partial^2 u}{\partial t^2} = a^2 \frac{\partial^2 u}{\partial x^2}.$$

If now as in § 26 we give for a certain instant the positions and velocities of the particles by putting $u = F(x)$, $\frac{\partial u}{\partial t} = G(x)$, by differentiation according to x we shall find the acceleration as $\frac{d^2 u}{dt^2} = F''(x)$. Now if the motion of either piston is arbitrarily prescribed, so that its acceleration and that the air in contact with it does not agree with that just found it is evident that there will be a discontinuity in the acceleration, and that waves will originate at the piston which would be quite different from those that would be propagated in a cylinder of infinite length. In fact we have treated the case of a fixed boundary, at which the acceleration of the air was obliged to be zero, while that due to the oncoming wave was not.
As a result we had a reflected wave, given by a quite different function of x and t. But we also observe cases in which not only the acceleration, but even the velocity and density become discontinuous at certain surfaces. We have a noteworthy case of this in the motion of a rifle bullet, which, moving faster than

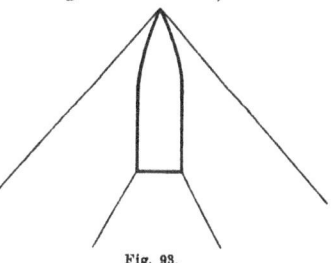

Fig. 93.

the velocity of sound, is accompanied by a so-called wave of shock (*onde de choc*, *Stoßwelle*) in which the sudden change of density is made evident by the change of index of refraction, Fig. 93 (made by an instantaneous electric spark). The sudden increase of density caused by a large shot may produce startling mechanical effects.

For greater convenience instead of the equations of hydrodynamics in the form associated with the name of Euler, Chapter I 106, 107, we shall adopt the form associated with the name of Lagrange, in which the coordinates x, y, z denote, not a fixed point in space, but the variable coordinates of a particular particle of fluid. The particle is identified by having its coordinates given as functions of three variable parameters (the method of curvilinear coordinates is described at length in the next chapter) a, b, c, which may be conveniently, though not necessarily, thought of as the values of x, y, z at an instant $t = 0$. We are then to put

$$x = x(a, b, c, t), \quad y = y(a, b, c, t), \quad z = z(a, b, c, t),$$

so that if for a given value of t we let a, b, c take all possible values we shall get the coordinates of all points in space, while if t varies the values of x, y, z for fixed a, b, c will give the varying coordinates of the same material particle. We shall then use the notation of ordinary derivatives to denote changes referring to a definite particle as it moves about, while we keep the sign of partial differentiation for its usual significance. Thus the components of velocity will be

$$\mathbf{v}_x = \frac{dx}{dt}, \quad \mathbf{v}_y = \frac{dy}{dt}, \quad \mathbf{v}_z = \frac{dz}{dt}$$

and of acceleration

$$\frac{d^2x}{dt^2}, \quad \frac{d^2y}{dt^2}, \quad \frac{d^2z}{dt^2},$$

instead of the four terms on the left of (106) Chapter I. Accordingly we may write the hydrodynamical equations as

$$(184) \quad \frac{d^2x}{dt^2} = F_x - \frac{1}{\varrho}\frac{\partial p}{\partial x}, \quad \frac{d^2y}{dt^2} = F_y - \frac{1}{\varrho}\frac{\partial p}{\partial y}, \quad \frac{d^2z}{dt^2} = F_z - \frac{1}{\varrho}\frac{\partial p}{\partial z},$$

which may be abbreviated into the single vector equation, in which $\mathbf{r}$ denotes the position vector of the particle,

$$(185) \qquad\qquad \frac{d^2\mathbf{r}}{dt^2} = \mathbf{F} - \frac{1}{\varrho}\operatorname{grad} p.$$

In order to obtain the equation of continuity consider the fluid that was originally contained in a rectangular parallelopiped whose extreme coordinates are a, b, c and $a + da$, $b + db$, $c + dc$. At any subsequent instant the first point has become x, y, z, the parallelopiped has become oblique, and the edge of length da which was parallel to the x-axis has obtained the projections

$$\frac{\partial x}{\partial a}\,da, \quad \frac{\partial y}{\partial a}\,da, \quad \frac{\partial z}{\partial a}\,da,$$

with corresponding results for the two others. Accordingly, by (15) Chapter I, the volume has become

$$\begin{vmatrix} \dfrac{\partial x}{\partial a}, & \dfrac{\partial y}{\partial a}, & \dfrac{\partial z}{\partial a} \\[2ex] \dfrac{\partial x}{\partial b}, & \dfrac{\partial y}{\partial b}, & \dfrac{\partial z}{\partial b} \\[2ex] \dfrac{\partial x}{\partial c}, & \dfrac{\partial y}{\partial c}, & \dfrac{\partial z}{\partial c} \end{vmatrix} \; da\,db\,dc$$

and since the mass is unchanged, and the ratio of the densities is in versely as that of the volumes, we have

$$(186) \qquad \omega = \frac{\varrho_0}{\varrho} = \begin{vmatrix} \dfrac{\partial x}{\partial a}, & \dfrac{\partial x}{\partial b}, & \dfrac{\partial x}{\partial c} \\[2mm] \dfrac{\partial y}{\partial a}, & \dfrac{\partial y}{\partial b}, & \dfrac{\partial y}{\partial c} \\[2mm] \dfrac{\partial z}{\partial a}, & \dfrac{\partial z}{\partial b}, & \dfrac{\partial z}{\partial c} \end{vmatrix}$$

We may call ω the ratio of dilatation, or ω/ϱ_0 the specific volume.

82. Discontinuity of the first order. Suppose that the displacement affects particles which were originally on a surface

$$F(a, b, c, t) = 0,$$

and let φ be a scalar function of the point which is continuous on crossing from the side 1 to the side 2 of the surface, that is $\varphi_2 = \varphi_1$. We shall use the sign δ to denote the discontinuity of a function, $\delta \psi = \psi_2 - \psi_1$.

Now if φ is continuous on the surface, we may differentiate it totally on either side in a direction tangential to the surface.

$$(187) \qquad \begin{aligned} d\varphi_1 &= \frac{\partial \varphi_1}{\partial a} \, da + \frac{\partial \varphi_1}{\partial b} \, db + \frac{\partial \varphi_1}{\partial c} \, dc, \\[2mm] d\varphi_2 &= \frac{\partial \varphi_2}{\partial a} \, da + \frac{\partial \varphi_2}{\partial b} \, db + \frac{\partial \varphi_2}{\partial c} \, dc. \end{aligned}$$

But since at all points on the surface $\varphi_2 = \varphi_1$, these two values are equal, so that by subtraction

$$\left(\frac{\partial \varphi_2}{\partial a} - \frac{\partial \varphi_1}{\partial a} \right) da + \left(\frac{\partial \varphi_2}{\partial b} - \frac{\partial \varphi_1}{\partial b} \right) db + \left(\frac{\partial \varphi_2}{\partial c} - \frac{\partial \varphi_1}{\partial c} \right) dc = 0,$$

that is

$$\delta \frac{\partial \varphi}{\partial a} \, da + \delta \frac{\partial \varphi}{\partial b} \, ab + \delta \frac{\partial \varphi}{\partial c} \, ac = 0,$$

for all points for which

$$\frac{\partial F}{\partial a} \, da + \frac{\partial F}{\partial b} \, db + \frac{\partial F}{\partial a} \, dc = 0.$$

That is, the coefficients of da, db, dc in these two equations must be proportional, or the vector whose components are $\delta \dfrac{\partial \varphi}{\partial a}$, $\delta \dfrac{\partial \varphi}{\partial b}$, $\delta \dfrac{\partial \varphi}{\partial c}$ must be normal to the surface.

We may call this the jump of grad φ.

Now since we have

$$\cos (nx) = \frac{\partial F}{\partial a} \Big/ h, \quad \cos (ny) = \frac{\partial F}{\partial b} \Big/ h, \quad \cos (nz) = \frac{\partial F}{\partial c} \Big/ h,$$

$$h^2 = \left(\frac{\partial F}{\partial a}\right)^2 + \left(\frac{\partial F}{\partial b}\right)^2 + \left(\frac{\partial F}{\partial c}\right)^2,$$

if λ is a factor of proportionality,

$$(188) \quad \delta \frac{\partial \varphi}{\partial a} = \lambda \cos (nx) = \frac{\lambda}{h} \frac{\partial F}{\partial a}, \quad \delta \frac{\partial \varphi}{\partial b} = \lambda \cos (nx) = \frac{\lambda}{h} \frac{\partial F}{\partial b},$$

$$\delta \frac{\partial \varphi}{\partial c} = \lambda \cos (nx) = \frac{\lambda}{h} \frac{\partial F}{\partial c},$$

which may be embraced in the single vector equation,

$$(188') \qquad\qquad \delta \nabla \varphi = \lambda \, \mathfrak{n}_1$$

where $\nabla \varphi$ stands for the gradient of φ (Chapter I, 23) and $\mathfrak{n}_1$ is a vector of unit length in the direction of the normal.

Similar reasoning may be applied to the components of a vector function, all of which are continuous at the surface, but whose derivatives are discontinuous. Corresponding to λ we may take three multipliers H_x, H_y, H_z, so that

$$\delta \frac{\partial x}{\partial a} = H_x \cos (nx), \quad \delta \frac{\partial x}{\partial b} = H_x \cos (ny), \quad \delta \frac{\partial x}{\partial c} = H_x \cos (nz),$$

$$(189) \; \delta \frac{\partial y}{\partial a} = H_y \cos (nx), \quad \delta \frac{\partial y}{\partial b} = H_y \cos (ny), \quad \delta \frac{\partial y}{\partial a} = H_y \cos (nz),$$

$$\delta \frac{\partial z}{\partial a} = H_z \cos (nx), \quad \delta \frac{\partial z}{\partial b} = H_z \cos (ny), \quad \delta \frac{\partial y}{\partial c} = H_z \cos (nz).$$

If x, y, z are the components of a vector $\mathbf{D}$, and $\mathbf{A}$ is an *arbitrary* vector these nine equations may be compressed into the single vector formula.

$$(190) \qquad\qquad \delta (\nabla \mathbf{D}_A) = \mathfrak{n}_1 H_A,$$

as is seen if we give $\mathbf{A}$ successively the directions of x, y, z. The equations (189) or (190), in which $\mathbf{D}$ represents the position vector of a particle, continuous at the surface, but with discontinuous derivatives of the first order, are called the *identical conditions*. Such a discontinuity is called a discontinuity of the first order. It is completely characterised by giving the vector $\mathbf{H}$ at each point in the surface.

The identical conditions follow from the mere fact that the discontinuities are distributed over some surface. If in addition as the time changes, there remains some single surface, not necessarily containing the same particles, nor remaining fixed in space, on which the discontinuities are still distributed, we shall obtain Hugoniot's conditions of *kinematic compatibility*. We may suppose that the equation

$$F(a, b, c, t) = 0,$$

denotes either a moving surface in three dimensional space, which we define as the wave of discontinuity, or as we have before done, a fixed hypersurface in the space of four dimensions a, b, c, t. Then as before for any continuous function $\delta\varphi = 0$,

$$(191) \qquad d\delta\varphi = \delta \frac{\partial\varphi}{\partial a}\, da + \delta \frac{\partial\varphi}{\partial b}\, db + \delta \frac{\partial\varphi}{\partial c}\, dc + \delta \frac{\partial\varphi}{\partial t}\, dt = 0$$

when

$$(192) \qquad \frac{\partial F}{\partial a}\, da + \frac{\partial F}{\partial b}\, db + \frac{\partial F}{\partial c}\, dc + \frac{\partial F}{\partial t}\, dt = 0$$

and as before we conclude that the hypergradient has a jump in the direction of the normal to the hypersurface, and accordingly

$$\delta \frac{\partial\varphi}{\partial a} = \frac{\lambda}{h} \frac{\partial F}{\partial a}, \quad \delta \frac{\partial\varphi}{\partial b} = \frac{\lambda}{h} \frac{\partial F}{\partial b}, \quad \delta \frac{\partial\varphi}{\partial c} = \frac{\lambda}{h} \frac{\partial F}{\partial c}, \quad \delta \frac{\partial\varphi}{\partial t} = \frac{\lambda}{h} \frac{\partial F}{\partial t}.$$

Applying this to the components of a vector, itself continuous, we find in addition to the equations (189)

$$(189a) \qquad \delta \frac{dx}{dt} = \frac{H_x}{h} \frac{\partial F}{\partial t}, \quad \delta \frac{dy}{dt} = \frac{H_y}{h} \frac{\partial F}{\partial t}, \quad \delta \frac{dz}{dt} = \frac{H_z}{h} \frac{\partial F}{\partial t}.$$

If x, y, z are the components of the displacement, this gives us the components of the jump of the velocity, which is a vector in the direction of $\mathbf{H}$, and equal to $\mathbf{H}$ multiplied by $\frac{1}{h} \frac{\partial F}{\partial t}$. The meaning of this factor we shall see if we put in (192) for da, db, dc the components of the distance dn that the wave has moved in the space a, b, c,

$$da = dn \cos(nx), \quad db = dn \cos(ny) \quad dc = dn \cos(nz).$$

Since we have

$$\frac{\partial F}{\partial a} = h \cos(nx), \quad \frac{\partial F}{\partial b} = h \cos(ny), \quad \frac{\partial F}{\partial c} = h \cos(nz),$$

this gives

$$(193) \qquad -\frac{1}{h} \frac{\partial F}{\partial t} = \frac{dn}{dt} = \theta,$$

but this is the velocity of propagation of the wave. Accordingly a discontinuity of the first order is completely defined by giving a vector $\mathbf{H}$ and a scalar θ, which result in the twelve equations of kinematic compatibility (189) (192). If the vector $\mathbf{H}$ is normal, the discontinuity of the first order is said to be longitudinal, if tangential it is said to be transverse.

The velocity θ is that of the propagation of the wave with respect to the particles of the fluid. In order to get its actual velocity of displacement with regard to fixed space, we must consider the actual coordinates x, y, z. Let the equation of the wave be

$$\Phi(x, y, z, t) = 0,$$

then as before the velocity of displacement will be found to be

(194)
$$T = - \frac{\dfrac{\partial \Phi}{\partial t}}{\sqrt{\left(\dfrac{\partial \Phi}{\partial x}\right)^2 + \left(\dfrac{\partial \Phi}{\partial y}\right)^2 + \left(\dfrac{\partial \Phi}{\partial z}\right)^2}}.$$

From these velocities we can find the discontinuity of the density in a wave of the first order. If S, Fig. 94, is the position of the wave at

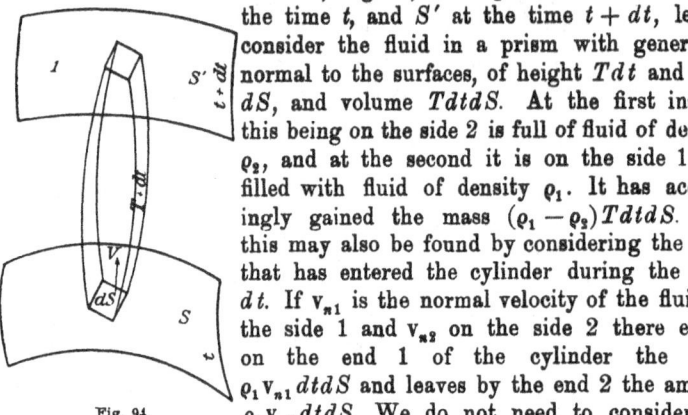

Fig. 94.

the time t, and S' at the time $t + dt$, let us consider the fluid in a prism with generators normal to the surfaces, of height $T dt$ and base dS, and volume $T dt dS$. At the first instant this being on the side 2 is full of fluid of density ϱ_2, and at the second it is on the side 1 and filled with fluid of density ϱ_1. It has accordingly gained the mass $(\varrho_1 - \varrho_2) T dt dS$. But this may also be found by considering the fluid that has entered the cylinder during the time dt. If v_{n1} is the normal velocity of the fluid on the side 1 and v_{n2} on the side 2 there enters on the end 1 of the cylinder the mass $\varrho_1 v_{n1} dt dS$ and leaves by the end 2 the amount $\varrho_2 v_{n2} dt dS$. We do not need to consider the amounts entering through the convex surface of the cylinder, which are infinitesimal in comparison. Equating the gain measured in the two ways we find

$$(\varrho_1 - \varrho_2)\, T dt dS = (\varrho_1 v_{n1} - \varrho_2 v_{n2}) dt dS,$$
$$\varrho_1(T - v_{n1}) = \varrho_2(T - v_{n2}).$$

But $T - v_{n1} = \theta_1$ is the velocity of propagation of the discontinuity with respect to the *substance* on the side 1, while $T - v_{n2} = \theta_2$ is the velocity with respect to the substance on the side 2, so that we have

(195)
$$\varrho_1 \theta_1 = \varrho_2 \theta_2.$$

83. Discontinuity of second order. Let us now suppose that all the derivatives of the first order are continuous, but that those of the second order are discontinuous. Treating the derivative $\partial x/\partial a$ as we before treated φ and differentiating along the surface $F = 0$, we find

$$\delta\, \frac{\partial^2 x}{\partial a^2}\, da + \delta\, \frac{\partial^2 x}{\partial a \partial b}\, db + \delta\, \frac{\partial^2 x}{\partial a \partial c}\, dc = 0$$

when
$$\frac{\partial F}{\partial a}\, da + \frac{\partial F}{\partial b}\, db + \frac{\partial F}{\partial c}\, dc = 0.$$

Accordingly
$$\frac{\delta \dfrac{\partial^2 x}{\partial a^2}}{\dfrac{\partial F}{\partial a}} = \frac{\delta \dfrac{\partial x}{\partial a \partial b}}{\dfrac{\partial F}{\partial b}} = \frac{\delta \dfrac{\partial^2 x}{\partial a \partial c}}{\dfrac{\partial F}{\partial c}}.$$

In like manner operating on $\partial x/\partial b$ and $\partial x/\partial c$ we find similar proportionalities for the other derivatives.

$$\frac{\delta \dfrac{\partial^2 x}{\partial a \partial b}}{\dfrac{\partial F}{\partial a}} = \frac{\delta \dfrac{\partial^2 x}{\partial b^2}}{\dfrac{\partial F}{\partial b}} = \frac{\delta \dfrac{\partial^2 x}{\partial b \partial c}}{\dfrac{\partial F}{\partial c}}, \text{ etc.}$$

These are all satisfied if we put

$$\delta \frac{\partial^2 x}{\partial a^2} = \frac{H_x}{h^2}\left(\frac{\partial F}{\partial a}\right)^2, \quad \delta \frac{\partial^2 x}{\partial b^2} = \frac{H_x}{h^2}\left(\frac{\partial F}{\partial b}\right)^2, \quad \delta \frac{\partial^2 x}{\partial c^2} = \frac{H_x}{h^2}\left(\frac{\partial F}{\partial c}\right)^2$$

(196) $\qquad \delta \dfrac{\partial^2 x}{\partial b \partial c} = \dfrac{H_x}{h^2}\dfrac{\partial F}{\partial b}\dfrac{\partial F}{\partial c}, \quad \delta \dfrac{\partial^2 x}{\partial c \partial a} = \dfrac{H_x}{h^2}\dfrac{\partial F}{\partial c}\dfrac{\partial F}{\partial a},$

$$\delta \frac{\partial^2 x}{\partial a \partial b} = \frac{H_x}{h^2}\frac{\partial F}{\partial a}\frac{\partial F}{\partial b}.$$

Similarly for the derivatives of y and z, if we replace the multiplier H_x by others H_y and H_z respectively. Accordingly the discontinuities in the derivatives according to a, b, c are defined by a vector H.

We come now to the derivatives of the velocity $\dfrac{dx}{dt}$, $\dfrac{dy}{dt}$, $\dfrac{dz}{dt}$ with respect to a, b, c. Since we suppose $\delta \dfrac{dx}{dt} = 0$, we have as before in (189)

(197) $\qquad \delta \dfrac{\partial}{\partial a}\dfrac{dx}{dt} = \dfrac{H_x'}{h}\dfrac{\partial F}{\partial a}, \quad \delta \dfrac{\partial}{\partial b}\dfrac{dx}{dt} = \dfrac{H_x'}{h}\dfrac{\partial F}{\partial b}, \quad \delta \dfrac{\partial}{\partial c}\dfrac{dx}{dt} = \dfrac{H_x'}{h}\dfrac{\partial F}{\partial c},$

$\qquad \delta \dfrac{\partial}{\partial a}\dfrac{dx}{dt} = \dfrac{H_y'}{h}\dfrac{\partial F}{\partial a}, \quad \delta \dfrac{\partial}{\partial b}\dfrac{dy}{dt} = \dfrac{H_y'}{h}\dfrac{\partial F}{\partial b}, \quad \delta \dfrac{\partial}{\partial c}\dfrac{dy}{dt} = \dfrac{H_y'}{h}\dfrac{\partial F}{\partial c}$ etc.,

so that these discontinuities are defined by a vector H'. The accelerations have discontinuities which constitute a third vector H''. These three vectors are not independent, but are connected by the equations of kinematic compatibility. As before, let us change a, b, c, t so that

$$\frac{\partial F}{\partial a}\, da + \frac{\partial F}{\partial b}\, db + \frac{\partial F}{\partial c}\, dc + \frac{\partial F}{\partial t}\, dt = 0.$$

We then obtain as before

$$\frac{\delta \dfrac{\partial^2 x}{\partial a^2}}{\dfrac{\partial F}{\partial a}} = -\frac{\delta \dfrac{\partial}{\partial a}\dfrac{dx}{dt}}{\dfrac{\partial F}{\partial t}}$$

and making use of the values of these ratios from the first of (196) and (197) we obtain

(198) $H_x'^i - \dfrac{H_x}{h}\dfrac{\partial F}{\partial t} = -\theta H_x,$ $H_y' = -\theta H_y,$ $H_z' = -\theta H_z.$

Accordingly the second vector $\mathbf{H}'$ is in the same direction as $\mathbf{H}$ and equal to it multiplied by the velocity.

Proceding in the same way with $H_x'' = \dfrac{d^2 x}{dt^2}$ we obtain similarly

(199) $H_z'' = -\theta H_z',$ $H_y'' = -\theta H_y',$ $H_z'' = -\theta H_z',$

and conclude that the third vector $\mathbf{H}$ representing the jump of acceleration is related to the second vector in the same manner. The three vectors $\mathbf{H}$, $\mathbf{H}'$, $\mathbf{H}''$ accordingly are in the same or contrary directions, and their magnitudes form a geometric progression of ratio θ equal to the velocity of propagation of the wave with respect to the fluid.

Just as we may obtain the jump in the divergence of a vector which is itself continuous in a wave of the first order by adding the diagonal terms in (189)

$$\delta \operatorname{div} D = \delta\left(\frac{\partial x}{\partial a} + \frac{\partial y}{\partial b} + \frac{\partial z}{\partial c}\right) = \mathbf{H}_x \cos(nx) + \mathbf{H}_y \cos(ny)$$
$$+ \mathbf{H}_z \cos(nz) = \mathbf{H}_n,$$

so in a wave of the second order we obtain the jump in the divergence of the velocity, itself continuous, from equations (197), which are the analogues of (189) for the velocity. Accordingly

(200) $$\delta \operatorname{div} v = \mathbf{H}_n'.$$

84. Hugoniot's Theorem for Velocity. A remarkable result was enunciated by Hugoniot, to the effect that the velocity of propagation of waves of acceleration may be found without integrating the differential equations of motion, by means of the so-called dynamical equations of compatibility. If the pressure is connected with the density by a physical relation $p = \varphi(\varrho)$, so that $dp/d\varrho$ is not a constant, as assumed in § 12, the equations of motion (184) become

(201) $$\frac{d^2 x}{dt^2} = F_x - \varphi'(\varrho)\frac{\partial \log \varrho}{\partial x}, \quad \frac{d^2 y}{dt^2} = F_y - \varphi'(\varrho)\frac{\partial \log \varrho}{\partial y},$$
$$\frac{d^2 z}{dt^2} = F_z - \varphi'(\varrho)\frac{\partial \log \varrho}{\partial z}.$$

Applying these to the two sides of a surface of discontinuity of the second order (the applied forces and $\varphi'(\varrho)$ being of course continuous) we obtain,

(202) $$\mathbf{H}'' = \delta\frac{dv}{dt} = -\varphi'(\varrho)\delta \nabla \log \varrho$$

Applying now to the scalar $\log \varrho$ the identical conditions, as in (188)

$$(203) \qquad\qquad \delta \nabla \log \varrho = \lambda \, \mathbf{n}_1$$

for the space derivatives, and the kinematical conditions of compatibility for the time derivative

$$(204) \qquad\qquad \delta \, \frac{d \log \varrho}{dt} = - \lambda \, \theta$$

the multiplier λ being the same in both cases, and θ the velocity of the wave, which we wish to find.

Now the equation of continuity, § 12, (107), combined with the meaning of the first two terms, as explained in § 8, (77), may be written

$$(205) \qquad\qquad \frac{d \log \varrho}{dt} = - \operatorname{div} \mathbf{v},$$

from which we obtain the jump

$$(206) \qquad\qquad \delta \, \frac{d \log \varrho}{dt} = - \delta \operatorname{div} \mathbf{v},$$

or by equation (200)

$$(207) \qquad\qquad \delta \, \frac{d \log \varrho}{dt} = - \mathbf{H}_n{}'.$$

Comparing this with (204) we determine as

$$\lambda = \frac{1}{\theta} \, \mathbf{H}_n{}',$$

inserting which in (203) gives

$$(208) \qquad\qquad \delta \nabla \log \varrho = \frac{1}{\theta} \, \mathbf{H}_n{}' \, \mathbf{n}_1 .$$

Now inserting this value in the dynamical equation of compatibility (202) with (199)

$$(209) \qquad\qquad \theta^2 \mathbf{H}' = \varphi'(\varrho) \mathbf{H}_n{}' \, \mathbf{n}_1 .$$

This vector equation shows that either $\mathbf{H}'$ is in the direction of the normal, when $\mathbf{H}_n{}' = |\, \mathbf{H}\, |$, the discontinuity is longitudinal, and is propagated with the velocity

$$(210) \qquad\qquad \theta = \sqrt{\varphi'(\varrho)} = \sqrt{\frac{dp}{d\varrho}}$$

or else $\mathbf{H}_n = 0$, when the discontinuity is transverse, and $\theta = 0$, that is the discontinuity is stationary, and always affects the same particles. The equation (210) is Laplace's celebrated formula for the velocity of sound, and agrees with what we have previously found for waves of infinitely small amplitude. We see that in general the velocity varies with the density, and increases with it.

85. Propagation in one Dimension. Characteristics. The propagation of motion parallel to a single direction, such as that of a gas moving by slices in a cylinder, has been thoroughly investigated by Hugoniot and Riemann. Suppose that u represents the displacement parallel to the axis of the cylinder and suppose that it satisfies the general equation (1) of this chapter.

$$(211) \qquad A\frac{\partial^2 u}{\partial x^2} + 2B\frac{\partial u}{\partial x}\frac{\partial u}{\partial t} + C\frac{\partial^2 u}{\partial t^2} = F\left(x, t, u, \frac{\partial u}{\partial x}, \frac{\partial u}{\partial t}\right).$$

Let there be a surface of discontinuity of the second order, on one side of which there is a motion given by an integral u_1 and on the other a different motion given by an integral u_2. In order that these motions may be propagated into each other, in the words of Hugoniot, they must satisfy the equations of compatibility. Now since at the surface of discontinuity,

$$\delta u = 0, \quad \delta\frac{\partial u}{\partial x} = 0, \quad \delta\frac{\partial u}{\partial t} = 0,$$

the equations of compatibility are

$$(212) \quad \delta\frac{\partial^2 u}{\partial x^2}\, dx + \delta\frac{\partial^2 u}{\partial x\partial t}\, dt = 0, \quad \delta\frac{\partial^2 u}{\partial x\partial t}\, dx + \delta\frac{\partial^2 u}{\partial t^2} dt = 0.$$

But since both integrals satisfy the equation (211), of which the second member, containing only continuous functions, is the same for each, we have

$$A\delta\frac{\partial^2 u}{\partial x^2} + 2B\delta\frac{\partial^2 u}{\partial x\partial t} + C\delta\frac{\partial^2 u}{\partial t^2} = 0.$$

Now these three equations for the three jumps cannot be satisfied unless the determinant of their coefficients vanishes, that is

$$(213) \qquad C\, dx^2 - 2B\, dx\, dt + A\, dt^2 = 0.$$

But this is the equation of the characteristics, derived just as in (6), (7) of this chapter.

Accordingly waves of acceleration are propagated along the characteristics, as we have seen before in a particular case. In fact this property may be used to define the characteristics. If as usual u denote the vertical coordinate, and x, t the coordinates on the floor, the integral is represented by a surface, and the characteristics may be defined as the projections on the floor of those curves on the integral surface along which it is joined by another integral surface so that the two are tangent. This is shown by the fact that not only the ordinates u are equal, $u = 0$ but that also the derivatives $p = \frac{\partial u}{\partial x}$, $q = \frac{\partial u}{\partial t}$, which determine the inclinations of the tangent plane, are also equal $\delta p = \delta q = 0$. The discontinuities of the second derivatives show that the curvature of the two integral surfaces is different.

In the present chapter we have generally supposed that the differential equation is linear, but all that has been said about the characteristics would hold even if the coefficients A, B, C, in equation (1) or (211) were functions of u, p, q, as well as of x, t. In general the characteristics would depend on the integral surface, but in the cases that we have considered they do not, but only on the differential equation, as we have seen. We shall now find that when we give up the hypothesis of infinitely small motions, the equation of propagation is not linear.

If we return to the notation of Lagrange, which is not necessary for one dimension, equation (186) becomes for one dimension

$$\omega = \frac{\varrho_0}{\varrho} = \frac{\partial x}{\partial a},$$

accordingly the equation of motion (184) or (201) becomes

$$(214) \qquad \frac{\partial^2 x}{\partial t^2} = - \varphi'(\varrho) \frac{\partial \log \varrho}{\partial x} = \frac{\varphi'(\varrho)}{\omega} \frac{\partial \omega}{\partial x} = \psi(\omega) \frac{\partial^2 x}{\partial a^2}$$

where

$$\psi(\omega) = \frac{\varphi'(\varrho)}{\omega^2}.$$

Accordingly the equation of the characteristics, (213) gives us the velocity of propagation

$$(215) \qquad \left(\frac{da}{dt}\right)^2 = \psi(\omega) = \psi\left(\frac{\partial x}{\partial a}\right).$$

On account of the occurrence of a function of $\frac{\partial x}{\partial a}$ in the coefficient of $\frac{\partial^2 x}{\partial a^2}$ the equation (214) is not linear. It may be made linear by a so-called Legendre transformation, but we shall not do this, but, with Hugoniot, differentiate along the characteristics, considering a as a function of t, defined by the equation (215).

$$(216) \qquad \begin{aligned} \frac{d}{dt}\left(\frac{\partial x}{\partial a}\right) &= \frac{\partial^2 x}{\partial t \partial a} + \frac{\partial^2 x}{\partial a^2} \frac{da}{dt} \\ \frac{d}{dt}\left(\frac{\partial x}{\partial t}\right) &= \frac{\partial^2 x}{\partial t^2} + \frac{\partial^2 x}{\partial t \partial a} \frac{da}{dt} \end{aligned}$$

Multiplying the first equation by $\frac{da}{dt}$ and subtracting from the second,

$$\frac{d}{dt}\left(\frac{\partial x}{\partial t}\right) - \frac{da}{dt} \frac{d}{dt}\left(\frac{\partial x}{\partial a}\right) = \frac{\partial^2 x}{\partial t^2} - \left(\frac{da}{dt}\right)^2 \frac{\partial^2 x}{\partial a^2}$$

inserting the value of $\left(\frac{da}{dt}\right)^2$ from (215) and taking account of our equation (214) the right hand member vanishes, so that we have

$$(217) \qquad \frac{d}{dt}\left(\frac{\partial x}{\partial t}\right) - \frac{da}{dt} \frac{d}{dt}\left(\frac{\partial x}{\partial a}\right) = 0.$$

Let us put for convenience

$$\int \sqrt{\psi(\omega)}\, d\omega = \chi(\omega), \quad \{\chi'(\omega)\}^2 = \psi(\omega)$$

so that the equation of the characteristics (215) becomes

$$(218) \qquad \frac{d\,a}{d\,t} = \pm\,\chi'(\omega) = \pm\,\chi'\!\left(\frac{\partial x}{\partial a}\right),$$

inserting which in (217) gives

$$(219) \qquad \frac{d}{d\,t}\!\left(\frac{\partial x}{\partial t}\right) \pm \chi'\!\left(\frac{\partial x}{\partial a}\right)\frac{d}{d\,t}\!\left(\frac{\partial x}{\partial a}\right) = 0.$$

This equation, or pair of equations can be at once integrated, for since the first member is, along the characteristics, a function only of t, it is an exact differential, and integrates into a constant. Accordingly we obtain

$$(220) \qquad
\begin{aligned}
\frac{\partial x}{\partial t} + \chi\!\left(\frac{\partial x}{\partial a}\right) &= \xi,\\[4pt]
\frac{\partial x}{\partial t} - \chi\!\left(\frac{\partial x}{\partial a}\right) &= \eta,
\end{aligned}$$

where ξ is constant along any characteristic of one family, but changes from one to another, while the same may be said of η for the other family. If we now take ξ, η for independent variables, we transform our equation (214) into the normal form for a hyperbolic equation, (14) which may be integrated by Riemann's method. It was in fact for this purpose that Riemann invented the method that we have described. We shall however not carry out this process, which is long, but consider merely some of the consequences of the equations (219). In the first place we may verify that all their integrals satisfy equation (214). For differentiating (220)

$$\frac{\partial^2 x}{\partial t^2} + \chi'\!\left(\frac{\partial x}{\partial a}\right)\frac{\partial^2 x}{\partial t \partial a} = 0,$$

$$\frac{\partial^2 x}{\partial t \partial a} + \chi'\!\left(\frac{\partial x}{\partial a}\right)\frac{\partial^2 x}{\partial a^2} = 0.$$

Multiplying the second by χ' and subtracting from the first, we find (214).

A developable surface may be defined as a surface generated by all the tangents to a twisted curve in space. Such a surface may be developed so as to lie on a plane without stretching. The reader not familiar with such surfaces may get an idea of their properties by cutting out a piece of paper along a curve (which must intersect the edge of the paper) drawing the tangents to the curve, and then lifting one corner so that the curve becomes twisted. Its tangents remain straight lines, and the curve becomes the so-called edge of regression or cuspidal edge of the developable surface. As a matter of fact there should be two sheets of paper, fastened together at the edge of regression by flexible hinges, so that when the bending takes place the two leaves will separate, and we. see that the surface has two sheets, one composed of the parts of the tangents on one side of the edge of regression, the other of the other parts. It will at once be seen that any tangent plane to a developable surface will be tangent along the whole line which is one of the tangents to the edge of regression. Accordingly such a surface, like a cylinder, has but a single infinity of tangent planes, instead of a double infinity, as usual. Along such a line, then, the angular coefficients of the tangent plane are constant, in other words $p = \dfrac{\partial x}{\partial t}$ and $q = \dfrac{\partial x}{\partial a}$ are functions of a single parameter, or there is a relation between them, not containing the coordinates. But either equation (220) is just such a relation, consequently the integrals of our equation (214) are represented by developable surfaces, instead of by cylinders, as in the case of small motions.

A motion compatible with rest, and proceeding from the left hand piston at $a = 0$, will be given, since for the air at rest $\dfrac{\partial x}{\partial t} = 0$ and $\dfrac{\partial x}{\partial a} = 1$, by putting $\xi = \chi(1)$, so that

$$(221) \qquad \qquad \frac{\partial x}{\partial t} + \chi\left(\frac{\partial x}{\partial a}\right) = \chi(1).$$

Physically this means that the velocity of the air particles depends on the density alone. The motion is propagated into the region of rest with the velocity of sound.

If through a fixed point, say the origin, we draw planes parallel to all the tangent planes of the developable surface, these planes envelope a cone whose generators are parallel to the tangents to the edge of regression. Hence all the integral surfaces have this cone in common, for the relation between p and q is given by the equation (221). This cone enables us to determine an integral surface (it is itself one) representing a wave of acceleration arising from a given motion of the piston at $a = 0$ represented by $x = f(t)$. For the curve in the plane $a = 0$ must at each of its points have its tangent lie in a tangent plane to the integral surface. Rotating a plane about this tangent line there will be a position where it will be parallel to one of the tangent planes to the cone just described (compare Fig. 30) and the generator of tangency will give the direction of the generator of the developable surface that is to be drawn through the given point of the curve. Thus all the generators are determined, and the surface is given.

It is plain that the motion will have a singularity at the edge of regression of the surface, for there successive tangents to the edge intersect, that is two waves of different velocities overtake each other.

In fact, if we give the piston a properly chosen acceleration, we may make the edge of regression reduce to a single point, the integral surface is then the cone spoken of, and all the waves overtake each other at the same time and place, and we have a discontinuity in the density, or a shock of compression. This is the so-called phenomenon of Riemann and Hugoniot, in which the waves of acceleration become waves of shock. It would be difficult to carry out the experiment with pistons, on account of the great acceleration required, but the great accelerations produced by explosions are sufficient, and the phenomenon has been observed by M. Vieille. Considerations of space forbid our proceeding farther with this important subject, but just as these sheets are going to press there has been received a very important paper by Professor A. E. H. Love, published in the Philosophical Transactions of the Royal Society of London, Series A, Vol. 222, on Lagrange's Ballistic Problem, in which, with the aid of Mr. F. B. Pidduck he has extended Hugoniot's researches and obtained important practical results on the propagation of waves of discontinuity in a gun.

CHAPTER VII.

SPHERICAL, CYLINDRICAL AND ELLIPSOIDAL HARMONICS OF LAPLACE, BESSEL AND LAMÉ.

86. Definition of Spherical Harmonics. Let us return to Laplace's equation and Dirichlet's Problem. This we have defined for any surface, but we have not actually carried out the solution. After the plane the simplest surface is a sphere, which we shall now consider.

A Spherical Harmonic function is a *homogeneous* function of x, y, z satisfying the usual conditions for a harmonic function, that is it is a simultaneous solution of the equations of Laplace and Euler,

(1) $$\Delta V = 0,$$

(2) $$x\frac{\partial V}{\partial x} + y\frac{\partial V}{\partial y} + z\frac{\partial V}{\partial z} = nV,$$

which is holomorphic, that is, finite, uniform, and continuous, and which vanishes at infinity. The *degree* of the harmonic is the n in the second equation. This is *usually* an integer. Determining a few rational polynomials of integral degrees, we obtain

$$V_0 = \text{constant},$$

(3) $$V_1 = ax + by + cz,$$

$$V_2 = ax^2 + by^2 + cz^2 + dxy + eyz + fzx$$

in which to satisfy $\Delta V = 0$ we must put $a + b + c = 0$, giving

(4) $V_2 = a(x^2 - z^2) + b(y^2 - z^2) + dxy + eyz + fzx$.

Consequently of the second degree there are five linearly independent harmonics, $x^2 - z^2$, $y^2 - z^2$, xy, yz, zx, and the *general* harmonic of the second degree is the sum of these each multiplied by an arbitrary constant.

$$V_3 = ax^3 + by^3 + cz^3 + dx^2y + ex^2z + fy^2x + gy^2z + hz^2x + kz^2y + lxyz,$$

$$\Delta V_3 = 6(ax + by + cz) + 2dy + 2ez + 2fx + 2gz + 2hx + 2ky = 0,$$

necessitating the relations,

$$6a + 2f + 2h = 0, \qquad a = -1/3\,(f + h),$$
$$6b + 2d + 2k = 0, \quad \text{or} \quad b = -1/3\,(d + k),$$
$$6c + 2e + 2g = 0, \qquad c = -1/3\,(e + g).$$

(5) $V_3 = d(x^2y - \frac{1}{3}y^3) + e(x^2z - \frac{1}{3}z^3) + f(y^2x - \frac{1}{3}x^3) + g(y^2z - \frac{1}{3}z^3)$
$\qquad\qquad + h(z^2x - \frac{1}{3}x^3) + k(z^2y - \frac{1}{3}y^3) + lxyz$.

There are seven arbitrary constants in V_3.

In a homogeneous polynomial of degree n there are the terms:

$$a_{n0}x^n + a_{n-1,0}x^{n-1}y^1 + a_{n-2,0}x^{n-2}y^2 + \cdots + a_{0_0}y^n$$
$$+ a_{n-1,1}x^{n-1}z + a_{n-2,1}x^{n-2}yz + \cdots + a_{0_1}y^{n-1}z$$
$$+ a_{n-2,2}x^{n-2}z^2 + \cdots + a_{0_2}y^{n-2}z^2$$
$$\cdot\ \cdot\ \cdot\ \cdot\ \cdot\ \cdot\ \cdot$$
$$\cdot\ \cdot\ \cdot\ \cdot\ \cdot\ \cdot$$
$$+ a_{0_n}z^n,$$

in number, $1 + 2 + 3 + \cdots + n + 1 = \dfrac{(n+1)(n+2)}{2}$. The sum of its second derivatives is a homogeneous polynomial of degree $n - 2$ and accordingly contains $n(n-1)/2$ terms. If this function is to vanish identically these coefficients must vanish, so that there are $n(n-1)/2$ relations between the $(n+1)(n+2)/2$ coefficients of the harmonic of degree n, leaving $2n + 1$ *arbitrary* coefficients.

If we insert spherical coordinates r the radius, θ the colatitude or polar distance, and φ, the longitude

(6) $x = r\sin\theta\cos\varphi, \quad y = r\sin\theta\sin\varphi, \quad z = r\cos\theta,$

a homogeneous function V_n contains r^n as factor,

(7) $$V_n = r^n Y_n(\theta, \varphi).$$

Y_n, a function of the geographical coordinates θ, φ on a spherical *surface*, is called a *surface* harmonic, as contrasted to V_n, which is called a *solid* harmonic.

The equation $Y_n = 0$ represents a cone, whose intersection with any sphere with center at 0 gives a geometrical representation of the harmonic. The contour lines $Y_n = $ const. do this still more graphically, e. g., $Y_2 = \dfrac{x^2 - y^2}{r^2} = $ const. gives a spherical ellipse.

Let u, v be any two continuous functions of x, y, z. Then

$$(8) \qquad \frac{\partial^2(uv)}{\partial x^2} = u\frac{\partial^2 v}{\partial x^2} + 2\frac{\partial u}{\partial x}\frac{\partial v}{\partial x} + v\frac{\partial^2 u}{\partial x^2}.$$

Changing x to y, z, and adding,

$$(9) \qquad \triangle(uv) = u\triangle v + 2\triangle(u, v) + v\triangle u.$$

Put $\qquad u = r^m$

$$\frac{\partial(r^m)}{\partial x} = mr^{m-1}\frac{\partial r}{\partial x} = mr^{m-2}x,$$

$$\frac{\partial^2(r^m)}{\partial x^2} = mr^{m-2} + m(m-2)r^{m-4}x^2,$$

$$(10) \qquad \triangle(r^m) = 3mr^{m-2} + m(m-2)r^{m-2} = m(m+1)r^{m-2}.$$

If V_n is a harmonic of degree n,

$$\triangle(r^m V_n) = r^m\triangle V_n + m(m+1)r^{m-2}V_n$$
$$(11) \qquad\qquad + 2mr^{m-2}\left(x\frac{\partial V_n}{\partial x} + y\frac{\partial V_n}{\partial y} + z\frac{\partial V_n}{\partial z}\right)$$
$$= r^{m-2}[m(m+1) + 2mn]V_n,$$

on account of the equations (1), (2) satisfied by V_n.

Accordingly if $m = -(2n+1)$, $r^m V_n$ is a spherical harmonic. Since V_n is of degree n, and r of degree 1, V_n/r^{2n+1} is of degree $-(n+1)$. Accordingly to any spherical harmonic of degree n, $V_n = r^n Y_n$, there corresponds another of degree $-(n+1)$; $V_{-(n+1)} = Y_n/r^{n+1}$. At a given spherical surface these two differ only by a const. factor. V_n vanishes for $r = 0$, and is ∞ for $r = \infty$,

$V_{-(n+1)}$ vanishes for $r = \infty$, and is ∞ for $r = 0$.

87. Dirichlet's Problem for Sphere. By means of spherical harmonics we can solve Dirichlet's problem for the sphere. Suppose that on the surface $r = R$, $V = V_s(\theta, \varphi)$ and that for $r < R$, $\triangle V = 0$.

Then *if* it is possible to develop the function $V_s(\theta, \varphi)$ in a convergent series of surface harmonics.

$$(12) \qquad\qquad V_s = Y_0 + Y_1 + Y_2 + \cdots$$

the series

$$(13) \qquad\qquad V = Y_0 + \frac{r}{R}Y_1 + \left(\frac{r}{R}\right)^2 Y_2 + \cdots$$

satisfies the conditions, for it converges faster than the above series, since $(r/R)^n < 1$ and when $r = R$ it takes the value V_s. Every term is harmonic, hence the sum is also.

The outer problem is stated similarly with the condition $V = 0$ for $r = \infty$. This is satisfied by

$$(14) \qquad V = \frac{R}{r} Y_0 + \left(\frac{R}{r}\right)^2 Y_1 + \left(\frac{R}{r}\right)^3 Y_2 + \cdots$$

each term of which is harmonic, and vanishes for $r = \infty$. The series converges for $r > R$.

88. Forms of Spherical Harmonics. Since we may permute the order of differentation, if V_n is harmonic

$$\triangle \frac{\partial V_n}{\partial x} = \frac{\partial}{\partial x} \triangle V_n = 0.$$

Thus we see that any derivative of a harmonic is a harmonic, so that

$$\frac{\partial^\alpha}{\partial x_\alpha} \frac{\partial^\beta}{\partial y^\beta} \frac{\partial^\gamma}{\partial z_\gamma} V_n$$

is a harmonic of degree $n - (\alpha + \beta + \gamma)$. Since $\frac{1}{r}$ is harmonic,

$$\frac{\partial^{\alpha + \beta + \gamma}}{\partial x^\alpha \partial y^\beta \partial z^\gamma} \left(\frac{1}{r}\right) = V_{-(1 + \alpha + \beta + \gamma)}.$$

If h_1 be a given constant direction with direction cosines, l, m, n,

$$\frac{\partial}{\partial h_1} = l_1 \frac{\partial}{\partial x} + m_1 \frac{\partial}{\partial y} + n_1 \frac{\partial}{\partial z}$$

and $\frac{\partial}{\partial h_1}\left(\frac{1}{r}\right)$ is a harmonic of degree $- 2$, to which corresponds the harmonic $r^3 \frac{\partial}{\partial h_1}\left(\frac{1}{r}\right)$ of degree 1. Multiplying this by an arbitrary constant A, we have

$$V_1 = A r^3 \frac{\partial}{\partial h_1}\left(\frac{1}{r}\right),$$

with *three* arbitrary constants, A, l, m (since $l^2 + m^2 + n^2 = 1$), and therefore the *general* harmonic of degree 1.

We have seen in § 57 that the Newtonian potential due to a point was $\frac{m}{r} = V_{-1}$, and that that due to a doublet of moment $M_1 = m d_1$ was $M_1 \frac{\partial}{\partial h}\left(\frac{1}{r}\right) = V_{-2}$, a harmonic of degree $- 2$. In like manner if two such doublets be placed parallel but with their centers at a distance apart d_2 in a direction h_2, the potential is easily seen to be

$$M_2 \frac{\partial}{\partial h_2} \frac{\partial}{\partial h_1}\left(\frac{1}{r}\right) = V_{-3}$$

where $M_2 = \lim M_1 d_2$. In like manner we may construct multiple points of any order, whose potential is the spherical harmonic

$$(14) \qquad V_{-(n+1)} = A \frac{\partial}{\partial h_1} \frac{\partial}{\partial h_2} \cdots \frac{\partial}{\partial h_n}\left(\frac{1}{r}\right).$$

This conception of the spherical harmonic is due to Maxwell. From the above we get

$$(15) \qquad V_n = A r^{2n+1} \frac{\partial}{\partial h_1} \frac{\partial}{\partial h_2} \cdots \frac{\partial}{\partial h_n}\left(\frac{1}{r}\right).$$

The directions $h_1, h_2, \ldots h_n$ are called those of the *axes* of the harmonic. Since for a direction two constants are necessary, these with A make $2n + 1$, therefore the above is the *most general* harmonic.

Examples:

$$V_1 = A r^3 \frac{\partial}{\partial h}\left(\frac{1}{r}\right) = A r^3 \left\{ l \frac{\partial}{\partial x}\left(\frac{1}{r}\right) + m \frac{\partial}{\partial y}\left(\frac{1}{r}\right) + n \frac{\partial}{\partial z}\right) \right\}$$

$$= A r^3 \left\{ -\frac{lx}{r^3} - \frac{my}{r^3} - \frac{nz}{r^3} \right\} = - A(lx + my + nz)$$

$$V_2 = A r^5 \left(l_1 \frac{\partial}{\partial x} + m_1 \frac{\partial}{\partial y} + n_1 \frac{\partial}{\partial z} \right) \left(l_2 \frac{\partial}{\partial x}\left(\frac{1}{r}\right) + m_2 \frac{\partial}{\partial y}\left(\frac{1}{r}\right) + n_2 \frac{\partial}{\partial z}\left(\frac{1}{r}\right) \right)$$

$$= - A r^5 \left(l_1 \frac{\partial}{\partial x} + m_1 \frac{\partial}{\partial y} + n_1 \frac{\partial}{\partial z} \right) \left(\frac{l_2 x + m_2 y + n_2 z}{r^3} \right)$$

$$= - A r^5 \left\{ l_1 \left(\frac{l_2}{r^3} - \frac{3 l_2 x^2}{r^5} - \frac{3 m_2 xy}{r^5} - \frac{3 n_2 xz}{r^5} \right) \right.$$

$$+ m_1 \left(\frac{m_2}{r^3} - \frac{3 m_2 y^2}{r^5} - \frac{3 l_2 xy}{r^5} - \frac{3 n_2 yz}{r^5} \right)$$

$$\left. + n_1 \left(\frac{n_2}{r^3} - \frac{3 n_2 z^2}{r^5} - \frac{3 l_2 xz}{r^5} - \frac{3 m_2 yz}{r^5} \right) \right\},$$

$$V_2 = A \left\{ -(l_1 l_2 + m_1 m_2 + n_1 n_2)(x^2 + y^2 + z^2) + 3(l_1 l_2 x^2 + m_1 m_2 y^2 + n_1 n_2 z^2) \right.$$

$$\left. + 3(l_1 m_2 + l_2 m_1)xy + 3(m_1 n_2 + m_2 n_1)yz + 3(n_1 l_2 + n_2 l_1)zx \right\}.$$

The coefficients are of course subject to the relations

$$l_1^2 + m_1^2 + n_1^2 = 1, \quad l_2^2 + m_2^2 + n_2^2 = 1.$$

89. Zonal Harmonics. If all the axes of the harmonic coincide, we may take its direction for one of the coordinate axes, and write $V_n = A r^{2n+1} \frac{\partial^n}{\partial z^n}\left(\frac{1}{r}\right)$. This will evidently be a polynomial in z and r, so that the *surface* harmonic will be simply a polynomial in $z/r = \cos(rz)$. This is called a Legendre's polynomial $P_n[\cos(rz)] = A r^{n+1} \frac{\partial^n}{\partial z^n}\left(\frac{1}{r}\right)$. The equation $P_n = 0$ may be shown to have n real roots between -1 and $+1$, and hence represents n circular cones of angles equal to the

$\cos^{-1}$ of these roots, intersecting the sphere in n parallels of latitude which divide the sphere into zones. These harmonics are accordingly called zonal harmonics, and $P_n(\cos(rz))$ is called the surface zonal harmonic.

90. Orthogonal Coordinates. It is important to deduce expressions for spherical harmonics in spherical coordinates. To do this we have to transform Laplace's equation into such coordinates. This being a case of a general transformation, we shall consider the latter.

If we have a set of three functions $\varrho_1(x, y, z)$, $\varrho_2(x, y, z)$, $\varrho_3(x, y, z)$ an equation $\varrho = $ const. represents a surface, two such represent the intersection of two surfaces, or a line, three equations represent the intersection of three, or a point. Hence the values of ϱ_1, ϱ_2, ϱ_3 determine a point, and are called its curvilinear coordinates. For instance

$$(16) \qquad \begin{aligned} \varrho_1 &= \sqrt{x^2 + y^2 + z^2} = r, \\ \varrho_2 &= \cos^{-1}\frac{z}{r} = \theta, \\ \varrho_3 &= \tan^{-1}\frac{y}{x} = \varphi. \end{aligned}$$

The level surfaces ϱ_1, ϱ_2, ϱ_3 are called the coordinate surfaces.

The normal to the surface $\varrho_r = $ const. has the direction cosines

$$\cos(n_r x) = \frac{\partial \varrho_r}{\partial x}\bigg/ h_r, \quad \cos(n_r y) = \frac{\partial \varrho_r}{\partial y}\bigg/ h_r, \quad \cos(n_r z) = \frac{\partial \varrho_1}{\partial z}\bigg/ h_r$$

where

$$h_r = |\operatorname{grad} \varrho_r|, \qquad h_r^2 = \left(\frac{\partial \varrho_r}{\partial x}\right)^2 + \left(\frac{\partial \varrho_r}{\partial y}\right)^2 + \left(\frac{\partial \varrho_r}{\partial z}\right)^2.$$

We have also in the direction n_1

$$(17) \qquad \frac{\partial \varrho_1}{\partial n_1} = \cos(n_1 x)\frac{\partial \varrho_1}{\partial x} + \cos(n_1 y)\frac{\partial \varrho_1}{\partial y} + \cos(n_1 z)\frac{\partial \varrho_1}{\partial z} = h_1$$

and

$$d n_1 = \frac{d\varrho_1}{h_1}$$

gives the distance along the normal in terms of $d\varrho_1$.

We are particularly concerned with the case in which at *every* point in space the normals to the three coordinate surfaces at that point are mutually perpendicular. The coordinates ϱ_1, ϱ_2, ϱ_3 are then said to form an orthogonal system, as in the example just given.

The conditions of perpendicularity are

$$(18) \qquad \begin{aligned} \frac{\partial \varrho_1}{\partial x}\frac{\partial \varrho_2}{\partial x} + \frac{\partial \varrho_1}{\partial y}\frac{\partial \varrho_2}{\partial y} + \frac{\partial \varrho_1}{\partial z}\frac{\partial \varrho_2}{\partial z} &= 0, \\ \frac{\partial \varrho_2}{\partial x}\frac{\partial \varrho_3}{\partial x} + \frac{\partial \varrho_2}{\partial y}\frac{\partial \varrho_3}{\partial y} + \frac{\partial \varrho_2}{\partial z}\frac{\partial \varrho_3}{\partial z} &= 0, \\ \frac{\partial \varrho_3}{\partial x}\frac{\partial \varrho_1}{\partial x} + \frac{\partial \varrho_3}{\partial y}\frac{\partial \varrho_1}{\partial y} + \frac{\partial \varrho_3}{\partial z}\frac{\partial \varrho_1}{\partial z} &= 0. \end{aligned}$$

Now differentiating ϱ_1, ϱ_2, ϱ_3 and dividing by h_1, h_2, h_3,

$$\frac{d\varrho_1}{h_1} = dn_1 = \frac{1}{h_1}\left(\frac{\partial \varrho_1}{\partial x}dx + \frac{\partial \varrho_1}{\partial y}dy + \frac{\partial \varrho_1}{\partial z}dz\right),$$

(19)
$$\frac{d\varrho_2}{h_2} = dn_2 = \frac{1}{h_2}\left(\frac{\partial \varrho_2}{\partial x}dx + \frac{\partial \varrho_2}{\partial y}dy + \frac{\partial \varrho_2}{\partial z}dz\right),$$

$$\frac{d\varrho_3}{h_3} = dn_3 = \frac{1}{h_3}\left(\frac{\partial \varrho_3}{\partial x}dx + \frac{\partial \varrho_3}{\partial y}dy + \frac{\partial \varrho_3}{\partial z}dz\right).$$

Now since the nine coefficients of dx, dy, dz are the nine direction cosines of n_1, n_2, n_3, by § 3, (7) these are the projections of the vector which has the rectangular components dx, dy, dz on the directions of the three normals. Accordingly dn_1, dn_2, dn_3 are the rectangular components in the new system of the line ds whose rectangular components in the old one are dx, dy, dz. Accordingly we have

(20)
$$ds^2 = dx^2 + dy^2 + dz^2 = \frac{d\varrho_1^2}{h_1^2} + \frac{d\varrho_2^2}{h_2^2} + \frac{d\varrho_3^2}{h_3^2},$$

a homogeneous quadratic function of $d\varrho_1$, $d\varrho_2$, $d\varrho_3$.

This may be shown directly as follows. We may express x, y, z, as functions of ϱ_1, ϱ_2, ϱ_3. We then have

$$dx = \frac{\partial x}{\partial \varrho_1}d\varrho_1 + \frac{\partial x}{\partial \varrho_2}d\varrho_2 + \frac{\partial x}{\partial \varrho_3}d\varrho_3,$$

(21)
$$dy = \frac{\partial y}{\partial \varrho_1}d\varrho_1 + \frac{\partial y}{\partial \varrho_2}d\varrho_2 + \frac{\partial y}{\partial \varrho_3}d\varrho_3,$$

$$dz = \frac{\partial z}{\partial \varrho_1}d\varrho_1 + \frac{\partial z}{\partial \varrho_2}d\varrho_2 + \frac{\partial z}{\partial \varrho_3}d\varrho_3.$$

Inserting these values in $d\varrho_1$, $d\varrho_2$, $d\varrho_3$,

$$d\varrho_1 = \frac{\partial \varrho_1}{\partial x}\left(\frac{\partial x}{\partial \varrho_1}d\varrho_1 + \frac{\partial x}{\partial \varrho_2}d\varrho_2 + \frac{\partial x}{\partial \varrho_3}d\varrho_3\right)$$

(22)
$$+ \frac{\partial \varrho_1}{\partial y}\left(\frac{\partial y}{\partial \varrho_1}d\varrho_1 + \frac{\partial y}{\partial \varrho_2}d\varrho_2 + \frac{\partial y}{\partial \varrho_3}d\varrho_3\right)$$

$$+ \frac{\partial \varrho_1}{\partial z}\left(\frac{\partial z}{\partial \varrho_1}d\varrho_1 + \frac{\partial z}{\partial \varrho_2}d\varrho_2 + \frac{\partial z}{\partial \varrho_3}d\varrho_3\right)$$

and equating coefficients of $d\varrho_1$,

(23)
$$\frac{\partial \varrho_1}{\partial x}\frac{\partial x}{\partial \varrho_1} + \frac{\partial \varrho_1}{\partial y}\frac{\partial y}{\partial \varrho_1} + \frac{\partial \varrho_1}{\partial z}\frac{\partial z}{\partial \varrho_1} = 1$$

and similarly changing 1 to 2 and 3.

Equating coefficients of $d\varrho_2$,

(24)
$$\frac{\partial \varrho_1}{\partial x}\frac{\partial x}{\partial \varrho_2} + \frac{\partial \varrho_1}{\partial y}\frac{\partial y}{\partial \varrho_2} + \frac{\partial \varrho_1}{\partial z}\frac{\partial z}{\partial \varrho_2} = 0$$

and similarly changing the suffixes.

Writing (23) and (24)

$$\frac{1}{h_1}\frac{\partial \varrho_1}{\partial x} h_1 \frac{\partial x}{\partial \varrho_1} + \frac{1}{h_1}\frac{\partial \varrho_1}{\partial y} h_1 \frac{\partial y}{\partial \varrho_1} + \frac{1}{h_1}\frac{\partial \varrho_1}{\partial z} h_1 \frac{\partial z}{\partial \varrho_1} = 1, \text{ etc.,}$$

we see that

$$\frac{1}{h_r}\frac{\partial \varrho_r}{\partial x} = h_r \frac{\partial x}{\partial \varrho_r}, \quad \frac{1}{h_r}\frac{\partial \varrho_r}{\partial y} = h_r \frac{\partial y}{\partial \varrho_r}, \quad \frac{1}{h_r}\frac{\partial \varrho_r}{\partial z} = h_r \frac{\partial z}{\partial \varrho_r} \quad r = 1,2,3$$

or the nine direction-cosines are otherwise expressed,

$$h_r \frac{\partial x}{\partial \varrho_r}, \quad h_r \frac{\partial y}{\partial \varrho_r}, \quad h_r \frac{\partial z}{\partial \varrho_r}. \qquad r = 1,2,3$$

Squaring and adding, putting equal to 1, and dividing by h_r^2,

$$(25) \qquad \frac{1}{h_r^2} = \left(\frac{\partial x}{\partial \varrho_r}\right)^2 + \left(\frac{\partial y}{\partial \varrho_r}\right)^2 + \left(\frac{\partial z}{\partial \varrho_r}\right)^2. \qquad r = 1,2.3$$

The conditions of orthogonality are also

$$(26) \qquad \frac{\partial x}{\partial \varrho_r}\frac{\partial x}{\partial \varrho_s} + \frac{\partial y}{\partial \varrho_r}\frac{\partial y}{\partial \varrho_s} + \frac{\partial z}{\partial \varrho_r}\frac{\partial z}{\partial \varrho_s} = 0, \qquad r,s = 1,2,3$$

and accordingly squaring and adding (21)

$$ds^2 = dx^2 + dy^2 + dz^2$$

$$(27) \quad \begin{aligned} &= d\varrho_1^2 \left\{ \left(\frac{\partial x}{\partial \varrho_1}\right)^2 + \left(\frac{\partial y}{\partial \varrho_1}\right)^2 + \left(\frac{\partial z}{\partial \varrho_1}\right)^2 \right\} \\ &+ d\varrho_2^2 \left\{ \left(\frac{\partial x}{\partial \varrho_2}\right)^2 + \left(\frac{\partial y}{\partial \varrho_2}\right)^2 + \left(\frac{\partial z}{\partial \varrho_2}\right)^2 \right\} \\ &+ d\varrho_3^2 \left\{ \left(\frac{\partial x}{\partial \varrho_3}\right)^2 + \left(\frac{\partial y}{\partial \varrho_3}\right)^2 + \left(\frac{\partial z}{\partial \varrho_3}\right)^2 \right\} \\ &= \frac{d\varrho_1^2}{h_1^2} + \frac{d\varrho_2^2}{h_2^2} + \frac{d\varrho_3^2}{h_3^2}. \end{aligned}$$

The element of the surface $\varrho_1 = $ const. is $dS_1 = \dfrac{d\varrho_2}{h_2}\dfrac{d\varrho_3}{h_3}$ and the element of the volume $d\tau = \dfrac{d\varrho_1}{h_1}\dfrac{d\varrho_2}{h_2}\dfrac{d\varrho_3}{h_3}$.

Let us now consider the divergence theorem, which in rectangular coordinates is

$$\iiint \text{div } A\, d\tau = -\iint A_n\, dS. \qquad (n \text{ internal})$$

Instead of the components A_x, A_y, A_z, let us consider the components of a vector A along the orthogonal directions of the normals to the coordinate surfaces $\varrho_1, \varrho_2, \varrho_3$. Call these A_1, A_2, A_3. Projecting along the normal n to S, we have

$$(28) \qquad A_n = A_1 \cos(nn_1) + A_2 \cos(nn_2) + A_3 \cos(nn_3).$$

If we divide the volume up into elementary *curved* prisms bounded by level surfaces of ϱ_1 and ϱ_2, we have at each case of cutting into or out of S respectively

$$\pm\, dS \cos(n n_1) = \frac{d\varrho_2}{h_2}\frac{d\varrho_3}{h_3}$$

and accordingly

$$-\int\int A_1 \cos(n n_1)\, dS_1 = -\int\int A_1 \frac{d\varrho_2}{h_2}\frac{d\varrho_3}{h_3} = \int\int\int \frac{\partial}{\partial \varrho_1}\Big(\frac{A_1}{h_2 h_3}\Big) d\varrho_1 d\varrho_2 d\varrho_3,$$

the change from the double to the triple integral involving the same considerations as in rectangular coordinates, § 50.

Transforming the other two integrals,

$$(29)\quad -\int\int A_n\, dS = \int\int\int\Big[\frac{\partial}{\partial \varrho_1}\Big(\frac{A_1}{h_2 h_3}\Big) + \frac{\partial}{\partial \varrho_2}\Big(\frac{A_2}{h_3 h_1}\Big) + \frac{\partial}{\partial \varrho_3}\Big(\frac{A_3}{h_1 h_2}\Big)\Big] d\varrho_1 d\varrho_2 d\varrho_3$$

$$= \int\int\int \operatorname{div} A \cdot d\tau = \int\int\int \operatorname{div} A \frac{d\varrho_1}{h_1}\frac{d\varrho_2}{h_2}\frac{d\varrho_3}{h_3}.$$

Since this is true for any volume, we find, by equating integrands,

$$(30)\qquad \operatorname{div} A = h_1 h_2 h_3 \Big\{\frac{\partial}{\partial \varrho_1}\Big(\frac{A_1}{h_2 h_3}\Big) + \frac{\partial}{\partial \varrho_2}\Big(\frac{A_2}{h_3 h_1}\Big) + \frac{\partial}{\partial \varrho_3}\Big(\frac{A_3}{h_1 h_2}\Big)\Big\}.$$

If the vector is lamellar, its projections are the projections of the gradient of a potential V, that is

$$(31)\quad A_1 = \frac{\partial V}{\partial n_1} = h_1 \frac{\partial V}{\partial \varrho_1},\quad A_2 = \frac{\partial V}{\partial n_2} = h_2 \frac{\partial V}{\partial \varrho_2},\quad A_3 = \frac{\partial V}{\partial n_3} = h_3 \frac{\partial V}{\partial \varrho_3}$$

and the above equation becomes

$$(32)\quad \varDelta V = h_1 h_2 h_3 \Big\{\frac{\partial}{\partial \varrho_1}\Big(\frac{h_1}{h_2 h_3}\frac{\partial V}{\partial \varrho_1}\Big) + \frac{\partial}{\partial \varrho_2}\Big(\frac{h_2}{h_3 h_1}\frac{\partial V}{\partial \varrho_2}\Big) + \frac{\partial}{\partial \varrho_3}\Big(\frac{h_3}{h_1 h_2}\frac{\partial V}{\partial \varrho_3}\Big)\Big\}$$

the result given by Lamé, by a laborious direct transformation.

91. Harmonics in Spherical Coordinates. For polar or spherical coordinates, $\varrho_1 = r$ the distance from the origin, $\varrho_2 = \theta$ the colatitude, $\varrho_3 = \varphi$ the longitude

$$(33)\qquad x = r \sin\theta \cos\varphi,\quad y = r \sin\theta \sin\varphi,\quad z = r \cos\theta.$$

The level surfaces for ϱ_1 are spheres, and since the normal is in the direction of the radius, $dn_1 = dr$, $h_1 = 1$, for ϱ_2 cones of semivertical angle θ, and since $dn_2 = r d\theta$, $h_2 = 1/r$, while for φ the level surfaces are planes through the polar axis of Z, $dn_3 = r \sin\theta d\varphi$ and $h_3 = 1/r \sin\theta$. The same results may be obtained from (25). Accordingly we obtain,

$$(34)\quad \varDelta V = \frac{1}{r^2 \sin\theta}\Big\{\frac{\partial}{\partial r}\Big(r^2 \sin\theta \frac{\partial V}{\partial r}\Big) + \frac{\partial}{\partial \theta}\Big(\sin\theta \frac{\partial V}{\partial \theta}\Big) + \frac{\partial}{\partial \varphi}\Big(\frac{1}{\sin\theta}\frac{\partial V}{\partial \varphi}\Big)\Big\}$$

$$= \frac{\partial^2 V}{\partial r^2} + \frac{2}{r}\frac{\partial V}{\partial r} + \frac{1}{r^2 \sin\theta}\frac{\partial}{\partial \theta}\Big(\sin\theta \frac{\partial V}{\partial \theta}\Big) + \frac{1}{r^2 \sin^2\theta}\frac{\partial^2 V}{\partial \varphi^2}.$$

Let us seek a solution of Laplace's equation which is the product of a function of r by a function of θ, φ, putting $V = R(r)\, Y(\theta, \varphi)$.

$$(35) \quad \frac{d^2R}{dr^2}Y + \frac{2}{r}\frac{dR}{dr}Y + \frac{R}{r^2\sin\theta}\frac{\partial}{\partial\theta}\left(\sin\theta\frac{\partial Y}{\partial\theta}\right) + \frac{R}{r^2\sin^2\theta}\frac{\partial^2 Y}{\partial\varphi^2} = 0$$

Multiplying by $r^2/R\,Y$ and transposing,

$$(36) \quad \frac{1}{Y}\frac{1}{\sin\theta}\frac{\partial}{\partial\theta}\left(\sin\theta\frac{\partial Y}{\partial\theta}\right) + \frac{1}{Y\sin^2\theta}\frac{\partial^2 Y}{\partial\varphi^2} = -\left\{\frac{r^2}{R}\frac{d^2R}{dr^2} + \frac{2r}{R}\frac{dR}{dr}\right\}$$

which must equal a constant, since on the right we have a function of r, on the left of θ, φ. If we call this c we have for R,

$$(37) \quad \frac{d^2R}{dr^2} + \frac{2}{r}\frac{dR}{dr} + \frac{c}{r^2}R = 0,$$

a linear equation with variable coefficients, and the singular point $r=0$. On attempting to solve it by means of a series, $R = \sum a_k r^k$ we find

$$\frac{dR}{dr} = \sum a_k k r^{k-1}, \quad \frac{d^2R}{dr^2} = \sum a_k k(k-1)r^{k-2}$$

$$\sum a_k\{k(k-1) + 2k + c\}r^{k-2} = 0,$$

which is satisfied only if $k(k+1) + c = 0$. This is quadratic in k, therefore instead of a series, there are only two *powers* of r which solve the equation, given by the roots of the quadratic. If for the arbitrary constant c we put $-n(n+1)$

$$k(k+1) - n(n+1) = 0,$$
$$(k-n)(k+n+1) = 0,$$
$$k = n; \quad k = -(n+1) \qquad \text{are the roots,}$$

and r^n and $1/r^{n+1}$ are the *only* solutions. But this is to say that $Ar^nY_n + BY_n/r^{n+1}$ is the general solution of Laplace's equation, where Y_n satisfies the equation.

$$(38) \quad \frac{1}{\sin\theta}\frac{\partial}{\partial\theta}\left(\sin\theta\frac{\partial Y_n}{\partial\theta}\right) + \frac{1}{\sin^2\theta}\frac{\partial^2 Y_n}{\partial\varphi^2} + n(n+1)Y_n = 0.$$

This is the equation as originally given by Laplace. The surface harmonic $Y_n(\theta, \varphi)$ is often called a Laplace's function. If we put as a variable, instead of θ,

$$\mu = \cos\theta, \quad d\mu = -\sin\theta\, d\theta$$

the equation (38) becomes

$$(39) \quad \frac{\partial}{\partial\mu}\left\{(1-\mu^2)\frac{\partial Y_n}{\partial\mu}\right\} + \frac{1}{1-\mu^2}\frac{\partial^2 Y_n}{\partial\varphi^2} + n(n+1)Y_n = 0.$$

92. Legendre's Polynomials. In the particular case that Y_n is independent of φ (39) becomes, Y_n becoming the zonal harmonic P_n,

$$(40) \qquad \frac{d}{d\mu}\left\{(1-\mu^2)\frac{dP_n}{d\mu}\right\} + n(n+1)P_n = 0,$$

or on differentiation,

$$(41) \qquad (1-\mu^2)\frac{d^2P_n}{d\mu^2} - 2\mu\frac{dP_n}{d\mu} + n(n+1)P_n = 0.$$

This is Legendre's differential equation (see Appendix), with the *two* singular points $\mu = \pm 1$. The point $\mu = 0$ is *not* singular, so that we may find a power-series as an integral. Putting $P_n = \sum_0^\infty a_k \mu^k$ we get an equation of recurrence to determine all the coefficients in terms of two,

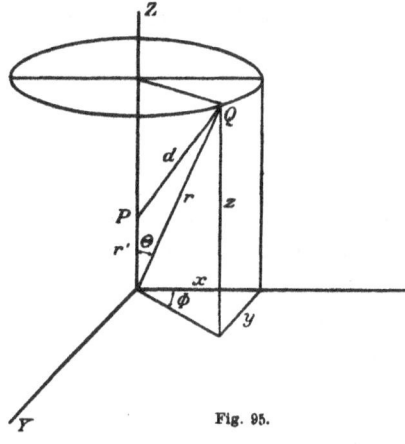

thus we get *two* series as particular integrals. *One* of these stops with μ^n, that is it is a polynomial, and this is called the Legendre polynomial $P_n(\mu)$. In order to determine the arbitrary coefficient of P_n we will define it as follows.

We know that the reciprocal distance from a fixed point P is a harmonic function of the coordinates x, y, z, and although it is not a *homogeneous* function unless P is at the origin, it may always be developed in a series of such, that

Fig. 95.

is of spherical harmonics. Also if P be taken for the pole of the spherical coordinates, the harmonics will be *zonal*. Call d the distance of the point Q, with coordinates x, y, z, from P, and reserve r for the polar coordinate of Q. Take the z-axis through P, and let r' be the distance OP, Fig. 95. Then we have

$$d^2 = r^2 + r'^2 - 2\mu rr', \quad \mu = \cos\theta,$$

$$(42) \qquad \frac{1}{d} = \{r^2 + r'^2 - 2\mu rr'\}^{-\frac{1}{2}} = [x^2 + y^2 + (z - r')^2]^{-\frac{1}{2}}$$

Considering this as a function of z let us develop by Taylor's theorem,

$$\frac{1}{d} = f(z - r') = f(z) - r'\left(\frac{\partial f}{\partial z}\right)_{r'=0} + \frac{1}{2!}r'^2\left(\frac{\partial^2 f}{\partial z^2}\right)_{r'=0} + \cdots$$

and since for

$$r' = 0, \quad \frac{1}{d} = \frac{1}{r}, \quad \frac{\partial f}{\partial z} = \frac{\partial}{\partial z}\left(\frac{1}{r}\right)$$

(43) $\frac{1}{d} = \frac{1}{r} - r'\frac{\partial}{\partial z}\left(\frac{1}{r}\right) + \frac{1}{2!}r'^2\frac{\partial^2}{\partial z^2}\left(\frac{1}{r}\right) + \cdots \frac{(-1)^n}{n!}r'^n\frac{\partial^n}{\partial z^n}\left(\frac{1}{r}\right) + \cdots$

Now multiplying and dividing each term by r^{n+1}

(44) $\frac{1}{d} = \frac{1}{r}\left\{P_0 + \frac{r'}{r}P_1 + \left(\frac{r'}{r}\right)^2 P_2 + \cdots + \left(\frac{r'}{r}\right)^n P_n + \cdots\right\}$

where we define

$$P_0 = 1, \quad P_n = \frac{(-1)^n}{n!}r^{n+1}\frac{\partial^n}{\partial z^n}\left(\frac{1}{r}\right)$$

as we have seen in § 89 for the harmonic with n coincident axes. The constant A has now been determined as $(-1)^n/n!$ for the reason that, since if $r' < r$, $\mu = 1$

$$\frac{1}{d} = (r - r')^{-1} = \frac{1}{r}\left(1 - \frac{r'}{r}\right)^{-1} = \frac{1}{r}\left\{1 + \frac{r'}{r} + \left(\frac{r'}{r}\right)^2 + \cdots\right\}$$

it makes for *every* value of n, $P_n(1) = 1$.

In order to find P_n as a polynomial in μ we may write

$$\frac{r}{d} = \left[1 - 2\frac{r'}{r}\left(\mu - \frac{r'}{2r}\right)\right]^{-\frac{1}{2}}$$

and develop by the binomial theorem.

(45) $\frac{r}{d} = \sum_{s=0}^{s=\infty}\frac{\frac{1}{2}\cdot\frac{3}{2}\cdot\frac{5}{2}\cdots(\frac{1}{2}+s-1)}{s!}\left(2\frac{r'}{r}\right)^s\left(\mu - \frac{r'}{2r}\right)^s.$

Devoloping the last factor,

(46) $\left(\mu - \frac{r'}{2r}\right)^s = \sum_{t=0}^{t=\infty}(-1)^t\frac{s(s-1)\cdots(s-t+1)}{t!}\mu^{s-t}\left(\frac{r'}{2r}\right)^t.$

(47) $\frac{r}{d} = \sum_{s=0}^{s=\infty}\sum_{t=0}^{t=\infty}(-1)^t 2^{s-t}\frac{\frac{1}{2}\cdot\frac{3}{2}\cdots(\frac{1}{2}+s-1)}{t!(s-t)!}\left(\frac{r'}{r}\right)^{s+t}\mu^{s-t}.$

Picking out all the terms for which $s + t = n$ we get for the coefficient of $\left(\frac{r'}{r}\right)^n$, writing $\frac{1\cdot3\cdot5\cdots(2n-1)}{n!} = \frac{(2n)!}{2^n(n!)^2}$,

(47a) $P_n = \frac{(2n)!}{2^n(n!)^2}\left[\mu^n - \frac{n(n-1)}{1\cdot2\cdot(2n-1)}\mu^{n-2} + \frac{n(n-1)(n-2)(n-3)}{2\cdot4\cdot(2n-1)(2n-3)}\mu^{n-4}\cdots\right].$

The first five $P's$ are

$$P_0(\mu) = 1,$$
$$P_1(\mu) = \mu,$$
$$P_2(\mu) = \tfrac{1}{2}(3\mu^2 - 1),$$
$$P_3(\mu) = \tfrac{1}{2}(5\mu^3 - 3\mu),$$
$$P_4(\mu) = \tfrac{1}{8}(35\mu^4 - 30\mu^2 + 3),$$
$$P_5(\mu) = \tfrac{1}{8}(63\mu^5 - 70\mu^3 + 15\mu).$$

In fact (see 44) the polynomials P_n may be defined as the coefficients of the powers of α^n in the development in powers of α of

$$(48) \qquad (1 - 2\alpha\mu + \alpha^2)^{-\frac{1}{2}} = 1 + P_1(\mu)\alpha + P_2(\mu)\alpha^2 + \cdots$$

This function is called the *generating function* for the polynomials $P_n(\mu)$.

93. Various Forms for Zonal Harmonic. If we develope the binomial $(\mu^2 - 1)^n = \mu^{2n} - n\mu^{2n-2} + \dfrac{n(n-1)}{2!}\mu^{2n-4}\cdots$ and differentiate n times, we find

$$(49)\ \frac{d^n}{d\mu^n}(\mu^2-1)^n = 2n(2n-1)\cdots(n+1)\mu^n - n(2n-2)(2n-3)\cdots(n-1)\mu^{n-2} + \cdots$$

$$(50)\quad \frac{1}{2^n n!}\frac{d^n}{d\mu^n}(\mu^2 - 1)^n = \frac{(2n)!}{2^n(n!)^2}\Big[\mu^n - \frac{n(n-1)}{1\cdot 2(2n-1)} + \cdots\Big] = P_n.$$

This is Rodrigues's expression for P_n and since $dy/dx = 0$ always has a root between two consecutive roots of $y = 0$, and since $(\mu^2 - 1)^n$ has n roots equal to 1 and n equal to -1, $P_n(\mu) = 0$ has n real roots between 1 and -1. Consequently it has no imaginary roots.

From the generating function,

$$(1 - 2\alpha\mu + \alpha^2)^{-\frac{1}{2}} = (1 - \alpha e^{i\theta})^{-\frac{1}{2}}(1 - \alpha e^{-i\theta})^{-\frac{1}{2}} = \sum_{n=0}^{n=\infty}\alpha^n P_n$$

$$= \Big[1 + \tfrac{1}{2}\alpha e^{i\theta} + \tfrac{1}{2}\cdot\tfrac{3}{2}\cdot\tfrac{1}{2!}\alpha^2 e^{2i\theta} + \tfrac{1}{2}\cdot\tfrac{3}{2}\cdot\tfrac{5}{2}\cdot\tfrac{1}{3!}\alpha^3 e^{3i\theta}\cdots\Big]$$

$$(51) \quad \times\Big[1 + \tfrac{1}{2}\alpha e^{-i\theta} + \tfrac{1}{2}\cdot\tfrac{3}{2}\cdot\tfrac{1}{2!}\alpha^2 e^{-2i\theta} + \tfrac{1}{2}\cdot\tfrac{3}{2}\cdot\tfrac{5}{2}\cdot\tfrac{1}{3!}\alpha^3 e^{-3i\theta}\cdots\Big]$$

$$= \sum_{k=0}^{k=\infty}\frac{1\cdot 3\cdot 5\cdots 2k-1}{2^k k!}\alpha^k e^{-ik\theta}\sum_{j=0}^{j=\infty}\frac{1\cdot 3\cdot 5\cdots 2j-1}{2^j j!}\alpha^j e^{ji\theta}.$$

Collecting the terms in α^n, with $k + j = n$

$$(52)\quad P_n(\mu) = \sum\frac{1\cdot 3\cdot 5\cdots 2k-1\cdot 1\cdot 3\cdot 5\cdots 2n-2k-1}{2^k k!\, 2^{n-k}(n-k)!}e^{(n-2k)i\theta}.$$

If we collect the terms in $k = p$ and $k = n - p$ we get

$$\sum \frac{1 \cdot 3 \cdot 5 \cdots 2p - 1 \cdot 1 \cdot 3 \cdot 5 \cdots 2n - 2p - 1}{2^n \, p! \, (n - p)!} \, 2 \cos (n - 2p) \theta$$

so that finally

$$(53) \quad P_n (\cos \theta) = 2 \frac{1 \cdot 3 \cdot 5 \cdots 2n - 1}{2 \cdot 4 \cdot 6 \cdots 2n} \left\{ \cos n\theta + \frac{1 \cdot n}{1 (2n - 1)} \cos (n - 2) \theta + \right.$$
$$\left. + \frac{1 \cdot 3 \cdot n \, (n - 1)}{1 \cdot 2 \cdot (2n - 1) (2n - 3)} \cos (n - 4) \theta \cdots \right\}$$

This form was given by Laplace and Legendre.

Consider the integral

$$(54) \qquad\qquad I = \int_0^\pi \frac{d\omega}{A - i B \cos \omega},$$

where A, B are real and $A > 0$.

$$(55) \qquad I = A \int_0^\pi \frac{d\omega}{A^2 + B^2 \cos^2 \omega} + i B \int_0^\pi \frac{\cos \omega \, d\omega}{A^2 + B^2 \cos^2 \omega}.$$

The second integral vanishes on account of the numerator taking equal and opposite signs for ω and $\pi - \omega$. Also

$$(56) \qquad \int_0^\pi \frac{d\omega}{A^2 + B^2 \cos^2 \omega} = 2 \int_0^{\frac{\pi}{2}} \frac{d\omega}{A^2 + B^2 \cos^2 \omega}.$$

Put

$$\tan \omega = u \sqrt{A^2 + B^2} / A, \qquad \frac{d\omega}{\cos^2 \omega} = du \frac{\sqrt{A^2 + B^2}}{A}$$

$$\frac{1}{\cos^2 \omega} = 1 + \frac{u^2 (A^2 + B^2)}{A^2}.$$

$$(57) \qquad I = \frac{2A}{\sqrt{A^2 (A^2 + B^2)}} \int_0^\infty \frac{du}{1 + u^2} = \frac{\pi}{\sqrt{A^2 + B^2}}.$$

If A is negative,

$$I = - \frac{\pi}{\sqrt{A^2 + B^2}}.$$

If we put

$$A = 1 - \alpha \mu,$$
$$B = \alpha \sqrt{1 - \mu^2}, \qquad\qquad 0 < \alpha < 1$$

$$(58) \qquad \int_0^\pi \frac{d\omega}{1 - \alpha (\mu + i \sqrt{1 - \mu^2} \cos \omega)} = \frac{\pi}{\sqrt{1 - 2\alpha\mu + \alpha^2}}.$$

The modulus of

$$\alpha (\mu + i \sqrt{1 - \mu^2} \cos \omega) \qquad\qquad < \alpha < 1$$

is

$$\alpha \sqrt{\mu^2 \sin^2 \omega + \cos^2 \omega}$$

Consequently

$$(59) \qquad \frac{1}{1 - \alpha(\mu + i\sqrt{1 - \mu^2} \cos \omega)} = \sum_{0}^{\infty} \alpha^n (\mu + i\sqrt{1 - \mu^2} \cos \omega)^n .$$

On the other hand, comparing with (48) we find

$$(60) \qquad P_n(\mu) = \frac{1}{\pi} \int_0^\pi (\mu + i\sqrt{1 - \mu^2} \cos \omega)^n \, d\omega .$$

This is Laplace's celebrated definite integral. This expression shows that $|P_n| < 1$ for all values of μ between $-1, +1$, for

$$|\mu + i\sqrt{1 - \mu^2} \cos \omega| = \sqrt{\mu^2 \sin^2 \omega + \cos^2 \omega} < 1 ,$$

$$|P_n| < \frac{1}{\pi} \int_0^\pi d\omega = 1 .$$

If on the other hand $\alpha > 1$,

$$(61) \quad (1 - 2\alpha\mu + \alpha^2)^{-\frac{1}{2}} = \frac{1}{\alpha}\left(1 - 2\frac{\mu}{\alpha} + \frac{\mu^2}{\alpha^2}\right) = \sum_0^\infty \frac{1}{\alpha^{n+1}} P_n(\mu)$$

Put $\qquad\qquad A = \alpha\mu - 1, \quad B = \alpha\sqrt{1 - \mu^2}$

$$\frac{1}{\sqrt{1 - 2\alpha\mu + \alpha^2}} = \frac{1}{\pi} \int_0^\pi \frac{d\omega}{(\alpha\mu - 1) - i\alpha\sqrt{1 - \mu^2} \cos \omega}$$

$$= \frac{1}{\pi} \int_0^\pi \frac{d\omega}{\alpha(\mu - i\sqrt{1 - \mu^2} \cos \omega)(1 - 1/\alpha(\mu - i\sqrt{1 - \mu^2} \cdot \cos \omega))} .$$

Now unless $\mu = 0$, α may be taken so large that

$$|1/\alpha(\mu - i\sqrt{1 - \mu^2} \cos \omega)| < 1, \quad 0 \leqq \omega \leqq \pi,$$

and developing

$$(62) \qquad P_n(\mu) = \frac{1}{\pi} \int_0^\pi \frac{d\omega}{(\mu - i\sqrt{1 - \mu^2} \cos \omega)^{n+1}}$$

or replacing ω by $\pi - \omega$

$$(63) \qquad P_n(\mu) = \frac{1}{\pi} \int_0^\pi \frac{d\omega}{(\mu + i\sqrt{1 - \mu^2} \cos \omega)^{n+1}} .$$

These formulae are due to Jacobi. From the first definite integral we may prove that $\lim\limits_{n=\infty} P_n(\mu) = 0$, for all values of μ *between* $-1, 1$.

For

$$(64) \qquad |\, P_n(\mu)\,| \leqq \frac{1}{\pi} \int_0^\pi |\, \mu + i\sqrt{1-\mu^2}\, \cos \omega \,|^n\, d\omega$$

$$\leqq \frac{1}{\pi} \int_0^\pi (\mu^2 \sin^2 \omega + \cos^2 \omega)^{\frac{n}{2}}\, d\omega\,.$$

Let ε be a positive quantity, then

$$\int_0^\pi = \int_0^\varepsilon + \int_\varepsilon^{\pi-\varepsilon} + \int_{\pi-\varepsilon}^\pi.$$

Since the integrand in (64) is less than unity,

$$\int_0^\varepsilon (\mu^2 \sin^2 \omega + \cos^2 \omega)^{\frac{n}{2}}\, d\omega < \int_0^\varepsilon d\omega = \varepsilon, \text{ and likewise} \int_{\pi-\varepsilon}^\pi < \varepsilon.$$

Replacing the integrand by its greatest value,

$$\int_\varepsilon^{\pi-\varepsilon} \{1 - (1-\mu^2)\sin^2 \omega\}^{\frac{n}{2}}\, d\omega < (\pi - 2\varepsilon)\,\{1 - (1-\mu^2)\sin^2 \varepsilon\}^{\frac{n}{2}}.$$

Accordingly,

$$|\, P_n(\mu)\,| < \frac{2\varepsilon}{\pi} + \frac{\pi - 2\varepsilon}{\pi}\,\{1 - (1-\mu^2)\sin^2 \varepsilon\}^{\frac{n}{2}}$$

where ε has a definite value. But $\lim\limits_{n=\infty} [1 - (1-\mu^2)\sin^2 \varepsilon]^{\frac{n}{2}} = 0$ so that if we make ε as small as we please, we make $|\, P_n\,|$ as small as we please.

In the formula (60) let us change the variable from ω to

$$\xi = \mu + i\sqrt{1-\mu^2}\, \cos \omega, \quad d\xi = -\, i\sqrt{1-\mu^2}\, \sin \omega\, d\omega,$$

$$\frac{(\xi-\mu)^2}{1-\mu^2} = -\cos^2 \omega, \ \sin^2 \omega = 1 + \frac{(\xi-\mu)^2}{1-\mu^2} = \frac{1-2\xi\mu+\xi^2}{1-\mu^2}$$

$$d\omega = \frac{i\, d\xi}{(1 - 2\mu\xi + \xi^2)^{\frac{1}{2}}},$$

$$(65) \qquad P_n(\mu) = -\frac{i}{\pi} \int_{\mu - i\sqrt{1-\mu^2}}^{\mu + i\sqrt{1-\mu^2}} \frac{\xi^n\, d\xi}{(1 - 2\mu\xi + \xi^2)^{\frac{1}{2}}}\,.$$

Put

$$\xi = e^{i\varphi}, \ d\xi = i e^{i\varphi} \delta\varphi, \ \cos \theta = \mu$$

$$1 - 2\mu\xi + \xi^2 = 1 - (e^{i\theta} + e^{-i\theta})\, e^{i\varphi} + e^{2i\varphi} = e^{i\varphi}\,\{e^{i\varphi} + e^{-i\varphi} - (e^{i\theta} + e^{-i\theta})\}$$

from which

$$P_n (\cos \theta) = \frac{1}{\pi} \int_{-\theta}^{\theta} \frac{e^{i(n+\frac{1}{2})\varphi} \, d\varphi}{\{2(\cos \varphi - \cos \theta)\}^{\frac{1}{2}}}$$

or since the imaginary part vanishes,

(66) $$P_n(\cos \theta) = \frac{2}{\pi} \int_{0}^{\theta} \frac{\cos\left(n+\frac{1}{2}\right)\varphi \, d\varphi}{\{2(\cos \varphi - \cos \theta)\}^{\frac{1}{2}}}.$$

This is the Mehler-Dirichlet integral.

94. Relations between neighbouring Harmonics. From the generating function

(48) $$(1 - 2\alpha\mu + \alpha^2)^{-\frac{1}{2}} = \sum_{n=0}^{n=\infty} \alpha^n P_n(\mu),$$

by partial differentiation with respect to α,

(67) $$(\mu - \alpha)(1 - 2\alpha\mu + \alpha^2)^{-\frac{3}{2}} = \sum_{n=0}^{n=\infty} n\alpha^{n-1} P_n(\mu).$$

Multiplying by
$$1 - 2\alpha\mu + \alpha^2$$

$$(\mu - \alpha) \sum_{n=0}^{n=\infty} \alpha^n P_n(\mu) = (1 - 2\alpha\mu + \alpha^2) \sum_{n=0}^{n=\infty} n\alpha^{n-1} P_n(\mu),$$

Equating the coefficients of α^n on both sides,

(68) $$\mu P_n(\mu) - P_{n-1}(\mu) = (n+1) P_{n+1}(\mu) - 2n\mu P_n(\mu) + (n-1) P_{n-1}(\mu)$$
$$(n + 1) P_{n+1}(\mu) - (2n + 1) \mu P_n(\mu) + n P_{n-1}(\mu) = 0$$

and for $n = 0$
$$P_1(\mu) - \mu P_0(\mu) = 0$$

an important recurrence-formula due to Ossian Bonnet.

Differentiating the generating function according to μ, and multiplying

$$\alpha(1 - 2\alpha\mu + \alpha^2)^{-\frac{3}{2}} = \sum_{n=0}^{n=\infty} \alpha^n \frac{dP_n}{d\mu},$$

$$\alpha \sum_{n=0}^{n=\infty} \alpha^n P_n(\mu) = (1 - 2\alpha\mu + \alpha^2) \sum \alpha^n \frac{dP_n}{d\mu},$$

from which

(69) $$P_n(\mu) = \frac{d\,P_{n+1}}{d\mu} + \frac{d\,P_{n-1}}{d\mu} - 2\,\mu\,\frac{d\,P_n}{d\mu}.$$

Multiplying the α-derivative of $\sum\limits_{n=0}^{n=n}\alpha\,P_n$ by α and the μ-derivative by $(\mu - \alpha)$, and equating,

$$(\mu - \alpha) \sum_{s=0}^{n=\infty} \alpha^n \frac{d\,P_n}{d\mu} = \sum_{n=0}^{n=\infty} n\,\alpha^n\,P_n(\mu),$$

from which

(70) $$\mu\,\frac{d\,P_n}{d\mu} - \frac{d\,P_{n-1}}{d\mu} = n\,P_n$$

and eliminating $\mu\,\dfrac{d\,P_n}{d\mu}$ from this and (69),

(71) $$\frac{d\,P_{n+1}}{d\mu} - \frac{d\,P_{n-1}}{d\mu} = (2n+1)\,P_n.$$

Making $n = 0, 1, 2, 3 \ldots$ and adding,

(71 a) $\quad 1 + 3\,P_1(\mu) + 5\,P_3(\mu) + \cdots + (2n+1)\,P_n(\mu) = \dfrac{d\,P_{n+1}}{d\mu} + \dfrac{d\,P_n}{d\mu},$

95. Associated Functions. Consider the integral

(72) $$\Phi(r) = \int\limits_{-1}^{1} (1 - \mu^2)^r \frac{d^r P_m}{d\mu^r} \frac{d^r P_n}{d\mu^r}\, d\mu$$

$P_n(\mu)$ satisfies Legendre's equation

(41) $$(\mu^2 - 1) \frac{d^2 P_n}{d\mu^2} + 2\,\mu\,\frac{d\,P_n}{d\mu} - n(n+1)P_n = 0,$$

differentiating with $r - 1$ times gives

(73) $(\mu^2 - 1)\dfrac{d^{r+1} P_n}{d\mu^{r+1}} + 2r\mu\,\dfrac{d^r P_n}{d\mu^r} - (n+r)\,(n-r+1)\dfrac{d^{r-1} P_n}{d\mu^{r-1}} = 0.$

Multiplying by $-(1 - \mu^2)^{r-1}$ we may write

(74) $\dfrac{d}{d\mu}\left[(1 - \mu^2)^r \dfrac{d^r P_n}{d\mu^r}\right] = -(n+r)(n-r+1)(1 - \mu^2)^{r-1}\dfrac{d^{r-1} P_n}{d\mu^{r-1}}.$

Integrating (72) by parts

(75)
$$\Phi(r) = \frac{d^{r-1} P_m}{d\mu^{r-1}} \frac{d^r P_n}{d\mu^r} (1 - \mu^2)^r \Big|_{-1}^{1}$$
$$- \int\limits_{-1}^{1} \frac{d^{r-1} P_m}{d\mu^{r-1}} \frac{d}{d\mu}\left[(1 - \mu^2)^r \frac{d^r P_n}{d\mu^r}\right] d\mu$$

and since the integrated part vanishes at both limits, from the differential equation (74)

(76) $$\Phi(r) = (n + r)\,(n - r + 1)\,\Phi(r - 1),$$

and in extension,

$$\Phi(r) = (n+r)(n-r+1)(n+r-1)(n-r+2)\,\Phi(r-2)$$
$$= (n-r+1)(n-r+2)\cdots(n+r)\,\Phi(0)$$
$$= \frac{(n+r)!}{(n-r)!}\,\Phi(0).$$

We have seen in Chapter III that the Legendre polynomials are normal functions for a rotating chain, and that they are consequently orthogonal. We may prove this from Legendre's equation. Putting in equation (275) Chapter III,

$$p(x) = 1 - x^2 \qquad q(x) = n(n+1)$$

we obtain Legendre's equation, and putting $u = P_m$, $v = P_n$, $\mu = x$, formula (278) becomes

(77)
$$(\mu^2 - 1)\,[P_n P_m' - P_m P_n']\Big|_{-1}^{1}$$
$$+ [n(n+1) - m(m+1)] \int_{-1}^{1} P_m(\mu)\,P_n(\mu)\,d\mu = 0.$$

Accordingly we have, if n is not equal m,

(78)
$$\int_{-1}^{1} P_m(\mu)\,P_n(\mu)\,d\mu = 0.$$

The function

(79)
$$(1 - \mu^2)^{\frac{r}{2}} \frac{d^r P_n}{d\mu^r} = \sin^r \theta \, \frac{d^r P_n(\cos\theta)}{d(\cos\theta)^r}$$

is called the associated function of order r and degree n and is denoted by $P_n^{(r)}(\mu)$, and by differentiation is found to be

(80)
$$P_n^{(r)}(\mu) = \frac{(2n)!}{2^n(n!)(n-r)!}(1-\mu^2)^{\frac{r}{2}}\Big[\mu^{n-r} - \frac{(n-r)(n-r-1)}{2(2n-1)}\mu^{n-r-2}$$
$$+ \frac{(n-r)(n-r-1)(n-r-2)(n-r-3)}{2\cdot4\cdot(2n-1)(2n-3)}\mu^{n-r-4}\cdots\Big].$$

We find then (72),

(81)
$$\Phi(r) = \int_{-1}^{1} P_n^{(r)} P_m^{(r)}\,d\mu = 0, \qquad\qquad m \neq n,$$

and it will be proved later that

(82)
$$\Phi(0) = \int_{-1}^{1} P_m(\mu)\,P_n(\mu)\,d\mu = \frac{2}{2n+1}, \qquad\qquad m = n,$$

so that we have

$$(83) \qquad \int_{-1}^{1} (P_n^{(r)}(\mu))^2 d\mu = \frac{2}{2n+1} \frac{(n+r)!}{(n-r)!}.$$

Putting $r = 0$ we get the formula for P_n.

Let us seek the differential equation satisfied by $P_n^{(r)}(\mu)$. Differentiating,

$$\frac{d P_n^{(r)}}{d\mu} = (1 - \mu^2)^{\frac{r}{2}} \frac{d^{r+1} P_n}{d\mu^{r+1}} - r\mu (1 - \mu^2)^{\frac{r}{2}-1} \frac{d^r P_n}{d\mu^r}$$

$$\frac{d^2 P_n^{(r)}}{d\mu^2} = (1 - \mu^2)^{\frac{r}{2}} \frac{d^{r+2} P_n}{d\mu^{r+2}} - 2r\mu (1 - \mu^2)^{\frac{r}{2}-1} \frac{d^{r+1} P_n}{d\mu^{r+1}}$$

$$- \left[r (1 - \mu^2)^{\frac{r}{2}-1} - r(r-2) \mu^2 (1 - \mu^2)^{\frac{r}{2}-2} \right] \frac{d^{r+1} P_n}{d\mu^{r+1}}.$$

From these we get

$$(1 - \mu^2) \frac{d^2 P_n^{(r}}{d\mu^2} - 2\mu \frac{d P_n^{(r)}}{d\mu} + n(n+1) P_n^{(r)}$$

$$(84) \qquad = (1 - \mu^2)^{\frac{r}{2}} \left[(1 - \mu^2) \frac{d^{r+2} P_n}{d\mu^{r+2}} - 2(r+1) \mu \frac{d^{r+1} P_n}{d\mu^{r+1}} \right.$$

$$\left. + \left(n(n+1) - r + \frac{r^2 \mu^2}{1 - \mu^2} \right) \frac{d^r P_n}{d\mu^r} \right].$$

Now if we differentiate Legendre's equation r times we have

$$(85) \qquad (1 - \mu^2) \frac{d^{r+2} P_n}{d\mu^{r+2}} - 2(r+1) \mu \frac{d^{r+1} P_n}{d\mu^{r+1}}$$

$$- [r(r+1) - n(n+1)] \frac{d^r P_n}{d\mu^r} = 0$$

so that the second member of (84) reduces to

$$(1 - \mu^2)^{\frac{r}{2}} \left[r(r+1) - r + \frac{r^2 \mu^2}{1 - \mu^2} \right] \frac{d^r P_n}{d\mu^r} = \frac{r^2}{1 - \mu^2} P_n^{(r)}.$$

Accordingly $P_n^{(r)}$ satisfies the differential equation

$$(86) \quad (1 - \mu^2) \frac{d^2 P_n^{(r)}}{d\mu^2} - 2\mu \frac{d P_n^{(r)}}{d\mu} + \left[n(n+1) - \frac{r^2}{1 - \mu^2} \right] P_n^{(r)} = 0.$$

96. General Harmonics. Let us now return to Laplace's equation,

$$(39) \qquad \frac{\partial}{\partial\mu} \left\{ (1 - \mu^2) \frac{\partial Y_n}{\partial\mu} \right\} + \frac{1}{1 - \mu^2} \frac{\partial^2 Y_n}{\partial\varphi^2} + n(n+1) Y_n = 0.$$

Suppose we undertake to find a particular solution of the form $Y_n = \Theta(\theta)\, \Phi(\varphi)$, we obtain by dividing by $\frac{Y_n}{1-\mu^2}$,

$$(87) \qquad \frac{1-\mu^2}{\Theta}\frac{d}{d\mu}\left\{(1-\mu^2)\frac{d\Theta}{d\mu}\right\} + n(n+1)(1-\mu^2) = -\frac{1}{\Phi}\frac{d^2\Phi}{d\varphi^2},$$

which as before must be a constant, say k^2, giving the two ordinary differential equations

$$(88) \qquad \frac{d^2\Phi}{d\varphi^2} + k^2\Phi = 0,$$

satisfied by

$$\Phi = A\cos k\varphi + B\sin k\varphi,$$

and

$$(89) \qquad \frac{d}{d\mu}\left\{(1-\mu^2)\frac{d\Theta}{d\mu}\right\} + n(n+1)\,\Theta = \frac{k^2\Theta}{1-\mu^2}$$

which is satisfied by $\Theta = P_n^{(k)}(\mu)$.

Thus we have the particular solutions

$$\cos k\varphi\,(\sin\theta)^k \frac{d^k P_n}{d\mu^k}, \quad \sin k\varphi\,(\sin\theta)^k \frac{d^k P_n}{d\mu^k}.$$

Since $P_n(\mu)$ is a polynomial of degree n, we may put

$$k = 0, 1, 2, 3 \ldots n,$$

which gives us the $2n+1$ harmonics

$$k = 0, \qquad P_n$$

$$k = 1, \qquad \cos\varphi\sin\theta\frac{dP}{d\mu}, \qquad \sin\varphi\sin\theta\frac{dP_n}{d\mu},$$

$$k = 2, \qquad \cos 2\varphi\sin^2\theta\frac{d^2P_n}{d\mu^2}, \quad \sin 2\varphi\sin^2\theta\frac{d^2P_n}{d\mu^2},$$

$$\cdots \cdots \cdots \cdots$$

$$k = n, \qquad \cos n\varphi\sin^n\theta\frac{d^nP}{d\mu^n}, \quad \sin n\varphi\sin^n\theta\frac{d^nP_n}{d\mu^n}.$$

Either of the harmonics

$$\cos k\varphi\sin^k\theta\frac{d^k P_n}{d\mu^k}, \quad \sin k\varphi\sin^k\theta\frac{d^k P_n}{d\mu^k},$$

vanishes for $\theta = 0$, $\theta = \pi$, and for $n - k$ other values, giving $n - k$ parallels of latitude. Also

$$\sin k\varphi = 0 \text{ for } \varphi = 0, \frac{\pi}{k}, \frac{2\pi}{k} \cdots \frac{k-1}{k}\pi, \frac{k+1}{\pi} \cdots 2\pi,$$

giving k meridians.

$$\cos k\varphi = 0 \text{ for } \varphi = \frac{\pi}{2k}, \frac{3\pi}{2k}, \cdots \frac{2k-1}{2k}2\pi,$$

also k meridians. Hence the zero values of these harmonics give k meridians making equal angles with each other, and $n - k$ parallels, which divide the sphere into spherical rectangles. Accordingly these are called *tesseral harmonics*, of degree n and type k. For $k = n$, since $\frac{d^n P_n}{d\mu^n}$ = constant there are no parallels, but only meridians, and the harmonic is called sectorial. Thus the zonal and sectorial harmonics are the extreme cases of the tesseral. As these $2n + 1$ harmonics are all independent, the general harmonic of degree n is the sum of each multiplied by an arbitrary constant, or

$$(90) \qquad V_n = r^n \sum_{k=0}^{k=n} (A_{kn} \cos k\varphi + B_{kn} \sin k\varphi) \sin^k \theta \frac{d^k P_n}{d\mu^k}.$$

97. Integral Relations. If we apply Green's theorem to two harmonic functions in a space bounded by a surface S,

$$(91) \qquad \iint \left\{ U \frac{\partial V}{\partial n} - V \frac{\partial U}{\partial n} \right\} dS = 0.$$

Let the space be that inside a sphere of radius r, and let

$$(92) \qquad \begin{array}{ll} U = V_n = r^n Y_n, & V = V_m = r^m Y_m \\ \dfrac{\partial U}{\partial n} = -nr^{n-1} Y_n, & \dfrac{\partial V}{\partial m} = -mr^{m-1} Y_m \end{array}$$

and the equation is

$$(93) \qquad r^{m+n-1}(m - n) \iint Y_m Y_n dS = 0.$$

Consequently, unless $m = n$

$$(94) \qquad \iint Y_m Y_n dS = 0$$

over any sphere, or the Y's are orthogonal. In fact it may be shown that they are normal functions for various spherical problems. If both Y's are zonal, we may take

$$dS = 2\pi \sin \theta d\theta$$

and

$$(95) \qquad \iint P_m P_n \sin \theta d\theta = 0,$$

or

$$\int_{-1}^{1} P_m(\mu) P_n(\mu) d\mu = 0, \qquad\qquad \mu = \cos \theta,$$

which is the same result as from Legendre's equation (78).

We may apply the theorem of Green, Chapter V, (70),

$$(96) \qquad V_P = \frac{1}{4\pi} \iint \left\{ V \frac{\partial \left(\frac{1}{d} \right)}{\partial n} - \frac{1}{d} \frac{\partial V}{\partial n} \right\} dS,$$

to obtain the development of a function in spherical harmonics. We have as above (44)

(44) $$\frac{1}{d} = \frac{1}{r} \sum_{s=0}^{s=\infty} \left(\frac{r'}{r}\right)^s P_s(\mu)$$

where

$$r' < r, \quad \mu = \cos (rr').$$

Putting in the above formula $V = r^m Y_m$,

(97) $$V_m(P) = \frac{1}{4\pi} \int_0^\pi \int_0^{2\pi} \left\{ r^m Y_m \sum_{s=0}^{s=\infty} (s+1) \frac{r'^s}{r^{s+2}} P_s(\mu) \right.$$

$$\left. + m r^{m-2} Y_m \sum \left(\frac{r'}{r}\right)^s P_s(\mu) \right\} r^2 \sin\theta \, d\theta \, d\varphi.$$

If the coordinates of P be r', θ', φ' we have $V_m(P) = r'^m Y_m(\theta', \varphi')$, while on the right we have an infinite series in powers of r', with definite integrals as coefficients. Since this is an identity for all $r' < r$, comparing coefficients we must have, in r'^s

(98) $$\int_0^\pi \int_0^{2\pi} Y_m(\theta, \varphi) P_s(\mu) \sin\theta \, d\theta \, d\varphi = 0 \qquad\qquad s \gtrless m$$

and if $s = m$

(99) $$\int_0^\pi \int_0^{2\pi} Y_m(\theta, \varphi) P_m(\mu) \sin\theta \, d\theta \, d\varphi = \frac{4\pi}{2m+1} Y_m(\theta', \varphi').$$

In forming this integral, we must put for μ the value from spherical trigonometry $\mu = \cos (rr') = \cos\theta \cos\theta' + \sin\theta \sin\theta' \cos(\varphi - \varphi')$. If we take for Y_m the zonal harmonic P_m and put $\theta' = 0$, $Y_m(\theta', \varphi') = P_m(1) = 1$, and we obtain (82)

$$\int_{-1}^1 P_m^2 d\mu = \frac{2}{2m+1},$$

the missing formula, (82).

98. Development in Spherical Harmonics. By means of the integral expressions (98), (99), we may find the development of a function of θ, φ in spherical harmonics, *assuming* that such a development is possible. Suppose that

(100) $$f(\theta, \varphi) = Y_0 + Y_1 + Y_2 + \cdots.$$

Multiply both sides by $P_n(\mu) \sin \theta \, d\theta \, d\varphi$ and integrate over the surface of the sphere. Every term vanishes except the nth, and we obtain

$$(101) \qquad \int\!\!\int f(\theta, \varphi) P_n(\mu) \sin \theta \, d\theta \, d\varphi = \frac{4\pi}{2n+1} Y_n(\theta', \varphi').$$

Accordingly to find the value of the nth term in the series at any point $P(\theta', \varphi')$ we find the value of the zonal harmonic with pole at P multiply its value at every point of the sphere by the value of f at that point and by $(2n+1)/4\pi$ times the element of area, and integrate over the sphere It is obvious that this is exactly the process described in Chapter III for development in series of normal functions.

It will be convenient to exhibit the development as a series of tesseral harmonics. Let us put

$$(102) \qquad \sum_{n=0}^{n=\infty} Y_n(\theta, \varphi) = \sum_{n=0}^{n=\infty} \sum_{k=0}^{k=n} P_n^{(k)}(\mu)[A_{nk} \cos k\varphi + B_{nk} \sin k\varphi],$$

and the coefficients A, B are to be determined. Multiply by $\cos m\varphi$ and integrate, and since

$$(103) \int_0^{2\pi} \cos m\varphi \sin k\varphi \, d\varphi = 0, \quad \int_0^{2\pi} \cos m\varphi \cos k\varphi \, d\varphi = \begin{cases} 0 \text{ if } m \neq k \\ \pi \text{ if } m = k > 0 \\ 2\pi \text{ if } m = k = 0 \end{cases}$$

we obtain

$$(104) \qquad \sum_{n=0}^{n=\infty} P_n^{(k)}(\mu) \lambda_k \pi A_{nk} = \int_0^{2\pi} f(\theta, \varphi) \cos k\varphi \, d\varphi \qquad \begin{aligned} \lambda_0 &= 2, \ k = 0. \\ \lambda_k &= 1, \ k \neq 0. \end{aligned}$$

Multiply by $P_m^{(k)}(\mu) d\mu$ and integrate, and since (81) (83)

$$\int_{-1}^{1} P_m^{(k)} P_n^{(k)} \, d\mu = \begin{cases} 0 \text{ if } m \neq n \\ \dfrac{(n+k)!}{(n-k)!} \dfrac{2}{2n+1}, \quad m = n \end{cases}$$

we have only one term on the left

$$(105) \qquad \frac{(n+k)!}{(n-k)!} \frac{2\lambda_k \pi}{2n+1} A_{nk} = \int_{-1}^{1}\!\!\int_0^{2\pi} f(\theta, \varphi) \cos k\varphi \, P_m^{(k)} \, d\mu \, d\varphi,$$

so that finally

$$(106) \qquad A_{nk} = \frac{(n-k)!}{(n+k)!} \frac{2n+1}{2\lambda_k \pi} \int_0^{\pi}\!\!\int_0^{2\pi} f(\theta, \varphi) \cos k\varphi \, P_n^{(k)} \sin \theta \, d\theta \, d\varphi$$

and in like manner

$$(107) \qquad B_{nk} = \frac{(n-k)!}{(n+k)!} \frac{2n+1}{2\lambda_k \pi} \int_0^{\pi}\!\!\int_0^{2\pi} f(\theta, \varphi) \sin k\varphi \, P_n^{(k)} \sin \theta \, d\theta \, d\varphi$$

Inserting, these values, we get the series

(108)
$$Y_n(\theta', \varphi') =$$

$$= \sum_{k=0}^{k=n} \left[\frac{(n-k)!}{(n+k)!} \frac{2n+1}{2\lambda_k\pi} \left(\cos k\varphi' \int_0^\pi \int_0^{2\pi} f(\theta, \varphi) \cos k\varphi \, P_n^{(k)} \sin \theta \, d\theta \, d\varphi \right. \right.$$

$$\left. \left. + \sin k\varphi' \int_0^\pi \int_0^{2\pi} f(\theta, \varphi) \sin k\varphi \, P_n^{(k)} \sin \theta \, d\theta \, d\varphi \right) \right] P_n^{(k)} \mu'.$$

Multiplying in the factors $\sin k\varphi'$, $\cos k\varphi'$, $P_n^{(k)}(\mu')$,

(109)
$$Y_n(\theta', \varphi')$$

$$= \int_0^\pi \int_0^{2\pi} \sum_{k=0}^{k=n} \frac{(n+k)!}{(n-k)!} \frac{2n+1}{2\lambda_k\pi} f(\theta, \varphi) \, P_n^{(k)}(\mu') \, P_n^{(k)}(\mu) \cos k(\varphi - \varphi') \sin \theta \, d\theta \, d\varphi.$$

Comparing with the previous formula (101), we find, using the value of

$$\mu = \cos (r\,r') = \cos \gamma$$

$$P_n\left(\cos \theta \cos \theta' + \sin \theta \sin \theta' \cos (\varphi - \varphi') \right)$$

(110)
$$= \sum_{k=0}^{k=n} \frac{(n-k)!}{(n+k)!} \frac{2}{\lambda_k} P_n^{(k)}(\cos \theta) \, P_n^{(k)}(\cos \theta') \cos k(\varphi - \varphi').$$

The zonal harmonic thus expressed as a symmetrical function of the coordinates of the two points P, Q is called a Laplace's coefficient, or biaxal harmonic. Considered as a function of θ, φ it contains two arbitrary constants, θ', φ'.

99. Convergence of Series. It is now important to show that the series obtained actually represents the function f, that is that if

(111) $$S_n = \frac{1}{4\pi} \sum_{s=0}^{s=n} \int_0^\pi \int_0^{2\pi} f(\theta, \varphi)(2s+1) P_s(\mu) \sin \theta \, d\theta \, d\varphi$$

we shall have

$$\lim_{n=\infty} S_n = f(\theta', \varphi').$$

This was demonstrated by Laplace, but without sufficient rigor, afterwards by Poisson, and finally in a rigorous manner by Lejeune-Dirichlet (Crelle, Journal der Mathematik Bd. 17) and Ossian Bonnet. (Journal des Mathématiques de Liouville 17). The following proof is due to Darboux.

Let us now introduce as new coordinates γ, ψ, so that P' is the new pole (Fig. 96). Then calling $f(\theta, \varphi) = F(\gamma, \psi)$

$$(112) \qquad S_n = \frac{1}{4\pi} \sum_{s=0}^{s=n} \int_0^\pi \int_0^{2\pi} F(\gamma, \psi) (2s+1) P_s(\cos\gamma) \sin\gamma \, d\gamma \, d\psi,$$

or putting

$$(113) \qquad \Phi(\gamma) =$$

$$= \frac{1}{2\pi} \int_0^{2\pi} F(\gamma, \psi) \, d\psi = \Psi(\mu),$$

that is the mean value of F on a circle whose distance from the pole is γ,

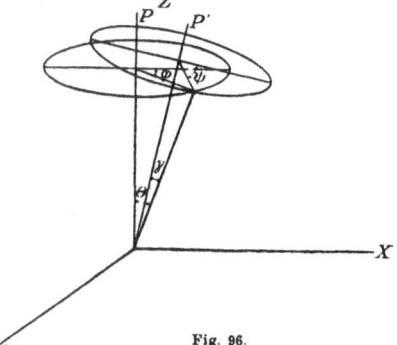

Fig. 96.

$$S_n = \sum_{s=0}^{s=n} \frac{2s+1}{2} \int_{-1}^1 \Psi(\mu) P_s(\mu) d\mu$$

$$(114) \qquad = \frac{1}{2} \int_{-1}^1 [1 + 3P_1 + 5P_2 + \cdots + (2s+1) P_s] \Psi(\mu) d\mu$$

$$= \frac{1}{2} \int_{-1}^1 \left(\frac{dP_{n+1}}{d\mu} + \frac{dP_n}{d\mu} \right) \Psi(\mu) d\mu,$$

by (71 a).

If $\Psi(\mu)$ is finite and continuous between -1, 1, we may integrate by parts

$$(115) \qquad S_n = \frac{1}{2} \Big[\Psi(\mu) (P_{n+1} + P_n) \Big]_{-1}^1 - \frac{1}{2} \int_{-1}^1 (P_{n+1} + P_n) \, \Psi'(\mu) d\mu$$

Now for

$$\mu = 1, \quad P_n = P_{n+1} = 1,$$
$$\mu = -1, \quad P_n = -P_{n+1} = (-1)^n,$$

so that the integrated part vanishes at -1, and

$$S_n = \Psi(1) - \frac{1}{2} \int_{-1}^1 (P_{n+1} + P_n) \, \Psi'(\mu) d\mu.$$

Now if $|\Psi'(\mu)| < M$ between -1, 1 the integral is less than

$$M \int_{-1}^1 (P_{n+1} + P_n) d\mu,$$

Now let ε be a positive quantity, the integral is the sum of the three

$$\int_{-1}^{1} = \int_{-1}^{-1+\varepsilon} + \int_{-1+\varepsilon}^{1-\varepsilon} + \int_{1-\varepsilon}^{1}.$$

But we have proved in § 93 that the middle term approaches the limit zero with increasing n. Also the other two vanish with ε. Consequently we have

$$\lim_{n=\infty} S_n = \Psi(1) = \Phi(\gamma = 0) = f(\theta'\varphi').$$

In order to integrate by parts, we have assumed $\Psi(\mu)$ continuous. If it has a finite number of discontinuities, we may break up the integrals on both sides of the discontinuities, and extend the demonstration to the general case. Hence any function satisfying Dirichlet's conditions can be developed in a series of spherical harmonics.

100. Definition of Bessel Functions. Let us consider the expression

(1) $$Z = e^{\frac{x}{2}(t - t^{-1})},$$

which may be developed in a series of positive and negative powers of t, since $e^{\frac{xt}{2}}$ is developable in positive powers of xt and $e^{-\frac{x}{2t}}$ in positive powers of x/t, except when $|t| = 0$. We have

$$e^{xt/2} = \sum_{k=0}^{k=\infty} \left(\frac{x}{2}\right)^k \frac{t^k}{k!}, \quad e^{-x/2t} = \sum_{j=0}^{j=\infty} (-1)^j \left(\frac{x}{2}\right)^j \frac{t^{-j}}{j!}$$

(2) $$Z = \sum_{j=0}^{j=\infty} \sum_{k=0}^{k=\infty} (-1)^j \left(\frac{x}{2}\right)^{j+k} \frac{t^{k-j}}{j!\,k!},$$

or putting $k - j = n$, $k = j + n$, $j + k = 2j + n$, and collecting in powers of t

(3) $$Z = \sum_{n=-\infty}^{n=\infty} t^n \sum_{j=0}^{j=\infty} \frac{(-1)^j \left(\frac{x}{2}\right)^{2j+n}}{j!\,(j+n)!} = \sum_{-\infty}^{\infty} t^n J_n(x)$$

if we write

(4) $$J_n(x) = \left(\frac{x}{2}\right)^n \sum_{s=0}^{s=\infty} \frac{(-1)^s \left(\frac{x}{2}\right)^{2s}}{s!\,(s+n)!}$$

Z is called the generating function for the functions $J_n(x)$. Compare § 93, (51).

If we form the partial derivatives of Z by x and t we obtain

$$\frac{\partial Z}{\partial x} = \frac{1}{2}(t - t^{-1})Z, \quad \frac{\partial^2 Z}{\partial x^2} = \frac{1}{4}(t - t^{-1})^2 Z,$$

$$\frac{\partial Z}{\partial t} = \frac{x}{2}\left(1 + \frac{1}{t^2}\right)Z, \quad \frac{\partial^2 Z}{\partial t^2} = \frac{x^2}{4}\left(1 + \frac{1}{t^2}\right)^2 Z - \frac{x}{t^3}Z,$$

from which it appears that Z satisfies the partial differential equation,

$$(5) \qquad x^2 \frac{\partial^2 Z}{\partial x^2} + x \frac{\partial Z}{\partial x} - t^2 \frac{\partial^2 Z}{\partial t^2} - t \frac{\partial Z}{\partial t} + x^2 Z = 0.$$

Substituting in this,

$$(3) \qquad Z = \sum_{-\infty}^{\infty} t^n J_n(x).$$

$$(6) \qquad \sum_{-\infty}^{\infty} t^n \left[x^2 \frac{d^2 J_n}{dx^2} + x \frac{d J_n}{dx} + (x^2 - n^2) J_n \right] = 0$$

or for each value of n

$$(6) \qquad \frac{d^2 J_n}{dx^2} + \frac{1}{x} \frac{d J_n}{dx} + \left(1 - \frac{n^2}{x^2}\right) J_n = 0.$$

This is Bessel's differential equation and the series $J_n(x)$ is called the Bessel function of order n. We have supposed the *parameter* n to be a real integer, while the argument x may be real or complex.

We had (3)

$$Z = J_0(x) + t J_1(x) + t^2 J_2(x) + \cdots$$
$$+ t^{-1} J_{-1}(x) + t^{-2} J_{-2}(x) + \cdots$$

and since Z is unchanged when we change t to $-t^{-1}$ we have

$$(7) \qquad Z = J_0(x) - t^{-1} J_1(x) + t^{-2} J_2(x) - \cdots$$
$$- t J_{-1}(x) + t_2 J_{-2}(x) - \cdots$$

so that by comparing coefficients we have

$$(8) \qquad J_n(x) = (-1)^n J_{-n}(x).$$

Accordingly we may write

$$(9) \qquad e^{\frac{x}{2}(t - t^{-1})} = J_0(x) + J_2(x)(t^2 + t^{-2}) + J_4(x)(t^4 + t^{-4}) + \cdots$$
$$+ J_1(x)(t - t^{-1}) + J_3(x)(t^3 - t^{-3}) + \cdots$$

and putting $t = e^{i\varphi}$,

$$(10) \qquad e^{ix \sin \varphi} = J_0(x) + 2 J_2(x) \cos 2\varphi + 2 J_4(x) \cos 4\varphi + \cdots$$
$$+ i \left[2 J_1(x) \sin \varphi + 2 J_3(x) \sin 3\varphi + \cdots \right]$$

which is the development of the exponential in a Fourier's series. If x and φ are both real, equating real and imaginary parts

(11)
$$\cos(x \sin \varphi) = J_0(x) + 2J_2(x) \cos 2\varphi + 2J_.(x) \cos 4\varphi + \cdots$$
$$\sin(x \sin \varphi) = 2J_1(x) \sin \varphi + 2J_3(x) \sin 3\varphi + \cdots$$

From these we get

(12)
$$J_{2n}(x) = \frac{1}{\pi} \int_0^\pi \cos(x \sin \varphi) \cos 2n\varphi \, d\varphi,$$

$$J_{2n+1}(x) = \frac{1}{\pi} \int_0^\pi \sin(x \sin \varphi) \sin(2n+1)\varphi \, d\varphi.$$

Changing φ to $\varphi + \pi/2$ in (11)

(13)
$$\cos(x \cos \varphi) = J_0(x) - 2J_2(x) \cos 2\varphi + 2J_4(x) \cos 4\varphi + \cdots$$
$$\sin(x \cos \varphi) = 2J_1(x) - 2J_3(x) \cos 3\varphi + \cdots$$

From any of these formulae the J's may be obtained as definite integrals.

Differentiating the generating function (3) according to t,

(14)
$$\frac{x}{2}(1 + t^{-2})Z = \sum_{-\infty}^{\infty} n t^{n-1} J_n(x), \quad \text{or}$$

$$\frac{x}{2}(1 + t^{-2}) \sum_{-\infty}^{\infty} t^n J_n(x) = \sum_{-\infty}^{\infty} n t^{n-1} J_n(x),$$

and by comparing coefficients of t^{n-1},

(15)
$$\frac{x}{2}[J_{n-1}(x) + J_{n+1}(x)] = nJ_n(x)$$

$$\frac{1}{2}[J_{n+1}(x) + J_{n-1}(x)] = \frac{nJ_n(x)}{x}.$$

From this formula J for any argument may be calculated if those for the next two values of n are known.

Differentiating Z according to x,

(16)
$$\frac{1}{2}(t - t^{-1}) \sum_{-\infty}^{\infty} t^n J_n(x) = \sum_{-\infty}^{\infty} t^n \frac{dJ_n(x)}{dx},$$

and as above

(17)
$$\frac{1}{2}(J_{n-1}(x) - J_{n+1}(x)) = \frac{dJ_n}{dx}.$$

By differentiating thus, and reapplying the formula, we may express all derivatives in terms of successive J's.

The equations (15) and (17) may be taken as functional equations to define any so-called *cylinder* function $C_n(x)$ for *any* value of n. Adding and subtracting we have

(18) $\qquad \dfrac{dC_n(x)}{dx} = -\dfrac{n}{x}C_n(x) + C_{n-1}(x), \qquad \dfrac{d}{dx}(x^n C_n) = x^n C_{n-1},$

or

(19) $\qquad \dfrac{dC_n(x)}{dn} = \dfrac{n}{x}C_n(x) - C_{n+1}(x), \qquad \dfrac{d}{dx}\left(\dfrac{C_n}{x^n}\right) = -\dfrac{C_{n+1}}{x^n}.$

Differentiating (19)

$$\frac{d^2 C_n}{dx^2} = -\frac{n}{x^2}C_n + \frac{n}{x}\frac{dC}{dx} - \frac{dC_{n+1}}{dx}$$

and taking $\dfrac{dC_n}{dx}$ from (19)

and $\dfrac{dC_{n+1}}{dx}$ from (18)

$$\frac{d^2 C_n}{dx^2} = -\frac{n}{x^2}C_n + \frac{n}{x}\left(\frac{n}{x}C_n - C_{n+1}\right) + \frac{n+1}{x}C_{n+1} - C_n.$$

Add to this (19) divided by x,

(20) $\qquad \dfrac{d^2 C_n}{dx^2} + \dfrac{1}{x}\dfrac{dC_n}{dx} + \left(1 - \dfrac{n^2}{x^2}\right)C_n = 0,$

which is Bessel's equation.

101. Bessel's equation. If we seek to determine the solution as a power series in ascending powers of x,

(21) $\qquad y = x^\varrho \displaystyle\sum_{k=0}^{k=\infty} a_k x^k, \quad \dfrac{dy}{dx} = \displaystyle\sum_{k=0}^{k=\infty}(\varrho + k)a_k x^{\varrho+k-1},$

$$\frac{d^2 y}{dx^2} = \sum_{k=0}^{k=\infty}(\varrho + k)(\varrho + k - 1)a_k x^{\varrho+k-2}$$

and substituting in (20)

(22) $\qquad \displaystyle\sum_{k=0}^{k=\infty} a_k\{[(\varrho + k)(\varrho + k - 1) + \varrho + k - n^2]x^{\varrho+k-2} + x^{\varrho+k}\} = 0.$

Equating to zero the coefficient of every power we get the recurrence equation,

(23) $\qquad (\varrho + k + n)(\varrho + k - n)a_k + a_{k-2}.$

Since there are to be no negative powers, $a_{-2} = 0$, which gives us the *indicial* equation,

(23) $\qquad (\varrho - n)(\varrho + n) = 0,$

with the roots $\varrho = n$ or $- n$. If we take $+ n$,

$$a_k = - \frac{a_{k-2}}{k(2n+k)}, \quad a_2 = - \frac{a_0}{2^2(n+1)}, \quad a_4 = - \frac{a_2}{4(2n+4)} = \frac{a_0}{2^4 2!(n+1)(n+2)}$$

and if we put

$$a_0 = \frac{1}{2^n \Pi(n)}, \quad \text{since} \quad \Pi(n) = n \Pi(n-1), \quad \text{(see (28) below)}$$

we have

$$a_{2k} = \frac{(-1)^k}{2^{n+2k} k! \Pi(n+k)},$$

and we have the particular integral

$$(24) \qquad y = \sum_{k=0}^{k=\infty} \frac{(-1)^k \left(\frac{x}{2}\right)^{n+2k}}{k! \Pi(n+k)} = J_n(x).$$

This is called the Bessel function of the first kind, agreeing with (4) if n is an integer.

If n is an integer and we take the root $- n$, the first n terms vanish, as they contain factorials of negative integers, (which are infinite) in the denominator, so that

$$(25) \qquad J_{-n}(x) = \sum_{k=0}^{k=\infty} \frac{(-1)^k \left(\frac{x}{2}\right)^{-n+2k}}{\Pi(-n+k) k!}$$

and putting $k = n + p$,

$$(26) \qquad J_{-n}(x) = \sum_{k=0}^{k=\infty} \frac{(-1)^{n+p} \left(\frac{x}{2}\right)^{n+2p}}{\Pi(p)(n+p)!} = (-1)^n J_n(x)$$

so that the two solutions are not linearly independent. The theory of linear differential equations shows that we must now have a solution $y = a J_n(x) + (\text{a power series})$. Instead of trying to find this directly we will pass to the limit. Let us put

$$(27) \qquad Y_n(x) = \frac{1}{\sin n\pi} (\cos n\pi J_n(x) - J_{-n}(x))$$

which is a solution, whatever n, and is indeterminate when n is an integer. We determine it by differentiation with respect to n.

If we write

$$(28) \qquad \Pi(z) = \lim_{n=\infty} \frac{1 \cdot 2 \cdot 3 \cdots n}{(1+z)(2+z)(3+z) \cdots (n+z)} n^z$$

from which[1])

(29) $$\Pi(z) = z\,\Pi(z-1), \quad \Pi(0) = 1,$$

so that when z is a whole number, $\Pi(z) = z!$. The equation (28) gives Π as a function of z for all values, so that it can be differentiated. If we put

(30) $$\Psi(z) = \frac{d\log\Pi(z)}{dz} = \frac{\Pi'(z)}{\Pi(z)},$$

we have from

$$\Pi(z+1) = (z+1)\,\Pi(z),$$

$$\Psi(z+1) = \Psi(z) + \frac{1}{z+1},$$

(30)

$$\Psi(1) = \Psi(0) + \frac{1}{1},$$

$$\Psi(2) = \Psi(0) + \frac{1}{1} + \frac{1}{2},$$

$$\cdots \cdots \cdots \cdots \cdots$$

$$\Psi(n) = \Psi(0) + \frac{1}{1} + \frac{1}{2} + \cdots + \frac{1}{n},$$

---

1) We have

$$\Pi(z-1) = \lim_{n=\infty} \frac{1\cdot 2\cdot 3\cdots n}{z(1+z)(1+2z)\cdots(n-1+z)}\,n^{z-1} = \lim_{n-1=\infty}\frac{1\cdot 2\cdot 3\cdots n-1}{z(1+z)\cdots(n-1+z)}\,n^{z}$$

whence

$$z\,\Pi(z-1) = \Pi(z).$$

We shall need the value of $\Pi(\tfrac{1}{2})$. Now from the definition,

$$\Pi\left(\frac{1}{2}\right) = \lim_{n=\infty}\frac{1\cdot 2\cdot 3\cdots n}{\frac{3}{2}\cdot\frac{5}{2}\cdot\frac{2n+1}{2}}\,n^{\frac{1}{2}} = \lim_{n=\infty}\frac{2\cdot 4\cdot 6\cdots 2n}{3\cdot 5\cdot 7\cdot(2n+1)}\,n^{\frac{1}{2}}$$

$$\Pi\left(-\frac{1}{2}\right) = \lim_{n=\infty}\frac{1\cdot 2\cdot 3\cdots n}{\frac{1}{2}\cdot\frac{3}{2}\cdot\frac{5}{2}\cdots\frac{2n-1}{2}}\,n^{-\frac{1}{2}} = \lim_{n=\infty}\frac{2\cdot 4\cdot 6\cdots 2n}{1\cdot 3\cdot 5\cdot(2n-1)}\,n^{-\frac{1}{2}}$$

and multiplying together,

$$\Pi\left(\frac{1}{2}\right)\Pi\left(-\frac{1}{2}\right) = \lim_{n=\infty}\frac{2^2\cdot 4^2\; 6^2\cdots(2n)^2}{3^2\cdot 5^2\cdot 7^2\cdots(2n-1)^2(2n+1)} = \frac{\pi}{2},$$

according to Wallis's value of π. But since $\Pi(\tfrac{1}{2}) = \tfrac{1}{2}\Pi(-\tfrac{1}{2})$ this gives

$$\Pi\left(\frac{1}{2}\right) = \frac{\sqrt{\pi}}{2}.$$

We also have in the limit,

$$\log\Pi(z) = \log 2 + \log 3\cdots + \log n + z\log n - \log(1+z) - \log(2+z)\cdots - \log(n+z)$$

and differentiating,

$$\Psi(z) = \frac{d}{dz}\log\Pi(z) = \log n - \left\{\frac{1}{1+z} + \frac{1}{2+z} + \frac{1}{3+z} + \cdots\frac{1}{n+z}\right\}$$

so that passing to the limit for $n=\infty$, and putting $z=0$,

$$\Psi(0) = -\lim_{n=\infty}\left\{1 + \frac{1}{2} + \frac{1}{3}\cdots + \frac{1}{n} - \log n\right\} = -C$$

which is the definition of Euler's constant.

and $\Psi(0) = -C = -0\cdot577215665$, Euler's constant.   Accordingly

$$\frac{\partial J_n(x)}{\partial n} = \sum_{k=0}^{k=\infty} \frac{(-1)^k \left(\frac{x}{2}\right)^{n+2k}}{k!\,\Pi(n+k)} \left\{\log\frac{x}{2} - \Psi(n+k)\right\},$$

$$\frac{\partial J_{-n}(x)}{\partial n} = \sum_{k=0}^{k=\infty} \frac{(-1)^k \left(\frac{x}{2}\right)^{-n+2k}}{k!\,\Pi(-n+k)} \left\{-\log\frac{x}{2} + \Psi(-n+k)\right\},$$

(31) $$\frac{\partial J_n}{\partial n} = J_n(x)\log\frac{x}{2} - \sum_{k=0}^{k=\infty} \frac{(-1)^k\,\Psi(n+k)\left(\frac{x}{2}\right)^{n+2k}}{k!\,\Pi(n+k)},$$

$$\frac{\partial J_{-n}}{\partial n} = -J_{-n}(x)\log\frac{x}{2} + \sum_{k=0}^{k=\infty} \frac{(-1)^k\,\Psi(-n+k)\left(\frac{x}{2}\right)^{-n+2k}}{k!\,\Pi(-n+k)}.$$

Evaluating (27) we have[1])

(32) $$Y_n(x) = \frac{-\pi\sin n\pi\, J_n + \cos n\pi\dfrac{\partial J_n}{\partial n} - \dfrac{\partial J_{-n}}{\partial n}}{\pi\cos n\pi}$$

and when n is an integer,

$$\pi Y_n = \frac{\partial J_n}{\partial n} - (-1)^n\frac{\partial J_{-n}}{\partial n},$$

so that since $$J_{-n} = (-1)^n J_n$$

we have

$$\pi Y_n(x) = 2J_n(x)\log\frac{x}{2} - \sum_{k=0}^{k=\infty} \frac{(-1)^k\,\Psi(n+k)\left(\frac{x}{2}\right)^{n+2k}}{k!\,\Pi(n+k)}$$

$$-\sum_{k=n}^{k=\infty} \frac{(-1)^{n+k}\,\Psi(-n+k)\left(\frac{x}{2}\right)^{-n+2k}}{k!\,\Pi(-n+k)} - \sum_{k=0}^{k=n-1} \frac{(-1)^{n+k}\,\Psi(-n+k)\left(\frac{x}{2}\right)^{-n+2k}}{k!\,\Pi(-n+k)}$$

Combining the two middle terms, and using (29) and (30)

$$\frac{\Psi(z)}{\Pi(z)} = \frac{\Psi(z+1) - \dfrac{1}{z+1}}{\Pi(z+1)/z+1} = \frac{(z+1)\left\{\Psi(z+1) - \dfrac{1}{z+1}\right\}}{\Pi(z+1)}$$

to reduce the last term, Ψ and Π being infinite for negative integers, putting $z = -n, -(n+1), \ldots$

$$\frac{\Psi(-n)}{\Pi(-n)} = -(-n+1)(-n+2)\ldots(-1) = (-1)^n\,\Pi(n-1),\ \&c.$$

---

[1]) The material between equations (31) and (33) was missing in the author's manuscript and has been supplied by the editor.

giving finally

$$(33) \quad \pi Y_n(x) = 2 J_n(x) \log \frac{x}{2} - \sum_{k=0}^{k=\infty} \frac{(-1)^k \left(\frac{x}{2}\right)^{n+2k}}{k!\,(n+k)!} \{\Psi(k+n) + \Psi(k)\}$$

$$- \sum_{k=0}^{k=n-1} \frac{(n-k-1)!}{k!} \left(\frac{2}{x}\right)^{n-2k}.$$

Thus the above form of Y_n is the form sought. It is called the Bessel function of the second kind. (It has already appeared in § 65.)

From our equation

$$(3) \quad e^{\frac{x}{2}(t-t^{-1})} = J_0(x) + \sum_{n=1}^{n=\infty} t^n J_n(x) + \sum_{n=1}^{n=\infty} (-1)^n t^{-n} J_n(x),$$

changing t to t^{-1},

$$(34) \quad e^{-\frac{x}{2}(t-t^{-1})} = J_0(x) + \sum_{n=1}^{n=\infty} t^{-n} J_n(x) + \sum_{n=1}^{n=\infty} (-1)^n t^n J_n(x),$$

and multiplying together we get an equation of the form

$$1 = A_0 + \sum_1^\infty A_n t^n + \sum_1^\infty A_{-n} t^{-n},$$

which being true of all values of t gives

$$A_0 = 1, \quad A_n = 0, \quad A_{-n} = 0.$$

$$(34) \quad A_0 = 1 = J_0^2(x) + 2 J_1^2(x) + 2 J_2^2(x) + \cdots$$

All the terms being positive, for real x the absolute values of J_0 must be less than 1, and of all others less than $1/\sqrt{2}$ for *all* real x.

102. Bessel's Functions as Definite Integrals. From the formula (9) with $t = e^{i\varphi}$

$$(35) \quad e^{ix\sin\varphi} = \sum_{n=-\infty}^{n=\infty} J_n(x) e^{in\varphi},$$

multiplying by $e^{-mi\varphi}$ and integrating,

$$(36) \quad \int_0^{2\pi} e^{ix\sin\varphi} e^{-mi\varphi} d\varphi = \sum_{n=-\infty}^{n=\infty} J_n(x) \int_0^{2\pi} e^{i(n-m)\varphi} d\varphi.$$

Now

$$\int_0^{2\pi} e^{iq\varphi} d\varphi = \begin{matrix} 0, & q \gtrless 0, \\ 2\pi, & q = 0, \end{matrix}$$

consequently

$$(37) \quad 2\pi J_n(x) = \int_0^{2\pi} e^{i(x\sin\varphi - n\varphi)} d\varphi,$$

and since for real x, $J_n(x)$ is real,

$$(38) \quad J_n(x) = \frac{1}{2\pi}\int_0^{2\pi}\cos(x\sin\varphi - n\varphi)\,d\varphi = \frac{1}{\pi}\int_0^{\pi}\cos(x\sin\varphi - n\varphi)\,d\varphi,$$

which is Bessel's original form. Expanding the cosine, we get the previous Fourier series integrals (12).

Consider the integral

$$(39) \quad I_{p,q} = \int_0^{\pi}\sin^{2p}\varphi\,\cos^{2q}\varphi\,d\varphi.$$

Integrating $\sin^{2p}\varphi\,\cos\varphi\,\cos^{2q-1}\varphi$ by parts,

$$I_{pq} = \frac{\sin^{2p+1}\varphi}{2p+1}\cos^{2q-1}\varphi\Bigg|_0^{\pi} + \frac{2q-1}{2p+1}\int_0^{\pi}\sin^{2(p+1)}\varphi\,\cos^{2(q-1)}\varphi\,d\varphi$$

$$= \frac{2q-1}{2p+1}\int_0^{\pi}\sin^{2p}\varphi\,(1-\cos^2\varphi)\cos^{2(q-1)}\varphi\,d\varphi = \frac{2q-1}{2p+1}\{I_{p,q-1} - I_{pq}\}.$$

$$(40) \quad I_{p,q} = \frac{2q-1}{2(p+q)}\,I_{p,q-1}.$$

In like manner

$$I_{p,q} = -\frac{\cos^{2q+1}\varphi\,\sin^{2p-1}\varphi}{2q+1}\Bigg|_0^{\pi} + \frac{2p-1}{2q+1}\int_0^{\pi}\cos^{2(q+1)}\varphi\,\sin^{2(p-1)}\varphi\,d\varphi$$

$$= \frac{2p-1}{2q+1}\int_0^{\pi}\cos^{2q}\varphi\,(1-\sin^2\varphi)\sin^{2(p-1)}d\varphi = \frac{2p-1}{2q+1}\{I_{p-1,q} - I_{p,q}\}.$$

$$(41) \quad I_{p,q} = \frac{2p-1}{2(p+q)}\,I_{p-1,q}.$$

By q successive applications of the reduction formula (40)

$$I_{p,q} = \frac{(2q-1)(2q-3)\ldots 3\cdot 1}{2(p+q)\,2(p+q-1)\ldots 2(p+1)}\,I_{p,0},$$

and by p applications of (41),

$$I_{p,0} = \frac{(2p-1)(2p-3)\ldots 3\cdot 1}{2p,\,2(p-1)\ldots 4\cdot 2}\,I_{0,0}$$

$$I_{p,q} = \frac{1\cdot 3\cdot 5\ldots(2q-1)\,1\cdot 3\cdot 5\ldots 2p-1}{2\cdot 4\cdot 6\ldots 2(p+q)}\,I_{0,0}$$

and since

$$I_{00} = \int_0^{\pi}d\varphi = \pi,$$

$$I_{p,q} = \frac{(2p)!\,(2q)!\,\pi}{2^{2(p+q)}\,p!\,q!\,(p+q)!}.$$

From this

$$\frac{2^{2n}\,n!}{(2n)!}\,\frac{1}{\pi}\int_0^{\pi}(-1)^k\,\frac{\sin^{2n}\varphi\,\cos^{2k}\varphi\cdot x^{2k}}{(2k)!}\,d\varphi=\frac{(-1)^k\left(\frac{x}{2}\right)^{2k}}{k!\,(n+k)!}.$$

Multiplying by $(x/2)^n$, and summing,

$$(42)\quad J_n(x)=\sum_{k=0}^{k=\infty}\frac{(-1)^k\left(\frac{x}{2}\right)^{2k+n}}{k!\,(n+k)!}$$

$$=\frac{2^{2n}\,n!}{(2n)!}\left(\frac{x}{2}\right)^n\frac{1}{\pi}\int_0^{\pi}\sin^{2n}\varphi\sum_{k=0}^{k=\infty}(-1)^k\frac{(x\cos\varphi)^{2k}}{(2k)!}\,d\varphi$$

$$=\frac{x^n}{1\cdot3\cdot5\ldots(2n-1)}\frac{1}{\pi}\int_0^{\pi}\sin^{2n}\varphi\,\cos(x\cos\varphi)\,d\varphi,$$

which has the advantage of plainly exhibiting the factor x^n. If we put $\cos\varphi=z$,

$$(43)\quad J_n(x)=\frac{x^n}{1\cdot3\cdot5\ldots(2n-1)\,\pi}\int_{-1}^{1}(1-z^2)^{\frac{2n-1}{2}}\cos xz\,dz.$$

The formulae (42) and (43) have been deduced under the assumption that n is an integer, but are true for any value of n if we replace the numerical factor in the denominator by $\sqrt{\pi}\,2^n\Pi(n-\tfrac{1}{2})$. If $(2n-1)/2$ is an integer, (43) may be integrated by parts, giving $J_n(x)$ in a finite number of terms involving trigonometric functions, when n is half an odd integer. These functions occur in the theory of spherical waves, as we shall see in Chapter VIII. We get the same result from the series, making use of the value of $\Pi(\tfrac{1}{2})$ that we have obtained. Perhaps the quickest way is to obtain $J_{\frac{1}{2}}(x)$ and $J_{\frac{3}{2}}(x)$ from (43) and then to obtain the rest from (15). We obtain

$$J_{\frac{1}{2}}(x)=\sqrt{\frac{2}{\pi x}}\sin x$$

$$J_{\frac{3}{2}}(x)=\sqrt{\frac{2}{\pi x}}\left(\frac{\sin x}{x}-\cos x\right)$$

$$J_{\frac{5}{2}}(x)=\sqrt{\frac{2}{\pi x}}\left\{\left(\frac{3}{x^2}-1\right)\sin x-\frac{3}{x}\cos x\right\}$$

$$\cdots\cdots\cdots\cdots\cdots\cdots$$

$$J_{-\frac{1}{2}}(x)=\sqrt{\frac{2}{\pi x}}\cos x$$

$$J_{-\frac{3}{2}}(x)=-\sqrt{\frac{2}{\pi x}}\left(\sin x+\frac{\cos x}{x}\right)$$

$$J_{-\frac{5}{2}}(x)=\sqrt{\frac{2}{\pi x}}\left\{\frac{3}{x}\sin x+\left(\frac{3}{x^2}-1\right)\cos x\right\}$$

$$\cdots\cdots\cdots\cdots\cdots\cdots$$

103. Cylindrical Coordinates. Equally important with spherical coordinates are cylindrical, or columnar coordinates, composed of the rectangular coordinate z, together with ϱ the distance from the Z-axis, and φ the azimuth or longitude, as before. Considering the normal distances to the coordinate surfaces, we easily find $h_\varrho = h_z = 1$, $h_\varphi = 1/\varrho$, which inserted in (32) § 90 gives

(44)
$$\Delta V = \frac{1}{\varrho}\left[\frac{\partial}{\partial \varrho}\left(\varrho \frac{\partial V}{\partial \varrho}\right) + \frac{\partial}{\partial \varphi}\left(\frac{1}{\varrho}\frac{\partial V}{\partial \varphi}\right) + \frac{\partial}{\partial z}\left(\varrho \frac{\partial V}{\partial z}\right)\right]$$
$$= \frac{\partial^2 V}{\partial \varrho^2} + \frac{1}{\varrho}\frac{\partial V}{\partial \varrho} + \frac{1}{\varrho^2}\frac{\partial^2 V}{\partial \varphi^2} + \frac{\partial^2 V}{\partial z^2}.$$

In order to find a particular solution of Laplace's equation we may try separating the variables, as in § 91. Let us put $V = R(\varrho)\,\Phi(\varphi)\,Z(z)$ and divide by V.

(45)
$$\frac{1}{R}\frac{d^2 R}{d\varrho^2} + \frac{1}{\varrho R}\frac{dR}{d\varrho} + \frac{1}{\varrho^2}\frac{1}{\Phi}\frac{d^2\Phi}{d\varphi^2} + \frac{1}{Z}\frac{d^2 Z}{dz^2} = 0.$$

By the same considerations as before we find that the last term must be constant. Calling it k^2 we get

(46)
$$\frac{d^2 Z}{dz^2} - k^2 Z = 0,$$

and multiplying by ϱ^2,

(47)
$$\frac{\varrho^2}{R}\frac{d^2 R}{d\varrho^2} + \frac{\varrho}{R}\frac{dR}{d\varrho} + k^2\varrho^2 + \frac{1}{\Phi}\frac{d^2\Phi}{d\varphi^2} = 0$$

As before the last term must be constant. Calling it $-n^2$ we get

(48)
$$\frac{d^2\Phi}{d\varphi^2} + n^2\Phi = 0$$

together with

(49)
$$\frac{d^2 R}{d\varrho^2} + \frac{1}{\varrho}\frac{dR}{d\varrho} + \left(k^2 - \frac{n^2}{\varrho^2}\right) R = 0.$$

But if we put $k\varrho = x$ this becomes Bessel's Equation.

(b)
$$\frac{d^2 R}{dx^2} + \frac{1}{x}\frac{dR}{dx} + \left(1 - \frac{n^2}{x^2}\right) R = 0.$$

We have thus separated the variables, and reduced the problem to the solution of three ordinary differential equations. We thus get the particular solution of Laplace's equation,

(50)
$$V = \sum\sum (A_{kn}\cosh kz + B_{kn}\sinh kz)(C_{kn}\cos n\varphi$$
$$+ D_{kn}\sin n\varphi)(E_{kn}J_n(k\varrho) + F_{kn}Y_n(k\varrho)).$$

For this reason the Bessel functions are sometimes called cylindrical harmonic functions. The equation (44) has already made its appearance in symmetrical distributions, independent of z and φ, in connection with the logarithmic potential.

104. Laplacian in Ellipsoidal Coordinates. A third set of important orthogonal coordinates are the socalled elliptic or ellipsoidal coordinates. The equation

$$(51) \qquad \frac{x^2}{a^2+\varrho} + \frac{y^2}{b^2+\varrho} + \frac{z^2}{c^2+\varrho} = 1,$$

in which ϱ is a parameter, represents a central surface of the second order, whose semi-axes are

$$\sqrt{a^2+\varrho}, \quad \sqrt{b^2+\varrho}, \quad \sqrt{c^2+\varrho}.$$

Let us suppose $\qquad a^2 > b^2 > c^2$.

The foci of the principal sections lie at distances from the center

$$\sqrt{a^2-b^2}, \quad \sqrt{a^2-c^2}, \qquad\qquad \text{on the X-axis,}$$

$$\sqrt{b^2-c^2} \qquad\qquad \text{on the Y-axis.}$$

That is the foci are the same for all values of ϱ, hence the family of surfaces obtained by giving ϱ all real values is *confocal*. For them to be *real* ϱ must be greater than $-a^2$.

If $-c^2 < \varrho < \infty$ all the terms are positive, we have an *ellipsoid*. If $-b^2 < \varrho < -c^2$, the first two terms are positive, the third negative, the section $z = 0$ is an ellipse, $y = 0$ is an hyperbola, $x = 0$ an hyperbola, we have a one-sheeted hyperboloid. If $-a^2 < \varrho < -b^2$ only the first term is positive, the sections $y = 0$, $z = 0$ are both hyperbolas, $x = 0$ is imaginary, we have an hyperboloid of two sheets.

Through any point x, y, z, we may draw one surface of each kind, for the above equation is a cubic in ϱ, which has three real roots lying as above. Let us call them $\lambda > \mu > \nu$. They determine the position of the point, and are called its *elliptic coordinates*.

$$(52) \quad \begin{aligned} \frac{x^2}{a^2+\lambda} + \frac{y^2}{b^2+\lambda} + \frac{z^2}{c^2+\lambda} &= 1, & \infty > \lambda > -c^2, \\ \frac{x^2}{a^2+\mu} + \frac{y^2}{b^2+\mu} + \frac{z^2}{c^2+\mu} &= 1, & -c^2 > \mu > -b^2, \\ \frac{x^2}{a^2+\nu} + \frac{y^2}{b^2+\nu} + \frac{z^2}{c^2+\nu} &= 1, & -b^2 > \nu > -a^2. \end{aligned}$$

To express x, y, z, in terms of λ, μ, ν we notice that the expression

$$(53) \qquad F(\varrho) \equiv \frac{x^2}{a^2+\varrho} + \frac{y^2}{b^2+\varrho} + \frac{z^2}{c^2+\varrho} - 1$$

vanishes for $\varrho = \lambda, \mu, \nu$, and that if reduced to the denominator $(a^2+\varrho)$ $(b^2+\varrho)(c^2+\varrho) = f(\varrho)$ the numerator is of the third degree. Hence

$$(54) \; F(\varrho) \equiv -\frac{(\varrho-\lambda)(\varrho-\mu)(\varrho-\nu)}{(a^2+\varrho)(b^2+\varrho)(c^2+\varrho)} \equiv \frac{x^2}{a^2+\varrho} + \frac{y^2}{b^2+\varrho} + \frac{z^2}{c^2+\varrho} - 1$$

Multiply by $\varrho + a^2$, and then put $\varrho = - a^2$, and we get

$$x^2 = \frac{(a^2 + \lambda)(a^2 + \mu)(a^2 + \nu)}{(a^2 - b^2)(a^2 - c^2)},$$

and in like manner

$$y^2 = \frac{(b^2 + \lambda)(b^2 + \mu)(b^2 + \nu)}{(b^2 - c^2)(b^2 - a^2)},$$

(55)

$$z^2 = \frac{(c^2 + \lambda)(c^2 + \mu)(c^2 + \nu)}{(c^2 - a^2)(c^2 - b^2)}.$$

Differentiating the above identity according to ϱ,

(56) $\left\{ \dfrac{x^2}{(a^2 + \varrho)^2} + \dfrac{y^2}{(b^2 + \varrho)^2} + \dfrac{z^2}{(c^2 + \varrho)^2} \right\} = \dfrac{(\varrho - \lambda)(\varrho - \mu)(\varrho - \nu)}{(a^2 + \varrho)(b^2 + \varrho)(c^2 + \varrho)}$

$\left\{ \dfrac{1}{\varrho - \lambda} - \dfrac{1}{\varrho + a^2} + \dfrac{1}{\varrho - \mu} - \dfrac{1}{\varrho + b^2} + \dfrac{1}{\varrho - \nu} - \dfrac{1}{\varrho + c^2} \right\}.$

Put $\varrho = \lambda$, and all terms on the right vanish, except the first, so that

(57) $$\frac{x^2}{(a^2 + \lambda)^2} + \frac{y^2}{(b^2 + \lambda)^2} + \frac{z^2}{(c^2 + \lambda)^2} = \frac{(\lambda - \mu)(\lambda - \nu)}{f(\lambda)},$$

and similarly

(58) $$\frac{x^2}{(a^2 + \mu)^2} + \frac{y^2}{(b^2 + \mu)^2} + \frac{z^2}{(c^2 + \mu)^2} = \frac{(\mu - \nu)(\mu - \lambda)}{f(\mu)}.$$

Forming the differences of the three equations (52)

$$(\mu - \nu)\left\{ \frac{x^2}{(a^2 + \mu)(a^2 + \nu)} + \frac{y^2}{(b^2 + \mu)(b^2 + \nu)} + \frac{z^2}{(c^2 + \mu)(c^2 + \nu)} \right\} = 0,$$

(59) $(\nu - \lambda)\left\{ \dfrac{x^2}{(a^2 + \nu)(a^2 + \lambda)} + \dfrac{y^2}{(b^2 + \nu)(b^2 + \lambda)} + \dfrac{z^2}{(c^2 + \nu)(c^2 + \lambda)} \right\} = 0,$

$$(\lambda - \mu)\left\{ \frac{x^2}{(a^2 + \lambda)(a^2 + \mu)} + \frac{y^2}{(b^2 + \lambda)(b^2 + \mu)} + \frac{z^2}{(c^2 + \lambda)(c^2 + \mu)} \right\} = 0.$$

Differentiating (55) by λ, μ, ν,

$$2x \frac{\partial x}{\partial \lambda} = \frac{(a^2 + \mu)(a^2 + \nu)}{(a^2 - b^2)(a^2 - c^2)} = \frac{x^2}{a^2 + \lambda}, \text{ etc., so that}$$

(60)
$$\frac{\partial x}{\partial \lambda} = \frac{1}{2}\frac{x}{a^2 + \lambda}, \quad \frac{\partial x}{\partial \mu} = \frac{1}{2}\frac{x}{a^2 + \mu}, \quad \frac{\partial x}{\partial \nu} = \frac{1}{2}\frac{x}{a^2 + \nu},$$

$$\frac{\partial y}{\partial \mu} = \frac{1}{2}\frac{y}{b^2 + \lambda}, \quad \frac{\partial y}{\partial \mu} = \frac{1}{2}\frac{y}{b^2 + \mu}, \quad \frac{\partial y}{\partial \nu} = \frac{1}{2}\frac{y}{b^2 + \nu},$$

$$\frac{\partial z}{\partial \lambda} = \frac{1}{2}\frac{z}{c^2 + \lambda}, \quad \frac{\partial z}{\partial \mu} = \frac{1}{2}\frac{z}{c^2 + \mu}, \quad \frac{\partial z}{\partial \nu} = \frac{1}{2}\frac{z}{c^2 + \nu}.$$

From this we have, from (59),

(61) $\dfrac{\partial x}{\partial \lambda}\dfrac{\partial x}{\partial \mu} + \dfrac{\partial y}{\partial \lambda}\dfrac{\partial y}{\partial \mu} + \dfrac{\partial z}{\partial \lambda}\dfrac{\partial z}{\partial \mu} = 0,$ unless $\lambda = \mu$, etc.,

so that the surfaces $\lambda = \text{const.}$, $\mu = \text{const.}$ are orthogonal by § 90 (26).

Similarly for the other pairs. We consequently have an orthogonal system. We have by § 90 (25), and (57) above

$$\frac{1}{h_\lambda^2} = \frac{1}{4}\left\{\frac{x^2}{(a^2+\lambda)^2}+\frac{y^2}{(b^2+\lambda)^2}+\frac{z^2}{(c^2+\lambda)^2}\right\} = \frac{1}{4}\frac{(\lambda-\mu)(\lambda-\nu)}{f(\lambda)},$$

so that

$$h_\lambda^2 = 4\frac{f(\lambda)}{(\lambda-\mu)(\lambda-\nu)},$$

(62)

$$h_\mu^2 = 4\frac{f(\mu)}{(\mu-\nu)(\mu-\lambda)},$$

$$h_\nu^2 = 4\frac{f(\nu)}{(\nu-\lambda)(\nu-\mu)},$$

and since

$$\Delta V = h_\lambda h_\mu h_\nu \left\{\frac{\partial}{\partial\lambda}\left(\frac{h_\lambda}{h_\mu h_\nu}\frac{\partial V}{\partial\lambda}\right)+\frac{\partial}{\partial\mu}\left(\frac{h_\mu}{h_\nu h_\lambda}\frac{\partial V}{\partial\mu}\right)+\frac{\partial}{\partial\nu}\left(\frac{h_\nu}{h_\lambda h_\mu}\frac{\partial V}{\partial\nu}\right)\right\}$$

and $\dfrac{h_\lambda}{h_\mu h_\nu} = \dfrac{1}{2}\sqrt{\dfrac{-f(\lambda)(\mu-\nu)^2}{f(\mu)f(\nu)}}$, $h_\lambda h_\mu h_\nu = \dfrac{-8\sqrt{-f(\lambda)f(\mu)f(\nu)}}{(\lambda-\mu)(\mu-\nu)(\nu-\lambda)}$

we obtain

(63) $\Delta V = \dfrac{-4\sqrt{-f(\lambda)f(\mu)f(\nu)}}{(\lambda-\mu)(\mu-\nu)(\nu-\lambda)}\left[\sqrt{\dfrac{-(\mu-\nu)^2}{f(\mu)f(\nu)}}\dfrac{\partial}{\partial\lambda}\left(\sqrt{f(\lambda)}\dfrac{\partial V}{\partial\lambda}\right)\right.$

$$\left.+\sqrt{\dfrac{(\nu-\lambda)^2}{f(\nu)f(\lambda)}}\dfrac{\partial}{\partial\mu}\left(\sqrt{-f(\mu)}\dfrac{\partial V}{\partial\mu}\right)+\sqrt{\dfrac{(\lambda-\mu)^2}{f(\lambda)f(\nu)}}\dfrac{\partial}{\partial\nu}\left(\sqrt{-f(\nu)}\dfrac{\partial V}{\partial\nu}\right)\right]$$

105. Lamé's Equation. If we put

(64) $\dfrac{d\lambda}{\sqrt{f(\lambda)}} = d\alpha$, $\dfrac{d\mu}{\sqrt{f(\mu)}} = d\beta$, $\dfrac{d\nu}{\sqrt{f(\nu)}} = d\gamma$,

so that λ, μ, ν are *elliptic functions* of α, β, γ we have

(65) $\Delta V = \dfrac{-4}{(\lambda-\mu)(\mu-\nu)(\nu-\lambda)}\left[(\mu-\nu)\dfrac{\partial^2 V}{\partial\alpha^2}+(\nu-\lambda)\dfrac{\partial^2 V}{\partial\beta^2}+(\lambda-\mu)\dfrac{\partial^2 V}{\partial\gamma^2}\right].$

Thus Laplace's equation becomes

(66) $(\mu-\nu)\dfrac{\partial^2 V}{\partial\alpha^2}+(\nu-\lambda)\dfrac{\partial^2 V}{\partial\beta^2}+(\lambda-\mu)\dfrac{\partial^2 V}{\partial\gamma^2} = 0.$

If we attempt to separate the variables and satisfy this by the product $V = L(\alpha)\,M(\beta)\,N(\gamma)$ we get

(67) $\dfrac{\mu-\nu}{L}\dfrac{d^2 L}{d\alpha^2}+\dfrac{\nu-\lambda}{M}\dfrac{d^2 M}{d\beta^2}+\dfrac{\lambda-\mu}{N}\dfrac{d^2 N}{d\gamma^2} = 0$

which is of the form,

$$(\mu-\nu)\,\varphi(\lambda)+(\nu-\lambda)\,\varphi(\mu)+(\lambda-\mu)\,\varphi(\nu) = 0.$$

This is satisfied by φ a constant, or $\varphi(x) = x$.

Accordingly we shall satisfy it if we put $\varphi(x) = gx + h$, obtaining the ordinary differential equations,

$$\frac{d^2L}{d\alpha^2} = (g\lambda + h)L,$$

(68)
$$\frac{d^2M}{d\beta^2} = (g\mu + h)M,$$

$$\frac{d^2N}{d\gamma^2} = (g\nu + h)N,$$

each function satisfying the same equation. The equation

(69)
$$\frac{d^2y}{dx^2} - (g\lambda(x) + h)y = 0$$

where $\lambda(x)$ is the elliptic function defined by

(70)
$$\left(\frac{d\lambda}{dx}\right)^2 = f(\lambda) \equiv (a^2 + \lambda)(b^2 + \lambda)(c^2 + \lambda),$$

is called Lamé's equation. If the constants g and h are arbitrary we get solutions of Laplace's equation, as above, but they are useful only if g and h are so determined that Lamé's equation has polynominals in λ as solutions. We have.

$$\frac{dy}{dx} = \frac{dy}{d\lambda}\frac{d\lambda}{dx} = \sqrt{f(\lambda)}\,\frac{dy}{d\lambda},$$

$$\frac{d^2y}{dx^2} = \frac{d\lambda}{dx}\frac{d}{d\lambda}\left(\sqrt{f(\lambda)}\,\frac{dy}{d\lambda}\right) = \sqrt{f(\lambda)}\,\frac{d}{d\lambda}\left(\sqrt{f(\lambda)}\,\frac{dy}{d\lambda}\right)$$

$$= f(\lambda)\frac{d^2y}{d\lambda^2} + \frac{1}{2}f'(\lambda)\frac{dy}{d\lambda},$$

so that Lamé's equation becomes

(71)
$$f(\lambda)\frac{d^2y}{d\lambda^2} + \frac{1}{2}f'(\lambda)\frac{dy}{d\lambda} - (g\lambda + h)y = 0$$

a linear equation similar to Legendre's and Bessel's. It has three singular points $\lambda = -a^2, \ -b^2, \ -c^2.$

Dividing by $f(\lambda)$,

(72)
$$\frac{d^2y}{d\lambda^2} + \frac{1}{2}\left\{\frac{1}{a^2+\lambda} + \frac{1}{b^2+\lambda} + \frac{1}{c^2+\lambda}\right\}\frac{dy}{d\lambda} - \frac{g\lambda+h}{(a^2+\lambda)(b^2+\lambda)(c^2+\lambda)}y = 0.$$

For either point the indicial equation is $r(r-1) + r/2 = 0$ with roots $r = 0, 1/2$, so that we have either a power series in λ, or such multiplied by $\sqrt{a^2 + \lambda}$. Consider λ as a function of x defined by the differential equation.

(70)
$$\lambda'^2 = \left(\frac{d\lambda}{dx}\right)^2 = f(\lambda) \equiv (\lambda + a^2)(\lambda + b^2)(\lambda + c^2) \equiv \lambda^3 + A\lambda^2 + B\lambda + C,$$

$$A = a^2 + b^2 + c^2, \quad B = b^2c^2 + c^2a^2 + a^2b^2, \quad C = a^2b^2c^2.$$

Differentiating repeatedly by x,

$2\lambda'\lambda'' = f'(\lambda)\lambda'$, and dividing by λ',

$$\lambda'' = \frac{1}{2}f'(\lambda) = \frac{3}{2}\lambda^2 + A\lambda + \frac{B}{2},$$

$$(72\,\mathrm{a})\ \lambda''' = \frac{1}{2}f''(\lambda)\lambda' = (3\lambda + A)\sqrt{f(\lambda)},$$

$$\lambda^{\mathrm{IV}} = \frac{1}{2}\{f''(\lambda)\lambda'^2 + f''(\lambda)\lambda''\} = \frac{1}{2}\{f'''(\lambda)f(\lambda) + \frac{1}{2}f''(\lambda)f'(\lambda)\},$$

. . . .

all successive derivatives are alternately polynomials in λ, or polynomials in λ multiplied by $\sqrt{f(\lambda)}$. In differentiating twice we increase the degree of the polynomial once, as follows.

n even			n odd		
Derivative order	Deg. of pol.		Deriv. order	Deg. of	Factor of $\sqrt{f(\lambda)}$
λ	0	1	λ'	1	0
λ''	2	2	λ'''	3	1
λ^{IV}	4	3	λ^{V}	5	2
			λ^{VII}	7	3
λ^{2n}	$2n$	$n+1$	λ^{2n+1} $2n+1$		n

Hence a linear function of derivatives of the same *parity*,

$$(72\,\mathrm{b})\qquad y = \lambda^{(n-2)} + a_1\lambda^{(n-4)} + a_2\lambda^{(n-6)} + \cdots$$

is either a polynomial in λ, of degree $n/2$ if n is even, or a polynomial multiplied by $\sqrt{f(\lambda)}$ and of degree $\dfrac{n-3}{2}$ if n is odd.

Considered as a polynomial in λ, if n is even, of degree $n/2$, let us put

$$y = \sum_{k=0}^{k=\frac{n}{2}} c_k \lambda^k$$

in the equation

$$(71)\qquad f(\lambda)\frac{d^2y}{d\lambda^2} + \frac{f'(\lambda)}{2}\frac{dy}{d\lambda} - (g\lambda + h)y = 0,$$

that is

$$(\lambda^3 + A\lambda^2 + B\lambda + C)\frac{d^2y}{d\lambda^2} + \left(\frac{3}{2}\lambda^2 + A\lambda + \frac{B}{2}\right)\frac{dy}{d\lambda} - (g\lambda + h)y = 0,$$

and the term of highest degree is

$$c_{\frac{n}{2}}\left\{\frac{n}{2}\left(\frac{n}{2} - 1\right) + \frac{3}{2}\cdot\frac{n}{2} - g\right\}\lambda^{\frac{n}{2}+1}$$

which, like every other term, must vanish. Accordingly we must take

$$g = \frac{n(n+1)}{4}.$$

The result of the substitution is

$$(73) \quad \sum_{k=0}^{k=\infty} c_k \left[k(k-1) \left\{ \lambda^{k+1} + A\lambda^k + B\lambda^{k-1} + C\lambda^{k-2} \right\} \right.$$

$$\left. + k \left\{ \frac{3}{2} \lambda^{k+1} + A\lambda^k + \frac{B}{2} \lambda^{k-1} \right\} - \left\{ g\lambda^{k+1} + h\lambda^k \right\} \right] = 0.$$

Picking out the coefficient of λ^{k+1},

$$(74) \quad c_k \left(k^2 + \frac{k}{2} - g \right) + c_{k+1} \left\{ (k+1)^2 A - h \right\}$$

$$+ c_{k+2} \left\{ (k+2)(k+1) + \frac{k+2}{2} \right\} B + c_{k+3}(k+2)(k+3) = 0$$

which is the recurrence formula. Since the polynomial of degree $n/2$ has $n/2 + 1$ terms, there are that number of equations of this sort, in each of which h appears multiplied by a c. Eliminating the c's, which appear linearly, we get an equation in h of degree $n/2 + 1$, which can be shown to have all roots different. There are thus $n/2 + 1$ polynomial solutions.

If n is odd, y is equal to $\lambda' = \sqrt{f(\lambda)}$ times a polynomial of degree $\frac{n-3}{2}$. Call this z,

$$y = z\lambda', \quad \frac{dy}{dx} = z\lambda'' + \lambda'\frac{dz}{dx},$$

$$\frac{d^2y}{dx^2} = z\lambda''' + 2\lambda''\frac{dz}{dx} + \lambda'\frac{d^2z}{dx^2} = (g\lambda + h)z\lambda', \qquad \text{by (69)}$$

Changing to λ as independent variable,

$$\frac{dz}{dx} = \frac{dz}{d\lambda}\lambda', \quad \frac{d^2z}{dx^2} = \frac{dz}{d\lambda}\lambda'' + \lambda'^2\frac{d^2z}{d\lambda^2},$$

and making use of (72a)

$$z\lambda'(3\lambda + A) + 3\frac{dz}{d\lambda}\lambda'\left(\frac{3}{2}\lambda^2 + A\lambda + \frac{B}{2}\right)$$

$$+ \frac{d^2z}{d\lambda^2}\lambda'(\lambda^3 + A\lambda^2 + B\lambda + C) = (g\lambda + h)z\lambda',$$

or dividing out λ',

$$(75) \quad \frac{d^2z}{d\lambda^2}(\lambda^3 + A\lambda^2 + B\lambda + C) + 3\frac{dz}{d\lambda}\left(\frac{3}{2}\lambda^2 + A\lambda + B\right)$$

$$+ z(3\lambda + A - g\lambda - h) = 0.$$

If z is a polynomial in λ of degree $(n-3)/2$, from the coefficient of the highest term we have as above

$$\left(\frac{n-3}{2}\right)\left(\frac{n-5}{2}\right) + \frac{9}{3}\left(\frac{n-3}{2}\right) + 3 - g = 0,$$

$$g = \frac{n(n+1)}{4}, \qquad \text{as before.}$$

Equating all coefficients to 0, we get $(n-3)/2 + 1 = (n-1)/2$ equations, and an equation for h of that degree, each root giving a polynomial. In this case (n odd) we also have solutions of the form

$$y = z\sqrt{\lambda + a^2}$$

where z is a polynomial.

$$\frac{dy}{d\lambda} = \frac{dz}{d\lambda}\sqrt{\lambda + a^2} + \frac{z}{2\sqrt{\lambda + a^2}},$$

$$\frac{d^2y}{d\lambda^2} = \frac{d^2z}{d\lambda^2}\sqrt{\lambda + a^2} + \frac{dz}{d\lambda}\frac{1}{\sqrt{\lambda + a^2}} - \frac{z}{4\sqrt{(\lambda + a^2)^3}},$$

and inserting in Lamé's equation, and dividing by $\sqrt{\lambda + a^2}$

$$(\lambda^3 + A\lambda^2 + B\lambda + C)\left(\frac{d^2z}{d\lambda^2} + \frac{dz}{d\lambda}\frac{1}{\lambda + a^2} - \frac{z}{4(\lambda + a^2)^2}\right)$$

$$+ \left(\frac{3}{2}\lambda^2 + A\lambda + \frac{B}{2}\right)\left(\frac{dz}{d\lambda} + \frac{z}{2(\lambda + a^2)}\right) - \left(\frac{n(n+1)}{4}\lambda + h\right)z = 0.$$

Multiplying by $(\lambda + a^2)$

$$(\lambda + a^2)(\lambda^3 + A\lambda^2 + B\lambda + C)\frac{d^2z}{d\lambda^2}$$

$$+ \left[\lambda^3 + A\lambda^2 + B\lambda + C + (\lambda + a^2)\left(\frac{3}{2}\lambda^2 + A\lambda + B\right)\right]\frac{dz}{d\lambda}$$

(76)

$$+ \left[\begin{array}{c} -\dfrac{\lambda^3 + A\lambda^2 + B\lambda + C}{4(\lambda + a^2)} + \dfrac{3}{4}\lambda^2 + \dfrac{A}{2}\lambda + \dfrac{B}{4} \\ -\left(\dfrac{n(n+1)}{4}\lambda + h\right)(\lambda + a^2) \end{array}\right]z = 0.$$

The highest term in z, $c_k\lambda^k$, which when substituted in (76) gives

$$c_k\left[k(k-1) + \frac{5}{2}k + \frac{1}{2} - \frac{n(n+1)}{4}\right]\lambda^{k+2},$$

is zero if $k = \dfrac{n-1}{2}$ which is accordingly the degree of the polynomial z. Equating its coefficients to zero gives $(n+1)/2$ equations with the same number of values for h. Substituting in like manner the factors $\sqrt{b^2 + \lambda}$, $\sqrt{c + \lambda}$ we get similar solutions, in all $\dfrac{3}{2}(n+1)$ which with the $\dfrac{n-1}{2}$ of the first kind gives $2n + 1$ different solutions.

If n is even, we may put

$$y = z\sqrt{(\lambda + b^2)(\lambda + c^2)}.$$

We get for z a polynomial of degree $\frac{n-2}{2}$, so that h is the root of an equation of degree $\frac{n}{2}$. There are thus $\frac{3n}{2}$ solutions of this type, which with the $\frac{n}{2}+1$ polynomial solutions make again $2n+1$ solutions.

If we make a linear transformation

$$\lambda = Ap + B$$

and determine the constants A and $B = -\frac{1}{3}(a^2+b^2+c^2)$ so that the sum of the roots is zero, and at the same time take $A = 4$, we obtain the Weierstrass normal form for $f(\lambda)$, and at the same time p becomes the Weierstrass elliptic function of its argument x, or as it is generally written u, Equation (70) then becomes

$$(p'(u^2) = \left(\frac{dp}{du}\right)^2 = 4p^3 - g_2 p - g_3$$

and Lamé's Equation (69) becomes

$$\frac{d^2y}{du^2} - (n(n+1)\, p(u) + h)\, y = 0$$

where the so-called invariants g_2, g_3 are easily found from a^2, b^2, c^2. The investigation is not changed in its nature, but is more elegantly carried on by the study of the properties of the elliptic functions. Lamé's equation has been the subject of thorough treatment by Hermite. For this mode of treatment the reader may consult Jordan, Cours d'Analyse, Tome III, Halphen, Traité des Fonctions Elliptiques, and Burkhardt, Elliptische Functionen, and for Lamé's equation, Hermite, Sur quelques applications des fonctions elliptiques (Paris, Gauthier-Villars).

106. Development in Lamé's Functions. If the function V satisfying Laplace's equation be developed in a series of homogeneous polynomials in x, y, z

(77) $$V = V_0 + V_1 + V_2 + \cdots$$

we have seen that each contains $2n + 1$ arbitrary constants. Substituting for x, y, z their values in λ, μ, ν, V_n becomes a homogeneous function of degree n in $\sqrt{\lambda + a^2}, \sqrt{\lambda + b^2}, \sqrt{\lambda + c^2}$ and it is of the same sort in μ and ν. Consequently if $V_n = LMN$ we have found the factors all satisfying the same equation. With each value of n there are $2n + 1$ values of h, accordingly just enough, and we have the most general harmonic.

Consider the two solutions $M_m^{h_p}(\mu)$, $M_n^{h_q}(\mu)$ We have

(78)
$$f(\mu)\frac{d^2 M_m}{d\mu^2} + \frac{f'(\mu)}{2}\frac{d M_m}{d\mu} - \left\{\frac{m(m+1)}{4}\mu + h_p\right\} M_m = 0,$$

$$f(\mu)\frac{d^2 M_n}{d\mu^2} + \frac{f'(\mu)}{2}\frac{d M_n}{d\mu} - \left\{\frac{n(n+1)}{4}\mu + h\right\} M_n = 0.$$

Multiply the first by M_n, the second by M_m and subtract, and dividing by $\sqrt{f(\mu)}$ We may write the result

$$
(79) \quad \frac{d}{d\mu}\left[\sqrt{f(\mu)}\left(M_m \frac{d M_n}{\cdot d\mu} - M_n \frac{d M_m}{d\mu}\right)\right]
$$
$$
= \frac{M_m M_n}{\sqrt{f(\mu)}}\left[\left\{m(m+1) - n(n+1)\frac{\mu}{4} + h_p - h_q\right\}\right].
$$

Integrating between limits μ_1, μ_2 which are both roots of $f(\mu) = 0$, the left-hand side disappears, and

$$
(80) \quad \frac{m(m+1) - n(n+1)}{4}\int_{\mu_1}^{\mu_2}\frac{M_m M_n}{\sqrt{f(\mu)}}\,\mu\,d\mu = (h_q - h_p)\int_{\mu_1}^{\mu_2}\frac{M_m M_n}{\sqrt{f(\mu)}}\,d\mu.
$$

Change μ to ν, M_m, M_n to N_m, N_n, with the same values of m, n, h, and integrating between, two other roots ν_1, ν_2

$$
(81) \quad \frac{m(m+1) - n(n+1)}{4}\int_{\nu_1}^{\nu_2}\frac{N_m N_n}{\sqrt{f(\nu)}}\,\nu\,d\nu = (h_q - h_p)\int_{\nu_1}^{\nu_2}\frac{N_m N_n}{\sqrt{f(\nu)}}\,d\nu.
$$

Multiplying these equations cross-wise and transposing, we get

$$
(82) \quad \frac{(h_p - h_q)}{4}\left[m(m+1) - n(n+1)\right]\int_{\mu_1}^{\mu_2}\int_{\nu_1}^{\nu_2}\frac{M_m N_m M_n N_n}{\sqrt{f(\mu)f(\nu)}}(\mu - \nu)\,d\mu\,d\nu = 0.
$$

Accordingly if either $m + n$ or $h_q + h_p$ the integral vanishes. The integration is over the whole surface of the ellipsoid λ.

$$
(83) \quad \int\int\frac{M_m N_m M_n N_n(\mu - \nu)}{\sqrt{f(\mu)f(\nu)}}\,d\mu\,d\nu = 0.
$$

This is the orthogonal property of the functions MN, and enables us to develop arbitrary function. The series

$$
(84) \quad V = \sum_{n=0}^{n=\infty}\sum_{p=1}^{p=2n+1}A_{np}L_n^{(p)}(\lambda)\,M_n^{(p)}(\mu)\,N_n^{(p)}(\nu)
$$

satisfies Laplace's equation, if convergent, and if it is to reduce to a given function $F(\mu, \nu)$ on the surface of the ellipsoid

$$
\frac{x^2}{a^2} + \frac{y^2}{b^2} + \frac{z^2}{c^2} = 1, \quad \text{that is } \lambda = 0,
$$

we must determine the coefficients

$$
(85) \quad A_{np}L_n^{(p)}(0) = C_{np}
$$

so that

$$
(86) \quad \sum\sum C_{no}M_n^{(p)}N_n^{p} = F(\mu, \nu).
$$

Multiply by $M_m M_n (\mu - \nu) \sqrt{f(\mu) f(\nu)}$ and integrate over the surface of the ellipsoid and all the terms vanish except when the products are the same, giving

$$(87) \qquad C_{np} = \frac{\displaystyle\iint \frac{F(\mu, \nu) M_m^{(p)} M_m^{(p)} (\mu - \nu) \, d\mu \, d\nu}{\sqrt{f(\mu) f(\nu)}}}{\displaystyle\iint \frac{M_m^{(p)2} M_n^{(p)2} (\mu - \nu) \, d\mu \, d\nu}{\sqrt{f(\mu) f(\nu)}}}$$

107. Convergence of Series. The coordinates of a point lying on the ellipsoid

$$\frac{x^2}{a^2 + \lambda} + \frac{y^2}{b^2 + \lambda} + \frac{z^2}{c^2 + \lambda} = 1$$

can be represented by two angular coordinates θ, φ,

$$(88) \qquad \begin{aligned} x &= \sqrt{a^2 + \lambda} \, \cos \theta \cos \varphi, \\ y &= \sqrt{b^2 + \lambda} \, \cos \theta \sin \varphi, \\ z &= \sqrt{c^2 + \lambda} \, \sin \theta, \end{aligned}$$

which with the values of $x \, y \, z$ in (55) give

$$(89) \qquad \begin{aligned} \cos \theta \cos \varphi &= \sqrt{\frac{(a^2 + \mu)(a^2 + \nu)}{(a^2 - b^2)(a^2 - c^2)}} \\ \cos \theta \sin \varphi &= \sqrt{\frac{(b^2 + \mu)(b^2 + \nu)}{(b^2 - c^2)(b^2 - a^2)}} \\ \sin \theta &= \sqrt{\frac{(c^2 + \mu)(c^2 + \nu)}{(c^2 - a^2)(c^2 - b^2)}} \end{aligned}$$

Let us relate with each point on the ellipsoid a point on a sphere of radius r with the angular coordinates θ, φ. Calling xyz the coordinates of the point on the sphere,

$$(90) \qquad \begin{aligned} x &= r \sqrt{\frac{(a^2 + \mu)(a^2 + \nu)}{(a^2 - b^2)(a^2 - c^2)}}, \quad y = r \sqrt{\frac{(b^2 + \mu)(b^2 + \nu)}{(b^2 - c^2)(b^2 - a^2)}}, \\ z &= r \sqrt{\frac{(c^2 + \mu)(c^2 + \nu)}{(c^2 - a^2)(c^2 - b^2)}} \end{aligned}$$

we find

$$x^2 + y^2 + z^2 = r^2,$$

$$(91) \qquad \begin{aligned} \frac{x^2}{a^2 + \mu} + \frac{y^2}{b^2 + \mu} + \frac{z^2}{c^2 + \mu} &= 0, \\ \frac{x^2}{a^2 + \nu} + \frac{y^2}{b^2 + \nu} + \frac{z^2}{c^2 + \nu} &= 0. \end{aligned}$$

The two latter equations represent cones, and it is easy to show, as before, that they are mutually orthogonal, so that r, μ, ν are orthogonal coordinates. In order to transform Laplace's equation we have $h_r = 1$,

$$(92) \qquad \begin{aligned} \frac{\partial x}{\partial \mu} &= \frac{1}{2} \frac{x}{a^2 + \mu}, \quad \frac{\partial y}{\partial \mu} = \frac{1}{2} \frac{y}{b^2 + \mu}, \quad \frac{\partial z}{\partial \mu} = \frac{1}{2} \frac{z}{c^2 + \mu} \\ \frac{1}{h_\mu^2} &= \frac{1}{4} \left\{ \frac{x^2}{(a^2 + \mu)^2} + \frac{y^2}{(b^2 + \mu)^2} + \frac{z^2}{(c^2 + \mu)^2} \right\} = -\frac{1}{4} F'(\mu) \end{aligned}$$

Now if we put

$$(93) \qquad F(\varrho) = \frac{x^2}{a^2 + \varrho} + \frac{y^2}{b^2 + \varrho} + \frac{z^2}{c^2 + \varrho},$$

we have

$$(94) \qquad F(\varrho) = \frac{(\varrho - \mu)(\varrho - \nu)r^2}{f(\varrho)},$$

since F vanishes for $\varrho = \mu$, $\varrho = \nu$, and multiplying by ϱ and putting $\varrho = \infty$ both sides become equal to r^2. Differentiating (94) logarithmically

$$(95) \qquad \frac{F'(\varrho)}{F(\varrho)} = \frac{1}{\varrho - \mu} + \frac{1}{\varrho - \nu} - \frac{f'(\varrho)}{f(\varrho)}.$$

Multiplying by $F(\varrho)$ and putting $\varrho = \mu$,

$$F'(\mu) = \frac{(\mu - \nu)r^2}{f(\mu)}$$

$$(96) \qquad h_\mu^2 = \frac{-4f(\mu)}{(\mu - \nu)r^2}, \quad \text{and similarly,} \quad h_\nu^2 = \frac{-4f(\nu)}{(\nu - \mu)r^2}.$$

Thus $\varDelta V = 0$ becomes

$$\frac{\partial}{\partial r}\left[\frac{r^2}{4}\frac{(\mu - \nu)}{\sqrt{f(\mu)f(\nu)}}\frac{\partial V}{\partial r}\right] + \frac{\partial}{\partial \mu}\left[\sqrt{\frac{-f(\mu)}{f(\nu)}}\frac{\partial V}{\partial \mu}\right] + \frac{\partial}{\partial \nu}\left[\sqrt{\frac{f(\nu)}{-f(\mu)}}\frac{\partial V}{\partial \nu}\right] = 0,$$

or

$$(97)\ (\mu - \nu)\frac{\partial}{\partial r}\left[\frac{r^2}{4}\frac{\partial V}{\partial r}\right] - \sqrt{f(\mu)}\frac{\partial}{\partial \mu}\left[\sqrt{f(\mu)}\frac{\partial V}{\partial \mu}\right] + \sqrt{f(\nu)}\frac{\partial}{\partial \nu}\left[\sqrt{f(\nu)}\frac{\partial V}{\partial \nu}\right] = 0.$$

Lamé's equation (68) or (71) for the solution $M_n(\mu)$, may be written,

$$(98) \qquad \sqrt{f(\mu)}\frac{d}{d\mu}\left(\sqrt{f(\mu)}\frac{dM_n}{d\mu}\right) = \left\{\frac{n(n+1)}{4}\mu + h\right\}M_n$$

and for N $\quad \sqrt{f(\nu)}\frac{d}{d\nu}\left(\sqrt{f(\nu)}\frac{dN_n}{d\nu}\right) = \left\{\frac{n(n+1)}{4}\nu + h\right\}N_n,$

from which, eliminating h, by multiplication and subtraction,

$$(99) \qquad \begin{aligned} \sqrt{f(\mu)}\frac{\partial}{\partial \mu}\left(\sqrt{f(\mu)}\frac{\partial(M_nN_n)}{\partial \mu}\right) - \sqrt{f(\nu)}\frac{\partial}{\partial \nu}\left(\sqrt{f(\nu)}\frac{\partial(M_nN_n)}{\partial \nu}\right) \\ = \frac{n(n+1)}{4}(\mu - \nu)M_nN_n. \end{aligned}$$

Now if in the equation of Laplace (97) we put $V = r^n Y_n$ we have $Y_n(\theta, \varphi) = Y_n(\mu, \nu)$ and dividing by r^m

$$(100) \qquad \begin{aligned} \frac{n(n+1)}{4}(\mu - \nu)Y_n - \sqrt{f(\mu)}\frac{\partial}{\partial \mu}\left(\sqrt{f(\mu)}\frac{\partial Y_n}{\partial \mu}\right) \\ + \sqrt{f(\nu)}\frac{\partial}{\partial \nu}\left(\sqrt{f(\nu)}\frac{\partial Y_n}{\partial \nu}\right) = 0 \end{aligned}$$

which is the same as (99). Consequently we see that M_nN_n satisfies the differential equation of a surface harmonic Y_n. But there are

$2n + 1$ arbitrary constants in Y_n, and there are the same number of values of h, giving as many harmonics, so that we have

$$(101) \qquad Y_n = \sum_{p=1}^{p=2n+1} M_n^{(p)} N_n^{(p)}.$$

Now any function that may be developed on the surface of a sphere in spherical harmonics Y_n may be developed in ellipsoidal harmonics $M_n N_n$ and Lamé's development is justified.

108. Surfaces of Revolution. Suppose that our functions depend only on the cylindrical coordinates z and $\varrho = \sqrt{x^2 + y^2}$, but not on φ. Then if we put

$$(102) \qquad z + i\varrho = \psi(u + iv),$$

u and v will be orthogonal coordinates in the plane, and since (see Appendix)

$$(103) \qquad \frac{\partial u}{\partial z} = \frac{\partial v}{\partial \varrho}, \quad \frac{\partial u}{\partial \varrho} = -\frac{\partial v}{\partial z},$$

we shall have

$$(104) \qquad h_u^2 = h_v^2 = \frac{1}{|\psi'(u+iv)|^2} = h^2. \text{ We already have } h_\varphi = \frac{1}{\varrho},$$

$$(105) \qquad \varDelta V = \frac{h^2}{\varrho} \left\{ \frac{\partial}{\partial u} \left(\varrho \frac{\partial V}{\partial u} \right) + \frac{\partial}{\partial v} \left(\varrho \frac{\partial V}{\partial v} \right) \right\} = 0, \text{ since } \frac{\partial V}{\partial \varphi} = 0.$$

If this is to be satisfied by a product $V = L(u) M(v)$ we must have

$$(106) \qquad \frac{1}{L} \frac{\partial}{\partial u} \left(\varrho \frac{dL}{du} \right) + \frac{1}{M} \frac{\partial}{\partial v} \left(\varrho \frac{dM}{dv} \right) = 0.$$

This requires $\varrho = f(u) \varphi(v)$, when

$$(107) \qquad \frac{1}{f(u)L} \frac{d}{du} \left(f(u) \frac{dL}{du} \right) + \frac{1}{\varphi(v)M} \frac{d}{du} \left(\varphi(v) \frac{dM}{dv} \right) = 0,$$

which falls apart into two ordinary differential equations,

$$(108) \qquad \begin{aligned} \frac{d^2L}{du^2} + \frac{f'}{f} \frac{dL}{du} + cL &= 0, \\ \frac{d^2M}{dv^2} + \frac{\varphi'}{\varphi} \frac{dM}{dv} - cM &= 0. \end{aligned}$$

If we put

$$(109) \qquad \begin{aligned} z + i\varrho &= \cos(u + iv) = \cos u \cosh v - i \sin u \sinh v, \\ z &= \cos u \cosh v, \quad \varrho = -\sin u \sinh v. \end{aligned}$$

Eliminating u and v,

$$(110) \qquad \begin{aligned} \frac{z^2}{\cosh^2 v} + \frac{\varrho^2}{\sinh^2 v} &= 1, \\ \frac{z^2}{\cos^2 u} - \frac{\varrho^2}{\sin^2 u} &= 1. \end{aligned}$$

The curves $u =$ const. are ellipses with the square of the focal distance $\cosh^2 v - \sinh^2 v = 1$, and $v =$ const. are hyperbolas confocal therewith. The ellipsoids are *prolate*. In our previous notation put

$$\frac{z^2}{a^2 + \lambda} + \frac{\varrho^2}{b^2 + \lambda} = 1, \quad \frac{z^2}{a^2 + \mu} + \frac{\varrho^2}{b^2 + \mu} = 1,$$

(111)
$$\cosh v = \sqrt{a^2 + \lambda}, \quad \sinh v = \sqrt{b^2 + \lambda} \qquad a^2 - b^2 = 1$$
$$\cos u = \sqrt{a^2 + \mu}, \quad i \sin u = \sqrt{b^2 + \mu},$$

$$dv = \frac{d\lambda}{2\sqrt{(a^2 + \lambda)(b^2 + \lambda)}} \quad i\, du = \frac{d\mu}{2\sqrt{(a^2 + \mu)(b^2 + \mu)}}$$

so that the elliptic functions have degenerated to circular.

We have

(112)
$$\frac{1}{h^2} = |\sin(u + iv)|^2 = \sin^2 u \cosh^2 v + \cos^2 u \sinh^2 v$$
$$= \cosh^2 v (1 - \cos^2 u) + \cos^2 u (\cosh^2 v - 1)$$
$$= \cosh^2 v - \cos^2 u = \lambda - \mu.$$

These coordinates make ϱ the product of a function of u by one of v, and in addition $1/h^2$ the *difference* of two such.

We have $f(u) = \sin u$, $\varphi(v) = - \sinh v$ and Laplace's equation reduces to

(113)
$$\frac{d^2 L}{du^2} + \operatorname{ctn} u \frac{dL}{du} + cL = 0,$$
$$\frac{d^2 M}{dv^2} + \operatorname{ctnh} v \frac{dM}{dv} - cM = 0.$$

If we put $\cos u = y$, $\cosh v = x$ we have

$$z = xy, \quad \varrho = \sqrt{(1 - y^2)(x^2 - 1)},$$
$$\frac{dL}{du} = - \frac{dL}{dy} \sin u, \quad \frac{d^2 L}{du^2} = - \frac{dL}{dy} \cos u + \frac{d^2 L}{dy^2} \sin^2 u,$$
$$\frac{d^2 L}{dy^2} \sin^2 u - \frac{dL}{dy} \cos u - \operatorname{ctn} u \sin u \frac{dL}{dy} + cL = 0.$$

(114)
$$(1 - y^2) \frac{d^2 L}{dy^2} - 2y \frac{dL}{dy} + cL = 0,$$

which is Legendre's equation, § 91, (41) if $c = n(n + 1)$,

$$L = P_n(y) = P_n(\cos u).$$

The equation for M is the same, so that $M = P_n(x)$, and finally

(115)
$$V = \sum_{n=0}^{n=\infty} A_n P_n(x) P_n(y).$$

If the ellipsoid is oblate, take $\varrho + iz = \cosh(v + iu)$

(116) $\quad \varrho = \cos u \cosh v = \sqrt{(1 + x^2)(1 - y^2)}, \quad x = \sinh v, \quad y = \sin u,$

$z = \sin u \sinh v = xy.$

(117)
$$\frac{\varrho^2}{1 + x^2} + \frac{z^2}{x^2} = 1, \quad \text{ellipse} \quad 0 < x < \infty,$$

$$\frac{\varrho^2}{1 - y^2} - \frac{z^2}{y^2} = 1, \quad \text{hyperbola}, \quad 0 < y < 1,$$

$$f(u) = \cos u, \quad \varphi(v) = \cosh v,$$

and

(118)
$$\frac{d^2 L}{du^2} - \tan u \frac{d L}{du} + c L = 0,$$

$$\frac{d^2 M}{dv^2} + \tanh v \frac{d M}{du} - c M = 0,$$

$$(1 - y^2) \frac{d^2 L}{dy^2} - 2 y \frac{d L}{dy} + n(n + 1) L = 0,$$

ane as before $L = P_n(y)$, while

(119) $\qquad (1 + x^2) \dfrac{d^2 M}{dx^2} + 2 x \dfrac{d M}{dx} - n(n + 1) M = 0.$

In this case we have $M = P_n(ix)$

(120)
$$V = \sum_{n=0}^{n=\infty} A_n P_n(y) P_n(ix).$$

All these developments have singularities at infinity. If we are to deal with problems outside of an ellipsoid, in which we have to consider the infinite region, we must consider the neighborhood of $x = \infty$. But $P_n(x)$ is a polynomial, and is infinite for $x = \infty$. We must then seek another solution of Legendre's equation.

In the general linear differential equation of the second order, if we have two solutions y_1 and y_2,

(121)
$$P \frac{d^2 y_1}{dx^2} + Q \frac{d y_1}{dx} + R y_1 = 0,$$

$$P \frac{d^2 y_2}{dx^2} + Q \frac{d y_2}{dx} + R y_2 = 0,$$

we have, eliminating the last terms,

$$P \left(y_2 \frac{d^2 y_1}{dx^2} - y_1 \frac{d^2 y_2}{dx^2} \right) + Q \left(y_2 \frac{d y_1}{dx} - y_1 \frac{d y_2}{dx} \right) = 0,$$

or as we may write it,

$$\frac{d}{dx} \log \left(y_2 \frac{d y_1}{dx} - y_1 \frac{d y_2}{dx} \right) = -\frac{Q}{P}.$$

Integrating we obtain

$$\log\left(y_2\frac{dy_1}{dx} - y_1\frac{dy_2}{dx}\right) = -\int\frac{Q}{P}\,dx + B,$$

$$y_2\frac{dy_1}{dx} - y_1\frac{dy_2}{dx} = Ce^{-\int\frac{Q}{P}\,dx},$$

and dividing by y_1^2

$$\frac{d}{dx}\left(\frac{y_2}{y_1}\right) = -\frac{C}{y_1^2}e^{-\int\frac{Q}{P}\,dx}$$

$$(122)\qquad y_2 = Cy_1\int\frac{e^{-\int\frac{Q}{P}\,dx}}{y_1^2}\,dx + \text{const.}$$

In the case of Legendre's equation,

$$-\frac{Q}{P} = \frac{2x}{1-x^2} = \frac{1}{1-x} - \frac{1}{1+x}$$

$$-\int\frac{Q}{P}\,dx = \log\frac{1}{1-x^2}; \quad e^{-\int\frac{Q}{P}\,dx} = \frac{1}{1-x^2}$$

and if we put $y_1 = P_n(x)$ and determine the constant so that

$$(123)\qquad Q_n(x) = P_n(x)\int_x^\infty\frac{dx}{\{P_n(x)\}^2(x^2-1)},$$

we have the so-called zonal harmonic of the second kind, as for Bessel's equation we have had a function of the second kind. This evidently *vanishes* for $x = \infty$ but has a singular point for $x = \pm 1$, where it is infinite like $\log(x\pm 1)$. We find since $P_0(x) = 1$, $P_1(x) = x$

$$(124)\qquad
\begin{aligned}
Q_0(x) &= \int_x^\infty\frac{dx}{x^2-1} = \frac{1}{2}\log\frac{x-1}{x+1}\\[2mm]
Q_1(x) &= x\int_x^\infty\frac{dx}{x^2(x^2-1)} = \frac{1}{2}x\log\frac{x+1}{x-1} - 1.
\end{aligned}$$

In fact we shall find for any n,

$$Q_n(x) = \frac{1}{2}P_n(x)\log\frac{x+1}{x-1} + \text{a series of } P\text{'s of lower order.}$$

These formulae are appropriate for a prolate ellipsoid. If it is oblate we have $Q_n(ix)$. Then we find instead of $\frac{1}{2}\log\frac{x+1}{x-1}$, $\text{ctn}^{-1}x$, and we have

$$q_0(x) = \text{ctn}^{-1}x, \quad q_1(x) = 1 - x\,\text{ctn}^{-1}x, \text{ etc.}$$

CHAPTER VIII.

APPLICATIONS OF SPHERICAL, CYLINDRICAL AND ELLIPSOIDAL HARMONICS.

109. Zonal Harmonics. Attraction of Circular Wire. Consider a wire in the shape of a circle of radius a, line-density δ, mass $m = 2\pi a\delta$, attracting according to the Newtonian law. Then its potential at a point at a distance z from its center on the axis of symmetry will be

$$(1) \qquad V = \frac{m}{d} = \frac{m}{\sqrt{a^2 + z^2}} = \frac{m}{a}\left(1 + \frac{z^2}{a^2}\right)^{-\frac{1}{2}} = \frac{m}{z}\left(1 + \frac{a^2}{z^2}\right)^{-\frac{1}{2}},$$

$$(2) \; z < a \qquad V = \frac{m}{a}\left\{1 - \frac{1}{2}\frac{z^2}{a^2} + \frac{1}{2}\cdot\frac{3}{4}\frac{z^4}{a^4}\cdots\right.$$
$$\left. + (-1)^s \frac{1\cdot3\cdot5\ldots(2s-1)}{2\cdot4\cdot6\ldots2s}\frac{z^{2s}}{a^{2s}} + \cdots\right\},$$

$$(3) \; z > a \qquad V = \frac{m}{z}\left\{1 - \frac{1}{2}\frac{a^2}{z^2} + \frac{1}{2}\cdot\frac{3}{4}\frac{a^4}{z^4}\cdots\right.$$
$$\left. + (-1)^s \frac{1\cdot3\cdot5\ldots(2s-1)}{2\cdot4\cdot6\ldots2s}\frac{a^{2s}}{z^{2s}} + \cdots\right\}.$$

Now we know that a potential symmetrical with respect to an axis can be developed as

$$(4) \qquad A_0 P_0 + A_1\frac{r}{a}P_1 + A_2\frac{r^2}{a^2}P_2 \cdots + A_s\frac{r^s}{a^s}P_s + \cdots, \qquad r < a,$$

or

$$(5) \qquad A_0\frac{a}{r}P_0 + A_1\frac{a^2}{r^2}P_1 + A_2\frac{a^3}{r^3}P_2 \cdots + A_s\frac{a^{s+1}}{r^{s+1}}P_s + \cdots, \qquad r > a$$

When the attracted point is on the axis $r = z$, we have every $P_s = 1$, § 92 so that by comparison

$$A_{2s} = (-1)^s \frac{1\cdot3\cdot5\ldots2s-1}{2\cdot4\cdot6\ldots2s}, \qquad A_{2s+1} = 0.$$

Accordingly we have finally

$$(6)\,r < a \quad V = \frac{m}{a}\left\{P_0 - \frac{1}{2}\frac{r^2}{a^2}P_2 + \frac{1\cdot3}{2\cdot4}\frac{r^4}{a^4}P_4 \ldots (-1)^s \frac{1\cdot3\cdot5\ldots(2s-1)}{2\cdot4\cdot6\ldots2s}\frac{r^{2s}}{a^{2s}}P_{2s} + \cdots\right\},$$

$$(7)\,r > a \quad V = \frac{m}{r}\left\{P_0 - \frac{1}{2}\frac{a^2}{r^2}P_2 + \frac{1\cdot3}{2\cdot4}\frac{a^4}{r^4}P_4 \ldots\right\}.$$

110. Potential of Disc and Current. By integration of the potential of a circular line we get that of a disc, as in Chapter V (55). On the axis,
$$V = 2\pi\sigma\left(\sqrt{a^2 + r^2} - r\right)$$

(8) $V = 2\pi\sigma\left\{-r + a\left(1 + \frac{1}{2}\frac{r^2}{a^2} - \frac{1}{2}\cdot\frac{1}{4}\frac{r^4}{a^4} + \frac{1\cdot1\cdot3}{2\cdot4\cdot6}\frac{r^6}{a^6}\cdots\right)\right\}$ $r < a$,

(9) $V = 2\pi\sigma\left\{-r + r\left(1 + \frac{1}{2}\frac{a^2}{r^2} - \frac{1}{2}\cdot\frac{1}{4}\frac{a^4}{r^4} + \frac{1\cdot1\cdot3}{2\cdot4\cdot6}\frac{a^6}{r^6}\cdots\right)\right\}$ $r > a$,

Accordingly as before, we have at points off the axis,

(10) $V = 2\pi\sigma\left\{a - rP_1 + \frac{1}{2a}r^2P_2 - \frac{1\cdot1}{2\cdot4}\frac{r^4}{a^3}P_4 + \frac{1\cdot1\cdot3}{2\cdot4\cdot6}\frac{r^6}{a^5}P_6\cdots\right\}$ $r < a$,

(11) $V = 2\pi\sigma\left\{\frac{1}{2}\frac{a^2}{r} - \frac{1\cdot1}{2\cdot4}\frac{a^4}{r^3}P_2 + \frac{1\cdot1\cdot3}{2\cdot4\cdot6}\frac{a^6}{r^5}P_4\cdots\right\}$ $r > a$.

We may find the potential due to a double distribution on the disc in the same way. According to a theorem of Ampère this will be the potential due to a circular current of electricity. According to § 57 (58) we have to differentiate the above potential according to the normal to the disc, obtaining

$$W = 2\pi\varkappa\left\{1 - \frac{r}{\sqrt{a^2 + r^2}}\right\}$$

(12) $W = 2\pi\varkappa\left\{1 - \frac{r}{a}\left(1 - \frac{1}{2}\frac{r^2}{a^2} + \frac{1\cdot3}{2\cdot4}\frac{r^4}{a^4} - \frac{1\cdot3\cdot5}{2\cdot4\cdot6}\frac{r^6}{a^6}\cdots\right)\right\}, r < a$,

(13) $W = 2\pi\varkappa\left\{1 - \left(1 - \frac{1}{2}\frac{a^2}{r^2} + \frac{1\cdot3}{2\cdot4}\frac{a^4}{r^4} - \frac{1\cdot3\cdot5}{2\cdot4\cdot6}\frac{a^6}{r^6}\cdots\right)\right\}$, $r > a$,

so that as before

(14) $W = 2\pi\varkappa\left\{1 - \frac{r}{a}P_1 + \frac{1}{2}\frac{r^3}{a^3}P_3 - \frac{1\cdot3}{2\cdot4}\frac{r^5}{a^5}P_5 + \cdots\right\}$, $r < a$

(15) $W = 2\pi\varkappa\left\{\frac{1}{2}\frac{a^2}{r^2}P_1 - \frac{1\cdot3}{2\cdot4}\frac{a^4}{r^4}P_3 + \frac{1\cdot3\cdot5}{2\cdot4\cdot6}\frac{a^6}{r^6}P_5\cdots\right\}$. $r > a$.

From these results we get many of use in electromagnetism.

111. Potential of thin Spherical Shell. Let us find the potential of a thin spherical shell, whose surface density is given as a function of the geographical coordinates θ, φ, and developed in surface harmonics.

(16) $\sigma = Y_0 + Y_1 + Y_2 \cdots$

Now we know that inside the sphere the potential is developable in a series of solid harmonics § 87

(17) $V = A_0 Y_0' + A_1\frac{r}{a}Y_1' + A_2\frac{r^2}{a^2}Y_2' + \cdots + A_s\frac{r^s}{a^s}Y_s'\cdots$ $r < a$

and outside

(18) $V = B_0\frac{a}{r}Y_0' + B_1\frac{a^2}{r^2}Y_1' + B_2\frac{a^3}{r^3}Y_2' + \cdots + B_s\frac{a^{s+1}}{r^{s+1}}Y_s'\cdots$ $r > a$.

Since the potential is continuous at the surface $r = a$ we must have $B_s = A_s$. But we also have at the surface Poisson's surface equation

(19) $$\frac{\partial V}{\partial n_1} + \frac{\partial V}{\partial n_2} = -4\pi\sigma \qquad \S 56\ (52)$$

that is

$$-\left(\frac{A_1}{a}Y_1' + \frac{2r}{a^2}A_2 Y_2' + \cdots + \frac{s r^{s-1}}{a^s}A_s Y_s' + \cdots\right)_{(r=a)}$$

$$-\left(A_0 \frac{a}{r^2}Y_0' + 2A_1\frac{a^2}{r^3}Y_1' + \cdots + (s+1)A_s\frac{a^{s+1}}{r^{s+2}}Y_s' + \cdots\right)_{(r=a)}$$

(20) $$= -4\pi\{Y_0 + Y_1 + Y_2 + \cdots\}$$

so that putting $r = a$, $Y_s^1 = Y_s$, we have

(21) $$A_s = \frac{4\pi a}{2s+1}.$$

Accordingly the potential at an external point due to the surface layer is, using (44), Chapter VII § 92

$$V = \iint \frac{\sigma\,dS}{d} = \frac{a^2}{r}\iint \sigma\left\{\sum_{s=0}^{s=\infty}\frac{a^s}{r^s}P_s(\mu)\right\}\sin\theta\,d\theta\,d\varphi$$

(21) $$= \frac{a^2}{r}\iint \sum_{n=0}^{n=\infty}Y_n(\theta,\varphi)\sum_{s=0}^{s=\infty}\frac{a^s}{r^s}P_s(\mu)\sin\theta\,d\theta\,d\varphi$$

$$= \sum_{s=0}^{s=\infty}\frac{4\pi}{2s+1}\frac{a^{s+2}}{r^{s+1}}Y_s(\theta',\varphi')$$

from which we get the important integral theorem,

(22) $$\iint Y_n(\theta,\varphi)P_s(\mu)\sin\theta\,d\theta\,d\varphi = \frac{4\pi}{2n+1}Y_n(\theta',\varphi'), \text{ if } n = s$$

or 0, if $n \neq s$ which is the same as (98) and (99) of chapter VII § 97, where it was proved for the interior of a spherical shell.

112. Potential of Body of any Shape. The potential of any body may be developed in spherical harmonics. If the coordinates of the attracting point are x', y', z', r', and of the attracted x, y, z, r.

(23) $$V = \iiint \frac{\rho\,d\tau'}{d} = \frac{Y_0}{r} + \frac{Y_1}{r^2} + \cdots \frac{Y_s}{r^{s+1}} + \cdots$$

and since when $r > r'$, (§ 91, 44)

$$\frac{1}{d} = \frac{1}{r}\left\{P_0 + \frac{r'}{r}P_1(\mu) + \frac{r'^2}{r^2}P_2(\mu) + \cdots\right\}$$

where

$$\mu = \cos(POP') = \frac{xx' + yy' + zz'}{rr'}$$

we obtain by insertion in the integral,

$$(24) \qquad Y_n = \iiint \varrho \, r'^n P_n(\mu) \, d\tau'.$$

Since we have

$$P_0(\mu) = 1, \qquad P_1(\mu) = \mu, \qquad P_2(\mu) = \tfrac{1}{2}(3\mu^2 - 1),$$

$$r' P_1(\mu) = r'\mu = \frac{xx' + yy' + zz'}{r},$$

$$r'^2 P_2(\mu) = \frac{1}{2}\left\{ \frac{3(xx' + yy' + zz')^2 - r'^2 r^2}{r^2} \right\},$$

we obtain

$$
\begin{aligned}
V = {}& \frac{1}{r}\iiint \varrho \, dx' \, dy' \, dz' + \frac{1}{r^3}\left\{ x\iiint \varrho \, x' \, dx' \, dy' \, dz' \right. \\
& + y\iiint \varrho \, y' \, dx' \, dy' \, dz' + z\iiint \varrho \, z' \, dx' \, dy' \, dz' \Big\} \\
& + \frac{1}{2r^5}\iiint \{ 3(x^2 x'^2 + y^2 y'^2 + z^2 z'^2 + 2xx'yy' + 2yy'zz' \\
& + 2zz'xx') - r^2(x'^2 + y'^2 + z'^2) \} \varrho \, dx' \, dy' \, dz' + \cdots
\end{aligned}
$$

(25)

The integral in the first term is evidently the mass of the body. The next three integrals are the products of the mass by the coordinates of the center of mass of the body, and if this point be taken for the origin they disappear. To recognize the meaning of the next integral we will define the moment of inertia of the body about one of the axes as the integral of each element of mass multiplied by the square of its distance from the axis. Thus we have the three

$$A = \iiint \varrho \, (y'^2 + z'^2) \, d\tau', \qquad B = \iiint \varrho \, (z'^2 + x'^2) \, d\tau',$$

$$C = \iiint \varrho \, (x'^2 + y'^2) \, d\tau'$$

about the axes of X, Y, Z respectively. Also such an integral as

$$D = \iiint \varrho \, y'z' \, d\tau'$$

and the two similar ones is called a product of inertia, and it is always possible to find three axes through the center of mass such as to make these three integrals vanish. There then remain the terms

$$(26) \quad V = \frac{M}{r} + \frac{1}{2}\frac{(B + C - 2A)x^2 + (C + A - 2B)y^2 + (A + B - 2C)}{r^5} + \cdots$$

of which the first and most important shows that at a distance the body acts as if concentrated at its center of mass (like a homogenous sphere) as we have proved in Chapter V, while the next term gives the most important correction. In dealing with the heavenly bodies these two terms are quite sufficient.

The same method may be applied to develop the potential of a magnet, which is defined as a body, every volume element of which acts like a doublet, § 57, whose moment per unit volume is a vector function with components I_x, I_y, I_z. Accordingly

$$(27) \quad V = \int\int\int \left\{ I_x \frac{\partial\left(\frac{1}{d}\right)}{\partial x'} + I_y \frac{\partial\left(\frac{1}{d}\right)}{\partial y'} + I_z \frac{\partial\left(\frac{1}{d}\right)}{\partial z'} \right\} dx'\, dy'\, dz'.$$

We easily find that there is no term of order -1, and that the term of order -2 represents the potential of a single doublet.

113. Hydrokinetic Applications. By means of spherical harmonics we may solve the problem of Neumann § 60, for a sphere. The external problem is that of finding the motion in a fluid reaching to infinity and at rest there, when the normal velocity is given all over the surface of a sphere. If the fluid is incompressible its velocity is solenoidal, and if it is derived from a velocity-potential φ this satisfies Laplace's equation. If the velocity is prescribed at the sphere, and developed in surface harmonics, we proceed exactly as in § 111. Supposing

$$(28) \qquad \left(\frac{\partial\varphi}{\partial r}\right)_{r=a} = Y_0 + Y_1 + Y_2 \cdots$$

the solution outside will be

$$(29) \qquad \varphi = A_0 \frac{Y_0}{r} + A_1 \frac{Y_1}{r^2} + A_2 \frac{Y_2}{r^3} + \cdots + A_s \frac{Y_s}{r^{s+1}} + \cdots$$

from which

$$(30) \qquad \frac{\partial\varphi}{\partial r} = -\left\{ A_0 \frac{Y_0}{r^2} + 2A_1 \frac{Y_1}{r^3} + 3A_2 \frac{Y_2}{r^4} + \cdots (s+1)\frac{A_s Y_s}{r^{s+2}} \cdots \right\}$$

and to give the surface values,

$$(31) \qquad A_s = -\frac{a^{s+2}}{s+1}.$$

Consequently the solution is

$$(32) \qquad \varphi = -a\left\{ \frac{a}{r} Y_0 + \frac{a^2}{2r^2} Y_1 + \frac{a^3}{3r^3} Y_2 + \cdots \right\}.$$

A simple but important application is that of a sphere moving with velocity ν in a straight line in the Z-axis. Resolving this normally at a point distant by an angle θ from the Z-axis,

$$(33) \qquad \left(\frac{\partial\varphi}{\partial r}\right)_{r=a} = \nu \cos\theta.$$

But this is the zonal harmonic P_1. Accordingly the solution is,

$$(34) \qquad \varphi = -\frac{\nu}{2}\frac{a^3}{r^2}\cos\theta,$$

or the action of the sphere is like that of a doublet, that is it produces a flow which is geometrically identical with the field of force produced by a doublet.

The similar problem for an ellipsoid of revolution may be solved by means of the elliptic coordinates of § 108, and the zonal harmonics of the second kind Q_n. If the ellipsoid is prolate, and moving with constant velocity ν in the direction of its axis, we may conveniently suppose it to be at rest, and the water to have a translation $-\nu$ superposed on its actual motion. Then we may take

$$(34) \qquad \varphi = A P_1(y) Q_1(x) - \nu z = A y Q_1(x) - \nu x y$$

where x and y are the elliptic coordinates. At the surface of the ellipsoid

$$(35) \qquad \frac{\partial \varphi}{\partial n} = (A y Q_1'(x) - \nu y) \frac{\partial x}{\partial n} = 0$$

from which

$$(36) \qquad A = \frac{\nu}{Q_1'(x_0)} = \frac{\nu}{\left\{ \dfrac{1}{2} \log \dfrac{x_0 + 1}{x_0 - 1} - \dfrac{x_0}{x_0^2 - 1} \right\}} \; .$$

Thus the velocity imparted to the water by the motion of the ellipsoid may be calculated. During the war there was some slight practical interest in this problem in connection with submarines. A table of the functions Q_1 and Q_1' was published by the author and Willard Fisher in the Proceedings of the National Academy of Sciences, March, 1919.

In cases of rotational symmetry like these where the lines of flow are in meridian planes it is convenient to introduce the so-called flux-function. Let us find the amount of fluid crossing a ribbon bounded by any curve AB in such a meridian plane and the curve formed by rotating it about the Z-axis through an angle $d\theta$. We find

$$(37) \qquad \int_A^B v_n \varrho \, d\theta \, ds = d\theta \int \frac{\partial \varphi}{\partial n} \varrho \, ds,$$

where n is the normal to the plane curve AB. Now if the fluid is incompressible this must be the same for all such curves joining A and B, consequently $\varrho \dfrac{\partial \varphi}{\partial n} ds$ must be an exact differential,

$$(38) \qquad \varrho \frac{\partial \varphi}{\partial n} ds = d\psi = \frac{\partial \psi}{\partial \varrho} d\varrho + \frac{\partial \psi}{\partial z} dz.$$

But, as in § 70 (3)

$$(39) \qquad \begin{aligned} \cos(n\varrho) &= -\cos(ds, z) = -\frac{dz}{ds}, \\ \cos(nz) &= \cos(ds, \varrho) = \frac{d\varrho}{ds}, \end{aligned}$$

therefore

$$(40) \qquad \varrho \, \frac{\partial \varphi}{\partial n} \, ds = \varrho \left(\frac{\partial \varphi}{\partial \varrho} \cos (n\varrho) + \frac{\partial \varphi}{\partial z} \cos (nz) \right) ds = \varrho \left(- \frac{\partial \varphi}{\partial \varrho} \, dz + \frac{\partial \varphi}{\partial z} \, d\varrho \right).$$

Consequently we must have

$$(41) \qquad \frac{\partial \psi}{\partial \varrho} = \varrho \, \frac{\partial \varphi}{\partial z}, \qquad \frac{\partial \psi}{\partial z} = - \varrho \, \frac{\partial \varphi}{\partial \varrho}.$$

The lines $\psi = $ const. are the stream-lines, and are orthogonal with the equipotentials.

Now since by (38) we have for any directions ds_1 and ds_2 which are at right angles to each other,

$$(42) \, | \qquad \varrho \, \frac{\partial \varphi}{\partial s_1} = \frac{\partial \psi}{\partial s_2}, \quad \text{we find} - \varrho \, \frac{\partial \varphi}{\partial s_2} = \frac{\partial \psi}{\partial s_1},$$

by the condition of orthogonality, and we may take for them the directions of the coordinate lines, giving

$$\varrho \, \frac{\partial \varphi}{\partial n_u} = \frac{\partial \psi}{\partial n_v}, \quad - \varrho \, \frac{\partial \varphi}{\partial n_v} = \frac{\partial \psi}{\partial n_u},$$

since $dn_u = du/h$, $dn_v = dv/h$, we have

$$(43) \qquad \varrho \, \frac{\partial \varphi}{\partial u} = \frac{\partial \psi}{\partial v}, \quad - \varrho \, \frac{\partial \varphi}{\partial v} = \frac{\partial \psi}{\partial u}.$$

Now changing from the variables u, v to x, y by means of the equations (116) § 108 these become

$$(44) \qquad (1 - y^2) \frac{\partial \varphi}{\partial y} = \frac{\partial \psi}{\partial x}, \quad - (1 + x^2) \frac{\partial \varphi}{\partial x} = \frac{\partial \psi}{\partial y}.$$

If we put $\qquad \varphi = P_n(y) \, Q_n(ix)$

we have

$$(45) \quad \frac{\partial \psi}{\partial x} = (1 - y^2) \, P_n'(y) \, Q_n(ix), \quad \frac{\partial \psi}{\partial y} = - i(1 + x^2) \, P_n(y) \, Q_n'(ix),$$

and replacing $P_n(y)$ by its value derived from Legendre's equation,

$$P_n(y) = - \frac{1}{n(n+1)} \frac{d}{dy} \left((1 - y^2) \frac{dP_n}{dy} \right)$$

from which

$$(46) \qquad \psi = \frac{1}{n(n+1)} (1 - y^2) \frac{dP_n}{dy} (1 + x^2) \frac{d \, Q_n(ix)}{dx}.$$

The simplest case is $n = 0$ when we cannot make use of (46). But we then have

$$(47) \qquad \varphi = A \operatorname{ctn}^{-1} x, \quad \psi = Ay,$$

which satisfy (44) The flow represented is that through a circular

hole in an infinite plane wall. The velocity at points in the plane of
the wall, inside the hole, is

$$\frac{\partial \psi}{\partial \varrho} = \frac{A}{\sqrt{1 - \varrho^2}}$$

and is infinite at the edge of the hole. The same solution gives us the
field due to electricity on a conducting circular disc.

114. Helmholtz's Equation. All the applications previously
treated have had to do with Laplace's equation. We have however seen
in Chapter III the extremely instructive nature of the study of vibrations, and the normal functions connected with them. We shall accordingly now consider the wave-equation,

$$(48) \qquad \frac{\partial^2 u}{\partial t^2} = a^2 \varDelta u,$$

which has been exhaustively studied in Chapter III for one dimension,
leading to sines and cosines as normal functions. In Chapter V we have
considered the relation of the normal functions to the integral equations for two and three dimensions, but have not explicitly found the
normal functions. We will now consider the case of circular and
spherical symmetry. Proceeding as in Chapter III, (128) and putting
$u = \psi(t)\, \varphi(x, y, z)$ we obtain

$$(49) \qquad \frac{d^2 \psi}{d t^2} + n^2 \psi = 0,$$

$$(50) \qquad \varDelta \varphi + k^2 \varphi = 0, \quad k = \frac{n}{a},$$

which is Helmholtz's equation. Let us first consider the case of two
dimensions, which will give us the case of a circular membrane, or of
the vibrations of a gas in a thin layer bounded by two parallel planes
and a circular cylinder normal to them, or of the vibrations of a liquid
in a circular tank with a flat bottom, § 13. Transforming the Laplacian
into cylindrical coordinates, § 103, (44)

$$(51) \qquad \frac{\partial^2 \varphi}{\partial \varrho^2} + \frac{1}{\varrho} \frac{\partial \varphi}{\partial \varrho} + \frac{1}{\varrho^2} \frac{\partial^2 \varphi}{\partial \theta^2} + k^2 \varphi = 0.$$

In order to separate the variables, we put as usual $\varphi = R(\varrho)\, \Theta(\theta)$
obtaining

$$(52) \qquad \frac{d^2 \Theta}{d \theta^2} + s^2 \Theta = 0,$$

$$(53) \qquad \frac{d^2 R}{d \varrho^2} + \frac{1}{\varrho} \frac{d R}{d \varrho} + \frac{k^2 \varrho^2 - s^2}{\varrho^2} R = 0.$$

These are (48) and (49), § 103. The first gives sines and cosines, the second Bessel's functions reducing to Bessel's equation if we put $k\varrho = x$. Accordingly any sum of products

$$(54) \qquad \varphi_s = (A_s \cos s\theta + B_s \sin s\theta) J_s(k\varrho)$$

is a solution of the equation 50.

Let us consider a membrane bounded by a circle of radius a, at which it is fastened, so that $u(a) = 0$. We must then have for any normal vibration

$$(55) \qquad J_s(ka) = 0,$$

which is the equation for the characteristic numbers. If k_{r_s} is a root of this equation, the normal functions are

$$(56) \qquad J_s(k_{r_s}\varrho) \cos s\theta, \quad J_s(k_{r_s}\varrho) \sin s\theta.$$

In order to develop a function $f(\varrho, \theta)$ in a series of normal functions

$$(57) \qquad f(\varrho, \theta) = \sum_r \sum_s J_s(k_{r_s}\varrho)(A_{r_s} \cos s\theta + B_{r_s} \sin s\theta),$$

let us multiply by $J_t(k_{q_t}\varrho) \cos t\theta$ and by the element of area $\varrho\, d\varrho\, d\theta$ and integrate over the circle

$$(58) \qquad \int_0^a \int_0^{2\pi} f(r, \theta) J_t(k_{q_t}\varrho) \cos t\theta\, \varrho\, d\varrho\, d\theta$$

$$= \sum_r \sum_s \iint J_t(k_{q_t}\varrho) J_s(k_{r_s}\varrho)(Ars \cos s\theta + Brs \sin s\theta) \cos t\theta\, \varrho\, d\varrho\, d\theta.$$

It is evident that functions having different values of s, t are orthogonal, while the orthogonality of those with $s = t$ and different k may be proved as in (159) Chapter V, or immediately from Bessel's equation as in Chapter III, (309) (311). We obtain

$$(58\text{a}) \qquad 2\int_0^a J_s^2(k_{r_s}\varrho)\varrho\, d\varrho = \varrho^2 J_s'^2(k_{r_s}a) + a^2\left(1 - \frac{s^2}{k_{r_s}^2 a^2}\right) J_s^2(ka)$$

and since if the boundary is fixed $J_s(k_{r_s}a) = 0$,

$$(59) \qquad \int_0^a J_s^2(k_{r_s}\varrho)\varrho\, d\varrho = \frac{1}{2} a^2 J_s'^2(k_{r_s}a).$$

Thus we may carry out the development, and by the methods of Chapter III we may also find the forced vibrations.

The forms of the vibrations for the different normal oscillations are easily seen. If $s = 0$ the displacement is the same at all points equally distant from the center, and there are a certain number of nodal circles. If k is the smallest root the membrane vibrates everywhere in the same direction, if the second root there is one circle, and so on, as for a string. The frequencies are not harmonic, but are proportional to the roots $k_{r,s}$. For other values of s there are also nodal radii, so that the circle is divided up into sectors of rings so that parts separated by a nodal line vibrate oppositely.

If the membrane does not extend as far as the center, but is fastened at an internal edge, we require a second term in the solution in the form of $Y_s(k\varrho)$, when the normal functions become entirely different.

115. Vibrations of Circular Plate. If instead of a membrane we consider a plate we have the same sort of changes as in the analogous cases of the string and bar, treated in Chapter III. The equation is now of the fourth order, involving the double Laplacian, § 14, (148'), Chapter I. Let us write

$$(60) \qquad \frac{\partial^2 u}{\partial t^2} + a^4 \Delta\Delta u = 0,$$

and separating the variables,

$$\frac{d^2\psi}{dt^2} + n^2\psi = 0,$$

$$(61) \qquad \Delta\Delta\varphi - k^4\varphi = 0, \qquad k^4 = \frac{n^2}{a^4}.$$

We may now write the equation, bearing in mind the significance of operators, as explained in Chapter I, § 5,

$$(62) \qquad (\Delta + k^2)(\Delta - k^2)\varphi = 0,$$

or in polar coordinates

$$(63) \quad \left(\frac{\partial^2}{\partial\varrho^2} + \frac{1}{\varrho}\frac{\partial}{\partial\varrho} + \frac{1}{\varrho^2}\frac{\partial^2}{\partial\theta^2} + k^2\right)\left(\frac{\partial^2}{\partial\varrho^2} + \frac{1}{\varrho}\frac{\partial}{\partial\varrho} + \frac{1}{\varrho^2}\frac{\partial^2}{\partial\theta^2} - k^2\right)\varphi = 0.$$

Accordingly we must add the solutions of both equations,

$$(64) \qquad \frac{\partial^2\varphi}{\partial\varrho^2} + \frac{1}{\varrho}\frac{\partial\varphi}{\partial\varrho} + \frac{1}{\varrho^2}\frac{\partial^2\varphi}{\partial\theta^2} + k^2\varphi = 0,$$

$$(65) \qquad \frac{\partial^2\varphi}{\partial\varrho^2} + \frac{1}{\varrho}\frac{\partial\varphi}{\partial\varrho} + \frac{1}{\varrho^2}\frac{\partial^2\varphi}{\partial\theta^2} - k^2\varphi = 0.$$

(65) being the same as (64) with k changed to ik. Compare (328), § 41 for the bar, as for the membrane. If the plate extends to the center, the functions $Y_s(k\varrho)$, $Y_s(ik\varrho)$ do not occur since the logarithm would make them infinite.

The nature of the normal functions will depend upon the boundary conditions, as in the case of the bar. We shall consider only a plate clamped at the outer edge, as in the telephone or phonograph. In that case we have, if we consider only the symmetrical case, $s = 0$

$$(66) \qquad \varphi = AJ_0(k\varrho) + BI_0(k\varrho), \quad \text{where } I_0(x) = J_0(ix)$$

and the conditions of clamping are

(67) $$\varphi = 0, \quad \frac{d\varphi}{d\varrho} = 0, \quad \varrho = a,$$

(68) $$A J_0(ka) + B I_0(ka) = 0,$$
$$A J_0'(ka) + B I_0'(ka) = 0.$$

This gives us the equation for the frequencies,

(68) $$J_0(ka) I_0'(ka) - I_0(ka) J_0'(ka) = 0.$$

The second frequency is nearly two octaves above the lowest. This explains how the equilibrium theory may do very well for such a diaphragm.

116. Spherical Vibrations. We might have arrived at Helmholtz's equation in considering the flow of heat in two dimensions, and applied it to cylindrical problems. But the results will be as well illustrated if we consider the three-dimensional problem of the cooling of a sphere. The equation of Fourier is

(69) $$\frac{\partial u}{\partial t} = c^2 \Delta u,$$

and if we suppose that the temperature is initially distributed in an arbitrary manner, and that the surface is maintained constantly at a fixed temperature, we have

(70) $$u = 0, \qquad r = a, \quad t > 0,$$
$$u = f(r, \theta, \varphi), \quad r < a, \quad t = 0.$$

Let us put $u = T(t) V(r, \theta, \varphi)$ when the variables separate, and we have

(71) $$\frac{1}{T} \frac{dT}{dt} = \frac{c^2}{V} \Delta V = -k^2 c^2,$$

(71a) $$T = A e^{-c^2 k^2 t}$$
$$\Delta V + k^2 V = 0,$$

so that we obtain Helmholtz's equation. If we consider the vibrations of air in a hollow sphere, we may find the normal vibrations, and consider the problem of the motion when the surface of the sphere executes a prescribed normal motion, so that in either problem we have to develop a function in a series on the surface of a sphere.

To solve Helmholtz's equation put $V = R(r) Y(\theta, \varphi)$, giving as in § 91, (36)

(72) $$\frac{1}{Y} \left(\frac{\partial^2 Y}{\partial \theta^2} + \operatorname{ctn} \theta \frac{\partial Y}{\partial \theta} + \frac{1}{\sin^2 \theta} \frac{\partial^2 Y}{\partial \varphi^2} \right)$$
$$= -\left(\frac{r^2}{R} \frac{d^2 R}{dr^2} + \frac{2r}{R} \frac{dR}{dr} + k^2 r^2 \right) = \text{const.}$$

and if we take for the constant the value $- n(n + 1)$, Y_n will be a surface spherical harmonic, while R_n satisfies

$$(73) \qquad \frac{d^2 R_n}{dr^2} + \frac{2}{r}\frac{d R_n}{dr} + \left\{ k^2 - \frac{n(n+1)}{r^2} \right\} R_n = 0,$$

a new linear differential equation. If we put $R = r^m z$

$$\frac{d R}{dr} = r^m \frac{dz}{dr} + m r^{m-1} z,$$

$$\frac{d^2 R}{dr^2} = r^m \frac{d^2 z}{dr^2} + 2 m r^{m-1}\frac{dz}{dr} + m(m-1) r^{m-2} z,$$

so that we obtain

$$(74) \qquad \frac{d^2 z}{dr^2} + \frac{2(m+1)}{r}\frac{dz}{dr} + \left\{ \frac{m(m+1) - n(n+1)}{r^2} + k^2 \right\} z = 0.$$

If we now take

$$2(m+1) = 1, \quad m = -\frac{1}{2},$$

$$m(m+1) - n(n+1) = - s^2, \quad s = \pm \left(n + \frac{1}{2} \right), \quad x = kr,$$

and the equation becomes

$$\frac{d^2 z}{dx^2} + \frac{1}{x}\frac{dz}{dx} + \left(1 - \frac{s^2}{x^2} \right) z = 0,$$

or that of Bessel, with s a half an odd number. We therefore make use of the functions at the end of § 102, and have the solution

$$(75) \qquad V_n = Y_n(\theta, \varphi) J_{n+\frac{1}{2}}(kr)/\sqrt{r}.$$

Accordingly after the normal functions are determined by the boundary conditions we shall have for the cooling sphere

$$(76) \qquad u = \sum_n \sum_p A_{np} e^{- c^2 k^2 t} Y_n(\theta, \varphi) J_{n+\frac{1}{2}}(k_{np} r)/\sqrt{r},$$

and for the vibrating gas

$$(77) \qquad u = \sum_n \sum_p (A_{np} \cos kct + B_{np} \sin kct) Y_n(\theta, \varphi) J_{n+\frac{1}{2}}(k_{np} r)/\sqrt{r}.$$

For the initial conditions in either case

$$(78) \qquad u = \sum_n \sum_p A_{np} Y_n(\theta, \varphi) J_{n+\frac{1}{2}}(k_{np} r)/\sqrt{r}.$$

The boundary conditions may be various. If the temperature of the cooling sphere is maintained constant at the surface $r = a$, calling u the temperature minus that constant, we have

$$(79) \qquad \sum_n \sum_p A_{np} Y_n(\theta, \varphi) J_{n+\frac{1}{2}}(k_{np} a)/\sqrt{a} = 0$$

so that we have for every term

$$(80) \qquad J_{n+\frac{1}{2}}(k_{np}a) = 0$$

which is the transcendental equation for the characteristic numbers. In the sound problem, if the boundary is rigid, which is the analogue of the membrane problem, we have at the surface

$$(81) \qquad \frac{\partial u}{\partial r} = 0, \qquad \left[\frac{d}{dr}\left\{\frac{1}{\sqrt{r}}J_{n+\frac{1}{2}}(k_{np}r)\right\}\right]_{r=e} = 0$$

which determines the characteristic numbers.

Let us then consider the development of a function $f(r, \theta, \varphi)$ as

$$(82) \qquad f(r, \theta, \varphi) = \frac{1}{\sqrt{r}}\sum_n\sum_p A_{np}\,Y_n(\theta, \varphi)J_{n+\frac{1}{2}}(k_{np}r).$$

Multiplying by $P_m(\cos\gamma)\sin\theta\,d\theta\,d\varphi$ and integrating over the surface of a sphere, r being constant, and all terms vanishing, except that for which $n = m$, we obtain, by (99) § 97,

$$(83) \qquad \begin{aligned} &\frac{Y_m(\theta', \varphi')}{\sqrt{r}}\sum_p A_{mp}J_{m+\frac{1}{2}}(k_{mp}r) \\ &= \frac{2m+1}{4\pi}\int_0^\pi\int_0^{3\pi} f(r, \theta, \varphi)\,P_m(\cos\gamma)\sin\theta\,d\theta\,d\varphi. \end{aligned}$$

Accordingly we must be able to develop a function of r as a series,

$$(84) \qquad \sqrt{r}f(r, \theta, \varphi) = \sum_p C_{np}J_{n+\frac{1}{2}}(k_{np}r),$$

where the k's are roots of the equation (80) or (81). In any of the ways previously described we find the orthogonal condition

$$(85) \qquad \int_0^a J_{n+\frac{1}{2}}(k_{np}r)J_{n+\frac{1}{2}}(k_{nq}r)r\,dr = 0 \qquad p \neq q.$$

Multiplying (84) by $rJ_{n+\frac{1}{2}}(k_{nq}r)$ and integrating we find

$$(86) \qquad C_{nq}(\theta, \varphi) = \frac{\displaystyle\int_0^a r^{\frac{1}{2}}f(r, \theta, \varphi)J_{n+\frac{1}{2}}(k_{nq}r)dr}{\displaystyle\int_0^a J_{n+\frac{1}{2}}^2(k_{nq}r)r\,dr} \qquad\qquad \text{cf. 315 § 40}$$

and the integrated squares may be found as above. Inserting in (83)

$$(87) \qquad \begin{aligned} &Y_n\sum_p A_{np}J_{n+\frac{1}{2}}(k_{np}r) \\ &= \frac{2n+1}{4\pi}\sum_p\int_0^\pi\int_0^{2\pi} J_{n+\frac{1}{2}}(k_{np}r)C_{np}(\theta, \varphi)P_n(\cos\gamma)\sin\theta\,d\theta\,d\varphi, \end{aligned}$$

and identifying terms in $J_{n+\frac{1}{2}}(k_{np}r)$ we have

$$(88) \qquad A_{np} Y_n = \frac{2n+1}{4\pi} \int_0^\pi \int_0^{2\pi} C_{np}(\theta, \varphi) P_n(\cos \gamma) \sin \theta \, d\theta \, d\varphi.$$

For the simplest vibration in the sphere, the motion being radial $n = 0$, u is independent of the angular coordinates θ, φ, and since

$$(89) \qquad J_{\frac{1}{2}}(x) = \sqrt{\frac{2}{\pi x}} \sin x, \quad \frac{1}{\sqrt{r}} J_{\frac{1}{2}}(kr) = \sqrt{\frac{2}{\pi k}} \frac{\sin kr}{r}$$

the equation (81) for the frequency is

$$(90) \qquad \tan ka = ka,$$

of which the roots may be discussed by a graphical construction. Since $k = 2\pi/\lambda$, where λ is the wave-length, the values of ka/π are the ratios of the diameter of the sphere to the wave-length. The first six values are

$$k_{op} a/\pi = 1.4303, \quad 2.4590, \quad 3.4709, \quad 4.4747, \quad 5.4818, \quad 6.4844.$$

The next simplest solution is that for $n = 1$, and since Y_1 is a zonal[1]) harmonic, this is independent of φ, and the motion is in meridional planes. Since

$$(91) \qquad J_{\frac{3}{2}} = \sqrt{\frac{2}{\pi x}} \left(\frac{\sin x}{x} - \cos x \right)$$

we have the frequency equation

$$(92) \qquad \tan ka = \frac{2ka}{2 - k^2 a^2}.$$

The lowest root gives

$$(93) \qquad k_n a/\pi = \cdot 6626,$$

so that the period is more than an octave lower than that of the lowest radial vibration above. This is in fact the lowest of all the vibrations possible in the sphere. The air surges to and fro somewhat as in a pipe.

The solution (89) is one of the elementary solutions of Helmholtz's equation that we have used in § 65. The other solutions may be derived from it by a process not unlike that by which spherical harmonics are derived from the elementary solution of Laplace's equation. If, instead of proceeding as we did with equation (74), we put $m = n$, $r^n Y_n$ becomes a solid spherical harmonic, which is multiplied by z_n, which becomes a solution of the equation, with $kr = x$,

$$(94) \qquad \frac{d^2 z_n}{dx^2} + \frac{2(n+1)}{x} \frac{dz_n}{dx} + z_n = 0.$$

This can readily be solved by series in ascending integral powers. We

---

1) See Rayleigh Theory of Sound Vol. I § 207 u. § 331.

shall not find them, but shall show that two successive z's are related by the equation

(95)
$$z_{n+1} = \frac{1}{x}\frac{dz_n}{dx}.$$

For if we put

(96)
$$\frac{dz_n}{dx} = x z_{n+1}, \quad \frac{d^2z_n}{dx^2} = z_{n+1} + x\frac{dz_{n+1}}{dx},$$

substitute in (94) and differentiate, we obtain

$$\frac{d^2z_{n+1}}{dx^2} + \frac{2(n+2)}{x}\frac{dz_{n+1}}{dx} + z_{n+1} = 0,$$

which is (94) with $n+1$ instead of n. But for $n = 0$ we have

(97)
$$\frac{d^2z_0}{dx^2} + \frac{2}{x}\frac{dz_0}{dx} + z_0 = 0,$$

which we have already shown has the solutions (§ 65)

$$\frac{\cos x}{x}, \quad \frac{\sin x}{x}, \quad \frac{e^{ix}}{x}, \quad \frac{e^{-ix}}{x}.$$

Applying the relation (95) n times we find as solutions

$$\Psi_n(x) = (-1)^n \left(\frac{1}{x}\frac{d}{dx}\right)^n \frac{\sin x}{x}$$

(98)
$$= \frac{1}{1\cdot 3\cdots(2n+1)}\left(1 - \frac{x^2}{2(2n+3)} + \frac{x^4}{2\cdot 4(2n+3)(2n+5)} + \cdots\right)$$

$$\Psi(x) = (-1)^n \left(\frac{1}{x}\frac{d}{dx}\right)^n \frac{\cos x}{x}$$

$$= \frac{1\cdot 3\cdots(2n-1)}{x^{2n+1}}\left(1 - \frac{x^2}{2(1-2n)} + \frac{x^4}{2\cdot 4(1-2n)(3-2n)} + \cdots\right).$$

We find

$$\Psi_0(x) = \frac{\sin x}{x}$$

(99)
$$\Psi_1(x) = \frac{\sin x}{x^2} - \frac{\cos x}{x^2}$$

$$\Psi_3(x) = \left(\frac{3}{x^5} - \frac{1}{x^3}\right)\sin n - \frac{3\cos x}{x^4}$$

which may be compared with the values of $J_{n+\frac{1}{2}}(x)$ in § 102.

117. Progressive Spherical Waves. In order to get progressive waves we shall need a solution of the wave-equation of the form

(100)
$$u = e^{i(pt-kr)}v(r)Y_n,$$

consequently we shall attempt to find a solution of the equation (73) of the form $R = e^{-ikr}v(r)$. Putting $-ikr = x$ the equation becomes

(101)
$$\frac{d^2R}{dx^2} + \frac{2}{x}\frac{dR}{dx} - \left\{1 + \frac{n(n+1)}{x^2}\right\}R = 0$$

or with $R + e^x v$,

(102) $\qquad \dfrac{d^2 v}{dx^2} + 2\left(1 + \dfrac{1}{x}\right)\dfrac{dv}{dx} + \left(\dfrac{2}{x} - \dfrac{n(n+1)}{x^2}\right)v = 0.$

Now putting

(103) $\qquad\qquad\qquad v = \sum a_s x^s$

we obtain

(104) $\quad \sum[a_s\{s(s-1) + 2s - n(n+1)\} + 2sa_{s-1}]x^{s-2} = 0$

which gives the recurrence formula

(105) $\qquad\qquad\qquad a_{s-1} = \dfrac{(n-s)(n+s+1)}{2s}\, a_s,$

from which we obtain

$a_{-2} = -\dfrac{n(n+1)}{2}\, a_{-1}$

$a_{-3} = -\dfrac{(n+2)(n-1)}{2\cdot 2}\, a_{-2} = \dfrac{(n+1)(n+2)n(n-1)}{2\cdot 4}\, a_{-1}$

$a_{-4} = -\dfrac{(n+3)(n-2)}{2\cdot 3}\, a_{-3} = -\dfrac{(n+1)(n+2)(n+3)n(n-1)(n-2)}{2\cdot 4\cdot 6}\, a_{-1}.$

Since n is an integer, we finally come to a zero coefficient, so that we have the polynomial solution

(106) $\quad v = \dfrac{c_{-1}}{x}\left(1 - \dfrac{n(n+1)}{2x} + \dfrac{(n+1)(n+2)n(n-1)}{2\cdot 4\, x^2}\right.$

$\left.\qquad\qquad - \dfrac{(n-2)(n-1)\ldots(n+3)}{2\cdot 9\cdot 6\, x^3}\ldots + \dfrac{1\cdot 2\cdot 3\ldots 2n}{2\cdot 4\cdot 6\ldots 2n\, x^n}\right)$

and replacing x by $-ikr$ we have

(107) $\quad R_n = \dfrac{e^{-ikr}}{ikr}\left\{1 + \dfrac{n(n+1)}{2ikr} + \dfrac{(n-1)n(n+1)(n+2)}{2\cdot 4\,(ikr)^2} + \cdots\right\}.$

The polynomial in the parenthesis is denoted by $f_n(ikr)$. It was introduced by Stokes.[1]

Accordingly we have the solution

(108) $\qquad\qquad u = \sum Y_n(\theta, \varphi)\dfrac{e^{i(pt - kr)}}{r} f_n(ikr).$

If $n = 0$ we have the symmetrical source, which we have already considered

(109) $\qquad\qquad\qquad u = \dfrac{e^{i(pt - hr)}}{r}.$

This motion could be produced by a sphere alternately expanding and contracting, like the pulsating spheres of C. A. Bjerknes.[2]

---

1) Stokes, On the Communication of Vibrations from a Vibrating Body to a Gas. Phil. Trans. 1868.

2) C. Bjerknes. Die Hydrodynamischen Fernkräfte. Leipzig, Barth, 1900.

If $n = 1$, Y_1 is a zonal harmonic, and we have

(110) $$u = \frac{\cos \theta}{r} e^{i(pt - kr)} \left\{ 1 + \frac{2}{ikr} \right\}.$$

This is the effect of a doublet, corresponding to the case of the moving sphere in the incompressible fluid already treated in § 113 This may be realized by a sphere oscillating to and fro. The gas will flow alternately from one side to the other of the sphere. The case of the tuning-fork is roughly analogous to a pair of doublets, for the gas flows out on both sides and in at the middle, and is represented by the case $n = 4$. For all these matters the reader should consult Rayleigh, Theory of Sound.

118. Waves from Ellipsoids of Revolution. If we have to do with surfaces of revolution, by (105) § 108, Helmholtz's equation becomes

(111) $$\frac{h^2}{\varrho} \left\{ \frac{\partial}{\partial u} \left(\varrho \frac{\partial V}{\partial u} \right) + \frac{\partial}{\partial v} \left(\varrho \frac{\partial V}{\partial v} \right) \right\} + k^2 V = 0.$$

If as before $\varrho = f(u)\,\varphi(v)$ and we put $V = L(u)\,M(v)$,

(112) $$\frac{1}{Lf(u)} \frac{d}{du} \left(f(u) \frac{dL}{du} \right) + \frac{1}{M\varphi(v)} \frac{d}{dv} \left(\varphi(v) \frac{dM}{dv} \right) + \frac{k^2}{h^2} = 0,$$

the variables do not separate unless $1/h^2$ is the sum of two functions each containing but one variable u or v. In the case of ellipsoids this is true, and if prolate $f(u) = \sin u$, $\varphi(v) = \sinh v$

$$\frac{1}{h^2} = \cosh^2 v - \cos^2 u, \quad \text{cf. § 108, (112)}.$$

(113) $$\frac{1}{L} \left\{ \frac{d^2 L}{du^2} + \operatorname{ctn} u \frac{dL}{du} \right\} + \frac{1}{M} \left\{ \frac{d^2 M}{dv^2} + \operatorname{ctnh} v \frac{dM}{dv} \right\}$$
$$+ k^2 (\cosh^2 v - \cos^2 u) = 0,$$

which separates into the two

(114) $$\frac{d^2 L}{du^2} + \operatorname{ctn} u \frac{dL}{du} + (C - k^2 \cos^2 u)\, L = 0,$$
$$\frac{d^2 M}{dv^2} + \operatorname{ctnh} v \frac{dM}{dv} + (k^2 \cosh^2 v - C)\, M = 0,$$

or in terms of the elliptic coordinates $x = \cosh v$, $y = \cos u$,

(115) $$(1 - y^2) \frac{d^2 L}{dy^2} - 2y \frac{dL}{dy} + (C - k^2 y^2)\, L = 0,$$
$$(x^2 - 1) \frac{d^2 M}{dx^2} + 2x \frac{dM}{dx} + (k^2 x^2 - C)\, M = 0$$

Putting $C = k^2 + \lambda$ we have

(116) $$\frac{d^2 L}{dy^2} - \frac{2y}{1 - y^2} \frac{dL}{dy} + \left(k^2 + \frac{\lambda}{1 - y^2} \right) L = 0$$

both equations being of the same form. This is a new differential equation, which has been studied by Maclaurin, Transactions of the Cambridge Philosophical Society, 1898, and Abraham, Mathematische Annalen, 52, 1899. The latter has used the equation in studying the electric waves radiated from an antenna in the form of a long ellipsoid. The study of this equation would take us too far.

119. Neumann's Addition Theorem for Bessel Functions.
It is evident from § 103 that a solution of Laplace's equation in cylindrical coordinates will be given by

$$(117) \qquad V = \sum_n \sum_\lambda A_{n\lambda} e^{-\lambda z} J_n(\lambda r) \cos n\varphi.$$

(In the remainder of this chapter we use r instead of ϱ). If d is the distance of a point with the coordinates r, φ, z from another r', 0, 0, we must be able to develop the reciprocal distance into such a series.

$$(118) \qquad \frac{1}{d} = \sum_n \sum_\lambda C_{n\lambda} e^{-\lambda z} J_n(\lambda r) \cos n\varphi.$$

When $z = 0$ and the two points are in the same plane perpendicular to the Z-axis, we have a solution[1])

$$J_0(\lambda \omega) \quad \text{where} \quad \omega^2 = r^2 + r'^2 - 2rr'.$$

Consequently we have by symmetry,

$$(118) \qquad J_0(\lambda \omega) = \sum_{n=0}^{n=\infty} A_n J_n(\lambda r) J_n(\lambda r') \cos n\varphi.$$

If we put $r' = 0$ all the terms vanish but the first, and since $\omega = r$, we find $A_0 = 1$.

Now from (10) § 100, changing φ to $\varphi + \frac{\pi}{2}$, or from (13)

$$(119) \qquad e^{ir\cos\varphi} = J_0(r) + 2\sum_{s=1}^{s=\infty} i^s J_s(r) \cos s\varphi,$$

and in like manner,

$$(120) \quad e^{ir\cos(\varphi-\theta)} = J_0(r) + 2\sum_{s=1}^{s=\infty} i^s J_s(r)(\cos s\varphi \cos s\theta + \sin s\varphi \sin s\theta).$$

Multiplying by $\cos n\varphi$ and integrating

$$(121) \qquad \int_0^{2\pi} e^{ir\cos(\varphi-\theta)} \cos n\varphi \, d\varphi = 2\pi i^n J_n(r) \cos n\theta.$$

---

1) For ω is the distance from a fixed point, and the solution $J_0(\lambda r)$ is no better than $J_0(\lambda \omega)$ where ω is the distance from any fixed point in the plane.

Now multiply by $e^{-ir'\cos\theta}/2\pi$ and integrate

$$(122) \quad \frac{1}{2\pi}\int_0^{2\pi} e^{-ir'\cos\theta}\,d\theta \int_0^{2\pi} e^{ir\cos(\varphi-\theta)}\cos n\varphi\,d\varphi = i^n J_n(r)\int_0^{2\pi} e^{-ir'\cos\theta}\cos n\theta\,d\theta.$$

The integral on the right becomes, by another application of (121), with $\theta = 0$, $2\pi i^n J_n(-r')$, so that the powers of i disappear, and we have

$$(123) \quad \frac{1}{2\pi}\int_0^{2\pi} e^{-ir'\cos\theta}\,d\theta \int_0^{2\pi} e^{ir\cos(\varphi-\theta)}\cos n\varphi\,d\varphi = 2\pi J_n(r)J_n(r').$$

Changing the order of integration gives

$$(124) \quad \frac{1}{2\pi}\int_0^{2\pi}\cos n\varphi\,d\varphi \int_0^{2\pi} e^{i[r\cos(\varphi-\theta)-r'\cos\theta]}\,d\theta = 2\pi J_n(r)J_n(r').$$

Put

$$r\cos\varphi - r' = a, \quad r\sin\varphi = b,$$

$$(125) \quad \int_0^{2\pi} e^{i(a\cos\theta+b\sin\theta)}\,d\theta = \int_0^{2\pi} e^{i\sqrt{a^2+b^2}\cos(\theta-\alpha)}\,d(\theta-\alpha), \quad \alpha = \tan^{-1}\frac{b}{a}$$

$$= 2\pi J_0(\sqrt{a^2+b^2}) = 2\pi J_0(\sqrt{r^2+r'^2-2rr'\cos\varphi} = 2\pi J_0(\omega).$$

Consequently we have

$$(126) \quad \frac{1}{\pi}\int_0^{2\pi} J_0(\omega)\cos n\varphi\,d\varphi = 2 J_n(r) J_n(r')$$

and all the coefficients in (118) except A_0 must be 2, and

$$(127) \quad J_0(\omega) = J_0(r) J_0(r') + 2\sum_{n=1}^{n=\infty} J_n(r) J_n(r')\cos n\varphi,$$

$$\omega^2 = r^2 + r'^2 - 2rr'\cos\varphi.$$

This is Neumann's addition theorem and putting $\varphi = 0$ or π,

$$J_0(r-r') = J_0(r) J_0(r') + 2\sum_{n=1}^{n=\infty} J_n(r) J_n(r')$$

$$J_0(r+r') = J_0(r) J_0(r') + 2\sum_{n=1}^{n=\infty}(-1)^n J_n(r) J_n(r').$$

120. Definite Integrals. We have § 102, (42)

$$(128) \quad J_0(bx) = \frac{1}{\pi}\int_0^{\pi}\cos(bx\cos\varphi)\,d\varphi.$$

Accordingly,

$$(129) \qquad \int_0^\infty e^{-ax} J_0(bx)\, dx = \frac{1}{\pi} \int_0^\pi dx \int_0^\infty e^{-ax} \cos(bx \cos \varphi)\, d\varphi,$$

and we may change the order of integration. Now we have

$$(129) \qquad \int_0^\infty e^{-mx} \cos nx\, dx = \frac{1}{2} \int_0^\infty \left(e^{(in-m)x} + e^{-(in+m)x} \right) dx$$

$$= \frac{1}{2} \left[\frac{e^{(in-m)x}}{in-m} - \frac{e^{-(in+m)x}}{in+m} \right]_0^\infty$$

which for real $m > 0$

$$= -\frac{1}{2} \left(\frac{1}{in-m} - \frac{1}{in+m} \right) = \frac{m}{m^2+n^2}.$$

From this

$$(130) \qquad \int_0^\infty e^{-ax} J_0(bx)\, dx = \frac{1}{\pi} \int_0^\pi \frac{a\, d\varphi}{a^2 + b^2 \cos^2 \varphi} = \frac{1}{\sqrt{b^2 + a^2}}.$$

It can be shown that this integral still holds if for a we put ia, so that

$$(131) \qquad \int_0^\infty e^{-iax} J_0(bx)\, dx = \frac{1}{\sqrt{b^2 - a^2}}.$$

Identifying reals and imaginaries,

$$(132) \qquad
\left.
\begin{aligned}
\int_0^\infty \cos ax\, J_0(bx)\, dx &= \frac{1}{\sqrt{b^2 - a^2}}, \\[2mm]
\int_0^\infty \sin ax\, J_0(bx)\, dx &= 0,
\end{aligned}
\right\} \quad b^2 > a^2$$

$$\left.
\begin{aligned}
\int_0^\infty \cos ax\, J_0(bx)\, dx &= 0, \\[2mm]
\int_0^\infty \sin ax\, J_0(bx)\, dx &= \frac{1}{\sqrt{a^2 - b^2}}
\end{aligned}
\right\} \quad a^2 > b^2$$

These integrals are due to H. Weber.

From the formula (130), writing

$$(133) \qquad \frac{1}{\sqrt{z^2 + r^2}} = \int_0^\infty e^{-tz} J_0(tr)\, dt, \quad \begin{matrix} z=0 \\ r=1 \end{matrix} \quad \int_0^\infty J_0(t)\, dt = 1,$$

can be deduced many others. Differentiating by z

$$(134) \qquad \frac{z}{(z^2 + r^2)^{3/2}} = \int_0^\infty e^{-zt} t J_0(tr)\, dt\,.$$

Differentiating (133) by r, since $J_0' = -J_1$ by (24)

$$(135) \qquad \frac{r}{(z^2 + r^2)^{3/2}} = \int_0^\infty e^{-zt} t J_1(tr)\, dt\,, \quad z = 0\,, \quad \frac{1}{r^2} = \int_0^\infty t J_1(rt)\, dt\,.$$

Integrating (135) by z from z to ∞,

$$(136) \qquad \frac{1}{r} - \frac{z}{r\sqrt{z^2 + r^2}} = \int_0^\infty e^{-tz} J_1(tr)\, dt\,.$$

Integrating again by z,

$$(137) \qquad \frac{\sqrt{z^2 + r^2}}{r} - \frac{z}{r} = \int_0^\infty \frac{e^{-tz}}{t} J_1(tr)\, dt\,, \quad z = 0\,, \quad \int_0^\infty \frac{J_1(tr)\, dt}{t} = 1\,.$$

Integrate (133) by z or (136) by r,

$$(138) \qquad \log\left(\frac{z + \sqrt{z^2 + r^2}}{r}\right) = \int_0^\infty \frac{1 - e^{-tz}}{t} J_0(tr)\, dt$$

From the addition theorem (127), (118)

$$(139) \qquad \frac{1}{2\pi} \int_0^{2\pi} J_0(t\omega)\, d\varphi = J_0(tr)\, J_0(tr')$$

$$(140) \qquad \frac{1}{2\pi} \int_0^{2\pi} J_0(t\omega) \cos n\varphi\, d\varphi = J_n(tr)\, J_n(tr')\,.$$

Since

$$\omega \frac{\partial \omega}{\partial r} = r - r' \cos\varphi\,, \qquad \omega \frac{\partial \omega}{\partial \varphi} = r r' \sin\varphi\,,$$

differentiating the addition theorem by r, $(J_0' = -J_1)$,

$$-t J_1(t\omega) \frac{(r - r' \cos\varphi)}{\omega} = -t J_1(tr) J_0(tr') + 2 \sum_{n=1}^{n=\infty} t J_n'(tr) J_n(tr') \cos n\varphi\,,$$

$$(141) \qquad \frac{1}{2\pi} \int_0^{2\pi} \frac{J_1(t\omega)(r - r' \cos\varphi)\, d\varphi}{\omega} = J_1(tr) J_0(tr')\,.$$

Differentiating by φ,

$$- t \frac{J_1(t\omega)}{\omega} rr' \sin\varphi = - 2 \sum_{n=1}^{n=\infty} n J_n(tr) J_n(tr') \sin n\varphi.$$

(142) $$\frac{trr'}{2\pi} \int_0^{2\pi} \frac{J_1(t\omega)}{\omega} \sin^2\varphi \, d\varphi = J_1(tr) J_1(tr').$$

Compare with (140).

121. Potentials of Symmetrical Bodies. For a circular line of density δ, radius a, at a point with coordinates r, z,

(143) $$V = \int \frac{\delta \, ds}{d} = \delta a \int_0^{2\pi} \frac{d\varphi}{\sqrt{z^2 + \omega^2}}, \quad \omega^2 = r^2 + a^2 - 2ar\cos\varphi.$$

Now by (133), this equals

(144) $$\delta a \int_0^{2\pi} d\varphi \int_0^{\infty} e^{-tz} J_0(t\omega) \, dt = \delta a \int_0^{\infty} e^{-tz} dt \int_0^{2\pi} J_0(t\omega) \, d\varphi,$$

and by (139)

(145) $$V = 2\pi a\delta \int_0^{\infty} e^{-zt} J_0(rt) J_0(at) \, dt.$$

For the flux-function, since by (41),

(41) $$\frac{\partial \psi}{\partial r} = r \frac{\partial V}{\partial z}, \quad \frac{\partial \psi}{\partial z} = - r \frac{\partial V}{\partial r},$$

(146) $$\psi = 2\pi a\delta r \int_0^{\infty} e^{-zt} J_1(rt) J_0(at) \, dt.$$

For a disc of radius a, density σ, calling (145) δV_a,

(147) $$V = \sigma \int_0^a V_a \, da = 2\pi\sigma \int_0^a a \, da \int_0^{\infty} e^{-zt} J_0(rt) J_0(at) \, dt.$$

Now from (18) § 100

$$\frac{d}{dx}(x J_1(x)) = x J_0(x), \quad \frac{d[at J_1(at)]}{d(at)} = at J_0(at)$$

(148) $$\int_0^a a J_0(at) \, da = \frac{a J_1(at)}{t}$$

(149) $$V = 2\pi a\sigma \int_0^{\infty} \frac{e^{-zt}}{t} J_1(at) J_0(rt) \, dt$$

(150) $$\psi = 2\pi\sigma a r \int_0^{\infty} \frac{e^{-zt}}{t} J_1(at) J_0(rt) \, dt.$$

For a double-distribution on a disc, or the solid angle subtended, § 57,

$$(151) \qquad W = \frac{\partial V}{\partial z} = 2\pi\varkappa a \int_0^\infty e^{-zt} J_0(rt) J_1(at) dt,$$

$$(152) \qquad \psi = 2\pi\varkappa ar \int_0^\infty e^{-zt} J_1(rt) J_1(at) dt.$$

This is the potential and flux function for an electrical current of radius a, unit strength.

122. Beltrami's Symmetrical Potentials. We shall devote the rest of this chapter to the exposition of some of the results obtained by Beltrami[1]) on potentials of bodies symmetrical about an axis of rotation, including the results just obtained.

We have seen that for $z > 0$, $V = Ae^{-zt}J_0(tr)$ is a solution of Laplace's equation, for which

$$\psi = A r e^{-zt} J_0'(tz)$$

is the flux-function.

For $z < 0$

$$(153) \qquad V = A e^{zt} J_0(tr)$$

$$\psi = - A r e^{zt} J_0'(tr).$$

For the surface density σ

$$(154) \qquad \left(\frac{\partial V}{\partial z}\right)_{+0} - \left(\frac{\partial V}{\partial z}\right)_{-0} = - 2 A t J_0(tr) = - 4\pi\sigma$$

consequently

$$\sigma(r) = \frac{At}{2\pi} J_0(tr).$$

The mass as far as r is

$$(155) \quad m(r) = 2\pi\int_0^r \sigma(r)r\, dr = At\int_0^r r J_0(tr)\, dr = Ar J_1(tr) = - \psi(r).$$

Multiplying by any function of t, $\varphi(t)$ and integrating, we still have a solution

$$(156) \qquad V = \int_0^\infty e^{-zt} J_0(rt)\varphi(t)\, dt,$$

$$(157) \qquad \psi = \int_0^\infty e^{-zt} J_1(rt)\varphi(t)\, dt,$$

1) E. Beltrami, Sulla Teoria delle Funzioni potenziali simmetriche. R. Accad. delle scienze di Bologna. Ser. 4, vol. 2, pp. 461—498, 1881.

for a distribution of density

(158)
$$\sigma(r) = \frac{1}{2\pi} \int_0^\infty t J_0(rt) \varphi(t) dt$$

of the first kind, Chapter III (288).

This is an integral equation for φ, σ being given.

Now the potential of a circular line (145)

(159)
$$V = v(a) = 2\pi a \delta \int_0^\infty e^{-zt} J_0(rt) J_0(at) dt$$

$$\psi = w(a) = 2\pi a r \delta \int_0^\infty e^{-zt} J_1(rt) J_0(at) dt$$

is of the above form, with

(160)
$$\varphi(t) = 2\pi a J_0(at),$$

so that from the formula for σ, (158)

(161)
$$\sigma(r) = a \int_0^\infty t J_0(at) J_0(rt) dt$$

must be 0 for $a \neq r$, ∞ for $a = r$.

For any distribution

(162)
$$V = \int_0^\infty v(a)\sigma(a) da, \qquad \psi = \int_0^\infty w(a)\sigma(a)$$

or

(163)
$$V = 2\pi \int_0^\infty e^{-zt} J_0(rt) dt \int_0^\infty J_0(at)\sigma(a) a \, da$$

(164)
$$\psi = 2\pi r \int_0^\infty e^{-zt} J_1(rt) dt \int_0^\infty J_0(at)\sigma(a) a \, da$$

for which

(165)
$$\varphi(t) = 2\pi \int_0^\infty J_0(at)\sigma(a) a \, da,$$

which is the solution of the integral equation (158). Inserting this in the integral equation,

(166)
$$\varphi(t) = \int_0^\infty J_0(st) s \, ds \int_0^\infty J_0(sr) r \varphi(r) dr,$$

which is Hankel's theorem similar to Fourier's, Chapter IV (52), § 43.

For instance, for a disc of radius a, $\sigma(r) = 0$, $r > a$

(167)
$$\sigma(r) = \int_0^\infty J_0(sr) s \, ds \int_0^a J_0(st) t\sigma(t) dt.$$

Now consider how to determine **V** from its values in the plane $z = 0$. From (156)

(168)
$$V(r) = \int_0^\infty J_0(rt)\,\varphi(t)\,dt.$$

Now this is of the same form as (158) if we put $V\,2\pi$ for σ, $\varphi(t)t$ for $\varphi(t)$, consequently the solution is

(169)
$$\frac{\varphi(r)}{r}\int_0^\infty J_0(rt)\,V(t)\,t\,dt,$$

which gives V,

(170)
$$V = \int_0^\infty\int_0^\infty e^{-zt}J_0(rt)\,J_0(st)\,V(s)\,st\,ds\,dt,$$

(171)
$$\psi = r\int_0^\infty\int_0^\infty e^{-zt}J_1(rt)\,J_0(st)\,V(s)\,st\,ds\,dt.$$

If we have a disc of radius a, $\sigma(r) = 0$, $r > a$, by (165)

(172)
$$\varphi(r) = 2\pi\int_0^a J_0(rt)\,\sigma(t)\,t\,dt.$$

For instance, if $\sigma = 1$

(173)
$$\varphi(r) = 2\pi\int_0^a J_0(rt)\,t\,dt = 2\pi a\,\frac{J_1(ar)}{r}.$$

If the potential is given, instead of the density, for $r < a$ only, the determination is more difficult, for then

(168)
$$V(r) = \int_0^\infty J_0(rt)\,\varphi(t)\,dt, \qquad\qquad \text{for } r < a,$$

and

(158)
$$\sigma(r) = \frac{1}{2\pi}\int_0^\infty J_0(rt)\,\varphi(t)\,t\,dt = 0 \qquad\qquad \text{for } r > a.$$

If we call $M(r)$ the mass of the ring outside r and inside a,

(174)
$$M(r) = 2\pi\int_r^a \sigma(r)\,r\,dr, \qquad M(a) = 0,$$

(175)
$$\frac{\partial M}{\partial r} = -2\pi r\sigma(r), \qquad \sigma = -\frac{1}{2\pi r}\frac{\partial M}{\partial r}$$

from which, by (165)

(176)
$$\varphi(t) = -\int_0^a J_0(rt)\frac{\partial M}{\partial r}\,dr = -M(r)J_0(rt)\Big|_0^a - t\int_0^a M(r)J_1(rt)\,dr$$

$$= M(0) - t\int_0^a M(r)J_1(rt)\,dr.$$

Introduce a function $F(r)$ such that

$$(177) \quad M(r) = \int_r^a \frac{F'(s)\,s\,ds}{\sqrt{s^2 - r^2}}, \quad M(0) = F(a), \quad F(0) = 0. \quad (177\,\mathrm{a})$$

We easily see that we may integrate first by r from 0 to s and then by s from 0 to a, (Dirichlet's integral formula),

$$
\begin{aligned}
(178) \quad \varphi(t) &= M(0) - t \int_0^a J_1(rt)\,dr \int_r^a \frac{F'(s)\,s\,ds}{\sqrt{s^2 - r^2}} \\
&= M(0) - t \int_0^a F'(s)\,s\,ds \int_0^s \frac{J_1(rt)}{\sqrt{s^2 - r^2}}\,dr.
\end{aligned}
$$

Now put $r = s \cdot \sin\theta, \; dr = s\cos\theta\,d\theta = \sqrt{s^2 - r^2}\,d\theta,$

$$(179) \quad \int_0^s \frac{J_1(rt)\,dr}{\sqrt{s^2 - r^2}} = \int_0^{\frac{\pi}{2}} J_1(st\sin\theta)\,d\theta = \int_0^{\frac{\pi}{2}} \sum_{n=0}^{n=\infty} \frac{(-1)^n \left(\frac{st\sin\theta}{2}\right)^{2n+1}}{n!\,(n+1)!}\,d\theta$$

and since

$$\int_0^{\frac{\pi}{2}} \sin^{2n+1}\theta\,d\theta = \frac{2^{2n}(n!)^2}{(2n)!\,(2n+1)}$$

$$(180) \quad \int_0^{\frac{\pi}{2}} J_1(st\sin\theta)\,d\theta = \sum_{n=0}^{n=\infty} (-1)^n \frac{(st)^{2n+1}}{(2n)!\,(2n+2)\,(2n+1)} = \frac{1 - \cos st}{st}$$

so that finally

$$
\begin{aligned}
(181) \quad \varphi(t) &= M(0) + \int_0^a (\cos st - 1)\, F'(s)\,ds \\
\varphi(t) &= \int_0^a F'(s)\cos(st)\,ds.
\end{aligned}
$$

Integrating by parts,

$$
\begin{aligned}
(182) \quad \varphi(t) &= F(s)\cos(st)\,\Big|_0^a + t \int_0^a F(s)\sin(st)\,ds \\
&= M(0)\cos(at) + t \int_0^a F(s)\sin(st)\,dt.
\end{aligned}
$$

We shall prove below that

$$(183) \quad F(r) = \frac{2}{\pi} \int_0^r \frac{V(s)\,s\,ds}{\sqrt{r^2 - s^2}}$$

and putting $\qquad r^2 - s^2 = t^2, \qquad - s\,ds = t\,dt,$

(184) $\qquad \dfrac{\pi}{2} F(r) = \displaystyle\int_0^r V(\sqrt{r^2 - s^2})\,dt$

and differentiating by r,

(185) $\dfrac{\pi}{2} F'(r) = V(a) + r \displaystyle\int_0^r \dfrac{V'(\sqrt{r^2 - t^2})}{\sqrt{r^2 - t^2}}\,dt = V(0) + r \displaystyle\int_0^r \dfrac{V'(s)\,ds}{\sqrt{r^2 - s^2}}$

From this in (181)

(186) $\qquad \varphi(t) = \dfrac{2V(0)}{\pi} \dfrac{\sin at}{t} + \dfrac{2}{\pi} \displaystyle\int_0^a r \cos(rt)\,dr \int_0^r \dfrac{V'(s)}{\sqrt{r^2 - s^2}}\,ds.$

Does the last equation (181) satisfy the equation (168), i. e.,

(187)
$$\int_0^\infty J_0(rt)\,dt \int_0^a F'(s) \cos(st)\,ds = V(r), \qquad\qquad r < a,$$

$$\int_0^\infty J_0(rt)t\,dt \int_0^a F'(s) \cos(st)\,ds = 0. \qquad\qquad r > a.$$

Changing the order, is

(188) $\qquad \displaystyle\int_0^a F'(s)\,ds \int_0^\infty J_0(rt) \cos(st)\,dt = V(r). \qquad\qquad r < a.$

Now

(132) $\qquad \displaystyle\int_0^\infty J_0(rt) \cos(st)\,dt = \dfrac{1}{\sqrt{r^2 - s^2}} \quad r > s, \; = 0, \; r < s.$

(189) $\qquad \displaystyle\int_0^r \dfrac{F'(s)\,ds}{\sqrt{r^2 - s^2}} = V(r). \qquad\qquad r < a.$

Now if F is defined as in (183)

$$\int_0^r \dfrac{F(t)t\,dt}{\sqrt{r^2 - t^2}} = \dfrac{2}{\pi} \int_0^r \dfrac{t\,dt}{\sqrt{r^2 - t^2}} \int_0^t \dfrac{V(s)s\,ds}{\sqrt{t^2 - s^2}}$$

which by Dirichlet's formula is

(190) $\qquad = \dfrac{2}{\pi} \displaystyle\int_0^r V(s)s\,ds \int_s^r \dfrac{t\,dt}{\sqrt{(r^2 - t^2)(t^2 - s^2)}} = \int_0^r V(s)s\,ds.$

Hence for $r < a$

(191) $\qquad \dfrac{d}{dr} \displaystyle\int_0^r \dfrac{F(t)t\,dt}{\sqrt{r^2 - t^2}} = rV(r).$

Performing the differentiation as in (183) (185), and since $F(0) = 0$

$$(192) \qquad r \int_0^r \frac{F'(t)\,dt}{\sqrt{r^2 - t^2}} = r\,V(r),$$

which is the equation (189) to be verified, so that F as defined in (183) satisfies the first condition.

If $r > a$, we have instead of the above

$$(193) \qquad V(r) = \int_0^a \frac{F'(t)\,dt}{\sqrt{r^2 - t^2}}, \qquad\qquad r > a,$$

which, with the formula for F, (183) gives

$$(194) \qquad V(r) = \frac{2}{\pi} \int_0^a \frac{dt}{\sqrt{r^2 - t^2}} \frac{d}{dt} \int_0^t \frac{V(s)\,s\,ds}{\sqrt{t^2 - s^2}},$$

gives the values outside the disc from those in it.

Now for the second equation of condition.

We have

$$(158) \qquad \sigma(r) = \frac{1}{2\pi} \int_0^\infty J_0(rs)\,\varphi(s)\,s\,ds,$$

$$(195) \qquad 2\pi r\sigma(r) = \frac{d}{dr}\left\{ r \int_0^\infty J_1(rs)\,\varphi(s)\,ds \right\},$$

and using the value of $\varphi(s)$ in (181)

$$(196) \qquad \begin{aligned} \int_0^\infty J_1(rs)\varphi(s)\,ds &= \int_0^\infty J_1(rs)\,ds \int_0^a F'(t)\cos(st)\,dt \\ &= \int_0^a F'(t)\,dt \int_0^\infty J_1(rs)\cos(st)\,ds. \end{aligned}$$

Now, it will be proved in (225)

$$(197) \qquad \int_0^\infty J_1(rs)\cos(st)\,ds = \frac{1}{r}, \qquad\qquad t < r,$$

$$= \frac{1}{r}\left(1 - \frac{t}{\sqrt{t^2 - r^2}}\right). \qquad\qquad t > r,$$

Therefore

$$(198) \qquad \int_0^\infty J_1(rs)\varphi(s)\,ds = \frac{1}{r}\int_0^a F'(t)\,dt = \frac{F(a)}{r}, \qquad\qquad r > a,$$

$$\int_0^\infty J_1(rs)\,\varphi(s)\,ds = \frac{1}{r}\int_0^r F'(t)\,dt + \frac{1}{r}\int_r^a \left(1 - \frac{t}{\sqrt{t^2 - r^2}}\right)F'(t)\,dt$$

$$(199)$$

$$= \frac{F(a)}{r} - \frac{1}{r}\int_r^a \frac{F'(t)\,t\,dt}{\sqrt{t^2 - r^2}}, \qquad\qquad r < a,$$

so that finally, by (195)

$$2\pi r\sigma(r) = 0, \qquad\qquad r > a$$

$$(200) \qquad\qquad 2\pi r\sigma(r) = -\frac{d}{dr}\int_r^a \frac{F'(t)\,t\,dt}{\sqrt{t^2 - r^2}}, \qquad\qquad r < a$$

and

$$(201) \qquad\qquad M(r) = 2\pi\int_r^a r\sigma(r)\,dr = -\int_r^a \frac{F'(t)\,t\,dt}{\sqrt{t^2 - r^2}}, \qquad\qquad r < a$$

as in (177).

As an application of formula (186) put $V = \text{const.} = 1$, $r < a$, then

$$\varphi(t) = \frac{2}{\pi}\frac{\sin at}{t},$$

and

$$V = \frac{2}{\pi}\int_0^\infty e^{-zt}J_0(rt)\frac{\sin at}{t}\,dt,$$

$$(202) \qquad\qquad \psi = -\frac{2}{\pi}\int_0^\infty e^{-zt}J_1(rt)\frac{\sin at}{t}\,dt,$$

$$\sigma(r) = \frac{1}{\pi^2}\int_0^\infty J_0(rt)\sin at\,dt, \quad = \frac{1}{\pi^2}\frac{1}{\sqrt{a^2 - r^2}} \qquad r < a$$

$$F(r) = \frac{2r}{\pi},$$

$$M(r) = \frac{2}{\pi}\sqrt{a^2 - r^2},$$

giving the potential and other functions for an electrified conducting disc.

Put in (165)

$$\sigma(r) = 1, \quad r < a, \quad \sigma(r) = 0, \quad r > a,$$

$$(203) \qquad\qquad \varphi(r) = 2\pi\int_0^a J_0(rt)t\,dt = 2\pi a\frac{J_1(ar)}{r}.$$

$$(204) \qquad\qquad V = 2\pi a\int_0^\infty e^{-zt}J_0(rt)J_1(at)\frac{dt}{t}$$

which is the same as (149) For the density, (158)

$$(205) \qquad \sigma(r) = a \int_0^\infty J_0(rt) J_1(at) dt$$

which integral accordingly is $1/a$ when $r < a$ or 0 when $r > a$.

123. Method for Definite Integrals. Let us put

$$(206) \qquad \int_0^\infty e^{-zt} J_n(rs) \cos st \, dt = P_n,$$

$$\int_0^\infty e^{-zt} J_n(rs) \sin st \, ds = Q_n,$$

$$(207) \qquad P_n + i Q_n = \int_0^\infty e^{-(z - it)s} J_n(rs) ds$$

where $z, r, t,$ are real and positive. From § 100, (10),

$$e^{irs \cos\varphi} = J_0(rs) + 2 \sum_{n=1}^{n=\infty} i^n J_n(rs) \cos n\varphi$$

$$(208) \quad e^{-(-z-it-ir\cos\varphi)s} = J_0(rs) e^{-(z-it)s} + 2 \sum_{n=1}^{n=\infty} i^n \cos n\varphi J_n(rs) e^{-(z-it)s},$$

and integrating with respect to s from 0 to ∞

$$(209) \quad \frac{1}{z - it - ir\cos\varphi} = P_0 + i Q_0 + 2 \sum_{n=1}^{n=\infty} i^n (P_n + i Q_n) \cos n\varphi.$$

Put this equal to

$$(210) \qquad \frac{\beta}{1 - 2\alpha \cos\varphi + \alpha^2}, \qquad \frac{1 + \alpha^2}{\beta} = z - it, \qquad ir = \frac{2\alpha}{\beta},$$

or eliminating β,

$$(211) \qquad \alpha^2 - 2\alpha \frac{z - it}{ir} + 1 = 0,$$

and solving for α,

$$(212) \qquad \alpha = \frac{ir}{\sqrt{r^2 + (z - it)^2} + z - it}$$

$$\beta = \frac{2}{\sqrt{r^2 + (z - it)^2} + z - it}$$

where the radical is taken with the $+$ sign, since for $r = 0$, $\alpha = 0$, $\beta = 1/(z - it)$. If we put

$$(213) \qquad \sqrt{r^2 + (z - it)^2} = Z - i T,$$

we are to have for $r = 0$, $Z = z$, $T = t$.

Squaring and comparing both sides,

(214) $$r^2 + z^2 - t^2 = Z^2 - T^2, \qquad zt = ZT,$$

(215) $$Z^2 - \frac{z^2 t^2}{Z^2} = \frac{z^2 t^2}{T^2} - T^2 = r^2 + z^2 - t^2$$

(216) $$Z = \sqrt{\tfrac{1}{2}\left(\sqrt{(r^2 + z^2 - t^2)^2 + 4z^2 t^2} + (r^2 + z^2 - t^2)\right)}$$
$$T = \sqrt{\tfrac{1}{2}\left(\sqrt{(r^2 + z^2 - t^2)^2 + 4z^2 t^2} - (r^2 + z^2 - t^2)\right)}.$$

For all radicals take their absolute values. The equation

(214) $$r^2 + z^2 - t^2 = Z^2 - T^2$$

can be written

(217) $$(z + Z)^2 + (t + T)^2 = r^2 + z(T^2 + z^2 + zZ + tT)$$

so that, when z, t, Z, T are positive,

(218) $$(z + Z)^2 + (t + T)^2 > r^2$$

unless $z = 0$, $T = 0$, when $t > r$.

Now we had

(212) $$\frac{1}{\alpha} = \frac{ir}{\sqrt{r^2 + (z - it)^2} + z - it} = \frac{ir}{Z + z - i(T + t)}$$

$$\left|\frac{1}{\alpha}\right| = \frac{r}{\sqrt{(Z + z)^2 + (T + t)^2}} < 1,$$

and consequently we may develop in a power-series,

$$\frac{\beta}{1 - 2\alpha \cos\varphi + \alpha^2} = \beta(1 - \alpha e^{i\varphi})^{-1}(1 - \alpha e^{-i\varphi})^{-1}$$

(219) $$= \beta \sum_{n=0}^{n=\infty} \alpha^n e^{in\varphi} \sum_{m=0}^{m=\infty} \alpha^m e^{-im\varphi} = \frac{\beta}{1 - \alpha^2}(1 + 2\alpha \cos\varphi + 2\alpha^2 \cos 2\varphi + \cdots).$$

But since, by

(210) $$\frac{1 + \alpha^2}{\beta} = z - it,$$

we have, subtracting this from $2/\beta$,

(220) $$\frac{1 - \alpha^2}{\beta} = \frac{2}{\beta} - (z - it) = \frac{ir}{\alpha} - (z - it) = \sqrt{r^2 + (z - it)^2},$$

by (212).

Accordingly

(215) $$\frac{1}{z - it - ir\cos\varphi} = \frac{1}{\sqrt{r^2 + (z - it)^2}}\left\{1 + 2\sum_{n=1}^{n=\infty} \alpha^n \cos n\varphi\right\}.$$

Comparing with (209),

$$i^n(P + iQ_n) = \frac{\alpha^n}{\sqrt{r^2 + (z - it)^2}}$$

$$P_n + iQ_n = \frac{1}{\sqrt{r^2 + (z - it)^2}} \left\{ \frac{r}{\sqrt{r^2 + (z - it)^2} + z - it} \right\}^n$$

(221)

$$= \frac{1}{\sqrt{r^2 + (z - it)^2}} \left\{ \frac{\sqrt{r^2 + (z - it)^2} - (z - it)}{r} \right\}^n$$

where

$$\sqrt{r^2 + (z - it)^2} = Z - iT.$$

For $n = 0, 1$

(222)
$$P_0 + iQ_0 = \frac{1}{z - iT},$$

(223)
$$P_1 + iQ_1 = \frac{1}{r}\left(1 - \frac{z - it}{Z - iT}\right).$$

If $z = 0$ we have

$$Z = \sqrt{\tfrac{1}{2}\{|r^2 - t^2| + (r^2 - t^2)\}}$$

(224)
$$T = \sqrt{\tfrac{1}{2}\{|r^2 - t^2| - (r^2 - t^2)\}}$$

$$Z = \sqrt{r^2 - t^2}, \qquad T = 0 \qquad\qquad t < r,$$

$$Z = 0 \qquad\qquad T = \sqrt{t^2 - r^2} \qquad\qquad t > r,$$

from which we get the following formulae.

$$\left.\begin{array}{l}
\displaystyle\int_0^\infty J_0(rs)\cos st\,ds = \frac{1}{\sqrt{r^2 - t^2}} \\[2.5ex]
\displaystyle\int_0^\infty J_0(rs)\sin st\,ds = 0 \\[2.5ex]
\displaystyle\int_0^\infty J_1(rs)\cos st\,ds = \frac{1}{r} \\[2.5ex]
\displaystyle\int_0^\infty J_1(rs)\sin st\,ds = \frac{t}{r\sqrt{r^2 - t^2}}
\end{array}\right\} \quad r > t$$

(225)

$$\left.\begin{array}{l}
\displaystyle\int_0^\infty J_0(rs)\cos st\,ds = 0 \\[2.5ex]
\displaystyle\int_0^\infty J_0(rs)\sin st\,ds = \frac{1}{\sqrt{t^2 - r^2}} \\[2.5ex]
\displaystyle\int_0^\infty J_1(rs)\cos st\,ds = \frac{1}{r}\left\{1 - \frac{t}{\sqrt{t^2 - r^2}}\right\} \\[2.5ex]
\displaystyle\int_0^\infty J_1(rs)\sin st\,ds = 0
\end{array}\right\} \quad t > r$$

The first two of these are Weber's integrals (132), while the last two give (197), which we have already used.

CHAPTER IX.

THEORY OF INTEGRAL EQUATIONS.

124. Integral Equations for Boundary Problems. The theory of integral equations has grown out of the attempt to solve the fundamental boundary problems of mathematical physics, in particular, the first boundary problem, or the problem of Dirichlet, that is, to find a function satisfying Laplace's equation, and taking on certain surface prescribed values.

Suppose that W stands for the potential of a double distribution

$$(1) \qquad W = \iint \sigma \frac{\partial \frac{1}{r}}{\partial n} dS = \iint \frac{\sigma \cos \varphi}{r^2} dS,$$

where φ is the angle made by the internal normal with the line from q the point of integration to the point for which we define W. We have proved in § 58 that such a potential has discontinuities in crossing the surface such that, if we characterize the limit approached on one side by the sign $+$ and on the other by $-$ we shall have

$$(2) \qquad \begin{aligned} W_+ - W &= 2\pi\sigma, \\ W - W_- &= 2\pi\sigma. \end{aligned}$$

The process that we are about to describe is a formulation by Poincaré of Carl Neumann's treatment of Dirichlet's problem.

Let us seek a double distribution which shall satisfy the following boundary condition

$$(3) \qquad \frac{1}{2}[W^+ - W^-]_S - \frac{\lambda}{2}\lfloor W^+ + W^-\rfloor = V_S$$

where λ is a constant parameter to be disposed of at pleasure. If we put $\lambda = 1$ we have

$$W^- = -V_S$$

and ($-$ denoting the outside) the function W solves the outer Dirichlet problem, while if $\lambda = -1$, $W^+ = V_S$, it solves the inner problem.

Now making use of the equations of discontinuity (2), the equation (3) becomes

$$(4) \qquad 2\pi\sigma - \lambda W = V_S$$

or putting in the value (1) of W,

$$(5) \qquad \sigma - \frac{\lambda}{2\pi} \iint \sigma \frac{\cos \varphi}{r^2} dS = \frac{V_S}{2\pi}.$$

If we denote the position of the point p on the surface by curvilinear coordinates u, v, and those of the point q of integration by u', v', we may write more explicitly

$$(6) \qquad \sigma(u, v) - \frac{\lambda}{2\pi} \iint \sigma(u', v') \frac{\cos \varphi(u', v') dS_q}{r_{pq}^2} = \frac{V}{2\pi}(u, v),$$

that is, V being a given function of u, v, we are to find the unknown function σ, from the above functional equation. We may make the problem general, while referring it to one independent variable if we associate the point p with a variable s, and that of integration q with t, both of which are to describe a fixed range of variation, which we may conveniently take from 0 to 1. For the unknown function put φ, for the given one f, and for the given function of *both* points let us write $K(s, t)$. Thus

$$(7) \qquad \varphi(s) - \lambda \int_0^1 K(s, t)\varphi(t)dt = f(s).$$

Such an equation, in which the unknown function φ enters under the integral sign as well as outside of it, is called by Hilbert an integral equation of the second kind. We have met it in quite a different problem in Chapter III, (202). The function $K(s, t)$ is called by Hilbert the "Kern" which we will translate as kernel or matrix, and which characterizes the integral equation.

The equation (7) was first solved by Fredholm, Stockholm-Öfversigt-Kongl. Vetenskaps. Akad. 1900. Acta-Mathematica XXVII, p. 365—1903. The theory has been much extended by Hilbert, in five papers in the Göttinger Nachrichten, 1904—5, and in his book on integral equations.

125. Algebraic Problem. The method of solving the integral equation (7), in which we shall assume that the kernel K is a symmetrical function,

$$K(s, t) = K(t, s),$$

is by considering this transcendental process as the limit of an algebraic process, as follows. Divide the field of integration into n equal parts $\delta = 1/n$ and replace the integral by a sum,

$$(8) \quad f(s) = \varphi(s) - \lambda\delta\left(K(s, \delta)\varphi(\delta) + K(s, 2\delta)\varphi(2\delta)\cdots + K(s, n\delta)\varphi(n\delta)\right).$$

Now since this holds for all values of s from 0 to 1 let us pick out the same points of subdivision, and put

$$(9) \quad \begin{aligned} f(\delta) &= \varphi(\delta) - \lambda\delta\left(K(\delta, \delta)\varphi(\delta) + K(\delta, 2\delta)\varphi(2\delta)\ldots\right), \\ f(2\delta) &= \varphi(2\delta) - \lambda\delta\left(K(2\delta, \delta)\varphi(\delta) + K(2\delta, 2\delta)\varphi(2\delta)\ldots\right)) \end{aligned}$$

. . . .

In order to save space, let us write

$$K_{pq} = K(p\delta, q\delta), \quad f_p = f(p\delta), \quad \varphi_p = \varphi(p\delta)$$

and $\lambda\delta = l$.

Then our n equations become

$$f_1 = \varphi_1 - l(K_{11}\varphi_1 + K_{12}\varphi_2 \cdots + K_{1n}\varphi_n),$$
$$f_2 = \varphi_2 - l(K_{21}\varphi_1 + K_{22}\varphi_2 \cdots + K_{2n}\varphi_n),$$

(10)

$$\cdot \quad \cdot \quad \cdot \quad \cdot \quad \cdot \quad \cdot \quad \cdot \quad \cdot \quad \cdot$$
$$\cdot \quad \cdot \quad \cdot \quad \cdot \quad \cdot \quad \cdot \quad \cdot \quad \cdot \quad \cdot$$

$$f_n = \varphi_n - l(K_{n1}\varphi_1 + K_{n2}\varphi_2 \cdots + K_{nn}\varphi_n),$$

for the $n\varphi's$, the $f's$ and $K's$ being given. Finally we are to make n increase without limit.

The determinant of the above equations, which will appear as the denominator of each φ, we call

(11)
$$d(l) = \begin{vmatrix} 1 - lK_{11}, & - lK_{12}, & \cdots & - lK_{1n} \\ - lK_{21}, & 1 - lK_{22}, & \cdots & - lK_{2n} \\ \cdot & \cdot \quad \cdot \quad \cdot & \cdot & \cdot \\ \cdot & \cdot \quad \cdot \quad \cdot & \cdot & \cdot \\ - lK_{n1}, & \cdot \quad lK_{n2}, & \cdots & 1 - lK_{nn} \end{vmatrix}$$

which as in § 30, (80), since $K_{rs} = K_{sr}$, is a symmetrical determinant, and is the discriminant of the quadratic form

(12) $x_1^2 + x_2^2 + \cdots + x_n^2 - l(K_{11}x_1^2 + K_{12}x_1x_2 + \cdots K_{nn}x_2^2).$

We shall make use of the following abbreviations.

$$Kxy = \sum_{p=1}^{p=n}\sum_{q=1}^{q=n} K_{pq}x_p y_q = K_{11}x_1y_1 + K_{12}x_1y_2 + K_{21}x_2y_1 + \cdots$$
$$Kx_1 = K_{11}x_1 + K_{12}x_2 + \cdots + K_{1n}x_n$$
$$Kx_2 = K_{21}x_1 + K_{22}x_2 + \cdots + K_{2n}x_n$$

(13)
$$\cdot \quad \cdot \quad \cdot \quad \cdot \quad \cdot \quad \cdot \quad \cdot \quad \cdot$$
$$Kx_n = K_{n1}x_1 + K_{n2}x_2 + \cdots + K_{nn}x_n$$
$$[xy] = x_1y_1 + x_2y_2 + \cdots + x_ny_n,$$

so that $d(l)$ is the discriminant of $[x, x] - lKxx$. It is at once seen that if $n = 3$ the equation $d(l) = 0$ is the cubic for the axes of a central quadric and that its roots in l are all real.

In order to solve the linear equations (10) we shall make use of another determinant which we shall call

(14)
$$D\left(l, \frac{x}{y}\right) \equiv \begin{vmatrix} 0, & x_1, & x_2, & \cdots & x_n \\ y_1, & 1 - lK_{11}, & - lK_{12}, & \cdots & - lK_{1n} \\ y_2, & - lK_{21}, & 1 - lK_{22}, & \cdots & - lK_{2n} \\ \cdot & \cdot \quad \cdot \quad \cdot & \cdot & \cdot \\ y_n, & \cdot \quad \cdot \quad \cdot \quad \cdot \quad \cdot & & 1 - lK_{nn} \end{vmatrix}$$

made by bordering $d(l)$ with $x's$ and $y's$, and symmetrical in these variables. If we replace in this each y_p by

$$Ky_p = K_{p1}y_1 + K_{p2}y_2 + \cdots + K_{pn}y_n,$$

we call it
$$D\left(l, \frac{x}{Ky}\right).$$

If we multiply $[xy]$ by $d(l)$ we may write

(15)
$$[xy]d(l) = \begin{vmatrix} [xy], & x_1, & x_2, \ldots x_n \\ 0, & 1 - lK_{11}, & -lK_{12}, \ldots \\ 0, & -lK_{21}, & 1 - lK_{22}, \ldots \\ & & \\ 0, & -lK_{n1}, & 1 - lK_{nn} \end{vmatrix}$$

the bordering with $x's$ in the first row being arbitrary.

Adding to this $D\left(l, \frac{x}{y}\right)$, which differs from it only in the first column, we get

(16)
$$\begin{vmatrix} [xy], & x_1, & x_2 \ldots x_n \\ y_1, & 1 - lK_{11}, & -lK_{12} \ldots \\ y_2, & -lK_{21}, & 1 - lK_{22} \ldots \\ \cdot & \cdot \quad \cdot \quad \cdot \quad \cdot \quad \cdot \quad \cdot \quad \cdot \quad \cdot \quad \cdot \end{vmatrix}$$

and this will be unchanged if we subtract from the first column each of the others multiplied, the r th column by y_r. Thus we get for the first column
$$0,$$
$$lKy_1,$$
$$lKy_2,$$
$$\cdot \quad \cdot \quad \cdot \quad \cdot$$
$$lKy_n,$$

so that the determinant is $lD\left(l, \frac{x}{Ky}\right)$. Accordingly we have the identity

(17)
$$d(l)[xy] + D\left(l, \frac{x}{y}\right) - lD\left(l, \frac{x}{Ky}\right) = 0$$

for *any* values of $x_1, x_2 \ldots x_n, y_1, y_2 \ldots y_n, l.$

Writing our equations (10)

(18)
$$f_r = \varphi_r - lK\varphi_r, \quad r = 1, 2, \ldots n$$

multiplying by arbitrary quantities y_r and adding, we obtain the linear form

(19)
$$[f, y] = [\varphi, y] - l[K\varphi, y],$$

and if we can find a form

(20)
$$[\varphi, y] = \varphi_1 y_1 + \varphi_2 y_2 \cdots + \varphi_n, y_n,$$

which identically satisfies the above, then its coefficients, considered as a linear form in y, are the $\varphi's$. But since $K_{rs} = K_{sr}$, referring to (13)

(21)
$$[K\varphi, y] = [\varphi, Ky],$$

hence we must have

(22) $$[f, y] = [\varphi, y] - l[\varphi, Ky]$$

and this will become an identity if we put

(23) $$[\varphi, y] = -\frac{D\left(l, {f \atop y}\right)}{d(l)},$$

for then (22) becomes

(24) $$[f, y] + \frac{D\left(l, {f \atop y}\right)}{d(l)} - l\,\frac{D\left(l, {f \atop Ky}\right)}{d(l)} = 0,$$

which we have shown to be an identity, (17). That is

(25)
$$[\varphi, y] = -\frac{1}{d(l)}
\begin{vmatrix}
0, & f_1, & f_2, \cdots & f_n \\
y_1, & 1 - lK_{11}, & -lK_{12}, \cdots & -lK_{1n} \\
y_2, & -lK_{21}, & 1 - lK_{22}, \cdots & -lK_{2n} \\
\cdot & \cdot \cdot \cdot \cdot \cdot \cdot \cdot & \cdot & \cdot \\
\cdot & \cdot \cdot \cdot \cdot \cdot \cdot \cdot & \cdot & \cdot \\
y_n & \cdot \quad \cdot \quad \cdot \quad \cdot \quad \cdot \quad \cdot & \cdot & 1 - lK_{nn}
\end{vmatrix}$$

If now l is not a root of $d(f) = 0$, the coefficients in the above form in y are the desired values of $\varphi_1, \varphi_2, \ldots, \varphi_n$.

126. Transcendental Problem. It now remains to pass to the limit for $n = \infty$. If we develop the determinant

(11)
$$d(l) = \begin{vmatrix}
1 - lK_{11}, & -lK_{12} \cdots \\
-lK_{21}, & 1 - lK_{22} \cdots \\
\cdot \quad \cdot \quad \cdot \quad \cdot \quad \cdot \quad \cdot \quad \cdot
\end{vmatrix}$$

in powers of l

(26) $$d(l) = 1 - d_1 l + d_2 l^2 \cdots \pm d_n l^n$$

in which $d_1 = K_{11} + K_{22} \cdots K_{nn}$,

(27)
$$d_2 = \begin{vmatrix} K_{11}, & K_{12} \\ K_{21}, & K_{22} \end{vmatrix} + \begin{vmatrix} K_{11}, & K_{13} \\ K_{31}, & K_{33} \end{vmatrix} \cdots + \begin{vmatrix} K_{22}, & K_{23} \\ K_{32}, & K_{33} \end{vmatrix} + \cdots$$

consisting of $n(n-1)/2!$ determinants, and in which any d_h is the sum of all minors of the h'th order whose diagonal row is taken out of the diagonal row of k's,

(28)
$$d_h = \Sigma \begin{vmatrix}
K_{p_1 p_1}, & K_{p_1 p_2}, & \ldots K_{p_1 p_h} \\
K_{p_2 p_1}, & K_{p_2 p_2}, & \ldots K_{p_2 p_h} \\
\cdot \quad \cdot & \cdot \quad \cdot & \cdot \quad \cdot \\
\cdot \quad \cdot & \cdot \quad \cdot & \cdot \quad \cdot \\
K_{p_h p_1}, & K_{p_h p_2}, & \ldots K_{p_h p_h}
\end{vmatrix}$$

where $\quad p_1 < p_2 < p_3 \cdots < p_h, \quad p_1, p_2, \ldots p_h = 1, 2, \ldots n.$

The sum contains as many determinants as can be made out of choices of h different elements out of n in the diagonal, that is $\frac{n(n-1)\ldots(n-h+1)}{h!} = \binom{n}{h}$. According to a theorem of Hadamard, Bull. des Sciences Math. (2) XVII (1893) the maximum value that a determinant of order n, containing $n!$ terms (which would be $n! \, K^n$ where K is the greatest value of any element if all the terms had the same sign, as they do not) is $n^{\frac{n}{2}} K^n$.

Consequently [1])

$$(29) \qquad |d_h| \leqq \binom{n}{h} h^{\frac{h}{2}} K^h \leqq \frac{h^{\frac{h}{2}}}{h!}(nK)^h \leqq \left(\frac{neK}{\sqrt{h}}\right)^h$$

that is

$$(30) \qquad \frac{|d_h|}{n^h} \leqq \left(\frac{eK}{\sqrt{h}}\right)^h.$$

Now since we had $\lambda/n = l$ we have

$$\lim_{n=\infty} l\, d_1 = \lambda \lim_{n=\infty} \frac{K_{11}+K_{22}\cdots+K_{nn}}{n} = \lambda \int_0^1 K(s_1, s_1)\, ds_1$$

$$\lim_{n=\infty} l^2 d_2 = \frac{\lambda^2}{2!} \sum \frac{\begin{vmatrix} K_{rr}, & K_{rs} \\ K_{sr}, & K_{ss} \end{vmatrix}}{n^2} = \frac{\lambda}{2!} \int_0^1\int_0^1 \begin{vmatrix} K(s_1, s_1), & K(s_1, s_2) \\ K(s_2, s_1), & K(s_2, s_2) \end{vmatrix} ds_1\, ds_2$$

and in like manner for any given l,

$$\lim_{n=0} \frac{d_h}{n^h} = h,$$

$$(31) \qquad \delta_h = \frac{1}{h!} \int_0^1 \cdots \int_0^1 \begin{vmatrix} K(s_1, s_1), K(s_1,s_2) \ldots K(s_1, s_h) \\ \cdot \quad \cdot \quad \cdot \quad \cdot \quad \cdot \quad \cdot \\ \cdot \quad \cdot \quad \cdot \quad \cdot \quad \cdot \quad \cdot \\ K(s_h, s_1). \quad \cdot \quad \cdot \quad \cdot \quad \cdot \; K s_h, s_h) \end{vmatrix} ds_1 \ldots ds_h.$$

The determinant $d(l)$ converges when n increases without limit toward the power series in λ,

$$(32) \qquad \delta(\lambda) = 1 - \delta_1 \lambda + \delta_2 \lambda^2 -$$

like the series,

$$(33) \qquad \sum_0^\infty \lambda^h \left(\frac{eK}{\sqrt{h}}\right)^h.$$

---

1) $e^h = 1 + h + \frac{h^2}{2!} \cdots + \frac{h^h}{h!} > \frac{h^h}{h!}$, $\frac{e^h}{h^{\frac{h}{2}}} > \frac{h^{\frac{h}{2}}}{h!}$.

This converges for all finite values of λ, since the ratio of the $(h+1)th$ to the hth term is

$$\frac{\lambda e K}{\left(1+\frac{1}{h}\right)^{\frac{h}{2}}\sqrt{h+1}}.$$

In like manner we develop the determinant,

$$(34) \qquad D\left(l,\begin{array}{c}x\\y\end{array}\right) = D_1\binom{x}{y} - lD_2\binom{x}{y} + l^2 D_3\binom{x}{y} \cdots \pm l^{n-1}D_n\binom{x}{y}$$

and putting $l = \lambda/n$ we find

$$\lim_{n=\infty}\frac{D_h\binom{x}{y}}{n^h} = \varDelta_h\binom{x}{y}$$

where

$$(35)\; \varDelta_h\binom{x}{y} = \frac{1}{h!}\int_0^1\cdots\int_0^1 \begin{vmatrix} 0, & x(s_1), & x(s_2)\ldots & x(s_h) \\ y(s_1), & K(s_1s_1), & K(s_1s_2)\ldots K(s_1s_h) \\ \cdot & \cdot & \cdots \cdot \\ y(s_h), & K(s_hs_1), & K(s_hs_2)\ldots K(s_hs_h) \end{vmatrix} ds_1 ds_2 \ldots ds_h,$$

and we shall have $\lim_{n=\infty}\left\{\frac{1}{n}D\left(\frac{\lambda}{n},\frac{x}{y}\right)\right\} = s\left(\lambda,\frac{x}{y}\right)$, where

$$(36) \qquad \varDelta\left(\lambda,\frac{x}{y}\right) = \varDelta_1\binom{x}{y} - \lambda\varDelta_2\binom{x}{y} + \lambda^2\varDelta_3\binom{x}{y}\cdots$$

which is a convergent series in λ, whose coefficients depend on the *arbitrary* functions $x(s)$, $y(s)$. We must now find the limit of the formula,

$$(17) \qquad d(l)[x,y] + D\left(l,\frac{x}{y}\right) - lD\left(l,\frac{x}{Ky}\right) = 0,$$

and we still need the expression for the last term. Now since

$$(13) \qquad Ky_p = K_{p1}y_1 + K_{p2}y_2 \cdots + K_{pn}y_n,$$

$$(37) \qquad \lim_{n=\infty}\frac{Ky_p}{n} = \int_0^1 K(s_p t)y(t)\,dt.$$

Dividing through above by n, we have

$$(38) \qquad \lim_{n=\infty}\frac{\lambda}{n^2}D\left(\frac{\lambda}{n},\frac{x}{Ky}\right) = \lim_{n=\infty}\frac{\lambda}{n}D\left(\frac{\lambda}{n},\frac{x}{\frac{Ky}{n}}\right)$$

$$= \lambda\left\{\varDelta\left(\lambda,\frac{x}{y}\right)\right\}_{\bar{y}(s)=\int_0^1 K(s,t)y(t)dt,}$$

or since the $y's$ enter linearly in each $\varDelta_h$ we may multiply by $y(t)dt$ and integrate outside the determinant,

$$(39) \qquad \lim_{n=\infty}\frac{\lambda}{n^2}D\left(\frac{\lambda}{n},\frac{x}{Ky}\right) = \lambda\int_0^1\left\{\varDelta\left(\lambda,\frac{x}{y}\right)\right\}_{\bar{y}(s)=K(s,t)}y(t)dt,$$

t disappears, and leaves the integral function of s. Thus our formula (17) divided by n, becomes

$$(40) \quad \delta(\lambda)\int_0^1 x(s)y(s)ds + \varDelta\left(\lambda, \frac{x}{y}\right) - \lambda\int_0^1 \left\{\varDelta\left(\lambda, \frac{x}{y}\right)\right\}_{\bar{y}(s)=K(s,\,t)} y(t)dt = 0.$$

This is an identity in λ, and holds when $x(s)$ and $y(s)$ are any continuous functions.

Now if we put $x(r) = K(s, r)$, $y(r) = K(t, r)$ the last and first terms become

$$(41) \quad -\int_0^1 \left[\lambda\varDelta\left(\lambda, \frac{x(r)}{y(r)}\right)_{\substack{x(r)=K(s,\,r)\\ y(r)=K(t,\,r)}} - \delta(\lambda)K(s, r)\right]K(t, r)dr$$

and put for brevity,

$$(42) \quad \varDelta(\lambda; s, t) = \lambda\left\{\varDelta\left(\lambda, \frac{x}{y}\right)\right\}_{\substack{x=K(s,\,r)\\ y=K(t,\,r)}} - \delta(\lambda)K(s, t),$$

then (41) is

$$-\int_0^1 \varDelta(\lambda; s, r)K(t, r)dr$$

while the middle term in (40) is, by (42)

$$(43) \quad \varDelta\left(\lambda\frac{x}{y}\right)_{\substack{x=K(s,\,r)\\ y=K(t,\,r)}} = \frac{1}{\lambda}\left\{\varDelta(\lambda; s, t) + \delta(\lambda)K(s, t)\right\}.$$

Consequently our equation (40) becomes

$$(44) \quad \delta(\lambda)K(s, t) + \varDelta(\lambda; s, t) - \lambda\int_0^1 \varDelta(\lambda; s, r)K(t, r)dr = 0$$

Dividing by $\delta(\lambda)$ and putting

$$(45) \quad \Gamma(s, t) = -\frac{\varDelta(\lambda; s, t)}{\delta(\lambda)},$$

(44) becomes

$$(46) \quad K(s, t) = \Gamma(s, t) - \lambda\int_0^1 \Gamma(s, r)K(t, r)dr.$$

This equation is symmetrical in $K(s, t)$ and $-\Gamma(s, t)$. The functions $\varDelta(\lambda; s, t)$ and $\Gamma(s, t)$ are symmetrical functions of s and t, containing the parameter λ, of which Γ is the ratio of two integral power-series in λ.

The equation (46) is an identity in s, t, and λ, and is called the *resolvent* for the kernel K. We have met it in § 40, (294). The function Γ is called the solving function, and enables us to solve the integral equation

(7) $$f(s) = \varphi(s) - \lambda \int_0^1 K(s, t)\varphi(t)dt,$$

as follows

(47) $$\varphi(s) = f(s) + \lambda \int_0^1 \Gamma(s, t)f(t)dt,$$

as we see by inserting the latter in the former.

Suppose we have two operations S_K and S_L

(48) $$S_K\varphi(s) = \varphi(s) - \lambda \int_0^1 K(s, t)\varphi(t)dt,$$
$$S_L\varphi(s) = \varphi(s) - \lambda \int_0^1 L(s, t)\varphi(t)dt,$$

and that we write the equation (7) as

(49) $$S_K\varphi(s) = f(s).$$

Then

(50) $$S_L S_K\varphi(s) = S_L f(s) = f(s) - \lambda \int_0^1 L(s, r)f(r)dr$$
$$= \varphi(s) - \lambda \int_0^1 K(s, t)\varphi(t)dt$$
$$- \lambda \int_0^1 L(s, r)dr\left[\varphi(r) - \lambda \int_0^1 K(r, t)\varphi(t)dt\right]$$

which will be equal to

(51) $$S_C\varphi(s) = \varphi(s) - \lambda \int_0^1 C(s, t)\varphi(t)dt$$

if

(52) $$C(s, t) = K(s, t) + L(s, t) - \lambda \int_0^1 L(s, r)K(r, t)dr.$$

Since, if both kernels are symmetrical, this is symmetrical in K and L, we have

(53) $$S_L S_K\varphi(s) = S_K S_L\varphi(s) = S_C\varphi(s)$$

or the two operations form a group and are commutative. If we put for $L(s, t) = -\Gamma(s, t)$, we have by (46) $C(s, t) = 0$ and $S_L f(s) = \varphi(s)$.

127. Fredholm's Formulae. We have by definition,(42),(36),(32)

(54)
$$\Delta(\lambda; s, t)$$

$$= -\sum_{h=1}^{h=\infty} \frac{(-\lambda)^h}{h!} \int_0^1 \cdots \int_0^1 \begin{vmatrix} K(ss_1), & K(ss_2) \ldots K(ss_h) \\ K(s_1 t), & K(s_1 s_1), K(s_1 s_2) \ldots \\ K(s_2 t), & K(s_2 s_1), K(s_2 s_2) \ldots \\ \cdot & \cdot \cdot \cdot \cdot \cdot \cdot \\ K(s_h t) & \cdots \cdots \cdots K(s_h s_h) \end{vmatrix} ds_1 ds_2 \ldots ds_h$$

$$- K(s,t) \sum_{h=0}^{h=\infty} \frac{(-\lambda)^h}{h!} \int_0^1 \cdots \int_0^1 \begin{vmatrix} K(s_1 s_1), K(s_1 s_2) \ldots K(s_1 s_h) \\ K(s_2 s_1), K(s_2 s_2) \ldots K(s_2 s_h) \\ \cdot \cdot \cdot \cdot \cdot \cdot \cdot \cdot \\ K(s_h s_1) \cdot \cdot \cdot \cdot K(s_h s_h) \end{vmatrix} ds_1 ds_2 \ldots ds_h$$

$$= -K(s,t) + \sum_{h=1}^{h=\infty} \frac{(-\lambda)^h}{h!} \int_0^1 \cdots \int_0^1 \begin{vmatrix} K(st), & K(ss_1) \ldots K(ss_h) \\ K(s_1 t), & K(s_1 s_1) \ldots K(s_1 s_h) \\ \cdot \cdot \cdot \cdot \cdot \cdot \cdot \cdot \\ K(s_h t), & K(s_h s_1) \ldots K(s_h s_h) \end{vmatrix} ds_1 ds_2 \ldots ds_h,$$

which we may compare with the series for $\delta(\lambda)$,

(55)
$$-\Delta(\lambda; s, t) = A_0(st) - \frac{\lambda}{1!} A_1(st) + \frac{\lambda^2}{2!} A_2(st) \ldots$$

(56)
$$\delta(\lambda) = 1 - \frac{\lambda}{1!} A_1 + \frac{\lambda^2}{2!} A_2 \ldots$$

(57)
$$A_n(st) = \int_0^1 \cdots \int_0^1 \begin{vmatrix} K(st), & K(ss_1) \ldots K(ss_n) \\ K(s_1 t), & K(s_1 s_1) \ldots K(s_1 s_n) \\ \cdot \cdot \cdot \cdot \cdot \cdot \cdot \cdot \\ K(s_n t) & \cdot \cdot \cdot \cdot K(s_n s_n) \end{vmatrix} ds_1 ds_2 \ldots ds_n.$$

If we put s and t equal, say to $s = s_{n+1}$ and integrate with respect to this variable, we shall obtain the relation

(58)
$$A_{n+1} = \int_0^1 A_n(ss) ds,$$

so that from (55) and (56) we obtain

(59)
$$\delta'(\lambda) = \frac{d\delta(\lambda)}{d\lambda} = \int_0^1 \Delta(\lambda; ss) ds.$$

These are Fredholm's formulae.

128. Solution by Iteration. We see by (45) that the solving function is the quotient of two power-series in λ. The calculation of the integrals involved in the coefficients is of enormous complication and in practice the solution in this manner is impossible. But if λ is small enough we may find a solution which is an entire function of λ, as follows.

(60) $$S\Gamma(st) = \lambda \int K(tr)\,\Gamma(rs)\,dr$$

we have (46)

(61) $$K(st) = (1 - S)\,\Gamma(st),$$

and if we treat this as an algebraic equation,

$$\Gamma(st) = (1 - S)^{-1}K(st) = (1 + S + S^2 + \cdots)\,K(st),$$

$$S K(st) = \lambda \int K(sr)\,K(rt)\,dr = \lambda\,K_2(st),$$

(62) $S^2 K(st) = S(SK(st)) = \lambda S K_2(st) = \lambda^2 \int K(tr)\,K_2(rs)\,dr = \lambda^2 K_3(st),$

$$\cdot\quad\cdot\quad\cdot\quad\cdot\quad\cdot\quad\cdot\quad\cdot\quad\cdot\quad\cdot\quad\cdot\quad\cdot\quad\cdot\quad\cdot\quad\cdot\quad\cdot\quad\cdot\quad\cdot$$

$$S^n K(st) = \lambda^n K_{n+1}(st)$$

$$K_1(st) = K(st)$$

where

(63) $$K_{n+1} = \int\!\int \cdots \int K_1(sr_1)\,K_{n-1}(r_1 r_2) \cdots K_n(r_n t)\,dr_1\,dr_2 \cdots dr_n,$$

is called the n th *iterated* matrix. Thus we have

(64) $$\Gamma(st) = K_1(st) + \lambda K_2(st) + \lambda^2 K_3(st) \ldots$$

This series converges only when $|\lambda|$ is less than the lowest root of $S(\lambda) = 0$. Let us try to express $\varDelta(\lambda; s, t)$ and $S(\lambda)$ by the iterated matrices $K_n(s, t)$.

Since

(59) $$\frac{dS(\lambda)}{d\lambda} = \int_0^1 \varDelta(\lambda; s, s)\,ds$$

or dividing by $S(\lambda)$,

(65) $$\frac{d\log S(\lambda)}{d\lambda} = -\int_0^1 \Gamma(ss)\,ds,$$

we have by using the series (64),

(66) $\dfrac{d\log S(\lambda)}{d\lambda} = -\left\{ \displaystyle\int_0^1 K_1(ss)\,ds + \lambda \int_0^1 K_2(ss)\,ds + \lambda^2 \int_0^1 K_3(ss)\,ds \cdots \right\}$

$$= -\{a_1 + a_2\lambda + a_3\lambda^2 + \cdots\}$$

where

(67) $$a_n = \int_0^1 K_n(ss)\,ds,$$

and integrating we obtain $\log S(\lambda)$. But we can find a series for $S(\lambda)$ itself

(56) $$S(\lambda) = 1 - \frac{\lambda}{1!}A_1 + \frac{\lambda^2}{2!}A_2 - \frac{\lambda^3}{3!}\cdots A_3$$

$$S'(\lambda) = -\left\{A_1 - \frac{\lambda}{1!}A_2 + \frac{\lambda^2}{2!}A_3\cdots\right\}$$

and using this in (66),

(68) $\qquad a_1 + a_2 \lambda + a_3 \lambda^2 + \cdots = \dfrac{A_1 - \dfrac{\lambda}{1!} A_2 + \dfrac{\lambda^2}{2!} A_3 - \cdots}{1 - \dfrac{\lambda}{1!} A_1 + \dfrac{\lambda^2}{2!} A_2 - \cdots}.$

Multiplying up and comparing coefficients in λ^n,

(69)
$$a_1 + (a_2 - a_1 A_1)\,\lambda + \left(a_3 - a_2 A_1 + \frac{a_1 A_2}{2!}\right)\lambda^2$$
$$+ \left(a_4 - a_3 A_1 + \frac{a_2 A_2}{2!} - \frac{a_1 A_3}{3!}\right)\lambda^3 \cdots = A_1 - \frac{A_2}{1!}\lambda + \frac{A_3}{2!}\lambda^2 - \frac{A_4}{3!}\lambda^3 + \cdots$$

(70)
$$A_1 = a_1,$$
$$a_1 A_1 - A_2 = a_2,$$
$$a_2 A_1 - \frac{a_1 A_2}{2!} + \frac{A_3}{2!} = a_3,$$
$$a_3 A_1 - \frac{a_2 A_2}{2!} + \frac{a_1 A_3}{3!} - \frac{A_4}{3!} = a_4,$$

from which we obtain the determinant

(71)
$$A_n = \begin{vmatrix} a_1, & n-1, & 0, & 0, & 0, \ldots 0 \\ a_2, & a_1, & n-2, & 0, & 0, \ldots 0 \\ a_3, & a_2, & a_1, & n-3, & 0, \ldots 0 \\ \cdot & \cdot & \cdot & \cdot & \cdot & \cdot & \cdot \\ \cdot & \cdot & \cdot & \cdot & \cdot & \cdot & \cdot \\ a_{n-1}, & a_{n-2}, & \cdot & \cdot & \cdot & \cdot & \cdot & 1 \\ a_n, & a_{n-1}, & \cdot & \cdot & \cdot & \cdot & \cdot & a_1 \end{vmatrix}$$

If we insert in the resolvent equation (44) the series (55), (56) we obtain

(72)
$$K(st)\left\{1 - \lambda A_1 + \frac{\lambda^2}{2!} A_2 - \frac{\lambda^3}{3!} A_3 + \cdots\right\} = A_0(st) - \frac{\lambda}{1!} A_1(st) + \frac{\lambda^2}{2!}(st)\cdots$$
$$- \lambda \int K(rt)\left[A_0(sr) - \frac{\lambda}{1} A_1(sr) + \frac{\lambda^2}{2!}(sr) - \cdots\right]dr$$

and comparing coefficients of λ^n,

(73) $\qquad A_n K(st) = A_n(st) + n \int K(rt) A_{n-1}(sr)\,dr,$

from which we may obtain

$$A_0(st) = K(st)$$

(74)
$$A_n(st) = \begin{vmatrix} K_1(st), & n, & 0, & 0, & 0, \ldots 0\; 0 \\ K_2(st), & a_1, & n-1, & 0, & 0, \ldots 0 \\ K_3(st), & a_2, & a_1, & n-2, & 0, \ldots 0 \\ \cdot & \cdot & \cdot & \cdot & \cdot & \cdot & \cdot \\ \cdot & \cdot & \cdot & \cdot & \cdot & \cdot & \cdot \\ K_n(st), & a_{n-1} & \cdot & \cdot & \cdot & \cdot & \cdot & a_1\; 1 \\ K_{n+1}(st), & a_n & \cdot & \cdot & \cdot & \cdot & \cdot & a_2\, a_1 \end{vmatrix}$$

These formulae were obtained by Plemelj. (Monatshefte für Math. u. Phys. 15 — 1904.)

The resolvent equation may be verified by developing according to the elements of the first row.

129. Characteristic Numbers and Normal Functions.

Let us go back and consider the case that λ is a root of the equation $S(\lambda) = 0$. Let the roots of the determinant

$$(11) \quad d(l) = \begin{vmatrix} 1 - l\,K_{11}, & -\,l\,K_{12}, & \cdots -l\,K_{1n} \\ -\,l\,K_{21}, & 1 - l\,K_{22}, & \cdots -lK_{2n} \\ \cdot\quad\cdot\quad\cdot\quad\cdot\quad\cdot\quad\cdot\quad\cdot \\ & & 1 - l\,K_{nn} \end{vmatrix}$$

be $l^{(1)}, l^{(2)} \ldots l^{(n)}$, and let us call the minors of the diagonal row $d_{11}, d_{22}, \ldots d_{nn}$.

The derivative of a determinant is the sum of n determinants obtained by replacing the rows in turn by their derivatives. Thus,

$$(75) \quad d'(l) = \begin{vmatrix} -K_{11}, & -K_{12}, & -K_{1n} \\ -K_{21}, 1 - K_{22}l, 1 - K_{2n} \\ \cdot\quad\cdot\quad\cdot\quad\cdot\quad\cdot\quad\cdot \end{vmatrix} + \begin{vmatrix} 1 - K_{11}l, & -K_{12}l, \cdots - K_{1n} \\ -K_{21}, & -K_{22}, \end{vmatrix} + \cdots$$

If this is multiplied by 1 and added to $d_{11}(l) + d_{22}(l) + \cdots + d_{nn}(l)$, we get $n\,d(l)$. Therefore

$$(76) \qquad d_{11}(l) + d_{22}(l) + \cdots + d_{nn}(l) = n\,d(l) - l\,d'(l) .$$

Suppose then $l^{(h)}$ is a root of $d(l) = 0$. Then

$$(77) \qquad d_{11}(l^{(h)}) + d_{22}(l^{(h)}) + \cdots + d_{nn}(l^{(h)}) = -l\,d'(l^{(h)}) .$$

If it is not a multiple root $d'(l^{(h)})$ is not zero, so that not all the minors $d_{rr}(l^{(h)})$ vanish, and the equations

$$(78) \qquad \begin{aligned} \varphi_1 - l(K_{11}\varphi_1 + K_{12}\varphi_2 + \cdots + K_{1n}\varphi_n) = 0, \\ \varphi_2 - l(K_{21}\varphi_1 + K_{22}\varphi_2 + \cdots + K_{2n}\varphi_n) = 0, \\ \cdot\quad\cdot\quad\cdot\quad\cdot\quad\cdot\quad\cdot\quad\cdot\quad\cdot\quad\cdot\quad\cdot\quad\cdot \end{aligned}$$

have solutions

$$(79) \qquad\qquad \varphi_1 = \varphi_1^{(h)}, \quad \varphi_2 = \varphi_2^{(h)}, \ldots \varphi_n = \varphi_2^{(h)} .$$

If we pass to the limit $n = \infty$, (78) become

$$(80) \qquad\qquad \varphi(s) - \lambda \int_0^1 K(s, t)\,\varphi(t)\,dt = 0,$$

which is the same as (215), § 38, and the function belonging to a root λ_λ of the equation $S(\lambda) = 0$, which we call φ_λ, satisfies the equation

$$(81) \qquad \varphi_\lambda(s) = \lambda_\lambda \int_0^1 K(s,t)\varphi_\lambda(t)\,dt.$$

These functions $\varphi_1(s)$, $\varphi(s)$, ... are called *Eigenfunction*, which we will translate characteristic or normal functions for the kernel $K(s,t)$ and the characteristic numbers. λ_1, λ_2, ...

Consider now the identity

$$(17) \qquad d(l)\,[x,y] + D\left(l,\,\genfrac{}{}{0pt}{}{x}{y}\right) - l\,D\left(l,\,\genfrac{}{}{0pt}{}{x}{Ky}\right) = 0,$$

which becomes, if we insert the root $l^{(\lambda)}$

$$(82) \qquad D\left(l^{(\lambda)},\,\genfrac{}{}{0pt}{}{x}{y}\right) = l_\lambda\,D\left(l^{(\lambda)}\,\genfrac{}{}{0pt}{}{x}{Ky}\right),$$

in which either side is a linear form in y_1, y_2 ... y_n whose coefficients are solutions of the homogeneous equations (78). Consequently $D\left(l^{(\lambda)},\,\genfrac{}{}{0pt}{}{x}{y}\right)$ contains as a factor the linear form

$$(83) \qquad [\varphi^{(\lambda)}, y] = \varphi_1^{(\lambda)}y_1 + \varphi_2^{(\lambda)}y_2 + \cdots + \varphi_n^{(\lambda)}y_n.$$

But an account of the symmetry in xy, we must have

$$D\left(l^{(\lambda)},\,\genfrac{}{}{0pt}{}{x}{y}\right) = [\psi^{(\lambda)}, x]\,[\varphi^{(\lambda)}, y]$$

and since the left-hand member is symmetrical in xy, we must have

$$(84) \qquad D\left(l^{(\lambda)},\,\genfrac{}{}{0pt}{}{x}{y}\right) = \pm\,[\varphi^{(\lambda)}, x]\,[\varphi^{(\lambda)}, y],$$

since all the $\varphi'(s)$ may be multiplied by any constant, so that the constant can be made ± 1.

If we develop the determinant, and compare the coefficients of $x_1 y_1$, $x_2 y_2$, ... we find

$$(85) \qquad d_{11}\left(l^{(\lambda)}\right) = \varphi_1^{(\lambda)2}, \quad d_{22}\left(l^{(\lambda)}\right) = \varphi_2^{(\lambda)2}, \ldots$$

$$(86) \qquad d_{11}(l^{(\lambda)}) + d_{22}(l^{(\lambda)}) + \cdots + d_{nn}(l^{(\lambda)}) = \mp\,[\varphi^{(\lambda)}, \varphi^{(\lambda)}]$$

but the left-hand member has been shown in (77) to be equal to $-\,l^{(\lambda)}d'(l^{(\lambda)})$.

Consequently

$$(87) \qquad \frac{D\left(l^{(\lambda)},\,\genfrac{}{}{0pt}{}{x}{y}\right)}{l^{(\lambda)}d'\left(l^{(\lambda)}\right)} = \frac{[\varphi^{(\lambda)}, x]\,[\varphi^{(\lambda)}, y]}{[\varphi^{(\lambda)}, \varphi^{(\lambda)}]}.$$

If we now proceed to the limit we obtain from (84) and (87) respectively,

$$(88) \qquad \varDelta\left(\lambda^{(h)}, \frac{x}{y}\right) = \pm \int_0^1 \varphi^{(h)}(s)\, x(s)\, ds \int_0^1 \varphi^{(h)}(s)\, y(s)\, ds,$$

$$(89) \qquad \frac{\varDelta\left(\lambda^{(h)}, \dfrac{x}{y}\right)}{\lambda^{(h)} s'\left(\lambda^{(h)}\right)} = \frac{\displaystyle\int_0^1 \varphi^{(h)}(s)\, x(s)\, ds \int_0^1 \varphi^{(h)}(s)\, y(s)\, ds}{\displaystyle\int_0^1 \left(\varphi^{(h)}(s)\right)^2 ds},$$

and putting $x(s) = K(r,s)$, $y(s) = K(r,t)$ and making use of (81)

$$(90) \qquad \frac{\varDelta\left(\lambda^{(h)}; s, t\right)}{s'\left(\lambda^{(h)}\right)} = \frac{\varphi^{(h)}(s)\, \varphi^h(t)}{\displaystyle\int_0^1 \left(\varphi^{(h)}(s)\right)^2 ds}.$$

130. Hilbert's Theorem. We have now to consider the problem of the transformation of a quadratic form into a sum or squares.

Since $D\left(l, \dfrac{x}{y}\right)$ is of degree $n-1$ in l we may develop in partial fractions,

$$(91) \qquad \frac{D\left(l, \dfrac{x}{y}\right)}{d(l)} = \frac{D\left(l^{(1)}, \dfrac{x}{y}\right)}{d^1(l^{(1)})} \frac{1}{l-l^{(1)}} + \frac{D\left(l^{(2)}, \dfrac{x}{y}\right)}{d^1(l^{(2)})} \frac{1}{l-l^{(2)}} + \cdots$$

or by (87)

$$(92) \qquad \frac{D\left(l, \dfrac{x}{y}\right)}{d(l)} = \frac{[\varphi^{(1)}, x][\varphi^{(1)}, y]}{[\varphi^{(1)}, \varphi^{(1)}]} \frac{l^{(1)}}{l-l^{(1)}} + \frac{[\varphi^{(2)}, x][\varphi^{(2)}, y]}{[\varphi^{(2)}, \varphi^{(2)}]} \frac{l^{(2)}}{l-l^{(2)}} + \cdots$$

which is an identity in l, x and y. For $l = 0$ we have by (17)

$$D\left(0, \frac{x}{y}\right) = -[x, y], \quad d(0) = 1,$$

$$(93) \qquad [x, y] = \frac{[\varphi^{(1)}, x][\varphi^{(1)}, y]}{[\varphi^{(1)}, \varphi^{(1)}]} + \frac{[\varphi^{(2)}, x][\varphi^{(2)}, y]}{[\varphi^{(2)}, \varphi^{(2)}]} + \cdots$$

Putting for y, Ky, since according to our equations (21) $[\varphi, Ky] = [K\varphi, y]$, and (78), $\varphi^{(h)} = l^{(h)} K\varphi^{(h)}$,

$$(94) \quad Kxy = [Kx, y] = [x, Ky] = \frac{[\varphi^{(1)}, x][\varphi^{(1)}, y]}{l^{(1)}[\varphi^{(1)}, \varphi^{(1)}]} + \frac{[\varphi^{(2)}, x][\varphi^{(2)}, y]}{l^{(2)}[\varphi^{(2)}, \varphi^{(2)}]} + \cdots$$

If we now put $x = y$, we have the canonical expression of a quadratic form as a sum of squares.

$$(95) \qquad Kxx = \frac{[\varphi^{(1)}, x]^2}{l^{(1)}[\varphi^{(1)}, \varphi^{(1)}]} + \frac{[\varphi^{(2)}, x]}{l^{(2)}[\varphi^{(2)}, \varphi^{(2)}]} + \cdots$$

The limiting form of the equation (94) is, since the double sum on the left becomes the double integral,

(96)
$$\int_0^1\int_0^1 K(s,t)x(s)y(t)ds\,dt = \frac{\frac{1}{\lambda_1}\int_0^1\varphi_1(s)x(s)ds\int_0^1\varphi_1(s)y(s)ds}{\int_0^1(\varphi_1(s))^2ds}$$
$$+ \frac{\frac{1}{\lambda_2}\int_0^1\varphi_2(s)x(s)ds\int_0^1\varphi_2(s)y(s)ds}{\int_0^1(\varphi_2(s))^2ds} + \cdots$$

If we multiply each function φ by a proper constant, putting

(97)
$$\frac{S_n(s)}{\sqrt{\left|\int_0^1(\varphi_n(s))^2ds\right|}} = \psi_n(s), \qquad \text{so that} \int_0^1(\psi_n s)^2 = 1,$$

the functions are normalized, and (90) and (94) become respectively

(98)
$$\frac{\Delta(\lambda^{(h)};s,t)}{s'(\lambda^{(h)})} = \psi_h(s)\psi_h(t),$$

(99)
$$\int_0^1\int_0^1 K(s,t)x(s)y(t)ds\,dt = \sum_n \frac{1}{\lambda_n}\int_0^1\psi_n(s)x(s)ds\int_0^1\psi_n(s)y(s)ds.$$

The last equation is Hilbert's fundamental theorem.

131. Development in Series of Normal functions. As an application of Hilbert's Theorem let us consider the development of an arbitrary function in normal functions. If $x(s)$ is any continuous function, and we put

(100)
$$f(r) = \int_0^1\int_0^1 K(s,t)K(r,t)x(s)ds\,dt,$$

we have, by twice applying (81)

(101)
$$\int_0^1 f(r)\psi_n(r)dr = \int_0^1\int_0^1 K(st)x(s)ds\,dt\int_0^1 K(r,t)\psi_n(r)dr$$
$$= \frac{1}{\lambda_n^2}\int_0^1\psi_n(s)x(s)ds$$

Let us now put in (99) $y(t) = K(r, t)$, when the left-hand member becomes $f(r)$,

$$(102) \qquad f(r) = \sum_n \frac{1}{\lambda_n} \int_0^1 \psi_n(s) x(s) ds \int_0^1 \psi_n(s) K(r, s) ds.$$

Again by the use of (81) we have

$$(103) \qquad f(r) = \sum_n \frac{\psi_n(r)}{\lambda_n^2} \int_0^1 \psi_n(s) x(s) ds,$$

which is the desired development in series,

$$(104) \qquad f(r) = \sum_n c_n \psi_n(r),$$

in which we find by comparison with (101)

$$(105) \qquad c_n = \int_0^1 f(r) \psi_n(r) dr$$

exactly as in the case of Fourier's series.

The treatment of integral equations that we have given in this chapter is that of Hilbert. We have given so many examples in Chapter III, where the subject has been treated in an entirely different manner, that we shall not pursue the subject farther.

APPENDIX.

1. Jacobians. If we have two functions of two independent variables $u(x, y)$ and $v(x, y)$ with a relation between them such as $v = \varphi(u)$ or $\psi(u, v) = 0$ it is evident that for all points for which u is constant, v is also constant, consequently the family of curves represented by the equation $u = $ const. is the same as that represented by $v = $ const. Consequently at every point the normals to the two lines $u = $ const. and $v = $ const. have the same direction. This is expressed by the condition,

$$\frac{\frac{\partial u}{\partial x}}{\frac{\partial v}{\partial x}} = \frac{\frac{\partial u}{\partial y}}{\frac{\partial v}{\partial y}}$$

or clearing of fractions

$$\frac{\partial u}{\partial x}\frac{\partial v}{\partial y} - \frac{\partial v}{\partial x}\frac{\partial u}{\partial y} = \begin{vmatrix} \dfrac{\partial u}{\partial x}, & \dfrac{\partial v}{\partial x} \\ \dfrac{\partial u}{\partial y}, & \dfrac{\partial v}{\partial y} \end{vmatrix} = 0.$$

The above determinant is called the Jacobian determinant, and its vanishing is the condition that the two functions u, v are not independent, but are connected by a functional relation. The above considerations may be reversed. If the Jacobian vanishes, the normals coincide at every point, so that the two families of curves coincide, and accordingly when u is constant v is constant, or there is a functional relation.

If we have three functions of three variables $u(x, y, z)$, $v(x, y, z)$, $w(x, y, z)$, between which there is a functional relation $w = \varphi(u, v)$ or $\psi(u, v, w) = 0$, it is evident that when u and v are constant w is also, that is the lines of intersection of the families of surfaces $u = $ const., $v = $ const. at every point touch one of the surfaces $w = $ const., or at every point the lines of intersection of the three surfaces u, v, w have a common direction. The three normals, being each perpendicular to this direction, are parallel to a plane. The condition for this is, by § 4, equation (15)

$$\begin{vmatrix} \dfrac{\partial u}{\partial x}, & \dfrac{\partial u}{\partial y}, & \dfrac{\partial u}{\partial z} \\ \dfrac{\partial v}{\partial x}, & \dfrac{\partial v}{\partial y}, & \dfrac{\partial v}{\partial z} \\ \dfrac{\partial w}{\partial x}, & \dfrac{\partial u}{\partial x}, & \dfrac{\partial w}{\partial z} \end{vmatrix} = 0.$$

Conversely, if at every point the Jacobian vanishes, the relations of the three surfaces will be as described, which implies a relation between the three functions.

2. Double Limits. Throughout the whole subject of the calculus we are confronted with the notion of a limit. If we have a set of numbers

$$a_1, a_2, a_3 \ldots a_n \ldots$$

having the property that to every positive number ε no matter how small there can be found a number μ such that for all integers n greater than μ we have

$$|a_n - A| < \varepsilon$$

then we say that the sequence of numbers a approaches A as a limit. The necessary and sufficient condition for a limit is that when ε is given we can find a number μ so large that

$$|a_n - a_{n+s}| < \varepsilon \qquad s = 1, 2, 3 \ldots$$

for all integers $n > \mu$. For the only way that the numbers a_n can approach all subsequent ones indefinitely closely is by approaching some constant value A.

We may also have a limit in which the different numbers are not characterized by discrete integers n, but vary continuously in an interval. Such a limit occurs in the definition of a continuous function. Suppose that $f(x)$ is a function of a variable x, then we say that the function is continuous in the neighborhood of a point x_0 if when ε is given however small we can find corresponding to it a number δ small enough so that

$$|f(x) - f(x_0)| < \varepsilon \quad \text{when} \quad |x - x_0| < \delta.$$

The condition for this is that for any two values x', x'' we have

$$|f(x') - f(x'')| < \varepsilon$$

whenever $|x' - x_0| < \delta$ and $|x'' - x_0| < \delta$. This evidently corresponds to the condition above, except that the largeness of the number μ is replaced by the smallness of the interval δ. In either case the essential idea is that the inequality applies to all values after a certain one.

When we have to do with functions of two variables, we have often to pass to two limits successively, and either of them may be of either sort, depending on discrete numbers n or on continuous variation. For instance consider the function

$$s_n(x) = \frac{nx}{nx + 1}$$

which depends upon the integer n and on the continuous variable x. We may consider the limit for n

$$\lim_{n \to \infty} \frac{nx}{nx + 1} = 1$$

while x remains fixed, or the limit for $x \longrightarrow 0$

$$\lim_{x \to 0} \frac{nx}{nx+1} = 0$$

while n remains fixed. So we may consider $f(x, y)$, a function of the two continuous variables x, y, in the neighborhood of a point $x = a$, $y = b$. Suppose first that for any fixed value of y we have

$$\lim_{x \to a} f(x, y) = \varphi(y)$$

and we then let y approach the value b. If $\varphi(y)$ approaches a limit we write

$$\lim_{y \to b} \varphi(y) = \lim_{y \to b} \left[\lim_{x \to a} f(x, y) \right].$$

On the other hand we may keep x fixed and let y approach b, and then let x approach a. The question arises, are the two limits

$$\lim_{x \to a} \left[\lim_{y \to b} f(x, y) \right] \quad \text{and} \quad \lim_{y \to b} \left[\lim_{x \to a} f(x, y) \right]$$

equal? That this is not in general true may be shown by the simplest examples. For instance let

$$f(x, y) = \frac{\alpha x + \beta y}{\gamma x + \delta y}, \quad \lim_{y \to 0} \left[\lim_{x \to 0} f(x, y) \right] = \frac{\beta}{\delta}, \quad \lim_{x \to 0} \left[\lim_{y \to 0} f(x, y) \right] = \frac{\alpha}{\gamma}$$

and these are not equal unless $\alpha \delta = \beta \gamma$.

We shall now consider a number of cases where the question of the possibility of interchanging the order of passing to the limit arises. The partial derivative of a function of two variables is defined

$$\frac{\partial f(x, y)}{\partial x} = \lim_{h \to 0} \frac{f(x+h, y) - f(x, y)}{h} \qquad \frac{\partial f}{\partial y} = \lim_{k \to 0} \frac{f(x, y+k) - f(x, y)}{k}$$

$$\frac{\partial^2 f}{\partial y \partial x} = \lim_{k \to 0} \frac{1}{k} \left[\lim_{h \to 0} \left\{ \frac{f(x+h, y+k) - f(x, y)}{h} \right\} - \lim_{h \to 0} \left\{ \frac{f(x+h, y) - f(x, y)}{h} \right\} \right]$$

$$\frac{\partial^2 f}{\partial x \partial y} = \lim_{h \to 0} \frac{1}{h} \left[\lim_{k \to 0} \left\{ \frac{f(x+h, y+k) - f(x, y)}{k} \right\} - \lim_{k \to 0} \left\{ \frac{f(x, y+k) - f(x, y)}{k} \right\} \right].$$

A definite integral is an example of a limit. If $f(x)$ is a function of x in the range of variation $a < x < b$, which we divide into a number of parts in any way by points $x_1, x_2, \ldots x_s,$ and call $\delta_s = x_s - x_{s-1}$ and if when we form the sum

$$\sum_{s=1}^{s=n} \delta_s f(\xi_s)$$

ξ_s being any point in the interval $x_{s-1} < \xi_s < x_s$, then if as we increase n, the number of subdivisions, the sum approaches a limit, it is the fundamental theorem of the integral calculus that the limit is independent of the method of subdivision into intervals or of the points taken in the intervals, and the limit defines the definite integral.

$$\lim_{n \to \infty} \sum_{s=1}^{s=n} \delta_s f(\xi_s) = \int_a^b f(x)\,dx.$$

Suppose now that the function f contains a variable parameter α and that for all values of α in a certain range the definite integral exists (that is the sum converges to a limit), the integral evidently is a function of α and my be differentiated according to α. Again the question arises whether the order of passing to the limits is indifferent, that is whether

$$\frac{\partial}{\partial \alpha} \int_a^b f(x, \alpha)\,dx = \int_a^b \frac{\partial f(x, \alpha)}{\partial \alpha}\,dx.$$

It has been supposed that the limits of integration a and b were finite. If one of them is not, say b, we have again to define the integral with limit of integration infinite as

$$\lim_{b \to \infty} \int_a^b f(x)\,dx = \int_a^\infty f(x)\,dx,$$

which is again a double limit. On the other hand if the integrand instead of being finite in the whole range of integration, becomes infinite at a point $x = x_0$, we must define the definite integral

$$\lim_{x' \to x_{0e}} \int_a^{x'} f(x)\,dx + \lim_{x'' \to x_{0e}} \int_{x''}^b f(x)\,dx = \int_a^b f(x)\,dx, \quad x' < x_0 < x''.$$

Finally in defining an infinite series, in which the terms are not constants, but functions of a variable x, $u_1(x), u_2(x), \ldots u_s(x) \ldots$

$$s_n(x) = u_1(x) + u_2(x) + \cdots + u_s(x) \cdots + u_n(x), \quad f(x) = \lim_{n \to \infty} s_n(x)$$

there arise several questions of double limits. Suppose that as x approaches a certain point x_0 every function $u_s(x)$ approaches a limit, is it then true that the function defined by the sum of the series is continuous, that is, can we write

$$\lim_{x \to x_0} f(x) = \lim_{x \to x_0} \left[\lim_{n \to \infty} s_n(x) \right] = \lim_{n \to \infty} \left[\lim_{x \to x_0} s_n(x) \right].$$

Furthermore, can we interchange the limits involved in differentiation and integration with that involved in summing the series,

$$\frac{df(x)}{dx} = \lim_{n \to \infty} \left[\frac{du_1(x)}{dx} + \frac{du_2(x)}{dx} + \cdots + \frac{du_n(x)}{dx} \right]$$

$$\int_a^b f(x)\,dx = \int_a^b \lim_{n \to 0} s_n(x)\,dx = \lim_{n \to \infty} \left[\int_a^b u_1(x)\,dx + \int_a^b u_2(x)\,dx + \cdots + \int_a^b u_n(x)\,dx \right]$$

or as we say can the series be differentiated and integrated term by term?

3. Uniform Convergence. All the cases enumerated above depend upon the consideration of what is called uniform convergence a phenomenon investigated by Stokes and Seidel, and may be investigated by means of a theorem due to E. H. Moore and W. F. Osgood. We shall illustrate the matter in the case of a series of functions. The series

$$u_1(x) + u_2(x) + \cdots + u_s(x) + \cdots = f(x)$$

converges for a certain value of x if, after assigning a number ϵ as small as we please, we can find a number μ such that, as soon as $n > \mu$

$$|r_n(x)| = |f(x) - s_n(x)| < \epsilon \quad \text{or} \quad |s_n(x) - s_{n+p}(x)| < \epsilon, \quad p = 1,2,3\ldots$$

Two cases may occur. As we vary x, when x approaches a certain value, it may be necessary to go farther on in the series, that is to take a larger value of μ, in order to have the necessary degree of convergence, we say the series, though convergent does not converge *uniformly*. On the contrary, if μ does not depend on the value of x in a certain interval, but the same value of μ answers, depending only on ε, for all values of x, we say the convergence is uniform in the interval. Under these circumstances we may prove the truth of the last three equations above. Suppose that in a certain interval each of the functions $u_s(x)$ is continuous, then the sum s_n of any finite number of them is also continuous. But since the series converges uniformly for any two values x', x'' in the interval we can find a number μ (the same for both) such that

$$|r_n(x')| < \frac{\varepsilon}{3}, \qquad |r_n(x'')| < \frac{\varepsilon}{3}, \qquad\qquad n > \mu.$$

Also because of the continuity of $s_n(x)$ we can find δ such that

$$|s_n(x') - s_n(x'')| < \frac{\varepsilon}{3}, \qquad |x' - x''| < \delta.$$

Accordingly we have

$$|f(x') - f(x'')| = |s_n(x') - s_n(x'') + r_n(x') - r_n(x'')| < \varepsilon, \qquad |x' - x''| < \delta,$$

which is the condition for continuity of $f(x)$.

In like manner a series of continuous functions may be integrated term by term if the series converges uniformly. For since $f(x) = s_n(x + r_n(x)$

$$\int_a^b f(x)\,dx - \int_a^b s_n(x)\,dx = \int_a^b r_n(x)\,dx$$

and since the series converges uniformly we can find for ε a fixed number μ such that $|r_n(x)| < \varepsilon$ when $n > m$. Accordingly

$$\left| \int_a^b f(x)\,dx - \int_a^b s_n(x)\,dx \right| = \left| \int_a^b r_n(x)\,dx \right| < \varepsilon\,(b-a)$$

or

$$\int_a^b f(x)\,dx = \lim_{n \to \infty} \int_a^b s_n(x)\,dx.$$

In the case of differentiation the matter is not so simple. For it may happen that the series of derivatives may not converge uniformly, or in fact may not converge at all. For instance the series,

$$\frac{\sin x}{1} + \frac{\sin 3x}{3} + \frac{\sin 5x}{5} + \cdots$$

which we have found in § 42 to represent a constant when $x \neq 0$, on being differentiated term by term gives

$$\cos x + \cos 3x + \cos 5x + \cdots$$

which does not converge at all, since we do not have for all x,

$$\lim_{s \to \infty} |u_s(x)| = 0.$$

But if the series of derivatives converges uniformly it represents the derivative of the original series. For if

$$f(x) = u_1(x) + u_2(x) +$$

and if the series of derivatives is

$$\varphi(x) = u_1'(x) + u_2'(x) + \cdots$$

and since $\varphi(x)$ is uniformly convergent, it may be integrated term by term, giving

$$\int_a^x \varphi(x)\,dx = u_1(x) - u_1(a) + u_2(x) - u_2(a) + \cdots = f(x) - f(a)$$

on differentiation of which we obtain,

$$\varphi(x) = f'(x).$$

The theorem for differentiation of a definite integral according to a parameter may be stated by saying that if $\partial f/\partial \alpha$ is a continuous function of both x and α the differentiation under the integral sign may be performed.

4. Definite Integrals. The evaluation of various definite integrals will illustrate some of these principles. The integral

$$\int_0^\infty \frac{\sin bx}{x}$$

which is one of those known by the name of Dirichlet, has been shown in § 42 to be convergent. Since changing the sign of b changes the sign of the integrand, it is evident that it changes the sign of the integral. Also the integral vanishes for $b = 0$. But otherwise it is evident by introducing the factor b into numerator and denominator and writing $bx = x'$,

$$\int_0^\infty \frac{\sin bx \, d(bx)}{bx} = \int_0^\infty \frac{\sin x'}{x'} \, dx'$$

that the integral does not depend on b, but is constant. Considered as a function of b, it has a discontinuity at $b = 0$.

Consider the integral

$$\int_0^\infty e^{-ax} dx = -\left.\frac{e^{-ax}}{a}\right|_0^\infty = \frac{1}{a}, \qquad a > 0.$$

It will still be convergent if we put for a a complex number,

$$\int_0^\infty e^{-(a+ib)x} dx = \frac{1}{a+ib} = \frac{a-ib}{a^2+b^2}, \qquad a > 0.$$

Separating the real and imaginary parts we obtain the two integrals,

$$\int_0^\infty e^{-ax}\cos bx \, dx = \frac{a}{a^2+b^2}, \qquad \int_0^\infty e^{-ax}\sin bx \, dx = \frac{b}{a^2+b^2}.$$

In the first of these let us now integrate again with respect to b.

$$\int_0^b db \int_0^\infty e^{-ax}\cos bx \, dx = \int_0^\infty e^{-ax} dx \int_0^b \cos bx \, db = \int_0^\infty e^{-ax} \frac{\sin bx}{x} \, dx$$

$$= a \int_0^b \frac{db}{a^2+b^2} = \tan^{-1}\frac{b}{a}$$

and putting $a = 0$ we obtain the desired result,

$$\int_0^\infty \frac{\sin bx}{x}\, dx = \frac{\pi}{2}, \qquad\qquad b > 0.$$

The integral

$$I_1 = \int_0^\infty e^{-x^2}\, dx,$$

which occurs in many parts of mathematical physics, including the Theory of Probability, is known by the name of Laplace. It represents the area of the so-called probability-curve $y = e^{-x^2}$. Inasmuch as the indefinite integral is not to be found except by a development in series, we are led to employ an artifice for the definite integral. Since the variable of integration is of no importance, let us write again

$$I_1 = \int_0^\infty e^{-x^2}\, dx = \int_0^\infty e^{-y^2}\, dy$$

and multiplying together

$$I_1^2 = \int_0^\infty e^{-x^2}\, dx \int_0^\infty e^{-y^2}\, dy = \int_0^\infty\int_0^\infty e^{-(x^2+y^2)}\, dx\, dy.$$

We are permitted to introduce e^{-x^2} under the sign of integration, since x and y are to be considered as independent variables. If we now consider them as coordinates on the floor, and z a vertical coordinate, the double integral will represent the volume of a solid of revolution bounded by the surface $z = e^{-(x^2+y^2)}$. We may find this by introducing polar coordinates, when the element of area on the floor will be $r\, dr\, d\theta$.

There may be some scruple in doing this, since the double-integral is the limit

$$\lim_{p = \infty} \int_0^p\int_0^p e^{-(x^2+y^2)}\, dx\, dy,$$

which represent the volume over square with side p. We easily see that this volume is greater than that of the figure of revolution over the circle of radius p, and less than that over the circle of radius $p\sqrt{2}$. Accordingly if the integral,

$$\int_0^{\frac{\pi}{2}}\int_0^p e^{-r^2}\, r\, dr\, d\theta$$

approaches limit for $p = \infty$ we shall have

$$I_1^2 = \int_0^{\frac{\pi}{2}} \int_0^\infty e^{-r^2} r \, dr \, d\theta.$$

The integration according to θ merely multiplies by $\pi/2$, while in the integral

$$\int e^{-r^2} r \, dr$$

on account of the factor r the integrand is an exact differential, and we obtain the indefinite integral

$$\int e^{-r^2} r \, dr = -\frac{1}{2} e^{-r^2}.$$

Passing to the limit we have

$$I_1^2 = \frac{\pi}{2} \lim_{p=\infty} \int_0^p e^{-r^2} r \, dr = \frac{\pi}{4} \lim (1 - e^{-p^2}) = \frac{\pi}{4}$$

so that we have the desired result,

$$I_1 = \int_0^\infty e^{-x^2} \, dx = \frac{\sqrt{\pi}}{2}.$$

In the above integral let us change the variable so that $x = \sqrt{\alpha} y$. Then

$$I_1 = \int_0^\infty e^{-\alpha y^2} \sqrt{\alpha} \, dy = \frac{\sqrt{\pi}}{2},$$

so that we obtain the integral with the parameter

$$I_2 = \int_0^\infty e^{-\alpha x^2} \, dx = \frac{1}{2} \sqrt{\frac{\pi}{\alpha}}.$$

If we differentiate this integral n times with respect to the parameter α we obtain

$$\frac{d^n I_2}{d\alpha^n} = (-1)^n \int_0^\infty e^{-\alpha x^2} x^{2n} \, dx = \frac{\sqrt{\pi}}{2} \left(-\frac{1}{2}\right) \left(-\frac{3}{2}\right) \cdots -\frac{(2n-1)}{2} \alpha^{-\frac{2n+1}{2}}.$$

The factor (-1) appears the same number of times on both sides, so that

$$I_3 = \int_0^\infty e^{-\alpha x^2} x^{2n} \, dx = \frac{1}{2} \sqrt{\frac{\pi}{\alpha}} \frac{1 \cdot 3 \cdot 5 \cdots (2n-1)}{2^n \alpha^n} = \frac{1}{2} \sqrt{\frac{\pi}{\alpha}} \frac{(2n)!}{n!} \frac{1}{(4\alpha)^n}.$$

From this integral we may construct

$$I_4 = \int_0^\infty e^{-\alpha x^2} \cos \beta x \, dx = \int_0^\infty dx \, e^{-\alpha x^2} \sum_{\beta=0}^{\beta=\infty} \frac{(-\beta^2 x^2)^n}{(2n)!}$$

$$= \sum_{n=0}^{n=\infty} \frac{(-\beta^2)^n}{(2n)!} \int_0^\infty e^{-\alpha x^2} x^{2n} \, dx = \sum_0^\infty \frac{(-\beta^2)^n}{(2n)!} \frac{1}{2} \sqrt{\frac{\pi}{\alpha}} \frac{(2n)!}{n!} \frac{1}{(4\alpha)^n}$$

$$= \frac{1}{2} \sqrt{\frac{\pi}{\alpha}} \sum_0^\infty \frac{1}{n!} \left(-\frac{\beta^2}{4\alpha}\right)^n = \frac{1}{2} \sqrt{\frac{\pi}{\alpha}} \, e^{-\frac{\beta^2}{4\alpha}}.$$

Consider the integral

$$I = \int_0^\infty e^{-\left(x^2 + \frac{a^2}{x^2}\right)} dx$$

which reduces to Laplace's when $a = 0$. We may differentiate with respect to a

$$\frac{dI}{da} = -2 \int_0^\infty \frac{a}{x^2} e^{-\left(x^2 + \frac{a^2}{x^2}\right)} dx$$

and if we change the variable by putting

$$x = \frac{a}{y}, \quad dx = -\frac{a \, dy}{y^2} = -\frac{x^2}{a} \, dy$$

this becomes

$$\frac{dI}{da} = -2 \int_0^\infty e^{-\left(y^2 + \frac{a^2}{y^2}\right)} dy = -2I.$$

This is a differential equation for I, integrating which we get

$$I = Ce^{-2a}.$$

Putting $a = 0$ we determine the constant as $\sqrt{\pi}/2$ so that we have finally

$$\int_0^\infty e^{-\left(x^2 + \frac{a^2}{x^2}\right)} dx = \frac{\sqrt{\pi}}{2} \, e^{-2a}.$$

If in this result we introduce for a a complex number

$$2a = (1 + i)\alpha, \quad a^2 = \frac{\alpha^2}{2} i$$

we have

$$\int_0^\infty e^{-x^2} e^{-\frac{a^2 i}{2x^2}} dx = \frac{\sqrt{\pi}}{2} \, e^{-\alpha} e^{-i\alpha}$$

which separates into the two real integrals

$$\int_0^\infty e^{-x^2} \cos \frac{\alpha^2}{2x^2}\, dx = \frac{\sqrt{\pi}}{2}\, e^{-\alpha} \cos \alpha$$

$$\int_0^\infty e^{-x^2} \sin \frac{\alpha^2}{2x^2}\, dx = \frac{\sqrt{\pi}}{2}\, e^{-\alpha} \sin \alpha.$$

In the integral

$$I = \int_0^{\frac{\pi}{2}} \frac{a\, d\varphi}{a^2 + b^2 \cos^2 \varphi},$$

let us in the denominator multiply a^2 by $\cos^2 \varphi + \sin^2 \varphi = 1$ and then divide numerator and denominator by $\cos^2 \varphi$, giving

$$I = \int_0^{\frac{\pi}{2}} \frac{a \sec^2 \varphi\, d\varphi}{a^2 + b^2 + a^2 \tan^2 \varphi}.$$

Let us now change the variable by putting

$$y = \frac{a}{\sqrt{a^2 + b^2}} \tan \varphi \quad dy = \frac{a}{\sqrt{a^2 + b^2}} \sec^2 \varphi\, d\varphi$$

when we obtain

$$I = \frac{1}{\sqrt{a^2 + b^2}} \int_0^\infty \frac{dy}{1 + y^2} = \frac{\pi}{2\sqrt{a^2 + b^2}}.$$

It is evident that if we extend the limits of integration as far as π the same values of the integrand appear again so that the integral is doubled, or

$$\int_0^\pi \frac{a\, d\varphi}{a^2 + b^2 \cos^2 \varphi} = \frac{\pi}{\sqrt{a^2 + b^2}}.$$

We have obtained in § 46, (132) for the Bessel function $J(x)$ the integral,

$$J(bx) = \frac{1}{\pi} \int_0^\pi e^{ibx \cos \omega}\, d\omega = \frac{1}{\pi} \int_0^\pi \cos (bx \cos \omega)\, d\omega,$$

(the field of integration is there from $-\pi$ to π, doubling the value and the imaginary part vanishing). If we multiply this by e^{-ax} and integrate with respect to x,

$$\int_0^\infty e^{-ax} J(bx)\, dx = \frac{1}{\pi} \int_0^\infty e^{-ax} dx \int_0^\pi \cos (bx \cos \omega)\, d\omega,$$

we may change the order of integration,

$$\int_0^\infty e^{-ax} J(bx)\, dx = \frac{1}{\pi} \int_0^\pi d\omega \int_0^\infty e^{-ax} \cos(bx \cos \omega)\, dx.$$

But the integral with respect to x has been found above, if we put $b \cos \omega$ instead of b. Accordingly

$$\int_0^\infty e^{-ax} J(bx)\, dx = \frac{1}{\pi} \int_0^\pi \frac{a\, d\omega}{a^2 + b^2 \cos^2 \omega} = \frac{1}{\sqrt{a^2 + b^2}}.$$

It may be shown that the integral remains convergent if for a we put ia, if a is real and positive. Thus we have

$$\int_0^\infty e^{-iax} J(bx) = \frac{1}{\sqrt{b^2 - a^2}}.$$

If $b^2 > a^2$, this is real, so that, separating the integral into its real and imaginary parts,

$$\int_0^\infty \cos ax J(bx)\, dx = \frac{1}{\sqrt{b^2 - a^2}}, \quad \int_0^\infty \sin ax J(bx)\, dx = 0, \qquad b^2 > a^2.$$

On the other hand if $b^2 < a^2$, the square root is imaginary, $i\sqrt{a^2 - b^2}$ and we have,

$$\int_0^\infty \cos ax J(bx)\, dx = 0, \quad \int_0^\infty \sin ax J(bx)\, dx = \frac{1}{\sqrt{a^2 - b^2}} \qquad b^2 < a_2.$$

These results have been otherwise found elsewhere.

5. The Complex Variable. All real numbers, positive and negative integral, fractional, and irrational, can be represented by points on a single line, whose distance from a fixed origin represents the number.

For the numbers $a\sqrt{-1}$, called imaginary, there is no place in this representation, hence we must find a new one. The number $p = a + bi$ where $i^2 = -1$, and a and b are real, may take on a doubly infinite system of values, for a and b may each take on an infinite set of values. It is therefore natural to represent these values by points on a plane, and we choose a point whose abscissa is a, and whose ordinate is b. This point will represent the complex number $a + bi$ which may lie anywhere in the plane, and all the points of the plane correspond to all possible complex numbers.

If for the number $z = x + yi$, represented by a point with co-ordinates x, y, we put in the polar coordinates

$$x = r \cos \varphi, \quad y = r \sin \varphi$$

we get

$$z = r(\cos \varphi + i \sin \varphi) = re^{i\varphi}$$

by replacing the functions of φ by their developments in series. The radius r is called the modulus, written $|z| = \sqrt{x^2 + y^2}$ and φ the argument of z. The sum of two complex quantities $p = a_1 + ib_1$ and $q = a_2 + ib_2$,

$$p + q = a_1 + a_2 + i(b_1 + b_2)$$

is represented by the diagonal of the parallelogram whose sides are equal to the moduli of p and q and make angles with the x-axis equal to the arguments of p and q. Hence

$$|p \pm q| \leqq |p| + |q|.$$

It is easily shown that the modulus of a product pq is equal to the product of the moduli, and that the argument of the product is the sum of the arguments for if

$$p = a_1 + ib_1 = r_1(\cos \varphi_1 + i \sin \varphi_1) = r_1 e^{i\varphi_1}$$

$$q = a_2 + ib_2 = r_2(\cos \varphi_2 + i \sin \varphi_2) = r_2 e^{i\varphi_2}$$

then

$$pq = r_1 r_2 e^{i(\varphi_1 + \varphi_2)} = r_1 r_2 \{\cos(\varphi_1 + \varphi_2) + ie(\varphi_1 + \varphi_2)\}.$$

Real quantities have the argument $\varphi = 0$, pure imaginaries $\varphi = \pi/2$.

A complex number cannot vanish unless both its real and imaginary points vanish, i. e., both of its coordinates vanish, and its modulus must vanish, its argument being arbitrary. A complex number is infinite when its modulus is infinite, irrespective of the argument. All infinite points are considered the same.

While a real variable may vary only by motion forward and back along a straight line, a complex variable may vary by its representative point describing any path whatever in the plane.

A function of a complex variable $z = x + iy$, if given as an analytic expression containing z, will be a certain function of the two variables x and y, and will contain a real part, which we will call $u(x,y)$ and an imaginary part $iv(x,y)$ where u and v are real functions of the two real variables x, y. Hence the study of functions of *an imaginary* variable maybe be made to depend on the study of function of *two* real variables.

$$f(z) = u(\mathbf{x}, y) + iv(x, y).$$

The representation of function and variable by means of abscissa and ordinate is not here applicable, for both variable and function have a doubly infinite variability.

We may not even represent the function by the length of a perpendicular to the plane of the complex variable, for while this would allow the variable all possible variation, the perpendicular could not at the same time represent the real and imaginary parts of the function. If, however, we should mark off two lengths on the same perpendicular, one equal to the real part u, and the other to v the coefficient of i in the imaginary part, these two parts would represent the value of the function, and as the variable moved, these points would describe two surfaces, which would completely represent the function.

The function may be otherwise represented by means of another plane in which u and v represent coordinates of a point representing the function. To every point x, y will then be associated a point u, v in the other plane. As the point x, y moves, so will the point u, v. As the point x, y, representing z, describes any curve, u, v, representing $f(z)$ describes another curve, if $f(z)$ is continuous, otherwise, the point may jump from one point to another.

The definition of continuity is then that two points on the function curve may be brought as near together as we please by taking the corresponding points on the variable curve sufficiently near. Or a function is continuous in a region of the z plane containing z, if given a positive quantity ε, as small as we please, we can find a corresponding δ so that

$$| f(z) - f(z_0) | < \varepsilon$$

whenever $| z - z_0 | < \delta$. Such a region may be called the neighborhood of the point z_0. By the representation by means of curves it is of importance to inquire whether, if the curve of z starting from a point z describing an arbitrary path, returns to it, the curve of $f(z)$ returns to the point it started from. If so, the function $f(z)$ within the region where this property holds, is said to be uniform, or singly-valued, for to every value z corresponds *one* value $f(z)$. A uniform function which is also finite and continuous, within a region, is said to be holomorphic. Let us examine the relation between an infinitely small change in z and the corresponding change in $f(z)$. The change dz has the modulus $| dz | = \sqrt{dx^2 + dy^2}$ and the argument $\omega = \tan^{-1} \dfrac{dy}{dx}$. The change

$$df(z) = d(u + iv) = du + i dv$$

has the modulus $| df(z) | = \sqrt{du^2 + dv^2}$ and the argument $\theta = \tan^{-1} \dfrac{dv}{du}$ now

$$du = \frac{\partial u}{\partial x} dx + \frac{\partial u}{\partial y} dy$$

$$dv = \frac{\partial v}{\partial x} dx + \frac{\partial v}{\partial y} dy$$

$$df = du + i dv = \frac{\partial u}{\partial x} dx + \frac{\partial u}{\partial y} dy + i \left\{ \frac{\partial v}{\partial x} dx + \frac{\partial v}{\partial y} dy \right\}$$

the ratio

$$\frac{df}{dz} = \frac{du + idv}{dx + idy} = \frac{\frac{\partial u}{\partial x}\,dx + \frac{\partial u}{\partial y}\,dy + i\left\{\frac{\partial v}{\partial x}\,dx + \frac{\partial v}{\partial x}\,dy\right\}}{dx + idy}$$

$$= \frac{\frac{\partial u}{\partial x} + i\frac{\partial v}{\partial x} + \left(\frac{\partial u}{\partial y} + i\frac{\partial v}{\partial y}\right)\frac{dy}{dx}}{1 + i\frac{dy}{dx}}$$

is in general dependent on dy/dx, that is on the direction in which we leave the point z. The value of the derivative will then not be determined for the point z irrespective of the direction of leaving it unless the numerator is a multiple of the denominator so that the expression containing dy/dx divides out. For this to be true we must have.

$$\left(\frac{\partial u}{\partial x} + i\frac{\partial v}{\partial x}\right) : 1 = \left(\frac{\partial u}{\partial y} + i\frac{\partial v}{\partial y}\right) : i$$

$$i\frac{\partial u}{\partial x} - \frac{\partial v}{\partial x} = \frac{\partial u}{\partial y} + i\frac{\partial v}{\partial y}$$

that is,

$$\frac{\partial u}{\partial x} = \frac{\partial v}{\partial y}, \quad \frac{\partial v}{\partial x} = -\frac{\partial u}{\partial y}.$$

In this case the function f has a definite derivative, and it is only where $u(x, y)$ and $v(x, y)$ satisfy these conditions that $u + iv$ is said to be an analytic function of z.

As examples of uniform functions of z are any power of z, z^n where n is a real integer, a sum of multiples of such powers, or an infinite series of such powers, if convergent.

Since the modulus of a product is the product of the moduli, and the argument of the product is the sum of the arguments, the argument of the nth power is n times the argument. If the modulus is 1, all powers lie on the circumference of a circle of radius unity.

Since

$$z = x + iy = r(\cos\varphi + i\sin\varphi) = re^{i\varphi}$$

increasing the argument by 2π does not change the value of z, and we may write

$$z = re^{i(\varphi + 2k\pi)}$$

(k is an *integer*), hence.

$$z^{\frac{1}{n}} = r^{\frac{1}{n}} e^{\frac{i(\varphi + 2k\pi)}{n}} = r^{\frac{1}{n}}\left\{\sin\left(\frac{\varphi}{n} + \frac{2\pi k}{n}\right) + i\cos\left(\frac{\varphi}{n} + \frac{2\pi k}{n}\right)\right\}$$

and besides a point with argument φ/n we have other points with arguments exceeding this by $2\pi/n$, $2(2\pi/n)$, $3(2\pi/2)$ ... $(n-1)(2\pi/n)$. There are then n *distinct* values of the nth root of a quantity, all having the same modulus $r^{\frac{1}{n}}$.

Accordingly, as z describes any curve, $z^{\frac{1}{n}}$ has n different representative points, each one describing a curve. These curves are all alike, if turned about the origin.

If z describes a closed path, in general, each branch $z^{\frac{1}{n}}$ describes a closed path, but if z describes a closed path about the origin, its argument increases by 2π, while the argument of $z^{\frac{1}{n}}$ increases only by $2\pi/n$, hence the function is not represented by a closed curve. This is the simplest case of a non-uniform function. As for $z = 0$ all the branches of $z^{\frac{1}{n}}$ unite at 0, which is called a branch-point or critical point of the function $z^{\frac{1}{n}}$.

Similarly a is a critical point for $(z - a)^{\frac{1}{n}}$. Since

$$z = r e^{i(\varphi + 2k\pi)}$$

$$\log z = \log r + i\varphi + 2k\pi i$$

the logarithm of z has an unlimited number of values differing by $2\pi i$. As z encircles the origin $\log z$ increases by $2\pi i$, so that the origin is a critical point.

If $f(z)$ is singly valued, but for certain points becomes infinite, as $1/(z - a)$ having everywhere else a definite value, and becoming infinite in such a manner that $1/f(z)$ remains finite for $z = a$, the point a is said to be a pole of $f(z)$, and $f(a)$ is said to be a meromorphic or fractional function. A remaining kind of point is furnished by the function $e^{\frac{1}{z}}$ for $z = 0$. If z is real $e^{\frac{1}{z}}$ approaches $+\infty$ if z approaches 0 from the right, but approaches 0 when z approaches 0 from the left. Hence the value of the function for $z = 0$ is not uniquely determined. It may be shown that it may take any value $\alpha + i\beta$, for let

$$\alpha + i\beta = e^{p + iq},$$

then

$$e^{\frac{1}{z}} = e^{\frac{1}{x + iy}}$$

must equal

$$e^{p + iq},$$

$$x + iy = \frac{1}{p + iq} = \frac{p - iq}{p^2 + q^2},$$

$$x = \frac{p}{p^2 + q^2}, \qquad y = -\frac{q}{p^2 + q^2}.$$

But $\alpha + i\beta$ is unchanged if we replace q by $q + 2k\pi$, hence the condition is satisfied by

$$x = \frac{p}{p^2 + (q + 2k\pi)^2}, \qquad y = \frac{-(q + 2k\pi)}{p^2 + (q + 2k\pi)^2}.$$

By taking k great enough, both x and y and hence $|z|$ may be made as small as we please, hence for $z = 0$, $e^{\frac{1}{z}}$ may be made to approach *any* value $\alpha + i\beta$. Such a point is called *essentially singular*.

An integral of a function of a complex variable $\int_{z_0}^{z} f(z) dz$ is defined if on any path representing the successive values of z varying from z_0 to z we take n points z_1, z_2, ... z_n and form the sum

$$\sum_{s=1}^{s=n} (z_s - z_{s-1}) f(z_s)$$

and take the limit that the sum approaches as the number of points increases indefinitely. This definition may be shown to be if $f(z) = u + iv$ equivalent to

$$\int f(z) dz = \int (u + iv)(dx + idy) = \int (u dx - v dy) + i \int (v dx + u dy).$$

The integral evidently depends, in general, upon the nature of the path from z_0 to z. In certain cases it does not, however. We have already examined the case where an integral involving two real functions $P(x, y)$ and $Q(x, y)$, $\int (P dx + Q dy)$ is independent of the path, and found the condition to be

$$\frac{\partial P}{\partial y} = \frac{\partial Q}{\partial x},$$

and applying this to

$$\int (u dx - v dy)$$

we find

$$\frac{\partial u}{\partial y} = -\frac{\partial v}{\partial x}$$

and to

$$\int (v dx + u dy)$$

we find

$$\frac{\partial v}{\partial y} = \frac{\partial u}{\partial x}.$$

These two conditions are however satisfied by every analytic function of z. It is further necessary, as will be seen by referring to the demonstration of the theorem on integrals, that between the two paths considered both u and v shall be always finite and continuous.

Hence we may state that if in any region $f(z)$ is a *holomorphic* function of z, $\int_{z_0}^{z} f(z) dz$ will be the same for all paths $z_0 z$ lying in the region, or the integral of a holomorphic function around any closed curve lying entirely in the region of holomorphism, and not surrounding any singular point, is *zero*. Consider $f(z) = z^n$ which is (for n integral) holomorphic if $n > 0$ but meromorphic if $s < 0$. Consider $\int z^n ds$ around any closed path about $s = 0$. The integral about any closed path C embracing the pole is evidently equal to the integral about any other also embracing it. For consider the integral from P around C, then along a line from point P' infinitely near P to a point on C', Q', then around C' in the reverse direction to Q, infinitely near to Q', and then back to P along a path infinitely near to $P'Q'$. (Fig. 97.)

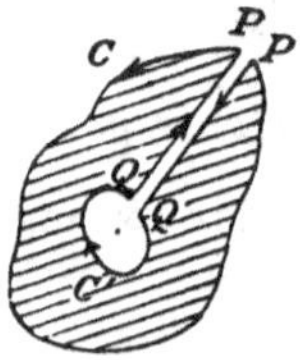

Fig. 97.

The region enclosed by the above closed path (shaded) is one in all of which the function is holomorphic, hence the integral is 0, i. e.,

$$\int_C + \int_{P'}^{Q'} - \int_{C'} + \int_Q^P = 0$$

where $\int_C$ means the integral around C in the direction opposite to the clockhands. But $\int_{P'}^{Q'}$ and $\int_Q^P$ are evidently equal and opposite, hence

$$\int_C = \int_{C'}.$$

Consider, therefore, for C' a circle about the origin, with radius ϱ on which
$$s = \varrho e^{i\varphi} \qquad ds = \varrho e^{i\varphi} i\, d\varphi, \qquad \text{since } \varrho \text{ is constant.}$$
$$z^n = \varrho^n e^{i n \varphi}$$
$$\int_{C'} z^n dz = i\varrho^{n+1} \int_{\varphi=0}^{\varphi=2\pi} e^{(n+1)i\varphi} d\varphi.$$

But
$$\int_0^{2\pi} e^{(n+1)i\varphi} d\varrho = \int_0^{2\pi} \cos(n+1)\varphi\, d\varphi + i\int_0^{2\pi} \sin(n+1)\varphi\, d\varphi$$
$$= \frac{1}{n+1}\left\{ \sin(n+1)\varphi \Big|_0^{2\pi} - i\cos(n+1)\varphi \Big|_0^{2\pi} \right\},$$

which vanishes for n a positive or negative integer, except $n = -1$.

If $n = -1$ we have
$$\int_{C'} \frac{ds}{z} = i\int_0^{2\pi} d\varphi = 2\pi i$$

irrespective of the value of ϱ, which may be taken as small as we please.

Now $\int_1^s \frac{dz}{z}$ is defined as $\log z$, hence we see again that $\log z$ increases by $2\pi i$ every time s revolves around the origin.

The integral $\int \dfrac{dz}{z-a}$ is shown, in like manner putting $z - a = \varrho e^{i\varphi}$, to be equal to $2\pi i$.

If

$$f(z) = \frac{A_{-n}}{(z-a)^n} + \frac{A_{-n+1}}{(z-a)^{n-1}} + \cdots + \frac{A_{-1}}{z-a} + A_0 + A_1(z-a) + A_2(z-a)^2 + \cdots$$

$$= \frac{1}{(z-a)^n} \{ A_{-n} + A_{-n+1}(z-a) + \cdots \}$$

a meromorphic function, the integral $\int f(z)\, dz$ around a closed curve embracing a is a sum of integrals of the form $A_s \int (z-a)^s ds$ all of which, by the above, are zero except $A_{-1} \int \dfrac{dz}{z-a} = 2\pi i A_{-1}$.

Hence for a meromorphic function, the integral $\int f(z)\, dz$ around its pole, is equal to $2\pi i A_{-1}$ where A_{-1} is the coefficient of $(z-a)^{-1}$ in the development. This coefficient is called the residue of the function for the pole a.

If $f(z)$ be a holomorphic function in a certain region, this integral $\int \dfrac{f(z)\, dz}{z-a}$ around a closed path in that region embracing a is equivalent to that around a circle c'

$$\int \frac{f(a)\, dz}{z-a} + \int \frac{f(z) - f(a)}{z-a}\, dz.$$

Now $|f(z) - f(a)| < \varepsilon$ where ε is as small as we please, if we take the radius of the circle small enough, for $f(z)$ is continuous. Hence

$$\int \frac{f(z)\, dz}{z-a} = f(a) \int \frac{dz}{z-a} + \varepsilon \int \frac{dz}{z-a}$$

and in the limit $\varrho = 0 \qquad \int \dfrac{f(z)\, dz}{z-a} = 2\pi i f(a)$

$$f(a) = \frac{1}{2\pi i} \int \frac{f(z)\, dz}{z-a}.$$

This is Cauchy's fundamental theorem. We may apply this formula to our linear differential equation

$$\frac{d^n y}{dt^n} + a_1 \frac{d^{n-1} y}{dt^{n-1}} + \cdots a_n y = 0.$$

Representing the n roots of the characteristic equation by points in the plane of z.

The integral around s_i

$$\frac{1}{2\pi i} \int \frac{C_i e^{zt}\, dz}{z - s_i} = C_i e^{s_i t}.$$

If now we take the integral around all the points s_i, it is evidently the sum of those around each and

$$C_1 e^{s_1 t} + C_2 e^{s_2 t} \ldots C_n e^{s_n t} = \frac{1}{2\pi i} \int \left\{ \frac{c_1}{z - s_1} \cdots \frac{c_n}{z - s_n} \right\} e^{zt} dz$$

$$= \frac{1}{2\pi i} \int \frac{P(z)}{F(z)} e^{zt} dz$$

where $F(z) = (z - s_1)(z - s_2) \ldots (z - s_n)$ and $P(z)$ is an *arbitrary* polynomial of degree less than n.

By the theorem of the residue the above integral taken around a pole is the residue of the function $\frac{P(z)}{F(z)} e^{zt}$ for that pole, and taken around severalpoles, is equal to the sum of the residues for those poles. But the roots of $F(z) = 0$ are poles for $\frac{P(z)}{F(z)} e^{zt}$.

Let these roots be $s_1 s_2 \ldots$ and their *multiplicity* $\mu_1 \mu_2 \ldots$ Then

$$\frac{P(z)}{F(z)} = \frac{\alpha_{\mu_1}}{(z - s_1)^{\mu_1}} + \frac{\alpha_{\mu_1 - 1}}{(z - s_1)^{\mu_1 - 1}} \cdots \frac{\alpha_1}{(z - s_1)}$$

$$+ \frac{\beta_{\mu_2}}{(z - s_2)^{\mu_2}} + \frac{\beta_{\mu_2 - 1}}{(z - s_2)^{\mu_2 - 1}} \cdots \frac{\beta_1}{(z - s_2)} + \cdots$$

the α's and β's being arbitrary constants. Also

$$e^{zt} = e^{s_k t} e^{(z - s_k)t} = e^{s_k t} \left[1 + (z - s_k)t + \frac{(z - s_k)^2 t}{1 \cdot 2} \cdots \right].$$

Multiplying together, the terms involving $\frac{1}{z - s_k}$ have the coefficient, which is the residue of $\frac{P(z)}{F(z)} e^{zt}$ for $z = s_k$

$$e^{s_k t} \left[y_1 + y_2 t + y_3 t^2 \ldots y_{\mu_k} \frac{t^{\mu_k - 1}}{(\mu_k - 1)!} \right] = P_k e^{s_k t}.$$

P_k being an arbitrary polynomial in t of degree $(\mu_k - 1)$. It has been shown that $C_k e^{s_k t}$ is a solution of the differential equation, when s_k is a root of $F(s) = 0$, hence we see that

$$\frac{1}{2\pi i} \int \frac{P(z)}{F(z)} e^{zt} dz$$

is a solution, and the above analysis has the advantage of giving the solution when there are equal roots.

The method of Cauchy may be applied to the equation with second member

$$\frac{d^n y}{dt^n} + a_1 \frac{d^{n-1} y}{dt^{n-1}} \cdots a_n y = T(t) \, .$$

put

$$y = \frac{1}{2\pi i} \int \frac{\xi(z) e^{zt}}{F(z)} dz$$

and substituting in dif. 29 first member

$$= \frac{1}{2\pi i}\int^c \frac{(z^n + a_1 z^{n-1}\ldots a_n)\xi e^{z't}dz}{F(z)} = \frac{1}{2\pi i}\int^c \xi e^{z't}dz.$$

If we can determine the function $\xi(z)$ and the line of integration so that

$$\frac{1}{2\pi i}\int^c \xi e^{z't}dz = T(t)$$

the above expression for y will be an integral of dif. 29. This may be done if $T = P(t)e^{\lambda t}$ where P is a polynomial in t, say

$$P \equiv \alpha_0 + \alpha_1 t + \alpha_2 t^2 \cdots + \alpha_m t^m.$$

Put

$$\xi = \frac{\beta_0}{z-\lambda} + \frac{\beta_1}{(z-\lambda)^2} \cdots \frac{\beta_m}{(z-\lambda)^{m+1}}$$

and integrate around a small circle about the pole λ.

The integral will be equal to the residue for the pole λ of

$$\xi(z)e^{z't} = \left[\frac{\beta_0}{z-\lambda} + \frac{\beta_1}{(z-\lambda)^2}\cdots\right]\left[1 + (z-\lambda)t + \frac{(z-\lambda)^2 t^2}{2!}\cdots\right]e^{\lambda t}$$

i. e.

$$e^{\lambda t}\left[\beta_0 + \beta_1 t + \cdots \frac{\beta_m}{m!}t^m\right]$$

which is equal to T if

$$\beta_0 = \alpha_0 \qquad \beta_1 = \alpha_1 \qquad \beta_m = m!\,\alpha_m.$$

The corresponding solution

$$\frac{1}{2\pi i}\int^c \frac{\xi e^{zt}}{F(z)}\,dt$$

is the residue of $\frac{\xi(z)e^{z't}}{F(z)}$ with respect to the pole $z = \lambda$. λ is a pole of order m for ξ and if λ is also a root of $F(z)$ of order μ. λ is a pole of order $m + \mu$ for the function considered and the residue will be $e^{z't}$ times a polynomial in t of deg. $m + \mu$.

In fact developing $\frac{1}{F(z)}$ in $z - \lambda$

$$\frac{1}{F(z)} = \frac{\gamma_\mu}{(z-\lambda)^\mu} + \cdots \frac{\gamma_1}{(z-\lambda)} + \delta_0 + \delta_1(z-\lambda)$$

$$e^{zt} = e^{\lambda t}\left[1 + (z-\lambda)t + \frac{(z-\lambda)^2 t^2}{2!}\cdots\right]$$

resp.

$$\frac{\xi(z)e^{z't}}{F(z)} = e^{\lambda t}(\text{Polynomial of degree } m + \mu).$$

As the terms of degree less than μ multiplied by $e^{z_k t}$ give a solution of the dif. eq. without second member if $s_k = \lambda$ is a μ'ple root of $F(s) = 0$ they may be omitted. Hence we see that when $T = P(t)e^{\lambda t}$ a particular solution is $t^\mu Q(t)e^{\lambda t}$ where Q is a polynomial of same degree as P and μ is the order of the root λ of $F(s) = 0$.

This method may be extended to the general case when T is an integral function of t, of exponentials $e^{\alpha t}e^{\beta t}$ and of sines and cosines $\sin \gamma t$, $\cos \gamma t \ldots$ for it may be put in the form $T' + T''$ of form $Pe^{\lambda t}$. To everyone of these terms may be found a particular solution, and the sum is the required.

The coefficients may be found as above, or by the method of undetermined coefficients.

Example 1.
$$\frac{d^2x}{dt^2} + n^2 x = \cos mt$$

we have found by undet. coef.

$$x = \frac{1}{n^2 - m^2} \cos mt \text{ if } m \neq n.$$

If $m = n$
$$\cos nt = \frac{e^{int} + e^{-int}}{2}$$

$$F(s) = s^2 + n^2$$

$$s_1 = in \quad s_n = -in$$

first consider
$$T_1 = \frac{1}{2} e^{int}$$

$$P = \frac{1}{2} \quad \lambda = in$$

and is a single root of $F(s)$ $\mu = 1$

$$\xi(z) = \frac{1}{2(z - in)}$$

$$\frac{1}{F(z)} = \frac{1}{(z - in)(z + in)}$$

$$= \frac{1}{z - in}\{(2in)^{-1} - (2in)^{-2}(z - in) + (2in)^{-3}(z - in)^2 \ldots\}$$

$$e^{zt} = e^{int}[1 + (z - in)t + (z - in)^2 t^2 \ldots]$$

residue
$$\frac{\xi(z)e^{zt}}{F(z)} = e^{int}\left[\frac{t}{4in} - \frac{1}{2(2in)^2}\right]$$

hence for $\lambda = in$ $y = e^{int}\left[\frac{t}{4in} + \frac{1}{8n^2}\right]$

and for $\lambda = -in$ $y = e^{-int}\left[-\frac{t}{4in} + \frac{1}{8n^2}\right]$

are particular solutions, adding we get

$$y = \frac{t}{2} \sin nt + \frac{1}{4n^2} \cos nt$$

the last term is a particular sol. of dif. eq. without second member. The general solution is then

$$y = c_1 \cos nt + c_2 \sin nt + \frac{t \sin nt}{2}.$$

Ex. 2.
$$\frac{d^2 y}{dt^2} + k\frac{dy}{dt} + n^2 y = a \cos mt$$

$$F(s) = s^2 + ks + n^2$$

$$s_1 = \frac{1}{2}\left(-k + i\sqrt{4n^2 - k^2}\right)$$

$$s_2 = \frac{1}{2}\left(-k - i\sqrt{4n^2 - k^2}\right)$$

$$a \cos mt = \frac{a}{2}\left(e^{mit} + e^{-mit}\right)$$

neither mit nor $-mit$ is a root of $F(s)$ hence the $P's$ are of deg. 0. Let us put

$$y = A \cos mt + B \sin mt$$

$$\frac{dy}{dt} = m\left[-A \sin mt + B \cos mt\right]$$

$$\frac{d^2 y}{dt^2} = -m^2\left[A \cos mt + B \sin mt\right].$$

Then
$$-m^2\left[A \cos mt + B \sin mt\right] + km\left[-A \sin mt + B \cos mt\right]$$
$$+ n^2\left[A \cos mt + B \sin mt\right] = a \cos mt$$

$$(n^2 - m^2)A + kmB = a \qquad B = \frac{km}{n^2 - m^2}A$$

$$(n^2 - m^2)B - kmA = 0 \qquad A\left[n^2 - m^2 + \frac{k^2 m^2}{n^2 - m^2}\right] = a$$

$$A = \frac{n^2 - m^2}{k^2 m^2 + (n^2 - m^2)^2}a$$

$$B = \frac{km}{k^2 m^2 + (n^2 - m^2)^2}a$$

$$y = e^{-\frac{kt}{2}}\left\{c_1 \cos\frac{\sqrt{4n^2 - k^2}}{2}t + c_2 \sin\frac{\sqrt{4n^2 - k^2}}{2}t\right\}$$
$$+ \frac{a}{k^2 m^2 + (n^2 - m^2)^2}\left\{(n^2 - m^2)\cos mt + km \sin mt\right\}.$$

The last term $= D \cos(mt - \alpha)$

if
$$D \cos\alpha = \frac{a(n^2 - m^2)}{k^2 m^2 + (n^2 - m^2)^2}$$

$$D \sin\alpha = \frac{akm}{k^2 m^2 + (n^2 - m^2)^2}$$

$$\text{tang } \alpha = \frac{km}{n^2 - m^2}$$

$$D = \frac{a}{\sqrt{k^2 m^2 + (n^2 - m^2)^2}}.$$

We shall need for the following the following theorems from the theory of functions.

We have seen that if $f(z)$ is holomorphic in a certain region enclosing a

$$f(a) = \frac{1}{2\pi i}\int \frac{f(z)}{z-a}\,dz$$

$$f'(a) = \frac{1}{2\pi i}\int \frac{f(z)}{(z-a)^2}\,dz$$

$$f^n(a) = \frac{n!}{2\pi i}\int \frac{f(z)}{(z-a)^{n+1}}\,dz.$$

The integrals being taken around any contour about a. If the contour be a circle of radius r

$$|z-a| = r \qquad |(z-a)^{n+1}| = r^{n+1}$$

and if M be the greatest value of $|f(z)|$ on the circumference

$$|f^n(a)| \leqq \frac{n!\,M2\pi r}{2\pi r^{n+1}} = \frac{Mn!}{r^n}$$

again in

$$f(a) = \frac{1}{2\pi i}\int \frac{f(z)\,dz}{z-a}$$

put

$$\frac{1}{z-a} = \frac{1}{z\left(1-\frac{a}{z}\right)} = \frac{1}{z} + \frac{a}{z^2} + \cdots + \left(\frac{a}{z}\right)^{n+1}\frac{1}{(z-a)}$$

$$f(a) = \frac{1}{2\pi i}\int \frac{f(z)\,dz}{z} + \frac{a}{2\pi i}\int \frac{f(z)\,dz}{z^2} \cdots \frac{t^n}{2\pi i}\int \frac{f(z)\,dz}{z^{n+1}}$$
$$+ \frac{1}{2\pi i}\int \frac{f(z)}{z-a}\left(\frac{a}{z}\right)^{n+1}dz$$

which is a series in powers of a with the remainder

$$R_n = \frac{1}{2\pi i}\int \frac{f(z)}{(z-a)}\left(\frac{a}{z}\right)^{n+1}dz.$$

If the contour is a circle of radius r about the origin and M is the maximum of $|f(z)|$ on the circumference, call $\varrho = |a|$

$$|R_n| \leqq M\left(\frac{\varrho}{r}\right)^{n+1}r$$

$$\lim_{n=\infty}|R_n| = 0 \ \text{ if } \varrho < r.$$

Hence if a is within the circle of radius r, $f(a)$ is developable in a convergent series, the coefficients being for $(z-a)^n$

$$\frac{1}{2\pi i}\int \frac{f(z)\,dz}{(z-a)^{n+1}} = \frac{f^{(n)}(0)}{n!}$$

so that the series is Maclaurin's. In like manner putting a for 0

$$f(z) = f(a) + (z-a) + (z-a)^2\frac{f'(a)}{2!}\cdots$$

so that a function which is holomorphic within a certain region can be developed in a series of ascending powers of $z - a$ in a circle reaching to the nearest point where the function ceases to be holomorphic. This circle is called the circle of convergence. Such a series is often represented by $P(z - a)$.

(Remarks on prolongation of functions). An important property of linear differential equations

$$\frac{d^n u}{dt^n} + p_1(t) \frac{d^{n-1} u}{dt^{n-1}} \cdots p_n(t) = 0$$

is that the singular points of their integrals are known.

As an equation of the n'th order has in its general solution n arbitrary constants, we may give for a given value t_0 arbitrary values to u and its $n - 1$ first derivatives

$$t = t_0 \quad u = u_0 \quad \frac{du}{dt} = u_0^1 \cdots \frac{d^{n-1} u}{dt^{n-1}} = u_0^{n-1}.$$

The question arises, whether the singular points of the integrals depend upon the arbitrary constants $u_0 \, u_0^1 \ldots u_0^{n-1}$. For instance, the diff. eq.

$$u \frac{du}{dt} = t$$

has the integral

$$u^2 - t^2 = c = u_0^2 - t_0^2$$

$$u = \sqrt{t^2 + u_0^2 - t_0^2}.$$

The function u is two-valued, and has the two branch-points

$$t = \pm \sqrt{t_0^2 - u_0^2}.$$

These are variable with the choice of u_0.

Consider on the other hand the linear equation

$$\frac{du}{dt} - \frac{u}{t^2} = 0.$$

$u = ct^{+\frac{1}{t}}$ which ceases to be holomorphic only for $t = 0$ independent of the value of the arbitrary constant. $t = 0$ is however a singular point of the coefficient of u, $-\frac{1}{t^2}$. We shall find this to be a property of all linear equations, their only singular points are those of the coefficients.

Suppose we write the equation

$$\frac{d^n u}{dt^n} = p_1 \frac{d^{n-1} u}{dt^{n-1}} + \cdots + p_n.$$

If we know that the integral u is holomorphic in any region, we may obtain the coefficients of the development, which are

$$u, \left(\frac{du}{dt}\right)_a, \frac{1}{2} \left(\frac{d^2 u}{dt^2}\right)_a \cdots$$

in terms of the first n, which are arbitrary, viz.

$$\frac{d^{n+1}u}{dt^{n+1}} = p_1 \frac{d^n u}{dt^n} + (p_1' + p_2) \frac{d^{n-1}u}{dt^{n-1}} + (p_2' + p_3) \cdots$$

$$\frac{d^{n+2}u}{dt^{n+2}} = p_1 \frac{d^{n+1}u}{dt^{n+1}} + (2p_1' + p_2) \frac{d^n u}{dt^n} \cdots$$

If then, all the functions p are holomorphic, necessitating their derivatives being so, all the coefficients will be finite. It remains to show that the development will be convergent in a circle reaching to the nearest singular point of any of the coefficients p.

For this purpose let us compare the dif. eq. with the following

$$\frac{d^n w}{dt^n} = \frac{M}{1 - \dfrac{t - t_0}{r}} \frac{d^{n-1}w}{dt^{n-1}} + \cdots \frac{M}{1 - \dfrac{t - t_0}{r}}$$

where r is the distance from t_0-to the nearest of the singular points of $p_1 p_2 \ldots p_n$ and M is the greatest value of any of their moduli. If we choose the values of $w \ldots w^{n-1}s$, greater than the corresponding values of u, $u' \ldots$, it can be shown that, for $t = t_0$ the moduli of the successive derivatives of w are greater than those of u, i. e., the coefficients of the development of w have moduli greater than those of u, hence if the series for w converges in a certain circle, that for u certainly does.

Put $\dfrac{t - t_0}{r} = z$ then to the circle with radius r in the t plane corresponds one of radius 1 in z-plane, and the equation for w is

$$(1 - z) \frac{d^n w}{dz^n} = Mr \frac{d^{n-1}w}{dz^{n-1}} + Mr^2 \frac{d^{n-2}w}{dz^{n-2}} \cdots Mr^n w.$$

This gives positive values for $\left(\dfrac{d^n w}{dz^n}\right)_0$ if the values of w and $\ldots w^{n-1}$ have been taken positive.

Let $\quad w = a_0 + a_1 z + a_2 z^2 + \cdots a_n z^n + \cdots a_{n+p} z^{n+p}$

of which $a_1 \ldots a_{n-1}$ are arbitrary, and all the a's are positive. Inserting in the equation, and making equal the coefficients of z^p

$$(n+p)(n+p-1) \cdots (p+1) a_{n+p} - (n+p-1)(n+p-2) \cdots$$
$$(p+1) p a_{(n+p-1)} = (n+p-1) \cdots (p+1) \, Mr a_{n+p-1}$$
$$+ (n+p-2) \cdots (p+1) Mr^2 a_{n+p-2} + \cdots$$

or $\qquad (n+p)(n+p-1) \cdots (p+1) a_{n+p}$
$$= (Mr + p)(n+p-1) \cdots (p+1) a_{n+p-1} + \cdots$$

The coefficients increase, for

$$\frac{a_{n+p}}{a_{n+p-1}} = \frac{Mr + p}{n+p} + \frac{Mr^2 a_{n+p-2}}{(n+p)(n+p-1) a_{n+p-1}} \cdots$$

all the terms beginning with the second are positive, now M being merely an upper limit, can be taken so that $Mr > n$ hence $a_{n+p} > a_{n+p-1}$. In spite of this increase of the coefficients, the series is convergent.

For $\dfrac{a_{n+p}}{a_{n+p-1}}$ contains a finite number of terms, n. The limit of this ratio as p increases, is hence the sum of the limits of the terms. Now $\underset{p=\infty}{\mathrm{Lim}}\ \dfrac{Mr+p}{n+p} = 1$ and of the other terms the limits are 0, since

$$\frac{a_{n+p-2}}{a_{n+p-1}} < 1.$$

Hence
$$\underset{p=\infty}{\mathrm{Lim}}\ \frac{a_{n+p}}{a_{n+p-1}} = 1.$$

In the series for w, the limit of the ratio of one term to the preceding is

$$\underset{p=\infty}{\mathrm{Lim}}\ \left| \frac{a_{n+p}}{a_{n+p-1}}\, z \right| = |\, z\, | < 1$$

hence the series for w is convergent when $|z| < 1$ or when $|t - t_0| < r$, hence the series for y is convergent in the same circle. Hence the integral y has no singular points except those of the coefficients p.

Hence it is important to study the form of the integrals in the neighbourhood of a singular point. Suppose $y_1 \ldots y_n$ are a fundamental system of integrals. If we make the variable t describe a circuit about a singular point a, when t has returned to its original value the functions $y_1 \ldots y_n$ will have, in general taken on new values $Y_1 Y_2 \ldots Y_n$. If $y_1 y_2 \ldots y_n$ form a fundamental system, evidently $Y_1 Y_2 \ldots Y_n$ must, for if they did not, there would be a linear relation between them, and making t turn about the singular point in the reverse direction, we should return with the linear relation still holding to $y_1 \ldots y_n$ which would hence not be a fundamental system.

Since $y_1 \ldots y_n$ are a fundamental system, $Y_1 \ldots Y_n$ must be linearly expressible in them, with const. coefficients.

$$Y_1 = a_{11} y_1 + a_{12} y_2 \cdots a_{1n} y_n$$
$$Y_2 = a_{21} y_1 + a_{22} y_2 \cdots a_{2n} y_n$$
$$\cdots \cdots \cdots \cdots$$
$$Y_n = a_{n1} y_1 + a_{n2} y_2 \cdots a_{nn} y_n.$$

The simplest singularity that a function can have is that when the variable turns once about the singular point the function is multiplied by a constant. e. g.
$$(x - a)^\lambda = R^\lambda e^{\lambda i \varphi}$$

and after a turn, φ has become $\varphi + 2\pi$
$$R^\lambda e^{\lambda i (\varphi + 2\pi)} = e^{2\pi i \lambda}(x - a)^\lambda$$

If λ is an integer $e^{2\pi i\lambda} = 1$ and the function $(x - a)^\lambda$ is singly valued, otherwise every turn multiplies it by $e^{2\pi i\lambda}$.

Any integral of our equation

$$u = A_1 y_1 + A_2 y_2 \cdots A_n y_n$$

becomes after the turn about a

$$U = A_1 Y_1 + A_2 Y_2 \cdots A_n Y_n$$

is it possible to find an integral u such that the turn multiplies it by a constant ω. If so $U = \omega u$

i. e.

$$A_1(a_{11} y_1 + a_{12} y_2 \cdots a_{1n} y_n)$$
$$+ A_2(a_{21} y_1 + a_{22} y_2 \cdots a_{2n} y_n)$$
$$\cdot \quad \cdot \quad \cdot \quad \cdot \quad \cdot \quad \cdot \quad \cdot \quad \cdot$$
$$A_n(a_{n1} y_1 + a_{n2} y_2 \cdots a_{nn} y_n)$$
$$= \omega(A_1 y_1 + A_2 y_2 + \cdots A_n y_n)$$

that is a linear homogeneous relation between $y_1 \ldots y_n$, but this is impossible, by hypothesis, hence every coefficient of ay must vanish.

$$A_1(a_{11} - \omega) + A_2 a_{21} + A_3 a_{31} \cdots A_n a_{n1} = 0$$
$$A_1 a_{12} + A_2(a_{22} - \omega) + A_3 a_{32} \cdots A_n a_{n2} = 0$$
$$\cdot \quad \cdot \quad \cdot \quad \cdot \quad \cdot \quad \cdot \quad \cdot \quad \cdot \quad \cdot \quad \cdot$$
$$A_1 a_{1n} + A_2 a_{2n} \quad \cdot \quad \cdot \quad A_n(a_{nn} - \omega) = 0.$$

These linear equations in A's can not be true for other than 0 values unless the characteristic equation

$$\begin{vmatrix} a_{11} - \omega, & a_{21} & \cdot & \cdot & \cdot & \cdot & a_{n1} \\ a_{12} &, & a_{22} - \omega, & a_{32} \cdots a_{n2} \\ \cdot & \cdot & \cdot & \cdot & \cdot & \cdot & \cdot & \cdot \\ a_{1n} &, & a_{2n} & \cdot & \cdot & \cdot & a_{nn} - \omega \end{vmatrix} = 0$$

i. e., w must be a root of the above equation of the nth degree.

Let the roots be $\omega_1 \omega_2 \ldots \omega_n$. Inserting any one of them in the above equations in A's, we may determine the A's and get a u. We have the n sets of A's, and hence n u's such that

$$U_1 = \omega_1 u_1$$
$$U_2 = \omega_2 u_2$$
$$\cdot \quad \cdot \quad \cdot \quad \cdot \quad \cdot$$
$$U_n = \omega_n u_n.$$

Furthermore these u's form a fundamental system, for if they did not, we should have

$$c_1 u_1 + c_2 u_2 \cdots c_n u_n = 0$$

turning once about the singular point

$$c_1 \omega_1 u_1 + c_2 \omega_2 u_2 \cdots c_n \omega_n u_n = 0$$

again

$$c_1 \omega_1^2 u_1 + c_2 \omega_2^2 u_2 \cdots$$

more times

$$c_1 \omega_1^{n-1} u_1 \cdots c_n \omega_n^{n-1} u_n = 0.$$

These equations in c cannot admit other than 0 values unless

$$\begin{vmatrix} w_1 & w_2 & w_n \\ w_1^2 & w_2^2 & w_n^2 \\ \cdot & \cdot & \cdot \\ w_1^{n-1} & w_2^{n-1} & w_n^{n-1} \end{vmatrix} = 0$$

which cannot be the case unless two ω's are equal. This property of the u's enables us to give their form in the case of no equal roots. The function $(u - a)^{r_1}$ as we have seen after a turn becomes multiplied by $e^{2\pi i r_1} \cdot u_1$ is multiplied by ω_1, hence if

$$w_1 = e^{2\pi i r_1} \quad \text{i. e.} \quad r_1 = \frac{1}{2\pi i} \log \omega_1$$

the quotient $\dfrac{u}{(u-a)^{r_1}}$ is unchanged, and is hence a uniform function $\varphi_1(u)$

$$u_1 = (u - a)^{r_1} \varphi_1(u)$$

in like manner

$$u_2 = (u - a)^{r_2} \varphi_2(u)$$
$$\cdot \quad \cdot \quad \cdot \quad \cdot \quad \cdot$$
$$u_n = (u - a)^{r_n} \varphi_n(u).$$

The functions $\varphi_1 \ldots \varphi_n$ are developable in positive and negative powers of $u - a$, and may have a for a pole or essentially singular point. Suppose that instead of the fundamental system $y_1 \ldots y_n$ we had begun with another $v_1 \ldots v_n$ so that instead of

$$Y_k = a_{k1} y_1 + a_{k2} y_2 \cdots a_{kn} y_n$$

we had

$$V_k = b_{k1} v_1 + b_{k2} v_2 \cdots b_{kn} v_n$$

but we may express the v's in y's

$$v = d_{11} y_1 + d_{22} y_2 \cdots d_{1n} y_n$$
$$\cdot \quad \cdot \quad \cdot \quad \cdot \quad \cdot \quad \cdot$$
$$v_n = d_{n1} y_1 \quad \cdot \quad \cdot \quad \cdot \quad d_{12} y_n$$

after a turn

$$V_k = d_{k1} Y_1 + d_{k2} Y_2 \cdots d_{kn} Y_n$$

and substitute the values of V_k in v_s and of those in $y's$ — also of $Y's$ in $y's$ and we get

$$b_{k1}(d_{11}y_1 + d_{12}y_2 \cdots d_{1n}y_n)$$
$$+ b_{k2}(d_{21}y_1 + d_{22}y_2 \cdots d_{2n}y_n)$$
$$\cdot \quad \cdot \quad \cdot \quad \cdot \quad \cdot \quad \cdot \quad \cdot \quad \cdot$$
$$b_{kn}(d_{n1}y_1 \cdot \quad \cdot \quad \cdot \quad d_{nn}y_n)$$
$$= d_{k1}(a_{11}y_1 + a_{12}y_2 \cdots a_{1n}y_n)$$
$$+ d_{k2}(a_{21}y_1 \quad \cdot \quad \cdot \quad \cdot \quad a_{2n}y_n)$$

the coefficient of y_p

$$b_{k1}d_{1p} + b_{k2}d_{2p} + \cdots b_{kn}d_{np}$$
$$= d_{k1}a_{1p} + d_{k2}a_{2p} \cdot \quad \cdot \quad d_{kn}a_{np} = c_{kp}.$$

The characteristic equations from the $y's$

$$F(\omega) = \begin{vmatrix} a_{11} - \omega & a_{21} & \cdots a_n \\ a_{12} & a_{22} - \omega & \cdots a_{n2} \\ \cdot & \cdot & \cdot \cdot \cdot \cdot \cdot \\ a_{1n} & \cdot & \cdot \cdot \cdot a_{nn} - \omega \end{vmatrix}$$

and from $v's$

$$G(\omega) = \begin{vmatrix} b_{11} - \omega & b_{21} & \cdots b_{n1} \\ b_{12} & b_{22} - \omega & \cdots b_{n2} \\ \cdot & \cdot & \cdot \cdot \cdot \cdot \cdot \cdot \\ b_{1n} & \cdot & \cdot \cdot \cdot b_{nn} - \omega \end{vmatrix}$$

multiplying these both by

$$\begin{vmatrix} d_{11} & d_{12} & d_{1n} \\ d_{21} & d_{22} & \cdot \cdot \\ \cdot & \cdot & \cdot \cdot \cdot \\ d_{n1} & \cdot & \cdot \quad d_{nn} \end{vmatrix}$$

(multiplying by horizontal rows)

$$F(\omega) \, | \, d \, | = \begin{vmatrix} c_{11} - d_{11}\omega, & c_{12} - d_{12}\omega, & \cdots c_{1n} - d_{1n}\omega \\ c_{21} - d_{21}\omega, & \cdot & \cdot \cdot \cdot \cdot \cdot \cdot \\ \cdot & \cdot & \cdot \cdot \cdot \cdot \cdot \cdot \end{vmatrix}$$

(by columns)

$$G(\omega) \, | \, d \, | = \begin{vmatrix} c_{11} - d_{11}\omega, & c_{12} - d_{12}\omega, & \cdots \\ c_{21} - d_{21}\omega, & c_{22} - d_{22}\omega, & \cdots \\ \cdot & \cdot & \cdot \cdot \cdot \end{vmatrix}$$

Now $| \, d \, |$ is not 0, hence $F(\omega) = G(\omega)$ or the characteristic equation is independent of the fundamental system chosen, and is an invariant of the differential equation.

In case we have equal roots in the characteristic equation $F(\omega) = 0$ the u's do not form a fundamental system. In this case take for a fundamental system u_1, and all the y's except 1. This can be done, for

$$u_1 = A_{11}y_1 + A_{12}y_2 \cdots$$

$$y_2 = \qquad\qquad y_2$$

$$\begin{vmatrix} A_{11} & A_{12} & & A_{1n} \\ 0 & 1 & 0 & 0 & 0 \\ 0 & 0 & 1 & 0 & 0 \\ \cdot & \cdot & \cdot & \cdot & \cdot \end{vmatrix} \neq 0$$

and let

$$U_1 = \omega_1 u_1$$

$$Y_2 = \beta_{21}u_1 + \beta_{22}y_2 + \beta_{23}y_3 \cdots \beta_{2n}y_n$$

$$Y_3 = \beta_{31}u_1 + \beta_{32}y_2$$

$$\cdot \quad \cdot \quad \cdot \quad \cdot \quad \cdot \quad \cdot \quad \cdot \quad \cdot \quad \cdot$$

$$Y_n = \beta_{n1}u_1 \quad \cdot \quad \cdot \quad \cdot \quad \cdot \quad \cdot \quad \cdot$$

the characteristic equation is

$$\begin{vmatrix} \omega_1 - \omega, & \beta_{21} & \beta_{31} & & \beta_{n1} \\ 0 & \beta_{22} - \omega & \beta_{3n} & & \beta_{n2} \\ 0 & \beta_{23} & & \beta_{33} - \omega & \\ 0 & & & & \end{vmatrix} = 0$$

$$\equiv (\omega_1 - \omega) \begin{vmatrix} \beta_{22} - \omega & \beta_{32} & \cdots \\ \beta_{23} & & \beta_{33} - \omega \cdots \end{vmatrix} = 0.$$

Now this has the same roots as the former char. eq. hence ω_1 is a root of

$$\begin{vmatrix} \beta_{22} - \omega, & \beta_{32} & \cdots \\ \beta_{23} & & \beta_{33} - \omega \cdots \\ \cdot & \cdot & \cdot \quad \cdot \quad \cdot \quad \cdot \quad \cdot \end{vmatrix} = 0.$$

Hence we may find quantities $B_2 \ldots B_n$ such that

$$(\beta_{22} - \omega_1)B_2 + \beta_{32}B_3 \quad \cdots \beta_{nn}B_n = 0$$

$$\beta_{23} \quad B_2 + (\beta_{33} - \omega_1) \cdots \beta_{n3}B_n = 0$$

$$\cdot \quad \cdot \quad \cdot \quad \cdot \quad \cdot \quad \cdot \quad \cdot \quad \cdot \quad \cdot \quad \cdot$$

Hence if we put

$$u_2 = B_2 y_2 + B_3 y_3 \cdots$$

$$U_2 = B_2 Y_2 + B_3 Y_3 \cdots$$

$$= B_2(\beta_{21}u_1 + \beta_{22}y_2 \cdots \beta_{2n}y_n) + B_3(\beta_{31}u_1 + \beta_{32}y_2 \cdots \beta_{3n}y_n) + \cdots$$

$$= (\beta_{21}B_2 + \beta_{31}B_3 \cdots \beta_{n1}B_n)u_1 + \omega_1(B_2 y_2 + B_3 y_3 \cdots B_n y_n)$$

$$= \omega_{21}u_1 + \omega_1 u_2.$$

By repeating the reasoning we may get if ω_1 is a λ — fold root

$$U_1 = \omega_1 u_1$$
$$U_2 = \omega_{21} u_1 + \omega_1 u_2$$
$$U_3 = \omega_{31} u_1 + \omega_{32} u_2 + \omega_1 u_3$$
$$\cdot \quad \cdot \quad \cdot \quad \cdot \quad \cdot \quad \cdot \quad \cdot \quad \cdot$$
$$U_\lambda = \omega_{11} u_1 + \omega_{12} u_2 \cdots \omega_2 u_2.$$

If there are various roots of multiplicity $\lambda_1 \lambda_2 \ldots \lambda_n$ we form for each a group of λ integrals all of which together form a fundamental system.

Consider now a λ fold root ω_1, there will be λ integrals $u_1 u_2 \ldots u_\lambda$ of which

$$u_1 = (x - a)^{r_1} \varphi_{11}(x)$$

where

$$r_1 = \frac{1}{2\pi i} \log \omega.$$

The second u_2 is such that $U_2 = \omega_{21} u_1 + \omega_1 u_2$ so that

$$\frac{U_2}{U_1} = \frac{\omega_{21}}{\omega_1} + \frac{u_2}{u_1}$$

that is the function $\frac{u_2}{u_1}$ has the singularity of being increased by a constant when n revolves about the singular point a, like a logarithm, hence

$$\frac{u_2}{u_1} - \frac{\omega_{21}}{\omega_1 \, 2\pi i} \log (x - a)$$

is unchanged, or is a uniform function in the neighborhood of $a = f(x)$

$$\frac{u_2}{u_1} = \frac{\omega_{21}}{\omega_1 \, 2\pi i} \log (x - a) + f(x)$$

$$u_2 = \frac{\omega_{21}}{2\pi i \omega_1} u_1 \log (x - a) + u_1 f(x).$$

But

$$u_1 = (x - a)^{r_1} \varphi_{11}(x)$$

hence

$$u_1 (x - a)^{-r_1} f(x)$$

is uniform, $= \varphi_{21}$ and we may take u_2

$$u_2 = (x - a)^{r_1} \{ \varphi_{21}(x) + \varphi_{22}(x) \log (x - a) \}$$

φ_{21} and φ_{22} are uniform, and φ_{22} differs only by a constant from φ_{11}. The third function u_3 is such that

$$U_3 = \omega_{31} u_1 + \omega_{32} u_2 + \omega_1 u_3$$

$$\frac{U_3}{U_1} = \frac{\omega_{31}}{\omega_1} + \frac{\omega_{32}}{\omega_1} \frac{u_2}{u_1} + \frac{u_3}{u_1}$$

inserting

$$\frac{u_2}{u_1} = \frac{\omega_{21}}{2\pi i \omega_1} \log (x - a) + f(x)$$

$$\frac{U_3}{U_1} = \frac{\omega_{31}}{\omega_1} + \frac{\omega_{32} \omega_{21}}{2\pi i \omega_1^2} \log (x - a) + \frac{u_3}{u_1} + \frac{\omega_{32}}{\omega_1} f(x)$$

hence $\frac{u_2}{u_1}$ acts in the region of a like

$$w = k\,[\log\,(x-a)]^2 + \psi(x)\,\log\,(x-a)$$

$$W = k\,[\log\,(x-a) + 2\pi i]^2 + \psi(x)\{\log\,(x-a) + 2\pi i\}$$

$$= k\,\{\log(x-a)\}^2 + 4\pi i k\,\log\,(x-a) - 4\pi^2 k + \psi(x)\,\log\,(x-a) + 2\pi i\psi(x)$$

$$= w + 4\pi i k\,\log\,(x-a) + 2\pi i\psi(x) - 4\pi^2 k$$

$$\frac{\omega_{31}}{\omega_1} = -4\pi^2 k; \quad \frac{\omega_{32}\,\omega_{31}}{2\pi i\,\omega_1^2} = 4\pi i k; \quad \frac{\omega_{32}}{\omega_1}f(x) = 2\pi i\psi(x)$$

hence

$$\frac{u_2}{u_1} - k\,(\log\,(x-a))^2 + \psi(x)\,\log\,(x-a) \quad\text{is uniform}$$

$$u_3 = (x-a)^{r_1}\{\varphi_{31} + \varphi_{32}\,\log\,(x-a) + \varphi_{33}\,\log^2\,(x-a)\}$$

$$u_\lambda = (x-a)^{r_1}\{\varphi_{\lambda 1} + \varphi_{\lambda 2}\,\log\,(x-a) + \varphi_{\lambda 3}\,\log^2\,(x-a)\cdots\varphi_{\lambda\lambda}\,\log^{\lambda-1}(x-a)\}.$$

All the functions φ may be expressed as linear functions of $\varphi_{11}\,\varphi_{21}\cdots\varphi_{\lambda 1}$.

It is important to consider what differential equation have integrals having the property that their products by $(x-a)^\lambda$ where λ is some finite number (0, pos. or neg.) remain finite for $x = a_1$. Such integrals are known as *regular*.

Since $(x-a)^p\{\log\,(x-a)\}^q$ is 0 for $x=a$ for all positive p's and q's, the preceding condition is equivalent to saying that the development of any function φ_{r_s} contains only a finite number of terms with negative powers of $(x-a)$. If r be the highest negative power we may take it out as a factor, and have

$$F = (x-a)^r\,[\varphi_0 + \varphi_1\,\log\,(x-a) + \varphi_2\,\log^2\,(x-a)\ldots\varphi_\lambda\,\log^\lambda\,(x-a)]$$

where the φ,s are all *holomorphic*. In this case $F(x-a)^{-r}$ becomes infinite only like a function

$$L = \alpha + \beta\,\log\,(x-a) + \gamma\,[\log\,(x-a)]^r\cdots$$

where α, β, γ ... are all const. In such a case F is said to belong to the exponent r.

It is evident that a regular expression F has a regular derivative

$$\frac{dF}{dx} = (x-a)^{r-1}\{\varphi_0 + \varphi_1\,\log\,(x-a) + \cdots + \varphi_\lambda\,(\log\,(x-a))^\lambda\}$$

$$+ (x-a)^r\{\varphi_0' + \varphi_1'\,\log\,(x-a)\cdots\varphi_\lambda'\,(\log\,(x-a))^\lambda + \frac{\varphi_1}{x-a}$$

$$+ \frac{\varphi_\lambda\,\lambda\,(\log\,(x-a))^{\lambda-1}}{x-a}$$

$$= (x-a)^{r-1}\{\varphi_0 + \varphi_1\,\log\,(x-a) + \cdots + \varphi_\lambda\,(\log\,(x-a))^\lambda$$

$$+ (x-a)\varphi_0' + (x-a)\varphi_1'\,\log\,(x-a)$$

$$+ \varphi_1 + \cdots \lambda\varphi_\lambda\,(\log\,(x-a))^{\lambda-1}\}$$

a regular expression with exponent $r-1$.

(Case of exception when $r = 0$ and $F(a)$ is not infinite.)
Also the product of two regular expressions is regular.
We may express the coefficients of the equation

$$\frac{d^n y}{d x^n} = p_1 \frac{d^{n-1} y}{d x^{n-1}} + p_2 \frac{d^{n-2}}{d x^{n-2}} \cdots p_n y$$

in terms of the n solutions $u_1 \ldots u_n$ for

$$\frac{d^n u_1}{d x^n} = p_1 u_1^{(n-1)} + p_2 u_1^{(n-2)} \cdots p_n u_1$$

$$u_2^{(n)} = p_1 u_2^{n-1} \cdot \quad \cdot \quad \cdot \quad \cdot \quad \cdot \quad \cdot$$

so that

$$p_\alpha = \frac{D_\alpha}{D}$$

where

$$D = \begin{vmatrix} u_1^{(n-1)} & u_1^{(n-2)} & \ldots & u_1 \\ u_2^{(n-1)} & u_2^{(n-2)} & \ldots & u_2 \\ \cdot & \cdot & & \cdot \\ u_n^{(n-1)} & \cdot & \cdot & u_n \end{vmatrix}$$

and D_α is the result of substituting $u_1^{(n)} u_2^{(n)} u_n^{(n)}$ in the α column of D.
Now when u turns about the critical point, $u_1 u_2 \ldots u_n$ undergo the
substitution

$$S = \begin{vmatrix} \omega_1 & 0 & 0 & 0 & 0 & 0 \\ \omega_{21} & \omega_1 & 0 & 0 & 0 & 0 \\ \omega_{31} & \omega_{32} & \omega_1 & & & \\ \cdot & \cdot & \cdot & \cdot & \cdot & \cdot \\ \omega_{\lambda 1} & \omega_{\lambda 2} & \ldots & \omega_1 & & \\ & & & 0 & \omega_{s+r} & 0 \\ & & & & \omega_{\lambda+2_1,\lambda} & \omega_2 \end{vmatrix}.$$

The derivatives of any order undergo the same substitution. Hence D
and D_α are multiplied by the determinant of the substitution and the
quotient $p_\alpha = \frac{D_\alpha}{D}$ is unchanged. Hence p_α must be a uniform function,
the logarithms must have disappeared. Also there can be in p_α only a
finite number of terms in negative powers of $x - a$. Hence p_α can
have at a only ordinary points or poles. We shall now show that p_α
can have a pole of no higher order than α.
Put $y = T\xi$ where

$$T = (x - a)^\beta \{ c_0 t - c_1 (x - a) + c_2 (x - a)^2 \cdots \}$$

$$\frac{d y}{d x} = \frac{d T}{d x} \xi + T \frac{d \xi}{d x}$$

$$\frac{d^k y}{d x^k} = \frac{d^k T}{d x^k} \xi + k \frac{d^{k-1} T}{d x^{k-1}} \xi' + \frac{k(k-1)}{2} T^{k-2} \xi'' \cdots$$

and our equation becomes.

$$T \frac{d^n \xi}{dx^n} = \left(- n T' + p_1 T\right) \frac{d^{n-1} \xi}{dx^{n-1}}$$
$$+ \left(- \frac{n(n-1)}{2} T'' + (n-1) p_1 T' + p_2 T\right) \frac{d^{n-2} \xi}{dx^{n-2}}$$

or dividing by T

$$\frac{d^n \xi}{dx^n} = \left(- n \frac{T''}{T} + p_1\right) \xi^{n-1} + \left[- \frac{n(n-1)}{2} \frac{T''}{T} + (n-1) p_1 \frac{T'}{T} \cdots\right].$$

If the original equation has regular integrals, this one in ξ has also, for

$$\xi = \frac{y}{T} = \frac{y}{(x-a)^\beta} \left\{\frac{1}{c_0} + \cdots\right\}.$$

Also since

$$T = (x-a)^\beta g(x-a); \quad T' = \beta (x-a)^{\beta-1} g(x-a) + (x-a)^\beta g'(x-a)$$

$$\frac{T'}{T} = \frac{1}{x-a} \left\{\beta + (x-a)^{\beta+1} \frac{g'}{g}\right\}$$

has a pole of order 1 at a; $\frac{T''}{T}$ of order 2, etc. Hence if p_α has a pole of order α or less in the original equation, the same property holds for $g_\alpha(\alpha)$ in the transformed equation and reciprocally, if it holds for the transformed, it will hold for the other, derived by the substitution $\xi = \frac{1}{T} y$. Now we have found that there is always one solution $u_1 = (x-a)^{r_1} \varphi_{11}$ without logarithms.

This integral being regular, will be of the form

$$(x-a)^\beta \{c_0 + c_1 (x-a) \cdots\}.$$

Take it for T then the transformed equation must admit the solution $\xi = 1$ and hence does not contain the term in ξ, and reduces to

$$\frac{d^n \xi}{dx^n} = y_1 \frac{d^{n-1} \xi}{dx^{n-1}} \cdots g_{n-1} \frac{d\xi}{dt}.$$

Put $\frac{d\xi}{dt} = \xi'$ and we have an equation of order $n-1$ in ξ' whose integrals being derivatives of the former ones, are regular.

Hence if the theorem is supposed true for eq.'s of order $n-1$ a will be a pole only of order α for g_α the theorem will be true for the equation in ξ and hence for that in y. We have then only to show it for $n = 1$

$$\frac{dy}{dx} + g_1 y = 0.$$

Put

$$y = (x-a)^\beta \{a_0 + a_1 (x-a) \cdots\} + \beta (x-a)^{\beta-1} \{a_0 + a_1 (x-a) + \cdots\}$$
$$+ q_1 (x-a)^\beta \{a_1 + 2 a_2 (x-a) + 3 a_3 (x-a)^2 \cdots\}$$
$$= (x-a)^{\beta-1} [\beta a_0 + \beta a_1 (x-a) \cdots + a_1 q_1 (x-a) \cdots] = 0.$$

Now if q_1, has a pole of order greater than 1, there will be a single term $a_1 g_1 (x - a)$ of degree less than the others, its coefficient must accordingly be 0.

Hence if a linear equation have regular integrals in the neighborhood of $x = a$ it must be of the form

$$\frac{d^n y}{d x^n} = \frac{P_1(x)}{x - a} \frac{d^{n-1} y}{d x^{n-1}} + \frac{P_2(x)}{(x - a)^2} \frac{d^{n-2} y}{d x^{n-2}} \cdots \frac{P_n(x)}{(x - a)^n} y$$

where $P_1 \ldots P_n$ are holomorphic in region $x = a$.

If the property holds for all the singular points $a_1 a_2 \ldots a_n$ the equation must have the form

$$\frac{d^n y}{d x^n} = \frac{P_1(x)}{\psi(x)} \frac{d^{n-1} y}{d x^{n-1}} + \frac{P_2(x)}{[\psi(x)]^2} \frac{d^{n-2} y}{d x^{n-2}} \cdots$$

$P_1 \ldots P_n$ being holomorphic in the whole plane and

$$\psi(x) = (x - a_1)(x - a_2) \cdots (x - a_\nu).$$

If the integrals are to be regular also for the neighborhood of $x = \infty$ i. e., $x^{-r} F$ is infinite only like $\log^2 x$ we must make the change $x = \frac{1}{t}$ and put the above condition for $t = 0$.

If the equation is

$$\frac{d^n y}{d x^n} + p_1 \frac{d^{n-1} y}{d x^{n-1}} \cdots p_n y = 0.$$

The transformed is

$$t^{2n} \frac{d^n y}{d t^n} + e_{01} t^{2n-1} \frac{d^{n-1} y}{d t^{n-1}} + e_{02} t^{2n-2} \frac{d^{n-2} y}{d t^{n-2}}$$

$$+ p_1 \left\{ t^{2n-2} \frac{d^{n-1} y}{d t^{n-1}} + e_{11} t^{2n-3} \frac{d^{n-2} y}{d t^{n-2}} \cdots \right\}$$

$$+ p_2 \left\{ t^{2n-4} \frac{d^{n-2} y}{d t^{n-2}} + e_{21} t^{2n-5} \frac{d^{n-3} y}{d t^{n-3}} \cdots \right\}$$

$$= t^{2n} \frac{d^n y}{d t^n}$$

$$+ \left\{ e_{01} t^{2n-1} + p_1 t^{2n-2} \right\} \frac{d^{n-1} y}{d t^{n-1}}$$

$$+ \left\{ e_{02} t^{2n-2} + e_{11} p_1 t^{2n-3} + p_2 t^{2n-4} \right\} \frac{d^{n-} y}{d t^{n-2}}$$

dividing by t^{2n}

$$\frac{d^n y}{d t^n} + \left\{ \frac{e_{01}}{t_n} + \frac{p_1}{t^2} \right\} \frac{d^{n-1} y}{d t^{n-1}} + \left\{ \frac{e_{02}}{t^2} + \frac{e_{11} p_1}{t^3} + \frac{p_2}{t^4} \right\} \frac{d^{n-2} y}{d t^{n-2}}$$

$$= \frac{d^n y}{d t^n} + q_1 + q_2 + \cdots q_k$$

$$q_k = \frac{e_{0k}}{t^k} + \frac{e_{1, k-1} p_1}{t^{k+1}} + \cdots \frac{p_k}{t^{2k}}$$

and for regular integrals for $t = 0$ we must have q_k have a pole of order not greater than k, i. e., $q_k t^k$ is finite for $t = 0$

$$= e_{0k} + \frac{e_{1,\,k-1}p_1}{t} + \frac{e_{2,\,k-2}p_2}{t^2} \cdots \frac{p_k}{t^k}$$

i. e. every $\frac{p_\alpha}{t^\alpha}$ must be finite for $t = 0$ or $p_\alpha x^\alpha$ finite for $x = \infty$.

But
$$p_\alpha = \frac{P_\alpha(x)}{[\psi(x)]^\alpha}$$

hence α must not be essentially negative for $P\alpha(x)$

$$\psi(x) = (x - a_1)(x - a_2) \cdots (x - a_\rho).$$

The denominator is of degree $\alpha\rho$. $P_\alpha x^\alpha$ must not be of greater degree than $\alpha\rho$; P_α not greater than $\alpha(\rho - 1)$. Hence that a linear differential equation shall have its integrals regular for every singular point, including $n = \infty$, it must have the form.

$$\frac{d^n y}{dx^n} + \frac{Q_{\rho-1}(x)}{\psi(x)} \frac{d^{n-1}y}{dx^{n-1}} + \frac{Q_2(\rho - 1)}{[\psi(x)]^2} \frac{d^{n-2}y}{dx^{n-2}} \cdots$$

where Q's are polynomials of the degree denoted by the suffixes and

$$\psi(x) = (x - a_1)(x - a_2) \cdots (x - a_\rho)$$

where $a_1 \ldots a_\rho$ are the critical points.

Suppose we transform the equation

$$\frac{d^n y}{dx^n} + \frac{P_1}{(x - a)} \frac{d^{n-1}y}{dx^{n-1}} + \frac{P_2}{(x - a)^2} \frac{d^{n-2}}{dx^{n-2}} \cdots \frac{P_n y}{(x - a)^n} = 0$$

as before by means of the substitution

$$y = \xi T$$

$$\frac{d^n \xi}{dx^n} + q_1 \frac{d^{n-1}\xi}{dx^{n-1}} + q_2 \frac{d^{n-2}\xi}{dx^{n-2}} \cdots q_n \xi = 0$$

$$q_k = n_k T^k + (n - 1)_{k-1} T^{k-1} \frac{P_1}{(x - a)} + (n - 2)_{k-2} T^{k-2} \frac{P_2}{(x - a)^2}$$

$$q_0 = T.$$

Now let us put

$$T = (x - a)^r$$

$$T' = r(x - a)^{r-1}; \quad T'' = r(r - 1)(x - a)^{r-2}$$

$$T_m = [\overset{m}{r}](x - a)^{r-m}$$

$$q_k = n_k [\overset{k}{r}](x - a)^{r-k} + (n - 1)_{k-1} [\overset{k-1}{r}](x - a)^{r-k+1} \frac{P_1}{x - a}$$

$$+ (n - 2)_{k-2} [\overset{k-2}{r}](x - a)^{r-k+2} \frac{P_2}{(x - a)^2} \cdots$$

$$= (x - a)^{r-k} \Big\{ n_k [\overset{k}{r}] + (n - 1)_{k-1} [\overset{k-1}{r}] P_1 + (n - 2)_{k-2} [\overset{k-2}{r}] P_2$$

$$+ (n - k)_0 [\overset{0}{r}] P_k \Big\}.$$

Hence we may write the equation in ξ, dividing through by $(x - a)^r$

$$\frac{d^n \xi}{d x^n} + \frac{Q_1}{(x - a)} \frac{d^{n-1} \xi}{d x^{n-1}} + \frac{Q_2}{(x - a)^2} \frac{d^{n-2} \xi}{d x^{n-2}} \cdots \frac{Q_n \xi}{(x - a)^n} = 0$$

where

$$Q_k = \left\{ n_k [\overset{k}{r}] + (n - 1)_{k-1} [\overset{k-1}{r}] P_1 \cdots \right\}.$$

The object of the transformation was to find an integral

$$y = (x - a)^r \{ c_0 + c_1 (x - a) \cdots \}.$$

Let us find the condition that the equation in ξ has an integral

$$c_0 + c_1 (x - a) + c_2 (x - a)^2 \cdots$$

in which c_0 is not 0.

Multiplying the differential equation in ξ by $(x - a)^n$, since all Q's are holomorphic, putting $x = a$ makes all terms vanish except the last

$$Q_n(a) \xi(a).$$

Now $\xi(a) = c_0$ hence we must have $Q_n(a) c_0 = 0$ and if c_0 is not 0.

$$Q_n(a) = [\overset{n}{n}][\overset{n}{r}] + (n - 1)_{n-2} [\overset{n-1}{r}] P_1 a + (n - 2)_{n-2} [\overset{n-2}{r}] P_2 a = 0.$$

Now

$$n_n = (n - 1)_{n-1} \cdots = 1.$$

Hence

$$r(r - 1)(r - 2) \cdots (r - n + 1) + r(r - 1) \cdots (r - n + 2) P_1 a$$
$$+ r(r - 1) \cdots (r - n + 3) P_2 a \cdots P_n(a) = 0.$$

This is an equation of the nth degree in r. $F(r) = 0$ and r in the integral

$$y = (x - a)^r \{ c_0 + c_1 (x - a) \cdots \}$$

must be one of its roots. It contains the values of the n functions $P_1 \ldots P_n$ for the point a, and is called the determining fundamental equation (Fuchs) for the singular point a.

It serves the same purpose as the characteristic equation

$$F(\omega) = \begin{vmatrix} \alpha_{11} - \omega & \alpha_{21} & \alpha_{n1} \\ \alpha_{12} & \alpha_{22} - \omega & \end{vmatrix} = 0$$

namely to determine the index in the integral $y = (x - a)^r \varphi$ which gave

$$r = \frac{1}{2 \pi i} \log \omega.$$

Now whereas the coefficients of the equation in ω have not yet been found, and depend in a very transcendental manner upon those of the coefficients of the differential equation, the coefficients in the equation for r are very simple, and contain besides integral numbers only the values of $P_1(a) \ldots P_n(a) \ldots$

The roots of the fundamental determining equation are equal to $\frac{1}{2\pi i}$ times the logarithms of the roots of the characteristic equation.

If any of the roots of the equation in r are equal, or differ by an integer, the logarithms of the corresponding ω's differ by $2\pi i$ times an integer, hence the ω's are the same, and the characteristic equation in ω has equal roots. In that case we get a set of integrals of the form previously found.

It may be shown that the series found are convergent by comparing the solutions with solutions of an auxiliary differential equation, in a manner analogous to that used for a non-singular point.

6. Linear Differential Equations. A *linear* differential equation is one of the form

$$X_0 \frac{d^n y}{dx^n} + X_1 \frac{d^{n-1} y}{dx^{n-1}} \cdots + X_{n}{}'_{y} = P,$$

where

$$X_0, \ X_1, \ \ldots X_n, \ P$$

are functions of x alone. If $P = 0$, the equation is called homogeneous, or without second member.

Those with second member can be made to depend upon those without, hence we shall first consider the latter.

The characteristic property of linear equations is the manner in which the arbitrary constants enter.

We need the following lemma.

If $y_1, y_2 \ldots y_{n+1}$ are functions of x for which the determinant

$$\begin{vmatrix} y_1, & y_2, & \cdots & y_{n+1} \\ \dfrac{dy_1}{dx}, & \dfrac{dy_2}{dx}, & \cdots & \dfrac{dy_{n+1}}{dx} \\ \dfrac{d^2 y}{dx^2}, & \dfrac{d^2 y_2}{dx^2}, & \cdots & \dfrac{d^2 y_{n+1}}{dx^2} \\ \cdot \ \cdot & \cdot \ \cdot & \cdots & \cdot \ \cdot \ \cdot \\ \dfrac{d^n y_1}{dx^n}, & \dfrac{d^n y_2}{dx^n}, & \cdots & \dfrac{d^n y_{n+1}}{dx^n} \end{vmatrix} = 0,$$

identically, then there is between the $n + 1$ functions a linear relation with constant coefficients,

$$c_1 y_1 + c_2 y_2 + \cdots + c_{n+1} y_{n+1} = 0.$$

Suppressing the last line and column we should have a determinant of the same nature, which we may suppose not equal to 0, for if it were we should have to demonstrate the property for n functions.

The minor being not zero we can find $n + 1$ functions λ such that

$$\lambda_1 y_1 + \lambda_2 y_2 + \cdots + \lambda_{n+1} y_{n+1} = 0,$$

$$\lambda_1 \frac{dy_1}{dx} + \lambda_2 \frac{dy_2}{dx} + \cdots + \lambda_{n+1} \frac{dy_{n+1}}{dn} = 0,$$

. . . .

. . . .

$$\lambda_1 \frac{d^n y_1}{dx^n} + \lambda_2 \frac{d^n y_2}{dx^n} + \cdots + \lambda_{n+1} \frac{d^n y_{n+1}}{dx^n} = 0.$$

Differentiating each (except the last) regarding the following,

$$y_1 \frac{d\lambda_1}{dx} + y_2 \frac{d\lambda_2}{dx} + \cdots + y_{n+1} \frac{d\lambda_{n+1}}{dx} = 0,$$

$$\frac{dy_1}{dx} \frac{d\lambda_1}{dx} + \frac{dy_2}{dx} \frac{d\lambda_2}{dx} + \cdots + \frac{dy_{n+1}}{dx} \frac{d\lambda_{n+1}}{dx} = 0,$$

. . . .

$$\frac{d^{n-1}y_1}{dx^{n-1}} \frac{d\lambda_1}{dx} + \frac{d^{n-1}y_2}{dx^{n-1}} \frac{d\lambda_2}{dx} + \cdots + \frac{d^{n-1}y_{n+1}}{dx^{n-1}} \frac{d\lambda_{n+1}}{dx} = 0,$$

which are of the same form for the $\frac{d\lambda_s}{dx}$ as for the $\lambda_s's$ (omitting the last). Hence

$$\frac{\frac{d\lambda_1}{dx}}{\lambda_1} = \frac{\frac{d\lambda_2}{dx}}{\lambda_2} \cdots; \quad \text{or} \quad \frac{d \log \lambda_1}{dx} = \frac{d \log \lambda_2}{dx}$$

$$\frac{\lambda_s}{\lambda_1} = \text{const.} = a_s; \quad a_1 y_1 + a_2 y_2 \cdots a_{n+1} y_{n+1} = 0.$$

Now if $y_1 y_2 \ldots y_{n+1}$ are solutions of the differential equation

$$X_0 \frac{d^n y}{dx} + X_1 \frac{d^{n-1}y}{dx^{n-1}} \cdots X_n y = 0$$

we have

$$X_0 \frac{d^n y_1}{dx_n} \quad \cdots \quad X_n y_1 = 0$$

$$X_0 \frac{d^n y_2}{dx_n} \quad \cdots \quad X_n y_2 = 0$$

. . . .

$$X_0 \frac{d^n y_{n+1}}{dx^{n+1}} \quad \cdots \quad X_n y_{n+1} = 0$$

and eliminating $X_0 \ldots X_n$ we have

$$\begin{vmatrix} y_1 & y_2 \cdots y_{n+1} \\ y_1' & y_2' \cdots y_{n+1}' \\ \cdots \cdots \cdots \end{vmatrix} = 0$$

so that there is a linear relation between the $n + 1$ y's.

Hence any integral y may be written

$$y = a_1 y_1 + a_2 y_2 \cdots a_n y_n$$

where $y_1 \ldots y_n$ are any particular solutions, not satisfying a linear relation. Such a set of integrals is called a fundamental system.

If we take

$$u_1 = \alpha_{11} y_1 + \alpha_{12} y_2 \cdots \alpha_{1n} y_m$$

$$u_2 = \alpha_{21} y_1 \quad \cdot \quad \cdot \quad \cdot \quad \cdot \quad \cdot$$

$$u_1 u_2 \ldots u_n$$

will form an fundamental system if $y_1 \ldots y_n$ do for if $\mid \alpha_m \mid \neq 0$

$$\begin{vmatrix} u_1 \ldots u_m \\ u_1' \ldots u_m' \\ \cdot \quad \cdot \quad \cdot \\ \cdot \quad \cdot \quad \cdot \end{vmatrix} = \begin{vmatrix} y_1 \cdots y_n \end{vmatrix} \quad \begin{vmatrix} \alpha_m \end{vmatrix}$$

Put

$$F(y) = X_0 y^{(n)} + X_1 y^{(n-1)} \cdots X_n y$$

$$F'(y) = n X_0 y^{(n-1)} + (n-1) X_1 y^{(n-2)} + X_{n-1} y$$

$$\cdot \quad \cdot \quad \cdot \quad \cdot \quad \cdot \quad \cdot \quad \cdot \quad \cdot \quad \cdot \quad \cdot \quad \cdot \quad \cdot$$

$$F^{(n-1)} y = 2 \cdot 3 \ldots n X_0 y' + (n-1)! \, X_1 y$$

$$F^n y = n! \, X_0 y$$

gives identically

$$F(yz) = z F(y) + \frac{z'}{1} F'(y) + \cdots \frac{z^n}{n!} F^n(y)$$

for

$$\frac{d}{dx}(zy) = zy' + yz'$$

$$\frac{d^2}{dx^2}(zy) = zy'' + 2y'z' + yz$$

$$\frac{d^n}{dx^n}(zy) = zy^n + \frac{n}{1} z' y^{n-1} \cdots yz^n.$$

If we choose y_1, so that

$$F^{(n-1)}(y_1) = 0 \quad \text{i. e.} \quad n X_0 y_1' + z F(y_1) = 0$$

giving

$$y_1 = e^{-\int \frac{z_1}{n_1} dx}$$

the equation $F(zy_1)$ will not contain $z^{(n-1)}$. If y is an integral of $F(y) = 0$ put $y = y_1 z$ the equation is

$$\frac{z^n F^n(y_1)}{n!} + \frac{z^{n-1} F^{(n-1)}}{(n-1)!} y' \cdots z F y_1 = 0$$

and since $F(y_1) = 0$

$$\frac{z^n F^n y_1}{n!} \cdots z' F'(y_1) = 0$$

and putting $z' = u \quad z = \int u\,dx$ the equation is of order $n - 1$ in u.

In general if we know k independent solutions, we can reduce the equation to one of order $n - k$, and k quadratures.

Put

$$y = \lambda_1 y_1 + \lambda_2 y_2 \cdots \lambda_k y_k$$

the λ's being k new functions connected by the $k - 1$ equations

$$y_1 \lambda_1' + y_2 \lambda_2' \cdots \lambda_k' y_k = 0$$

$$y_1' \lambda_1' + y_2' \lambda_2' \cdot \ \cdot \ \cdot = 0$$

$$\cdot \ \cdot \ \cdot \ \cdot \ \cdot \ \cdot \ \cdot \ \cdot$$

$$y_1^{k-2} \lambda_1' \ \cdot \ \cdot \ \cdot \ \cdot = 0$$

$$y' = \lambda_1 y_1' + \lambda_2 y_2' \cdots + \lambda_k y_k'$$

$$y_k'' = \lambda_1 y_1'' \ \cdot \ \cdot \ \cdot + \lambda_k y_k''$$

$$\cdot \ \cdot \ \cdot \ \cdot \ \cdot \ \cdot \ \cdot$$

$$y^{k-1} = \lambda_1 y_1^{k-1} \cdot \ \cdot \ \cdot + \lambda_k y_k^{k-1}$$

$$y^k = F_1 \ \cdot \ \cdot \ y^n = F_{n-k+1}.$$

F_r being a linear function of the λ's and their r first derivatives.

Substituting in the differential equation, whe shall have a linear equation containing λ's and their $n - k + 1$ first derivatives.

Now the differential equation is satisfied if the λ's are constants, since $y_1 \ldots y_k$ are solutions. Hence the terms in $\lambda_1 \ldots \lambda_k$ drop out and there remain $\lambda_1' \ldots \lambda_k'$ and their $n - k$ derivatives.

But from the above $k - 1$ equations of condition we may determine $\lambda_2' \lambda_3' \ldots \lambda_k'$ as linear functions of λ, and substituting we get an equation of the $n - k$th order for λ_1', we then obtain $\lambda_1' \ldots \lambda_k'$, and by quadrations $\lambda_1 \ldots \lambda_k$.

E. g., suppose we know a solution y, of

$$\frac{d^2 y}{dx^2} + p_1 \frac{dy}{dx} + p_2 y = 0$$

put

$$y = \lambda_1 y_1 \qquad k - 1 = 0$$

$$\frac{dy}{dx} z + \lambda_1 y_1'$$

$$\frac{d^2 y}{dx^2} = \lambda_1'' y_1 + 2 \lambda_1' y_1' + \lambda_1 y_1''$$

$$\lambda_1'' y_1 + 2 \lambda_1' y_1' + p_1 (\lambda_1' y_1) = 0$$

$$\lambda_1'' y_1 + (2 y_1' + p_1 y_1) \lambda_1' = 0.$$

$$\frac{\lambda_1''}{\lambda_1'} = -\frac{2 y_1'}{y_1} - p_1; \quad \log \lambda_1' = -2 \log y_1 - \int p_1 \, dx + c$$

$$\lambda_1' = A \frac{e^{-\int p_1 \, dx}}{y_1^2}; \quad \lambda_1 = A \int \frac{e^{-\int p_1 \, dx}}{y_1^2} \, dx + B$$

$$y = A y_1 \int \frac{e^{-\int p_1 \, dx}}{y_1^2} \, dx + B y_1.$$

Multiplying the equation

$$\frac{d^n y}{d x^n} + p_1 \frac{d^{n-1} y}{d x^n} \cdots p_n y = 0$$

by an undetermined function u and integrating by parts.

$$\int u \left\{ \frac{d^n y}{d x^2} + p_1 \frac{d^{n-1} y}{d x^{n-1}} \cdots p_n y \right\} dx = 0$$

$$u y^{n-1} - u' y^{(n-2)} + u'' (y)^{n-3} \cdots$$

$$+ p_1 u y^{(n-2)} - (p_1 u)' y^{(n-3)} \cdots$$

$$+ (-1)^n \int y [u^n - (p_1 u)^{n-1} + (p_2 u)^{n-2} \cdots (-1)^n p_n y] dx = 0.$$

If u is a solution of

$$\frac{d^n u}{d x^n} - \frac{d^{n-1}}{d x^{n-1}} (p_1 n) + \frac{d^{n-2}}{d x^{n-2}} (p_2 u) \cdots (-1)^n p_n y = 0$$

the integral disappears above, and we have an equation of order $n - 1$ for y, containing an arbitrary constant. The function u is called a multiplier for the differential equation, for it converts the expression.

$$y^{(n)} + p_1 y^{(n-1)} \cdots p_n y$$

into the derivative of a linear function of y and its $n - 1$ derivatives.

If the above equation admits the solution $u = $ const., the original differential equation is called an exact differential equation.

The above equation for the multiplier is called the adjunct equation. The general solution of the equation with second member

$$X_0 y^{(n)} + X_1 y^{(n-1)} \cdots = P$$

is obtained by adding to any particular solution of it u the general solution of the equation without second member. For if y be the solution of the latter, $u + y$ satisfies the equation with second member, and contains n constants, hence is the general solution.

To find the solution of the equation with second member we first find n independent solutions of the equation without second member, $y_1 \cdots y_n$ then put

$$y = \lambda_1 y_1 + \lambda_2 y_2 \cdots \lambda_n y_n$$

where the $\gamma's$ are defined by the conditions

$$\lambda_1' y_1 + \cdots + \lambda_n' y_n = 0$$
$$\lambda_1' y_1' + \cdots + \lambda_n' y_n' = 0$$
$$\lambda_1' y_1'' + \cdots + \lambda_n' y_n'' = 0$$
$$\cdot \quad \cdot \quad \cdot \quad \cdot \quad \cdot \quad \cdot \quad \cdot \quad \cdot$$
$$\lambda_1' y_1^{n-2} + \cdots + \lambda_n' y_n^{n-2} = 0$$

giving

$$y' = \lambda_1 y_1' + \cdots + \lambda_n y_n'$$
$$y'' = \lambda_1 y_1'' + \cdots + \lambda_n y_n''$$
$$\cdot \quad \cdot \quad \cdot \quad \cdot \quad \cdot \quad \cdot \quad \cdot \quad \cdot$$
$$y^{n-1} = \lambda_1 y_1^{(n-1)} + \cdots + \lambda_n y_n^{(n-1)}$$
$$y^n = \lambda_1 y_1^{(n)} + \cdots + \lambda_n y_n^{(n)} + \lambda_1' y_1^{(n-1)} + \cdots \lambda_n' y_n^{(n-1)}$$

substituting in the differential equation

$$\lambda_1[X_0 y_1^n + X_1 y_1^{(n-1)} \cdots X_n y_1] + \lambda_2[X_0 y_2^n + X_1 y_2^{(n-1)} \cdots X_n y_2] \cdots$$
$$+ \lambda_n[X_0 y_n^n + X_1 y_n^{(n-1)} \cdots X_n y_n] + X_0[\lambda_1' y_1^{(n-1)} + \lambda_2 y_2^{(n-1)} \cdots \lambda_n' y_n^{n-1}] = P$$

and all disappears except the last line. This together with the above $n - 1$ linear equations in λ's determines them, and the λ's are obtained by quadratures.

Ex. $$\frac{d^2 y}{dx^2} + u^2 y = \cos mx.$$

Equations with constant coefficients.

$$\frac{d^n y}{dx^n} + a_1 \frac{d^{n-1} y}{dx^{n-1}} \cdots a_n x = 0.$$

Integrated by Euler, as follows.

Put
$$y = e^{sx}$$

$$\frac{dy}{dx} = se^{sx} \quad \frac{d^2 y}{dx^2} = s^2 e^{sx} \quad \frac{d^n y}{dx^2} = s^n e^{sx}$$

$$e^{sx}[s^n + a_1 s^{n-1} \cdots a_n] = 0$$

satisfied if s is a root of

$$F(s) \equiv s^n + a_1 s^{n-1} \cdots + a_n = 0$$

which is called the characteristic equation. If this equation has no unequal roots $s_1 \ldots s_n$ we shall have n particular solutions $e^{s_1 x} e^{s_2 x} \ldots e^{s_n x}$.

They are independent, for the determinant

$$\begin{vmatrix} e^{s_1 x} & e^{s_2 x} & \cdots & e^{s_n x} \\ s_1 e^{s_1 x} & s_2 e^{s_n x} & \cdots & s_n e^{s_n x} \\ \cdot & \cdot & \cdot & \cdot \\ s_1^{n-1} e^{s_1 x} & & & s^{n-1} e^{s_n x} \end{vmatrix} = e^{(s_1 + s_2 \cdots s_n) x X} \begin{vmatrix} 1 & 1 & \cdots & 1 \\ s_1 & s_2 & \cdots & s_n \\ \cdot & \cdot & \cdot & \cdot \\ s_1^{n-1} & s_2^{n-1} & \cdots & s_n^{n-1} \end{vmatrix}$$

The latter determinant, being of the order $n - 1$ in any s, and vanishing if any two are equal, is equal to the product of all the differences of the roots.

$$(s_n - s_1)(s_n - s_2) \cdots (s_n - s_{n-1})(s_{n-1} - s_1)(s_{n-1} - s_2) \cdots$$

and vanishes only when two roots are equal.

The expression

$$\frac{d^n e^{sx}}{dx^n} + a_1 \frac{d^{n-1} e^{sx}}{dx^{n-1}} \cdots a_n e^{sx} = e^{sx} F(s)$$

differentiated $i - 1$ times with respect to s gives

$$\frac{d^n (x^{i-1} e^{sx})}{dx^n} + a_1 \frac{d^{n-1}(x^{i-1} e^{sx})}{dx^{n-1}}$$
$$= e^{sx} [x^{(i-1)} F(s) + (i-1) x^{(i-2)} F'(s) \cdots F^{(i-1)}(s)].$$

If s, is a multiple root of order i of $F(s) = 0$, we have

$$F'(s_1) = F''(s_1) \cdots = F^{(i-1)}(s_1) = 0$$

so that the right hand member vanishes for $s = s_1$ and $x^{i-1} e^{s_1 x}$ is a solution of the differential equation. It is *a fortiori* evident that

$$x^{i-2} e^{s_1 x}, \quad x^{i-3} e^{s_1 x}, \ldots x e^{s_1 x},$$

are solutions, so that the general solution is

$$y = e^{s_1 x} [c_1 + c_2 x + c_3 x^2 \cdots c_i x^{i-1}]$$
$$+ c_{i+1} e^{s_2 x} + c_{i+3} e^{s_3 x} \cdots c_n e^{s_{n-i} x}.$$

To solve a linear differential equation without second member with constant coefficients, then, we find the roots of the characteristic equation

$$s^n + a_1 s^{n-1} \cdots a_n = 0$$

where the roots are $s_1 s_2 \ldots$ and their multiplicities $\mu_1 \mu_2 \ldots$ and the integral is

$$P_1(x) e^{s_1 x} + P_2(x) e^{s_2 x} \cdots$$

where $P_1 P_2$ are arbitrary polynomials of degree $\mu_1 - 1$, $\mu_n - 1 \ldots$

Suppose that the equation has an imaginary root $s_1 = \alpha + \beta i$. It will also have the conjugate $s_2 = \alpha - \beta i$ if all the coefs. are real, and the order of multiplicity will be the same for s_1 and s_2. Now since

$$e^{\beta i x} = i \sin \beta x + \cos \beta x$$

the terms of the integral

$$P_1 e^{(\alpha + \beta i) x} + P_2 e^{(\alpha - \beta i) x}$$

$$= e^{\alpha x}[(P_1 + P_2) \cos \beta x + i(P_1 - P_2) \sin \beta x]$$

or as the coefficients in $P_1 P_2$ are entirely arbitrary, and real or imaginary

$$= e^{\alpha x}(Q_1 \cos \beta x + Q_2 \sin \beta x)$$

Q_1 and Q_2 being arbitrary polynomials of order $\mu - 1$. Equations of the form

$$a_0 (a x + b)^n \frac{d^n y}{d x^n} + a_1 (a x + b)^{n-1} \frac{d^{n-1} y}{d x^{n-1}} \cdots$$

can be reduced to equations with constant coefficients by putting

$$a x + b = a e^t$$

and changing variable

$$d x = d t e^t$$

$$\frac{d y}{d x} = \frac{d y}{d t} e^{-t}$$

$$\frac{d^2 y}{d x^2} = \frac{d^n y}{d t^2} e^{-2t} - \frac{d y}{d t} e^{-2t}$$

$$\cdot \quad \cdot \quad \cdot \quad \cdot \quad \cdot \quad \cdot \quad \cdot \quad \cdot \quad \cdot \quad \cdot$$

and multiplying by the coefficients the exponentials disappear.

INDEX

(*Prepared by Galoust M. Elgal*)

A CATALOG OF SELECTED
DOVER BOOKS
IN SCIENCE AND MATHEMATICS

Astronomy

CHARIOTS FOR APOLLO: The NASA History of Manned Lunar Spacecraft to 1969, Courtney G. Brooks, James M. Grimwood, and Loyd S. Swenson, Jr. This illustrated history by a trio of experts is the definitive reference on the Apollo spacecraft and lunar modules. It traces the vehicles' design, development, and operation in space. More than 100 photographs and illustrations. 576pp. 6 3/4 x 9 1/4. 0-486-46756-2

EXPLORING THE MOON THROUGH BINOCULARS AND SMALL TELESCOPES, Ernest H. Cherrington, Jr. Informative, profusely illustrated guide to locating and identifying craters, rills, seas, mountains, other lunar features. Newly revised and updated with special section of new photos. Over 100 photos and diagrams. 240pp. 8 1/4 x 11. 0-486-24491-1

WHERE NO MAN HAS GONE BEFORE: A History of NASA's Apollo Lunar Expeditions, William David Compton. Introduction by Paul Dickson. This official NASA history traces behind-the-scenes conflicts and cooperation between scientists and engineers. The first half concerns preparations for the Moon landings, and the second half documents the flights that followed Apollo 11. 1989 edition. 432pp. 7 x 10.
0-486-47888-2

APOLLO EXPEDITIONS TO THE MOON: The NASA History, Edited by Edgar M. Cortright. Official NASA publication marks the 40th anniversary of the first lunar landing and features essays by project participants recalling engineering and administrative challenges. Accessible, jargon-free accounts, highlighted by numerous illustrations. 336pp. 8 3/8 x 10 7/8. 0-486-47175-6

ON MARS: Exploration of the Red Planet, 1958-1978--The NASA History, Edward Clinton Ezell and Linda Neuman Ezell. NASA's official history chronicles the start of our explorations of our planetary neighbor. It recounts cooperation among government, industry, and academia, and it features dozens of photos from Viking cameras. 560pp. 6 3/4 x 9 1/4. 0-486-46757-0

ARISTARCHUS OF SAMOS: The Ancient Copernicus, Sir Thomas Heath. Heath's history of astronomy ranges from Homer and Hesiod to Aristarchus and includes quotes from numerous thinkers, compilers, and scholasticists from Thales and Anaximander through Pythagoras, Plato, Aristotle, and Heraclides. 34 figures. 448pp. 5 3/8 x 8 1/2.
0-486-43886-4

AN INTRODUCTION TO CELESTIAL MECHANICS, Forest Ray Moulton. Classic text still unsurpassed in presentation of fundamental principles. Covers rectilinear motion, central forces, problems of two and three bodies, much more. Includes over 200 problems, some with answers. 437pp. 5 3/8 x 8 1/2. 0-486-64687-4

BEYOND THE ATMOSPHERE: Early Years of Space Science, Homer E. Newell. This exciting survey is the work of a top NASA administrator who chronicles technological advances, the relationship of space science to general science, and the space program's social, political, and economic contexts. 528pp. 6 3/4 x 9 1/4.
0-486-47464-X

STAR LORE: Myths, Legends, and Facts, William Tyler Olcott. Captivating retellings of the origins and histories of ancient star groups include Pegasus, Ursa Major, Pleiades, signs of the zodiac, and other constellations. "Classic." – *Sky & Telescope.* 58 illustrations. 544pp. 5 3/8 x 8 1/2. 0-486-43581-4

A COMPLETE MANUAL OF AMATEUR ASTRONOMY: Tools and Techniques for Astronomical Observations, P. Clay Sherrod with Thomas L. Koed. Concise, highly readable book discusses the selection, set-up, and maintenance of a telescope; amateur studies of the sun; lunar topography and occultations; and more. 124 figures. 26 halftones. 37 tables. 335pp. 6 1/2 x 9 1/4. 0-486-42820-6

Browse over 9,000 books at www.doverpublications.com

Chemistry

MOLECULAR COLLISION THEORY, M. S. Child. This high-level monograph offers an analytical treatment of classical scattering by a central force, quantum scattering by a central force, elastic scattering phase shifts, and semi-classical elastic scattering. 1974 edition. 310pp. 5 3/8 x 8 1/2. 0-486-69437-2

HANDBOOK OF COMPUTATIONAL QUANTUM CHEMISTRY, David B. Cook. This comprehensive text provides upper-level undergraduates and graduate students with an accessible introduction to the implementation of quantum ideas in molecular modeling, exploring practical applications alongside theoretical explanations. 1998 edition. 832pp. 5 3/8 x 8 1/2. 0-486-44307-8

RADIOACTIVE SUBSTANCES, Marie Curie. The celebrated scientist's thesis, which directly preceded her 1903 Nobel Prize, discusses establishing atomic character of radioactivity; extraction from pitchblende of polonium and radium; isolation of pure radium chloride; more. 96pp. 5 3/8 x 8 1/2. 0-486-42550-9

CHEMICAL MAGIC, Leonard A. Ford. Classic guide provides intriguing entertainment while elucidating sound scientific principles, with more than 100 unusual stunts: cold fire, dust explosions, a nylon rope trick, a disappearing beaker, much more. 128pp. 5 3/8 x 8 1/2. 0-486-67628-5

ALCHEMY, E. J. Holmyard. Classic study by noted authority covers 2,000 years of alchemical history: religious, mystical overtones; apparatus; signs, symbols, and secret terms; advent of scientific method, much more. Illustrated. 320pp. 5 3/8 x 8 1/2.
0-486-26298-7

CHEMICAL KINETICS AND REACTION DYNAMICS, Paul L. Houston. This text teaches the principles underlying modern chemical kinetics in a clear, direct fashion, using several examples to enhance basic understanding. Solutions to selected problems. 2001 edition. 352pp. 8 3/8 x 11. 0-486-45334-0

PROBLEMS AND SOLUTIONS IN QUANTUM CHEMISTRY AND PHYSICS, Charles S. Johnson and Lee G. Pedersen. Unusually varied problems, with detailed solutions, cover of quantum mechanics, wave mechanics, angular momentum, molecular spectroscopy, scattering theory, more. 280 problems, plus 139 supplementary exercises. 430pp. 6 1/2 x 9 1/4. 0-486-65236-X

ELEMENTS OF CHEMISTRY, Antoine Lavoisier. Monumental classic by the founder of modern chemistry features first explicit statement of law of conservation of matter in chemical change, and more. Facsimile reprint of original (1790) Kerr translation. 539pp. 5 3/8 x 8 1/2. 0-486-64624-6

MAGNETISM AND TRANSITION METAL COMPLEXES, F. E. Mabbs and D. J. Machin. A detailed view of the calculation methods involved in the magnetic properties of transition metal complexes, this volume offers sufficient background for original work in the field. 1973 edition. 240pp. 5 3/8 x 8 1/2. 0-486-46284-6

GENERAL CHEMISTRY, Linus Pauling. Revised third edition of classic first-year text by Nobel laureate. Atomic and molecular structure, quantum mechanics, statistical mechanics, thermodynamics correlated with descriptive chemistry. Problems. 992pp. 5 3/8 x 8 1/2. 0-486-65622-5

ELECTROLYTE SOLUTIONS: Second Revised Edition, R. A. Robinson and R. H. Stokes. Classic text deals primarily with measurement, interpretation of conductance, chemical potential, and diffusion in electrolyte solutions. Detailed theoretical interpretations, plus extensive tables of thermodynamic and transport properties. 1970 edition. 590pp. 5 3/8 x 8 1/2. 0-486-42225-9

Browse over 9,000 books at www.doverpublications.com

Engineering

FUNDAMENTALS OF ASTRODYNAMICS, Roger R. Bate, Donald D. Mueller, and Jerry E. White. Teaching text developed by U.S. Air Force Academy develops the basic two-body and n-body equations of motion; orbit determination; classical orbital elements, coordinate transformations; differential correction; more. 1971 edition. 455pp. 5 3/8 x 8 1/2. 0-486-60061-0

INTRODUCTION TO CONTINUUM MECHANICS FOR ENGINEERS: Revised Edition, Ray M. Bowen. This self-contained text introduces classical continuum models within a modern framework. Its numerous exercises illustrate the governing principles, linearizations, and other approximations that constitute classical continuum models. 2007 edition. 320pp. 6 1/8 x 9 1/4. 0-486-47460-7

ENGINEERING MECHANICS FOR STRUCTURES, Louis L. Bucciarelli. This text explores the mechanics of solids and statics as well as the strength of materials and elasticity theory. Its many design exercises encourage creative initiative and systems thinking. 2009 edition. 320pp. 6 1/8 x 9 1/4. 0-486-46855-0

FEEDBACK CONTROL THEORY, John C. Doyle, Bruce A. Francis and Allen R. Tannenbaum. This excellent introduction to feedback control system design offers a theoretical approach that captures the essential issues and can be applied to a wide range of practical problems. 1992 edition. 224pp. 6 1/2 x 9 1/4. 0-486-46933-6

THE FORCES OF MATTER, Michael Faraday. These lectures by a famous inventor offer an easy-to-understand introduction to the interactions of the universe's physical forces. Six essays explore gravitation, cohesion, chemical affinity, heat, magnetism, and electricity. 1993 edition. 96pp. 5 3/8 x 8 1/2. 0-486-47482-8

DYNAMICS, Lawrence E. Goodman and William H. Warner. Beginning engineering text introduces calculus of vectors, particle motion, dynamics of particle systems and plane rigid bodies, technical applications in plane motions, and more. Exercises and answers in every chapter. 619pp. 5 3/8 x 8 1/2. 0-486-42006-X

ADAPTIVE FILTERING PREDICTION AND CONTROL, Graham C. Goodwin and Kwai Sang Sin. This unified survey focuses on linear discrete-time systems and explores natural extensions to nonlinear systems. It emphasizes discrete-time systems, summarizing theoretical and practical aspects of a large class of adaptive algorithms. 1984 edition. 560pp. 6 1/2 x 9 1/4. 0-486-46932-8

INDUCTANCE CALCULATIONS, Frederick W. Grover. This authoritative reference enables the design of virtually every type of inductor. It features a single simple formula for each type of inductor, together with tables containing essential numerical factors. 1946 edition. 304pp. 5 3/8 x 8 1/2. 0-486-47440-2

THERMODYNAMICS: Foundations and Applications, Elias P. Gyftopoulos and Gian Paolo Beretta. Designed by two MIT professors, this authoritative text discusses basic concepts and applications in detail, emphasizing generality, definitions, and logical consistency. More than 300 solved problems cover realistic energy systems and processes. 800pp. 6 1/8 x 9 1/4. 0-486-43932-1

THE FINITE ELEMENT METHOD: Linear Static and Dynamic Finite Element Analysis, Thomas J. R. Hughes. Text for students without in-depth mathematical training, this text includes a comprehensive presentation and analysis of algorithms of time-dependent phenomena plus beam, plate, and shell theories. Solution guide available upon request. 672pp. 6 1/2 x 9 1/4. 0-486-41181-8

Browse over 9,000 books at www.doverpublications.com

HELICOPTER THEORY, Wayne Johnson. Monumental engineering text covers vertical flight, forward flight, performance, mathematics of rotating systems, rotary wing dynamics and aerodynamics, aeroelasticity, stability and control, stall, noise, and more. 189 illustrations. 1980 edition. 1089pp. 5 5/8 x 8 1/4.   0-486-68230-7

MATHEMATICAL HANDBOOK FOR SCIENTISTS AND ENGINEERS: Definitions, Theorems, and Formulas for Reference and Review, Granino A. Korn and Theresa M. Korn. Convenient access to information from every area of mathematics: Fourier transforms, Z transforms, linear and nonlinear programming, calculus of variations, random-process theory, special functions, combinatorial analysis, game theory, much more. 1152pp. 5 3/8 x 8 1/2.   0-486-41147-8

A HEAT TRANSFER TEXTBOOK: Fourth Edition, John H. Lienhard V and John H. Lienhard IV. This introduction to heat and mass transfer for engineering students features worked examples and end-of-chapter exercises. Worked examples and end-of-chapter exercises appear throughout the book, along with well-drawn, illuminating figures. 768pp. 7 x 9 1/4.   0-486-47931-5

BASIC ELECTRICITY, U.S. Bureau of Naval Personnel. Originally a training course; best nontechnical coverage. Topics include batteries, circuits, conductors, AC and DC, inductance and capacitance, generators, motors, transformers, amplifiers, etc. Many questions with answers. 349 illustrations. 1969 edition. 448pp. 6 1/2 x 9 1/4.

0-486-20973-3

BASIC ELECTRONICS, U.S. Bureau of Naval Personnel. Clear, well-illustrated introduction to electronic equipment covers numerous essential topics: electron tubes, semiconductors, electronic power supplies, tuned circuits, amplifiers, receivers, ranging and navigation systems, computers, antennas, more. 560 illustrations. 567pp. 6 1/2 x 9 1/4.   0-486-21076-6

BASIC WING AND AIRFOIL THEORY, Alan Pope. This self-contained treatment by a pioneer in the study of wind effects covers flow functions, airfoil construction and pressure distribution, finite and monoplane wings, and many other subjects. 1951 edition. 320pp. 5 3/8 x 8 1/2.   0-486-47188-8

SYNTHETIC FUELS, Ronald F. Probstein and R. Edwin Hicks. This unified presentation examines the methods and processes for converting coal, oil, shale, tar sands, and various forms of biomass into liquid, gaseous, and clean solid fuels. 1982 edition. 512pp. 6 1/8 x 9 1/4.   0-486-44977-7

THEORY OF ELASTIC STABILITY, Stephen P. Timoshenko and James M. Gere. Written by world-renowned authorities on mechanics, this classic ranges from theoretical explanations of 2- and 3-D stress and strain to practical applications such as torsion, bending, and thermal stress. 1961 edition. 560pp. 5 3/8 x 8 1/2.   0-486-47207-8

PRINCIPLES OF DIGITAL COMMUNICATION AND CODING, Andrew J. Viterbi and Jim K. Omura. This classic by two digital communications experts is geared toward students of communications theory and to designers of channels, links, terminals, modems, or networks used to transmit and receive digital messages. 1979 edition. 576pp. 6 1/8 x 9 1/4.   0-486-46901-8

LINEAR SYSTEM THEORY: The State Space Approach, Lotfi A. Zadeh and Charles A. Desoer. Written by two pioneers in the field, this exploration of the state space approach focuses on problems of stability and control, plus connections between this approach and classical techniques. 1963 edition. 656pp. 6 1/8 x 9 1/4.

0-486-46663-9

Mathematics–Bestsellers

HANDBOOK OF MATHEMATICAL FUNCTIONS: with Formulas, Graphs, and Mathematical Tables, Edited by Milton Abramowitz and Irene A. Stegun. A classic resource for working with special functions, standard trig, and exponential logarithmic definitions and extensions, it features 29 sets of tables, some to as high as 20 places. 1046pp. 8 x 10 1/2.   0-486-61272-4

ABSTRACT AND CONCRETE CATEGORIES: The Joy of Cats, Jiri Adamek, Horst Herrlich, and George E. Strecker. This up-to-date introductory treatment employs category theory to explore the theory of structures. Its unique approach stresses concrete categories and presents a systematic view of factorization structures. Numerous examples. 1990 edition, updated 2004. 528pp. 6 1/8 x 9 1/4.   0-486-46934-4

MATHEMATICS: Its Content, Methods and Meaning, A. D. Aleksandrov, A. N. Kolmogorov, and M. A. Lavrent'ev. Major survey offers comprehensive, coherent discussions of analytic geometry, algebra, differential equations, calculus of variations, functions of a complex variable, prime numbers, linear and non-Euclidean geometry, topology, functional analysis, more. 1963 edition. 1120pp. 5 3/8 x 8 1/2.   0-486-40916-3

INTRODUCTION TO VECTORS AND TENSORS: Second Edition--Two Volumes Bound as One, Ray M. Bowen and C.-C. Wang. Convenient single-volume compilation of two texts offers both introduction and in-depth survey. Geared toward engineering and science students rather than mathematicians, it focuses on physics and engineering applications. 1976 edition. 560pp. 6 1/2 x 9 1/4.   0-486-46914-X

AN INTRODUCTION TO ORTHOGONAL POLYNOMIALS, Theodore S. Chihara. Concise introduction covers general elementary theory, including the representation theorem and distribution functions, continued fractions and chain sequences, the recurrence formula, special functions, and some specific systems. 1978 edition. 272pp. 5 3/8 x 8 1/2.   0-486-47929-3

ADVANCED MATHEMATICS FOR ENGINEERS AND SCIENTISTS, Paul DuChateau. This primary text and supplemental reference focuses on linear algebra, calculus, and ordinary differential equations. Additional topics include partial differential equations and approximation methods. Includes solved problems. 1992 edition. 400pp. 7 1/2 x 9 1/4.   0-486-47930-7

PARTIAL DIFFERENTIAL EQUATIONS FOR SCIENTISTS AND ENGINEERS, Stanley J. Farlow. Practical text shows how to formulate and solve partial differential equations. Coverage of diffusion-type problems, hyperbolic-type problems, elliptic-type problems, numerical and approximate methods. Solution guide available upon request. 1982 edition. 414pp. 6 1/8 x 9 1/4.   0-486-67620-X

VARIATIONAL PRINCIPLES AND FREE-BOUNDARY PROBLEMS, Avner Friedman. Advanced graduate-level text examines variational methods in partial differential equations and illustrates their applications to free-boundary problems. Features detailed statements of standard theory of elliptic and parabolic operators. 1982 edition. 720pp. 6 1/8 x 9 1/4.   0-486-47853-X

LINEAR ANALYSIS AND REPRESENTATION THEORY, Steven A. Gaal. Unified treatment covers topics from the theory of operators and operator algebras on Hilbert spaces; integration and representation theory for topological groups; and the theory of Lie algebras, Lie groups, and transform groups. 1973 edition. 704pp. 6 1/8 x 9 1/4.   0-486-47851-3

Browse over 9,000 books at www.doverpublications.com

A SURVEY OF INDUSTRIAL MATHEMATICS, Charles R. MacCluer. Students learn how to solve problems they'll encounter in their professional lives with this concise single-volume treatment. It employs MATLAB and other strategies to explore typical industrial problems. 2000 edition. 384pp. 5 3/8 x 8 1/2.　　0-486-47702-9

NUMBER SYSTEMS AND THE FOUNDATIONS OF ANALYSIS, Elliott Mendelson. Geared toward undergraduate and beginning graduate students, this study explores natural numbers, integers, rational numbers, real numbers, and complex numbers. Numerous exercises and appendixes supplement the text. 1973 edition. 368pp. 5 3/8 x 8 1/2.　　0-486-45792-3

A FIRST LOOK AT NUMERICAL FUNCTIONAL ANALYSIS, W. W. Sawyer. Text by renowned educator shows how problems in numerical analysis lead to concepts of functional analysis. Topics include Banach and Hilbert spaces, contraction mappings, convergence, differentiation and integration, and Euclidean space. 1978 edition. 208pp. 5 3/8 x 8 1/2.　　0-486-47882-3

FRACTALS, CHAOS, POWER LAWS: Minutes from an Infinite Paradise, Manfred Schroeder. A fascinating exploration of the connections between chaos theory, physics, biology, and mathematics, this book abounds in award-winning computer graphics, optical illusions, and games that clarify memorable insights into self-similarity. 1992 edition. 448pp. 6 1/8 x 9 1/4.　　0-486-47204-3

SET THEORY AND THE CONTINUUM PROBLEM, Raymond M. Smullyan and Melvin Fitting. A lucid, elegant, and complete survey of set theory, this three-part treatment explores axiomatic set theory, the consistency of the continuum hypothesis, and forcing and independence results. 1996 edition. 336pp. 6 x 9.　　0-486-47484-4

DYNAMICAL SYSTEMS, Shlomo Sternberg. A pioneer in the field of dynamical systems discusses one-dimensional dynamics, differential equations, random walks, iterated function systems, symbolic dynamics, and Markov chains. Supplementary materials include PowerPoint slides and MATLAB exercises. 2010 edition. 272pp. 6 1/8 x 9 1/4.　　0-486-47705-3

ORDINARY DIFFERENTIAL EQUATIONS, Morris Tenenbaum and Harry Pollard. Skillfully organized introductory text examines origin of differential equations, then defines basic terms and outlines general solution of a differential equation. Explores integrating factors; dilution and accretion problems; Laplace Transforms; Newton's Interpolation Formulas, more. 818pp. 5 3/8 x 8 1/2.　　0-486-64940-7

MATROID THEORY, D. J. A. Welsh. Text by a noted expert describes standard examples and investigation results, using elementary proofs to develop basic matroid properties before advancing to a more sophisticated treatment. Includes numerous exercises. 1976 edition. 448pp. 5 3/8 x 8 1/2.　　0-486-47439-9

THE CONCEPT OF A RIEMANN SURFACE, Hermann Weyl. This classic on the general history of functions combines function theory and geometry, forming the basis of the modern approach to analysis, geometry, and topology. 1955 edition. 208pp. 5 3/8 x 8 1/2.　　0-486-47004-0

THE LAPLACE TRANSFORM, David Vernon Widder. This volume focuses on the Laplace and Stieltjes transforms, offering a highly theoretical treatment. Topics include fundamental formulas, the moment problem, monotonic functions, and Tauberian theorems. 1941 edition. 416pp. 5 3/8 x 8 1/2.　　0-486-47755-X

Browse over 9,000 books at www.doverpublications.com

Mathematics–Logic and Problem Solving

PERPLEXING PUZZLES AND TANTALIZING TEASERS, Martin Gardner. Ninety-three riddles, mazes, illusions, tricky questions, word and picture puzzles, and other challenges offer hours of entertainment for youngsters. Filled with rib-tickling drawings. Solutions. 224pp. 5 3/8 x 8 1/2.   0-486-25637-5

MY BEST MATHEMATICAL AND LOGIC PUZZLES, Martin Gardner. The noted expert selects 70 of his favorite "short" puzzles. Includes The Returning Explorer, The Mutilated Chessboard, Scrambled Box Tops, and dozens more. Complete solutions included. 96pp. 5 3/8 x 8 1/2.   0-486-28152-3

THE LADY OR THE TIGER?: and Other Logic Puzzles, Raymond M. Smullyan. Created by a renowned puzzle master, these whimsically themed challenges involve paradoxes about probability, time, and change; metapuzzles; and self-referentiality. Nineteen chapters advance in difficulty from relatively simple to highly complex. 1982 edition. 240pp. 5 3/8 x 8 1/2.   0-486-47027-X

SATAN, CANTOR AND INFINITY: Mind-Boggling Puzzles, Raymond M. Smullyan. A renowned mathematician tells stories of knights and knaves in an entertaining look at the logical precepts behind infinity, probability, time, and change. Requires a strong background in mathematics. Complete solutions. 288pp. 5 3/8 x 8 1/2.
0-486-47036-9

THE RED BOOK OF MATHEMATICAL PROBLEMS, Kenneth S. Williams and Kenneth Hardy. Handy compilation of 100 practice problems, hints and solutions indispensable for students preparing for the William Lowell Putnam and other mathematical competitions. Preface to the First Edition. Sources. 1988 edition. 192pp. 5 3/8 x 8 1/2.   0-486-69415-1

KING ARTHUR IN SEARCH OF HIS DOG AND OTHER CURIOUS PUZZLES, Raymond M. Smullyan. This fanciful, original collection for readers of all ages features arithmetic puzzles, logic problems related to crime detection, and logic and arithmetic puzzles involving King Arthur and his Dogs of the Round Table. 160pp. 5 3/8 x 8 1/2.
0-486-47435-6

UNDECIDABLE THEORIES: Studies in Logic and the Foundation of Mathematics, Alfred Tarski in collaboration with Andrzej Mostowski and Raphael M. Robinson. This well-known book by the famed logician consists of three treatises: "A General Method in Proofs of Undecidability," "Undecidability and Essential Undecidability in Mathematics," and "Undecidability of the Elementary Theory of Groups." 1953 edition. 112pp. 5 3/8 x 8 1/2.   0-486-47703-7

LOGIC FOR MATHEMATICIANS, J. Barkley Rosser. Examination of essential topics and theorems assumes no background in logic. "Undoubtedly a major addition to the literature of mathematical logic." – *Bulletin of the American Mathematical Society.* 1978 edition. 592pp. 6 1/8 x 9 1/4.   0-486-46898-4

INTRODUCTION TO PROOF IN ABSTRACT MATHEMATICS, Andrew Wohlgemuth. This undergraduate text teaches students what constitutes an acceptable proof, and it develops their ability to do proofs of routine problems as well as those requiring creative insights. 1990 edition. 384pp. 6 1/2 x 9 1/4.   0-486-47854-8

FIRST COURSE IN MATHEMATICAL LOGIC, Patrick Suppes and Shirley Hill. Rigorous introduction is simple enough in presentation and context for wide range of students. Symbolizing sentences; logical inference; truth and validity; truth tables; terms, predicates, universal quantifiers; universal specification and laws of identity; more. 288pp. 5 3/8 x 8 1/2.   0-486-42259-3

Browse over 9,000 books at www.doverpublications.com